Cambridge Monographs in African Archaeology
70
Series Editors: John Alexander, Laurence Smith and Timothy Insoll

Archaeofaunal remains from the past 4000 years in Sahelian West Africa

Domestic livestock, subsistence strategies and environmental changes

Veerle Linseele

BAR International Series 1658
2007

Published in 2016 by
BAR Publishing, Oxford

BAR International Series 1658

Cambridge Monographs in African Archaeology 70
Archaeofaunal remains from the past 4000 years in Sahelian West Africa

ISBN 978 1 4073 0094 8

BAR Publishing is the trading name of British Archaeological Reports (Oxford) Ltd. British Archaeological Reports was first incorporated in 1974 to publish the BAR Series, International and British. In 1992 Hadrian Books Ltd became part of the BAR group. This volume was originally published by Archaeopress in conjunction with British Archaeological Reports (Oxford) Ltd / Hadrian Books Ltd, the Series principal publisher, in 2007. This present volume is published by BAR Publishing, 2016.

Printed in England

BAR
PUBLISHING

BAR titles are available from:

BAR Publishing
122 Banbury Rd, Oxford, OX2 7BP, UK
EMAIL info@barpublishing.com
PHONE +44 (0)1865 310431
FAX +44 (0)1865 316916
www.barpublishing.com

Preface

This volume is a slightly altered version of the doctoral dissertation that I defended in November 2005 at the Catholic University of Leuven (Belgium), where I am presently working as a postdoctoral fellow at the CAS (Center for Archaeological Sciences).

There have been many people in the course of time to whom I owe a special and warm thank you. First of all to Wim Van Neer, my main supervisor, who has been there for me from the first to the very last second of my PhD. I am especially grateful to him for reading through my texts so quickly and thoroughly. I am also very grateful to my other supervisor, Pierre Vermeersch, for being critical and encouraging, and always being there when I needed his help or advice. Then of course I should also thank Achilles Gautier, who initiated me into archaeozoology and has been supportive and helpful ever since.

Without the project "SFB 268" at the J.W. Goethe University (Frankfurt-am-Main, Germany), and the Deutsche Forschungsgemeinschaft who funded it, this work would never have been possible. I am especially indebted to Peter Breunig and Katharina Neumann. I hold good memories of my German cooperation, both professionally and on a more personal level. Sincere thanks are due to Detlef Gronenborn, Maya Hallier-von Czerniewicz, Alexa Höhn, Stephanie Kahlheber, Carlos Magnavita, Sonja Magnavita, Christoph Pelzer, Lucas Petit, Nicole Rohde, Nicole Rupp, Peter Wendt and Birgitt Wiesmüller. A very important thank you also to the Belgian Federal Science Policy Office, because without the scholarship they gave me, my adventures in archaeozoology might have been ended long ago.

I am also very grateful to Augustin Holl, who allowed me to include the material from the Cameroonian Blé sites in my study. I would equally like to acknowledge some foreign colleagues who have kindly sent me reprints of their publications: Shawn Badenhorst, Hubert Berke, Roger Blench, Isabelle Chenal-Vélardé, Hélène Jousse, Fiona Marshall and Ina Plug. I have learned a lot from discussions with them and other scholars. A warm thank you also to Sheila Hamilton-Dyer for assistance in turning my English into "proper" English.

I owe a great deal to Sven Lambrecht, whose work I have taken over. A sincere thank you also to the people who shared their knowledge with me and assisted with identifications, and that only for the honour of being mentioned in these acknowledgements: Dirk Van Damme for the molluscs, France de Broin-de Lapparent for the turtles and tortoises, Wim Wendelen for the small rodents and Joris Peters for the bullfrog and buffalo.

There is, furthermore, a long list of colleagues and former colleagues from the Royal Museum of Central Africa in Tervuren, who have all helped me in one way or another, sometimes just by being there and listening to me moaning: An Alen, Gert Boden, Ides "Bobo" Boone, Henk Breman, Garin Cael, Els Cornelissen, Mark Hanssens, Julie Herregods, Marc Herremans, Alexandre Livingstone-Smith, Michel Louette, Danny Meirte, Tobias Musschoot, Brigitte Osselaer, Miguel Parrent, Fabienne Pigière, Alain Reygel, Jos Snoeks, Sofie Vanpoucke, Elina Rijmenants, Alexander Vral, Wim Tavernier, Franck Theeten, Mircea Udrescu, Emmanuel Vreven and Wim Wouters, but especially Bea De Cupere. I also owe a tremendous debt to my parents. They have always supported me in every possible way. Thanks also to the rest of my family, especially my brother and my grandparents, and to the friends, An, Maaike, Marc, Sandrine and Simon, that have been closest to me during the preparation of my dissertation.

Leuven, November 2006

Veerle Linseele
Center for Archaeological Sciences
Katholieke Universiteit Leuven
Celestijnenlaan 200E
B-3001 Leuven
Belgium
E-mail: Veerle.linseele@arts.kuleuven.be

Contents

List of figures

List of tables

Chapter 1. Research goals and strategies

1.1. Situation of the research within the multidisciplinary project "SFB 268"

The archaeofauna that is the subject of this study was excavated in northern Burkina Faso and in north-eastern Nigeria, close to the southern shores of Lake Chad, by a team of archaeologists from the Johann Wolfgang Goethe-University at Frankfurt am Main (Germany) under the direction of Prof. Dr. P. Breunig. The team was part of the large multidisciplinary research project SFB 268 "Cultural development and language history in the West African savannah" that ran between 1988 and 2002 and also included archaeobotanists, botanists, ethnographers, geographers and linguists. The archaeologists' main subject of investigation was the interaction of prehistoric populations with their environment (Breunig *et al.*, 1993b). More precisely, archaeological research focused on the beginning of food production and, in a later stage, also on the transition between the Late Stone Age and the Iron Age. The excavated sites in Burkina Faso date approximately between 2000 BC and AD 1400, while those from Nigeria cover the whole chronological sequence between 2000 BC and present.[1] Excavations in Nigeria are continuing at present in the new project "Ecological and cultural change in West and Central Africa", also based at Frankfurt am Main.

Besides the material excavated by the team from Frankfurt, fauna from sites at Blé in the extreme north of Cameroon, bordering the Nigerian part of the research area, is also included in this study. Prof. Dr. A. Holl (2002), University of Michigan, Ann Arbor (Michigan, U.S.A.) excavated the sites in the framework of the Houlouf Project, which aimed to study the development of complex ranked societies in the southern Lake Chad area. The sites at Blé date from about the end of the first millennium BC to the sixteenth century AD.

1.2. Earlier archaeozoological studies in the research area and neighbouring regions

Geographically, this study thus concentrates mainly on the northernmost parts of Burkina Faso and the southern Lake Chad area, two regions in arid sub-Saharan West Africa. In northern Burkina Faso no studies on archaeological fauna were conducted prior to the present one. In neighbouring Mali and especially in the regions around the Niger bend, however, substantial archaeozoological data are available thanks to the work of K. MacDonald, W. Van Neer and a few other researchers (see Fig. 1 and references therein). Nonetheless, faunal research did not always result in the publication of full reports.

Although extensive archaeological excavations have been conducted in the region, data on archaeofauna from the Nigerian part of the research area and adjacent parts of the southern Lake Chad area in Cameroon and Chad are patchy and usually of poor standard (see Fig. 1 and references therein). In northern Nigeria, Connah (1981) carried out pioneering archaeological work in the 1960s. Connah also published data on the archaeofauna, studied by Fagan (mammals, birds) and Howes (fish), but the data are mainly limited to numbers of specimens of each species found by archaeological unit. There was a separate publication on the mollusc remains, however, with thorough interpretations (Connah and McMillan, 1995). Holl (2002) conducted extensive archaeological research in northern Cameroon and also integrated faunal data into his archaeological reports. Again, the data are mere lists of species found in each archaeological unit and the accuracy of some of the identifications also seems questionable. Catfish is, for example, systematically referred to with the Latin genus name "*Gymnarchus*", which is not a catfish at all. Marliac (1991) added annexes on the fauna, analysed by Columeau, to his study of a few sites in the Diamaré region of northern Cameroon, but again these are not much more than "laundry lists". Another Cameroonian site included in Fig. 1, Blabli, was never truly archaeozoologically analysed. The radiocarbon dates given for the latter site were obtained on ovicaprine bones, which appear to be the oldest indications for domestics in West Africa south of the Sahara (David and Sterner, 1989). Lebeuf (1962, 1969) has also carried out intensive archaeological research in the southern Lake Chad area, and especially the Chadian part, but faunal data were only published for Mdaga, studied by Thomas (Lebeuf *et al.*, 1980). It does not constitute a full report, however. Only from the Chadian part of the southern Lake Chad area, for Koyom, is such a report available (Rivallain and Van Neer, 1983, 1984).

1.3. Research questions

The faunal material available for this study has important scientific potency. First of all, results of its analysis can be integrated with and compared to data from other disciplines within the multidisciplinary project "SFB 268". Perhaps even more importantly, the faunal assemblages allow the tracing of diachronic and

[1] Unless otherwise specified, all dates cited in this study are calibrated BC or AD. Details on original radiocarbon dates are omitted from the actual text but can be found by cited location in Appendix A. Calibration has been carried out with OxCal 3.10, or with the calibration curve of Reimer *et al.* (2004) when radiocarbon dates with standard deviations were not available.

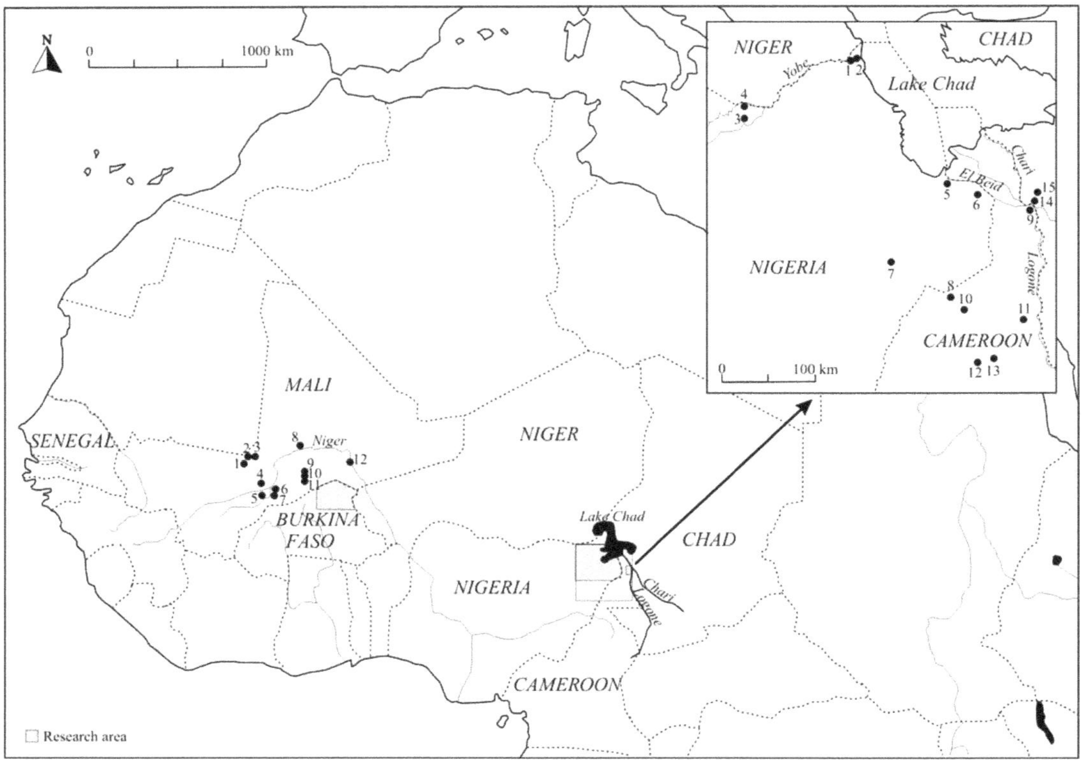

Site		Faunal study	Date
Niger bend			
1	Akumbu	MacDonald and Van Neer, 1993, 1994	AD 400-1400
2	Kobadi	Raimbault *et al*., 1987; Jousse and Chenal-Vélardé, 2001-2002	2nd-1st millenium BC
3	Kolima-Sud	MacDonald, 1994, 100-105, Tables 5.7-5.10; MacDonald and Van Neer, 1994	1700-500 BC
4	Dia-Shoma	MacDonald and MacDonald, 2000; Bedaux *et al*., 2001	800 BC-AD 1700
5	Jenné-Jeno	MacDonald, 1995; Van Neer, 1995	250 BC-AD 1400
6	Toguéré Galia	Bedaux *et al*., 1978	End 1st-beginning 2nd mill. AD
6	Toguéré Doupwil	Bedaux *et al*., 1978	End 1st-beginning 2nd mill. AD
7	Hamadallahi	Vélardé, 1995; Chenal-Vélardé, 1996	19th cent. AD
8	Mouyssam II (KNT 2)	Guérin and Faure, 1991	1st mill. AD
9	Windé Koroji Sud-Ouest	MacDonald, 1994, 231-232, Table 7.3, 1996	1700-1500 BC
9	Windé Koroji Ouest	MacDonald, 1994, 236, Table 7.4, 1996	2100-900 BC
10	Zampia	MacDonald, 1994, 238, 1996	800 BC
11	Tongo Maaré Diabel	MacDonald *et al*., 1994	1st mill. AD
12	Gao Saney	MacDonald and MacDonald, 1996	AD 900-1200
12	Gao Ancien	MacDonald and MacDonald, 1996; Barrett-Jolley, 2000	6th-16th cent. AD
12	Gadei	Cook, 2000; Milner, 2000; Stangroome, 2000	7th-16th cent. AD
Southern Lake Chad area			
1	Yau	Connah, 1981, Histogram 9.1, 211-212; Connah and McMillan, 1995	9/10th-13/15th cent. AD
2	Ajere	Connah, 1981, Histogram 9.2, 213; Connah and McMillan, 1995	2nd mill. AD
3	Birnin Gazargamo	Connah, 1981, Histogram 9.3-9.4, 218-219; Connah and McMillan, 1995	AD 1470-1812
5	Kursakata	Connah, 1981, Histogram 5.2, 98; Connah and McMillan, 1995	1000 BC-early 2nd mill. AD
6	Daima	Connah, 1981, 137-139, 160, 192-194, Table 8.3; Connah and McMillan, 1995	550 BC-AD 1150
7	Borno 38	Connah, 1981, 91-92; Connah and McMillan, 1995	End 2nd mill. BC-middle 1st mill. BC
8	Blabli	David and Sterner, 1987, 1989	2500 BC-AD 0
9	Deguesse[1]	Holl, 2002, 35-43	1900 BC-AD 1800
9	Krenak[1]	Holl, 2002, 43-47	1900 BC-AD 1400
9	Hamei[1]	Holl, 2002, 47-53	AD 500-1800
9	Mishiskwa[1]	Holl, 2002, 55-60	AD 500-1800
9	Madaf[4]	Holl, 2002, 60-65	AD 500-1800
9	Amachita[1]	Holl, 2002, 65-68	AD 1400-1800
9	Sororo[1]	Holl, 2002, 68-72	AD 1600-1800
9	Blé[1]	Holl, 2002, 111-120	AD 250-1400
9	Krenak-Sao[1]	Holl, 2002, 130-138	AD 500-1000
9	Houlouf[4]	Holl, 2002, 139-187	AD 0-1800
10	Aissa Dugjé	Bourges *et al*., 1999; MacEachern *et al*., 2001	Middle 1st-early 2nd mill. AD
11	Mongossi	Marliac, 1991, 237-241	5/6-15/16th cent. AD
12	Salak	Marliac, 1991, 30, Fig. 42	6-13th cent. AD
13	Goray	Marliac, 1991, 380-382	9-15/16th cent. AD
14	Koyom	Rivallain and Van Neer, 1983, 1984	18-19th cent. AD
15	Mdaga	Lebeuf et al., 1980, 21-96; Thomas, 1980	425 BC-AD 1780

[1]see Fig. 10 for more precise location

Fig. 1: Sites in the research area and neighbouring regions where earlier archaeozoological research has been carried out

geographic trends because of their origin from a relatively large area and time period. Chronologically, the fauna covers almost the entire time span between 2000 BC and the present. Geographically it comes from two areas situated at about the same latitude, but which are very distinct because of the presence of Lake Chad in the Nigerian and Cameroonian part. In the following chapter, it will moreover become clear that within the southern Lake Chad area several different geographic zones also have to be discerned. Finally, an important reason to study the faunal material is its origin from areas where very little archaeozoological research has been conducted so far (see 1.2.). For northern Burkina Faso there are simply no data available at all, while the problem in northern Nigeria and the southern Lake Chad area in general is more qualitative. Archaeozoological studies are also scanty in sub-Saharan West Africa as a whole. The only relatively well-investigated areas are Senegal (Linares de Sapir, 1971; Thilmans and Ravisé, 1980, 111-123; Chavane, 1985, 112-114; Van Neer and Bocoum, 1991; Van Neer, 1997, unpublished report; MacDonald and MacDonald, 2000, unpublished report; McIntosh and Bocoum, 2000) and especially the Malian Niger bend (see 1.2.).

The research questions in this study partly follow those formulated by the Frankfurt archaeologists. As mentioned, one of their major research themes was the beginning of food production in western Africa. The period in which this major innovation took place has been named "Final Stone Age" (Breunig and Neumann, 2002a) and "West African Neolithic" (McIntosh, 2001), among others. It was decided here not to use "Neolithic" and derived terms, because they are overly associated with the situation in Europe and the Near East where agriculture, livestock keeping, sedentism and ceramics appear almost simultaneously (the so-called Neolithic package). A more in-depth discussion on the use of the term "Neolithic" in Africa is given in Sinclair *et al.* (1993) and Ambrose (1997). The oldest form of food production in Africa is livestock keeping (Gautier, 1984b, 2002; Marshall, 1998; Marshall and Hildebrand, 2002), but there is considerable variation in the date and mode of its introduction between different parts of the continent (MacDonald and MacDonald, 2000; Marshall, 2000; Smith, 2000; Van Neer, 2000). It is evident that archaeozoology is one of the prime sources for the study of the history of livestock keeping, although a lot of information can also be deducted from genetics, iconography and linguistics for example. A survey of the possible sources with an evaluation of their potential and weaknesses can be found in Blench (2000a).

With the exception of the guineafowl (*Numida meleagris*[2]) and very recent experiments with the domestication of local wild species, none of the domestic animals of western Africa are indigenous to the region. A major subject of this study is therefore the establishment of the timing of the appearance of the domestic animals in the research area and an evaluation of the possible routes through which they arrived there. A *status quaestionis* for Africa west of Lake Chad, mainly from archaeological bone remains, is given in MacDonald and MacDonald (2000). For guineafowl no domestication centre is documented yet and a possible local domestication in western Africa is therefore also a subject of investigation. During their spread into western Africa, most domestic species seem to have undergone a size decline and dwarf types developed. This study makes new metrical data available that can be used to trace dwarfism back in time, but it can also document other size variations of domestic species.

Contrary to domestic animals, the wild forms of many cultivated crops do occur in West Africa. In the 1970s Harlan (1971) proposed a West African origin for the most important African crops solely based on the present distribution of their wild progenitors. In the Lake Chad area the wild ancestors of pearl millet (*Pennisetum americanum*), sorghum (*Sorghum bicolor*) and rice (*Oryza glaberrima*) can for example be found. One of the goals of the Frankfurt project was to test Harlan's hypothesis with direct archaeological evidence. Data gathered by the project's archaeobotanists on the appearance and possible local domestication of cultivated crops are summarised here and compared against the data on the history of domestic livestock.

Besides gathering information on the history of the different domestic animals in the research area, a major aim of this study is the reconstruction of the palaeoeconomy and palaeoecology of the investigated sites. Wilkinson and Stevens (2003, 136) defined "economy" as the production and consumption of goods and services. One of the palaeoeconomic questions that will be addressed is the composition of the former diet, firstly the importance of animal versus plant food and secondly, at a more detailed level, the relative importance of domestic, collected, hunted, fowled and fished animals. At present, the research area consists of a mosaic of ethnic groups that all have their own economic specialisation, such as fishing, nomadic pastoralism and agriculture, or any combination of these, but that often work together and exchange food products. This means that the dietary composition of a human group does not necessarily directly reflect its own economic activities (cf. production versus consumption). In this study it is tried to determine when, why and how economic specialisation developed. In connection with this aim, possible evidence for trade in animals or animal products is also looked for. All animal-related economic activities, collecting, fishing, hunting, fowling and herding, are studied in more detail. Stock keeping strategies are

[2] Throughout this work, Latin names are added to the vernacular names of plants and animals the first time they are mentioned in a chapter. In addition, they are used in all headings containing names of animal taxa.

analysed, besides the purposes why domestics were kept, either mainly for their meat or for secondary products (e.g., skin, wool, milk, power). Hunting and fishing gear and techniques are investigated, as well as techniques for food processing and food preservation. Where possible, it is tried to go beyond a pure economic interpretation of the faunal data and gather data on past religion or ritual, cultural or ethnic identity and social status of the sites' inhabitants.

In the 1950s "Cultural Ecology" defended that all aspects of human culture are determined by the natural environment.[3] Although this degree of environmental determination seems to be a step too far, human culture, including economy, is to a certain extent determined by the environment, hence the importance of palaeoecological studies for archaeology. The studied faunal remains, especially but not exclusively those of wild animals, allow reconstruction of the former environment around the sites, which is to a large extent determined by the climate at that time. The general outline of the Holocene climatological evolution in (West) Africa is increasingly well documented (Vernet, 2002), but smaller events and the local impact of climate changes are still poorly known. The fauna can thus also provide proxy data for diachronic changes in the environment, which may be confronted with information from other disciplines. Besides climate, anthropogenic factors can also induce environmental changes. Differentiation of the two is often difficult, however, and it is tried to establish if this is perhaps possible from the faunal remains. Parallel with the seasonal changes that environments undergo throughout a year, human (economic) activities shift, indicated with the term "seasonal round". Depending on their main economic activity and the environment they live in, people will be able to stay at the same locality year-round (sedentism), or will move according to the seasonal availability of resources (nomadism, transhumance). Archaeological visibility is usually inversely related to the degree of mobility and this study aims at evaluating what kind of economies or economic activities may be missing from the archaeological record.

1.4. Research strategies

The main source to answer the research questions formulated, are the data obtained from the archaeozoological analysis of about ten sites in northern Burkina Faso and twenty-five in the southern Lake Chad area. The stress is on intersite comparison rather than detailed study of individual locations. In the Chapter 2 a description is given of both regions as they are today (climate, geography, ethnography), followed by an overview of what is known on their past. In the same chapter the studied sites are described. The methodology used for the archaeozoological analysis is discussed in Chapter 3, while Chapter 4 contains the description of the faunal remains by animal taxon. Before the actual interpretation of the fauna can be done, a taphonomical analysis needs to be carried out in order to better understand how and why the remains were preserved at, and recovered from, the sites. This subject is tackled in Chapter 5, where the archaeozoological productivity and preservation by site is also discussed. The actual interpretation of the data follows in Chapter 6, with the formulation of possible answers on the research questions. Data obtained through new faunal research are thereby put in a wider context by comparing them with data from other archaeological sites, in sub-Saharan West Africa and beyond. This study moreover aims at synthesising these data with what is known from iconography, textual evidence, genetics, animal production, ethnography and linguistics. To this end an extensive survey of the relevant literature has been carried out. Every type of source used has its own constraints (Blench, 2000a), which are discussed more substantially where appropriate.

[3] See Wilkinson and Stevens (2003) for theoretical approaches in archaeozoology and other disciplines within environmental archaeology.

Chapter 2. The research area

2.1. Today: climate, geography, the people and their subsistence

Fig. 2 illustrates the geographic position of both parts of the research area near the northern edge of the West African savannah, an almost 1500 km wide zone reaching from the Sahara in the north to the rainforest in the south. Many different systems of subdivision of the savannah zone have been proposed because its vegetation is a continuum between the extremes at the two ends. In this study the subdivisions by le Houérou (1989) and White (1983) are followed, after the example of the archaeobotanists of the Frankfurt project (Breunig and Neumann, 2004). According to these systems the savannah consists of the Sahel zone, corresponding climatologically with a yearly rainfall of 100-600 mm, the Sudan zone, with a rainfall of 600-1000 mm, and the Guinean zone, with a rainfall of more than 1000 mm. Some aspects of the savannah make it particularly favourable for human habitation, e.g., ease of movement in an open landscape stimulating the development of trade and exchange networks (Connah, 1987) and a natural richness in wild plant resources (Neumann, 2003). There are, however, some limiting ecological factors which are at least of equal importance, including a highly variable rainfall regime, poor soil fertility and the occurrence of several endemic diseases, such as malaria and sleeping sickness or trypanosomiasis (Connah, 1987; Breman, 1992). The latter is transmitted by the tsetse fly (*Glossina* sp.), whose distribution zone was, in the 1960s, between 15°N and 29°S, its northern limit corresponding approximately with the 600 mm isohyet, or the border between the Sahel and Sudan zone (Lambrecht, 1964). Although arid West Africa has many similarities with the East African arid zone, there is one major difference; the latter has an equatorial rainfall regime with two rainy seasons, while the former has only one rainy season (Le Houérou, 1989, 3). This has consequences on the grass cover, putting more nutritional stress on West African than on East African herbivores.

Fig. 2: Rainfall isohyets and eco-climatic zones of West Africa (after Le Houérou, 1989, 7)

2.1.1. Burkina Faso

All studied sites from Burkina Faso are situated in the country's most northerly province, the Oudalan, which borders Mali in the north and Niger in the east. The Oudalan has a typical Sahelian climatic regime with extremely variable rainfall. The short rainy season lasts from June to September and the long dry season from November to May (Krings, 1980, 15). The 400 mm isohyet runs through the line Oursi-Markoye and corresponds with the northern limit of permanent villages (Gallais, 1975, 167-168; Krings, 1980, 15). Temperatures can be very high during the dry season (average maxima: 38.3-40.8° C), with a cooler period between December and February (average minima: 15.1-17.5° C) (Grouzis, 1988, 16-17). The dry season is also characterised by north-eastern winds blowing from the Sahara (harmattan). During the rainy season the sky is often overcast which keeps temperatures down (Krings, 1980, 15). The prevailing wind direction is then south-west (mousson).

Geographically, the Oudalan is characterised by stabilised sand dunes of Pleistocene origin, which are east-west trending and 1-10 km wide, alternated by crystalline plains with isolated granitic inselbergs and lateritic crusts (Albert *et al.*, 1997). Because the dunes block northwards water drainage, lakes are formed on their southern edges. Some lakes, like the lake of Oursi and Darkoye can contain water year-round, while most other lakes dry up completely by February/March (Krings, 1980, 30).

The Oudalan is presently inhabited by a multitude of ethnic groups, which all have their own economic specialisation. It is a crossroads of cultural influences from sedentary agriculturalists of the Sudan zone and Saharan nomadic pastoralists (Krings, 1980, 49-50). The present ethnic structure of the Oudalan would have been formed around the middle of the nineteenth century AD. Krings (1980, 49-50) claims that the Gulmanceba (syn. Gourmantché, Gulmance) and Kurumba, both sedentary agriculturalists and speaking Songhay, are the oldest ethnic groups of the region, but does not give any arguments to substantiate this view. In the northern Oudalan, the Kel Tamashek (syn. Touareg) are the dominant ethnic group. They are Islamised Berbers originating from the Central Sahara and have an extremely hierarchic society (Barral, 1967). As nomadic pastoralists, they cover large distances in the course of one year (Gallais, 1975, 63-75). The animal species they keep differ according to caste; people from the highest castes, for example, keep dromedaries (*Camelus thomasi* f. dromedarius) (Krings, 1980, 53). The Bella (syn. Iklan, Bouzou), the former slaves of the Kel Tamashek are, with 44 % of the total population, the largest ethnic group of the Oudalan (Krings, 1980, 96). They have a long tradition of goat (*Capra aegagrus* f. hircus) keeping, the only animal the Kel Tamashek allowed them to keep, but since their liberation they have widened their range of livestock species (Krings, 1980, 57). The Islamic Peul (syn. Fulbe, Fula, Fellah, Fellata, Felaata, Fulani, Woodabe), a non-negroid group, started to infiltrate the Oudalan in the seventeenth century AD (Krings, 1980, 49). They speak Fulfulde, a language of the West Atlantic branch. Like the Kel Tamashek, they are pastoral nomads, but their annual movements are limited to between 50 and 200 km per year (Gallais, 1975, 161). The central and southern part of the Oudalan is dominated by sedentary agriculturalists (Krings, 1980, 72). The Songhay (syn. Sonraï, Songoi, Sonrai, Sonroi) settled in the Oudalan during the sixteenth century AD, coming from the Niger area (Barral, 1967; Krings, 1980, 49, 67). They are a Negroid ethnic group and principally live along the banks of the Niger between Timbuktu (Mali) and northern Nigeria (MacLean and Insoll, 1999). Although the majority of the Songhay are agriculturalists, fishing is an important element in their economy. They have specialist sub-groups; the Sorko, who mainly practice fishing and the Gow, who are hunters (Insoll, 1996, 3). The Rimaybe, descending from the Gulmanceba, are former slaves of the Peul (Krings, 1980, 65). They are chiefly farmers but also keep ovicaprine flocks near their villages, up to about 100 animals per family. The few head of cattle (*Bos primigenius* f. taurus) they own are entrusted to the Peul to herd. The Mallebe are sedentary farmers assimilated by the Peul (Krings, 1980, 66). Their main occupation is the cultivation of pearl millet (*Pennisetum americanum*), but they also keep sheep (*Ovis ammon* f. aries) and goats and in some groups cattle are also relatively important. The Mossi are nowadays an ethnic minority in the Oudalan, but in the eighteenth century AD they formed the main section of the population of Sahelian Burkina Faso together with the Gulmanceba (Krings, 1980, 68). In all of West Africa the Mossi are considered as specialised pearl millet farmers. Their language belongs to the Gur branch within the Niger-Congo phylum.

With 6 % percent of the total population of Burkina Faso, the Oudalan possesses about 20 % of the countries livestock (Grouzis, 1988, 11). A few geographic features make the region particularly favourable for stock keeping (Barral, 1967). First of all, it is situated outside the tsetse-zone and although it has an extremely irregular rainfall regime, rains are still sufficient to sustain year round watering points for livestock (e.g., wells, lakes, springs). The rainfall also allows the formation of pasture areas of better quality than in the Sudan zone. Sahelian herders stress that the finer grasses from the Sahel zone are significantly better fodder than the hard, woody, grass species from the Sudan zone (Krings, 1980, 44-45). The main constraint for livestock keeping in the Oudalan is the limited availability of drinking places during the dry season; pasture areas are useless when situated more than 20-25 km away from such a watering spot (Krings, 1980, 45). The cattle in the area are of the zebu (*indicus*) type, while both sheep and goat are of the large savannah forms (Grouzis, 1988, 11) (see 4.6.10. for more details on breeds). The estimated proportion of cattle versus ovicaprines in the Oudalan is about 1 to 3, with an average of 41 animals by herd. The sandy dune soils of the Oudalan yield good pasture areas and are also favourable spots for agriculture (Vogelsang *et al.*, 1999). About ninety percent of the regions cultivated surface is taken by pearl millet (Grouzis, 1988, 10), which is grown during the rainy season. After the droughts of the early 1970s, the cultivation of sorghum, thriving better on clay soils, also expanded. Cultivation usually takes place in agroforestry systems, defined as land-use units combining the cultivation of crops with the systematic exploitation of woody plants growing on the same plot (Sturm, 1997). Besides agriculture, the gathering of wild plants also contributes significantly to obtaining vegetable food (Grouzis, 1988, 14). Barral (1977, 20-22) stresses the limited economic importance of wild terrestrial animals in the Oudalan, which are severely threatened by population growth and competition with growing numbers of domestic animals.

Fig. 3: African language families (Blench, 1993)

2.1.2. Nigeria

The Nigerian part of the research area is situated in the country's Borno State, bordering Niger in the north, Chad in the north-west and Cameroon in the west. Its climate is very similar to the Sahelian climate in northern Burkina Faso (Tschierschke, 1998). The average yearly rainfall varies between 600 mm in the south and 200 mm in the north. The dry season lasts from November to March and rains fall between April and October with most precipitation in July and August. The mean annual temperature is around 27° C; the beginning and end of the rainy season are the hottest periods of the year. Winds can be strong and wind directions are similar to the ones described for northern Burkina Faso, with the harmattan blowing north-east from October to April and south-westerly winds, bringing rain, from May until September.

Geographically, the area is largely dominated by the presence of Lake Chad. During the Early and Middle Holocene climatic optimum (see 2.2.1.) it was about as large as the Caspian Sea is today and is referred to as Lake Mega-Chad. The Bama Ridge, a former shoreline of Lake Mega-Chad that reaches a height of up to 12 m at some places, demarcates its maximal extension (Thiemeyer, 2004). Closer to the lake, the Ngelewa Beach Ridge, which has not been precisely dated but is probably from slightly before 2000 BC, marks another lake level that was slightly higher than the recent one (*Ibid.*). After the mid-Holocene, the lake gradually retreated because of increasing aridity, be it with many large and small oscillations, until it reached its present size (Brunk and Gronenborn, 2004). Present-day Lake Chad is remarkable for its shallowness. Together with high evaporation, this is one of the causes of its constantly fluctuating shape and size, on seasonal, mid-term and long-term time scales (Servant and Servant-Vildary, 1980; Franke-Scharf *et al.*, 2004). The "average" Lake Chad covers a surface of about 21000 km^2, with depths ranging between 2.5 and 6.5 m (Carmouze and Lemoalle, 1983). The lake reaches its highest annual level in December-January and is at its lowest mid-October, at the end of the rainy season (Tschierschke, 1998). About 90 % of the lake's water supply stems from the rivers Chari and Logone. The sources of these rivers are situated more then 500 km to the south, in a region with an annual precipitation of around 1500 mm. Correlations have been shown for rainfall in the source area of the Chari and Logone and the level of Lake Chad on the one hand, and local rainfall in the Lake Chad area on the other (*Ibid.*). Brunk and Gronenborn (2004) do not agree with this, however, and stress that there is not always a good correlation between lake levels and local climatic conditions, because the catchment area of the lake lies in the Sudan and Guinean zone, while the central parts of its basin are part of the Sahel zone.

The first ground that became available after the retreat of the lake, around 2000 BC, is the geographical unit called the Bama Deltaic Complex. Its landscape has a low relief formed by previous deltas of the rivers Yedseram and Ngadda (Breunig *et al.*, 1992). In the central section, sand plains and clay depressions predominate, whereas flat, stabilized longitudinal dunes occur in the north and south-east. The clay depressions are believed to be the remnants of an ancient lagoon formed by the backwaters of Lake Mega-Chad (Breunig *et al.*, 1993b). During the rainy season they are filled up with water and some can contain water year-round. About a millennium after the Bama Deltaic Complex the Chad Lagoonal Complex also became free. This geographical unit is known as the *firgi* area, after the Kanuri name for the flat lagoonal clay plains that dominate its landscape. The plains are almost treeless and are covered with perennial grasses in the wet season. Sand islands interrupt the heavy, dark clays extending over a surface of about 5000 km^2 (Breunig *et al.*, 1992). The clay plains continue in the neighbouring area of Cameroon where they are known under the name *yaéré*. Large parts of the *firgi* plains are inundated annually, both by rainwater and by water coming form the major rivers (Bawden, 1972). Beadle (1981, 223) specifies that floodwater comes from the Chari and Logone, while excess water is evacuated both by the Logone and the El Beid. Permanent settlement in the *firgi* and *yaéré* is only possible on land above the annual average flood level (Holl, 2003, 303). Semi-permanent villages are more widespread, located on the edge of the land above flood level, while dry season camps are situated in the hinterland depressions below the flood levels, where the presence of water and high-quality grazing attract pastoralists from an extensive area.

Due to the particular geographic circumstances of the plains south of Lake Chad Basin there are no stones in the area (Connah and Freeth, 1989; Rupp, 2004). The nearest outcrops are situated at a distance of about 70 km.

Like the Oudalan of Burkina Faso, the Nigerian part of the research area is mainly Islamic and has a mixture of various ethnic groups. Linguistically they can be divided into two main groups, those speaking Saharan languages and those speaking Chadic languages, besides two smaller language groups, Semitic and West-Atlantic speakers. The Kotoko (Bouquet, 1990, 258-261) belong to the Chadic speaking group. They claim to be the descendants of the Sao or So giants, which are legendary fishermen and hunters. They are probably the oldest inhabitants of the southern Lake Chad area and are mainly fishing people and also usually city dwellers. Today they are mostly confined to the Cameroonian side of the lake. Another Chadic speaking group, the Gamergu, lives more in the western part of the study area, near the shores of the Yedseram (Cyffer *et al.*, 1996). A third Chadic speaking group are the Buduma (syn. Yedina) (Bouquet, 1990, 237-245). These extremely mobile people inhabit the islands of Lake Chad, adapting to the smallest changes in its levels, and are notorious for cattle and slave raiding. Their main economic activities are, in order of importance, cattle herding, cultivation of millet and fishing. The Kuri are another ethnic group living on the lake's islands and speaking the same language as the Buduma, but their economic stress is on agriculture (Bouquet, 1990, 248-250). The majority of the inhabitants of the Nigerian southern Lake Chad area today refer to themselves as Kanuri. Kanuri belongs to the Saharan language group and seems to have displaced the Chadic languages in the region during the course of the present millennium (Connah, 1981, 35). The Kanuri are farmers and sometimes own cattle too, which they usually give to the Shuwa Arabs to herd (Blench, 1995; Breunig, 1995; Sturm *et al.*, 1996). The latter are the only Semitic-speaking group in the southern Lake Chad area (Connah, 1981, 29). They form the majority of its Arab population and started penetrating the area in the mid seventeenth century AD (Braukämper, 2004). Initially they were Saharan camel nomads but under the constraint of the new ecological circumstances they faced in the Lake Chad area, they adopted a mixed economy of cattle-breeding and cultivation of millet. The Nigerian part of the research area also has a minority of Peul (syn. Fulbe, Fula, Fellah, Fellata, Felaata, Fulani, Woodabe) speaking Fulfulde, a West Atlantic language, and who started migrating into the area from Central Niger around the beginning of the twentieth century AD (Shareika, 2004). In contrast to their relatives in Burkina Faso, they are highly mobile cattle herders.

North-eastern Nigeria is situated in a zone that is now recorded to be free of tsetse flies, but was apparently only cleared after 1967 (http://ergodd.zoo.ox.ac.uk/teseweb/all_species.htm) Goats seem to be ubiquitous throughout Nigeria, while cattle are mostly confined to the northern two-thirds of the country, where herds are often accompanied by small flocks of sheep (Bourn *et al.*, 1994). In the north zebu and Kuri are the prevailing cattle types, Balami and Uda the most important sheep types and Sokoto reds the main type of goats, besides Sahel goats which are restricted to a strip along the frontier with Niger (see 4.6.10. for more on details breeds). Cultivation in the southern Lake Chad area concentrates on two crops: millet, rain-fed and only on sandy soils, and sorghum, often on seasonally inundated clay soils using a particular technique of dry season cultivation, locally called *masakwa*[4] (Zach *et al.*, 1996). Besides cultivated crops, wild plants are also intensively used (Neumann, 2003).

2.2. The past 4000 years: Climate, archaeology and history

Archaeology is almost the only source of knowledge on human life in the West African past, up to around the end of the first millennium AD, but after that date, historical information is also available. The oldest written sources on West Africa are Arabic. Arabs gradually began to penetrate West Africa after their invasion of northern Africa during the seventh century AD (Mones, 1992), coinciding roughly with the beginning of what has been called the West African medieval period. Most of the surviving Arabic sources on West Africa are external, i.e. written by Arabs who lived in the Muslim world north and east of the Sahara (Levtzion and Hopkins, 1981, 1). Some of the most important sources, such as Al-Bakri writing in 1068 AD, had never even set foot in western Africa (Levtzion and Hopkins, 1981, 62). The corpus of Arabic sources has been assembled in a single volume of translations (Levtzion and Hopkins, 1981), while all data related to food are synthesised in Lewicki (1974). European descriptions of the coast and hinterland appear in the fifteenth century AD and from the nineteenth century AD there are also texts by early European travellers who explored the African continent (DeCorse, 1997). These are of great interest because of the information they contain on West Africa before the radical changes that set in during the colonial era, between about AD 1880 and 1935 (Boahen, 1990).

2.2.1. Climate

The climate in sub-Saharan West Africa is governed by seasonal migrations of the Intertropical Convergence Zone (ITCZ; the meteorological equator), which are in turn responses to changes in the location of maximal solar heating. This results in belts of monsoonal climate regimes with summer rains and winter droughts and with more seasonal extremes and decreasing precipitation, the further from the equator. In sub-Saharan West Africa, like everywhere in the world, the climate has undergone considerable changes in the course of time, on a millennial to decadal scale (Hoelzmann *et al.*, 2004). Nevertheless, its seasonality has remained a constant since as far back as the Miocene (Maley, 1996). The archaeology and history of the research area must be seen against the background of climatic changes, as these often appear to have functioned as a trigger for changes in human adaptive strategies (e.g., Brooks, 1998). Abrupt dry intervals seem to have particularly influenced human behaviour (e.g., Hassan, 1996). Nonetheless, Vernet (2002) has argued that a more precise chronological level, ideally 25 years or a human generation, of climatological reconstructions is necessary in order to find causal relations with cultural changes.

---

[4] In the dry season, dams and ditches are constructed around the clay fields in order to keep as much water as possible on them once the rains start (Sturm *et al.*, 1996). During the rainy seasons, the clay farms are usually weeded and hoed because the clay is then soft (*Ibid.*). In August, certain sorghum varieties are sown into small seedbeds, and after a growth period of about six weeks, at the end of the rainy season, the seedlings are removed and transplanted into the clay fields, where they are harvested around January (*Ibid.*).

One of the most important techniques applied in order to study the past African climate is the use of records of lake level fluctuations, gathered mainly by means of geomorphology, sedimentology, stable isotope contents of primary lacustrine carbonates and organic matter, and biological remains (e.g., diatoms, ostracods). Overviews are given, for example, in Servant and Servant-Vildary (1980), Street-Perrott and Perrott (1993), Jolly *et al.* (1998), Gasse (2000) and Street-Perrott *et al.* (2000). Another important source is the record of vegetation through pollen analysis, e.g., Salzmann (1999, 2000) or Salzmann and Waller (1998). For more recent periods these records often face the problem, however, of distinguishing between anthropogenic and climatological factors influencing vegetation. Lézine and Casanova (1989) and Hoelzmann *et al.* (2004) have confronted both lacustrine and vegetation records for the reconstruction of past climates, the former for tropical West Africa and the latter for the Sahara, Sahel and Arabian Peninsula. Vegetation responses to climatic changes may be delayed, and the two types of data are therefore not always consistent, sometimes even contradictory (e.g., Salzmann, 1999).

All evidence points to a climatic optimum during the Early and Middle Holocene and it is estimated that the ITCZ moved about 500 km northwards during this period (Gasse, 2000). Northerly zones witnessed a higher rainfall than at any time since and this has presumably led to a "greening" of the present Saharan desert. Fish remains point to the existence of lakes in the area with permanent, well aerated, water (Van Neer and Gayet, 1988). Large savannah fauna, like elephants (*Loxodonta africana*) and giraffes (*Giraffa camelopardalis*), populated the Sahara during this early-mid Holocene climatic optimum (Le Quellec, 1999). Based on animal bone remains from the period, Peters and Pöllath (2003) were able to detect an east-west gradient, with more humid conditions in the western parts of the Sahara. Rock art specialist Muzzolini (1995, 25) discerns two humid phases in the Saharan Early and Middle Holocene (ca. 10000-5500 BC and ca. 4200-2500 BC) interrupted by an arid spell (ca. 5500-4500 BC). He connects the second humid phase with the appearance of pastoral communities and the earliest phases of rock art, depicting large wild fauna as well as domestic animals. The pastoral groups left archaeological traces, in the form of hearths for example (Gabriel, 1987), but bone remains of their flocks have been recovered as well (e.g., Gautier, 1987a).

Dry conditions prevailed again in the whole Sahara around 2000 BC, but in its eastern parts, they had already begun at least a millennium earlier (Gasse, 2000). Increased aridification initiated large-scale population movements southwards from the Sahara, into the present savannah zones, archaeologically traceable from Mauritania to Niger (McIntosh, 1994). This is also the period in both parts of the research area from which the first archaeological traces have been found (see 2.2.2. and 2.2.3.). Southward migrations, as a reaction to prolonged periods of aridity, have also been attested more recently during the great drought in 1913-1914 (Brunk and Gronenborn, 2004). Due to the increasingly dry climate, vegetation belts must have retreated southwards, although Hoelzmann *et al.* (2004) have proven that the model of shifting vegetation zones is too simplistic. Nonetheless, the Saharan people moving to the south may have been following their resources (Breunig and Neumann, 2004). The population movements seem to have caused considerable cultural changes in both parts of the research area. In later periods, cultural innovations often seem to have also come from the north (see 2.2.2. and 2.2.3.).

In sub-Saharan West Africa the general trend towards aridification seems to have reached its height during the first millennium BC (Breunig and Neumann, 2004). Brooks (1998) considers the period between ca. 300 BC and AD 300 as a transitional dry phase, before ecological conditions improved again during the following centuries. Between AD 300 and 1100 relatively benign climate circumstances reigned, while there are several historical indications for climatic deterioration in the thirteenth and fourteenth centuries AD, causing conflicts between human groups (Brunk and Gronenborn, 2004). There have also been several periods during the past few centuries in which the climates of northern hemisphere Africa differed from those of today. Between the sixteenth and eighteenth centuries AD a period of wetter circumstances is documented, for example, while several major droughts occurred in the eighteenth century AD (Nicholson, 1980; Brunk and Gronenborn, 2004).

The local impact of climatic fluctuations in the research area has been studied as part of the "SFB 268" at the University of Frankfurt. For Burkina Faso, data from pollen and diatom analysis are available (Morczinek, 1995; Ballouche and Neumann, 1995; Andres *et al.*, 1996; Ballouche, 2001; Neumann *et al.*, 2004). For Nigeria vegetation changes have been studied through pollen analysis (Salzmann, 1996, 1999, 2000; Salzmann and Waller, 1998) and another important data set for the area comes from diachronic reconstructions of levels of Lake Chad both through the Frankfurt project (Brunk and Gronenborn, 2004) and older studies (Maley, 1981). Lake Mega-Chad can be connected with the climatic optimum of the Early and Middle Holocene. Higher lake levels were mainly due to higher rainfall, but also to less evaporation because of higher humidity and lower temperatures (Maley, 1973). Subsequent shrinking of Lake Chad after the mid-Holocene is linked to the increasingly dry climate. However, the lake has known many size oscillations, which are better documented for more recent periods, when historical sources and direct observations become available (Brunk and Gronenborn,

2004). In addition to pollen analysis and fluctuating lake sizes, results from archaeobotanical analysis have also been used as proxy data for climatological reconstructions. They are especially important on a localised geographic level.

2.2.2. Burkina Faso

2.2.2.1. General

Late Stone Age

Despite extensive archaeological survey, no sites older than 2200 BC have been discovered thus far in the north of Burkina Faso. Like many other regions in West Africa, the area must have been only sparsely populated before the Late Holocene. The oldest discovered sites date between 2200 and 1000 BC and seem to be the remnants of occupation by highly mobile groups (Vogelsang *et al.*, 1999; Vogelsang, 2000). The sites are concentrated in the dunes, they are shallow and do not contain indications for settlement structures, apart from some pits with unknown function. The cultural material is composed of comb- and roulette decorated pottery, polished stone axes, some grinding equipment, a microlithic stone industry with segments as the dominating type and sometimes bifacially retouched arrowpoints of a Saharan type. The composition of the material has allowed the distinction of two regional groups, the "Dori" facies and the "Tin Akof" facies. The recovered plant remains consist mainly of fruits of wild trees and shrubs, but some sparse domesticated pearl millet was also present (Neumann *et al.*, 2000). AMS-dating run on the grains yielded a date of around 1000 BC (Vogelsang *et al.*, 1999; Vogelsang, 2000), but the crop was already present in the oldest of the site's occupational phases (Kahlheber and Neumann, in press), based on the archaeological inventory beginning around 1800 BC (Kühltrunk, 2000). The domestication process probably took place elsewhere, since the wild types were not found in archaeobotanical samples, which evidence ought to have been present if the people from northern Burkina Faso had experimented with their cultivation (Neumann *et al.*, 2000).

Iron Age

With the exception of the pottery site Kissi 49 (Magnavita *et al.*, 2002), no archaeological traces are known in the north-east of Burkina Faso for the entire first millennium BC. Breunig and Neumann (2002a) have argued that the apparent discontinuity was a reaction to environmental instability, which induced cultural innovations. The innovations would have become clear around the beginning of our era, with the appearance of large Iron Age sites, marking a shift to a sedentary way of life.

The Iron Age in north-eastern Burkina Faso has been divided into Early (AD 0-500), Middle (AD 500-1000) and Late Iron Age (AD 1000-1400) based on changes in ceramic decorations combined with radiocarbon dates and stratigraphies of excavated sites (von Czerniewicz, 2004, 124). After the fourteenth century AD the archaeological sequence for the region stops, although the very last occupation phases at the mounds are missing because of erosion. The settlement mounds were probably abandoned as a consequence of political instability in the region, and an ethnical change took place whereby nomadic groups replaced the sedentary ones (Albert *et al.*, 2000). Pelzer *et al.* (2004) see a relationship with the fall of the Songhay Empire[5] in AD 1591. Other causes may include environmental deterioration and perhaps also conflicts between agriculturalists and pastoralists (Kahlheber and Neumann, in press). Parallels for the Iron Age ceramics and settlement pattern in the Oudalan can be found at contemporary sites in southern Mali, in the Inland Niger Delta and the Méma (von Czerniewicz, 2004, 151). South of the Oudalan, ceramic traditions, like the one recorded in the Chaîne de Gobnangou in south-eastern Burkina Faso, seem to be different (von Czerniewicz, 2004, 159).

Besides 21 surface sites, a total of 116 Iron Age mounds, 1 to 8 m high and with a circumference of 20 to 200 m, have been mapped for the Oudalan (von Czerniewicz, 2004, 126, 137). Like the Late Stone Age sites, these are mostly found in the dune areas, which are still favoured for human habitation. The determining factors for the choice of settlement areas seem to have been the availability of fertile soils and water. The settlement mounds usually occur in small groups and are interpreted as villages, or at least large hamlets (Breunig and Neumann, 2002a; von Czerniewicz, 2004, 138). Most sites are large accumulations, up to 8 m thick, of cultural remains, without much stratification and with only indirect evidence for settlement structures in the form of pisoliths[6] (Breunig and Neumann, 2002a). Excavations at Oursi hu-beero (BF 97/30), however, have revealed the burnt remains of a large two-storied house, consisting of at least 14 rooms, on top of a settlement mound (Hallier and Petit, 2001).

With the beginning of the Iron Age, iron tools were introduced into daily life, for example for food preparation (knifes, awls) or hunting (arrow- and lance

[5] At the height of its power, in the early sixteenth century AD, Songhay was the largest empire ever created in Tropical Africa (Crowder, 1977, 47-52). In the second half of the fifteenth century AD it had replaced the Empire of Mali as the most important power in the western savannah (Crowder, 1977, 26-35). Together with the Empire of Ghana and the Kanem-Borno, the latter was part of the largest tree empires that medieval West Africa has known before AD 1450.

[6] Pisoliths are small, round iron concretions and remains of the laterite crust that was formed in the region during the Tertiary and Pliocene part of the Quaternary (Albert *et al.*, 1997). They presumably arrived at the sites as temper for the mud used in house construction (von Czerniewicz, 2004, 9).

points) (von Czerniewicz, 2004, 137). The Iron Age also seems to have marked a change to full farming. Pearl millet was already known since the end of the Late Stone Age and from the second half of the first millennium AD onwards there is also evidence for sorghum (Höhn *et al.*, 2004). Cultivation of pearl millet probably took place in agroforestry systems, as it does today (Kahlheber, 1999). Growing of domestic crops never seems to have totally replaced the harvesting of wild plants (Neumann, 2003). Evidence for the sheabutter tree (*Vitellaria paradoxa*) at several of the Iron Age sites, a tree that can no longer thrive in the area, indicates that rainfall during the Iron Age was higher than at present (Neumann *et al.*, 1998; Höhn *et al.*, 2004).

From the sixth-seventh century AD onwards there is archaeological evidence for long distance trade from West to North Africa, as for example at the cemetery of Kissi in Burkina Faso, where cowry shells and glass pearls have been found amongst other exotic goods (Magnavita *et al.*, 2002). The earliest finds of sorghum have been seen as a possible indication for trade as well, and the crop may have been an imported luxury food or a remnant of passing caravans (Kahlheber, 2003, 224). Finds of weapons at the cemetery of Kissi point to conflicts during the later Iron Age (Magnavita *et al.*, 2002). Contacts with North African Arabs initiated Islamisation of western Africa, although the first Arabs in the region were traders, not interested in imposing their religion on the local population (Diallo, 1990). By the sixteenth century AD, Islam was well established in the region, but remained a cult of the courts and trade centres for a long time (El Fasi and Hrbek, 1988; Masonen, 1997). The rural hinterland stayed only slightly touched by Islam until the nineteenth and twentieth century AD when large-scale conversions took place (de Benoist, 1983).

2.2.2.2. Description of the studied sites

The excavation strategies of the Frankfurt team concentrated mainly on establishing a chronological framework for northern Burkina Faso since it was virtually archaeological "Terra Incognita" up to their arrival. Most trenches dug were therefore only a few square metres in size. Because stratification was usually not clear, excavations were carried out by square metre in spits of 10 cm down to sterile levels. All sediment was dry sieved on 10 mm meshes. Sediments for archaeobotanical investigation were additionally sieved on 2.5, 1 and 0.5 mm meshes and fractions flotated. For most of the sites, archaeological and archaeobotanical finds from only one square metre were studied, in contrast to the faunal remains that were usually analysed for all squares. This was not done out of diligence but because it had been proven necessary to study samples from a larger volume for the faunal analysis than for that of other disciplines. The radiocarbon dates of the sites are summarised in Appendix A. In Appendix B, tables for each site indicate the weight of the faunal remains by square and spit (see 2.2). A more in-depth discussion on the amounts of faunal remains recovered by locality is given in 5.6.

Sites of the "Tin Akof" facies (BF 94/133) (Vogelsang *et al.*, 1999; Kühltrunk, 2000; Vogelsang, 2000)

The site of Tin Akof, in the very north of Burkina Faso, near the Malian border, has given its name to the northern Late Stone Age facies of the Oudalan. The material culture of the Tin Akof facies is characterised by retouched arrowheads, the absence of microliths and a predominance of ceramics with organic temper. The excavated trench at the site measured 17 m^2 and was investigated by quarters of a square metre and in spits of 5 cm instead of the usual 10 cm. The stratigraphy of the site consisted of an almost sterile upper sand layer of around 30 cm thick, underneath which a cultural layer of up to 40 cm thick was present. Ten pits of between 0.2 and 0.7 m deep, relatively rich in finds, were recorded. Their function remains unknown, however. Inside the pits organic preservation was relatively good and some charcoal and bones (350 g) have been retrieved from them. Outside the pits no organic material was preserved. The study of the ceramics found in the pits combined with radiocarbon dates suggest an occupation of the site during three phases, dated between 1800 and 1000 BC.

Sites of the "Dori" facies (BF 94/40 and BF 94/96) (Vogelsang *et al.*, 1999; Vogelsang, 2000)

The Dori facies is confined to the southern part of the Oudalan. Stone tools at the sites, mainly microliths, are almost exclusively made out of quartz. Two sites near Dori, BF 94/40 and BF 94/96, were excavated in a 6 m^2 trench, by square metre and by 5 cm spit. The stratigraphy at both sites was similar to Tin Akof (BF 94/133), with an anthropogenic layer of 40 cm thick underneath a layer of aeolian sand of around 30 cm thick. Radiocarbon dates from the site cover a large time span, between about 2000 and 1000 BC, probably as a result of perturbation (unspecified; Kühltrunk, in prep.). BF 94/96 yielded very few animal remains (14 g) and only one bone was recovered from BF 94/40 (1 g).

Sites near Oursi

Four sites were excavated in the dunes in the vicinity of the present village of Oursi, that was founded in the seventeenth century AD by Songhay and still has a predominantly Songhay population (Froment, 1988, 35). The village is situated at the edge of one of the

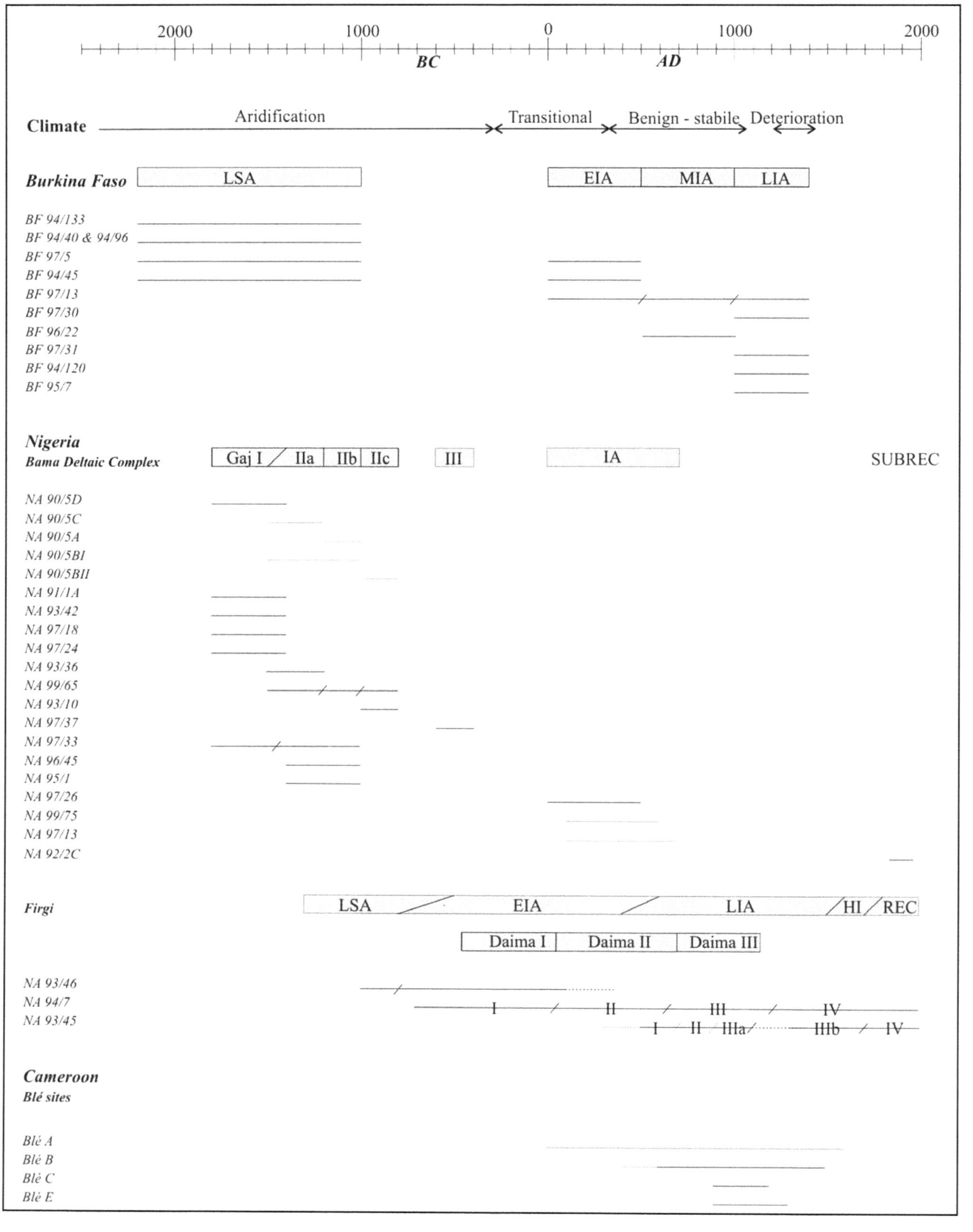

Fig. 4: Time scale with studied sites and summary climatological data

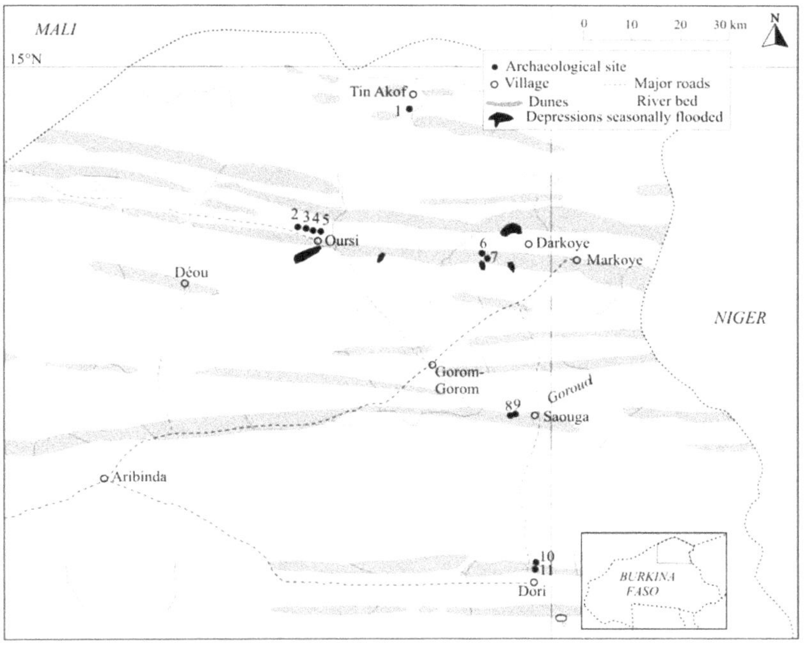

Fig. 5: Studied sites in northern Burkina Faso.
1: Tin Akof (BF 94/133), 2: Oursi (BF 94/45), 3: Corcoba (BF 97/5), 4: Oursi hu-beero (BF 97/30), 5: Oursi village (BF 97/13), 6: Kissi 22 (BF 96/22), 7: Kissi 40 (BF 97/31), 8: Saouga 94/120 (BF 94/120), 9: Saouga 95/7 (BF 95/7), 10: Dori 94/40 (BF 94/40), 11: Dori 94/96 (BF 94/96)

largest lakes of the Oudalan to contain water year round in years with good rainfall (Krings, 1980, 30). When water is at its highest, in August/September, depths of up to 5 m can be measured.

Corcoba (BF 97/5) (Breunig and Neumann, 2002a; Breunig, pers. comm.)

Corcoba is a flat site situated on an old dune, a few kilometres north-west of the lake of Oursi. It has Late Stone Age deposits, with the earliest date of all sites near Oursi, as well as Early Iron Age deposits. Eight trenches of 1 m^2 were dug at the locality. Trenches I to VI were dated to the Late Stone Age, trench VIII to the Early Iron Age, and trench VII covered both phases. Trenches II to VI were excavated to investigate the contents of a large pit, which had a depth of up to one metre and covered an area of several square metres. Thanks to good preservation conditions in the pit, a relatively large sample of organic remains, including bones, could be recovered. The other Late Stone Age contexts yielded no (trench I), or very few, bone remains (trench VII), while more bones were present in the Iron Age context. With the exception of trench I and VII, all trenches were archaeobotanically sampled, but domestic plants were only identified from the Iron Age context (Kahlheber, 2003, 99, Table 20). For the Late Stone Age a temporary occupation by a relatively large group is proposed (Kühltrunk, in prep.).

Oursi (BF 94/45) (von Czerniewicz, 2004, 19-47)

The archaeological site Oursi is situated about two kilometres west from the village, near the shores of the lake. It is visually the most marked mound of a group of twelve. The excavated trench was 4 by 3 m in size and sterile layers were encountered at a depth of 6.0 m. Only 6 of the squares were excavated to the maximum depth and below -5.0 m, only one half of squares E7 and F7 was dug out. The layers below -4.3 m were attributed to the Late Stone Age, those from -4.0 m upwards to the Early Iron Age. Late Stone Age and Early Iron Age are separated by a chronological hiatus of about 1000 years. Nevertheless, the faunal distribution does not show a hiatus between the levels of the two periods. For the faunal analysis all material from layers above -4.0 m was therefore considered as Early Iron Age, and remains from below this level as Late Stone Age. Compared to the faunal sample from the Early Iron Age, that from the Late Stone Age is very poor. Archaeological and archaeobotanical analysis concentrated on square G8, while for the fauna all squares were studied. In addition, the animal remains retrieved from the archaeobotanical samples were analysed.

Oursi village (BF 97/13) (von Czerniewicz, 2004, 48-71)

Oursi village is the largest mound of a group of 25, situated about 1 km north of the present village of Oursi and further away from the lake than the aforementioned site Oursi, which lies 2 km more to the west. A 3 by 3 m trench was opened on the mound top and excavated to a depth of 8.1 m, which could only be reached in 2 out of the 9 squares (L6 and M6) for practical and safety reasons. The site's stratigraphy covers all Iron Age phases: Early Iron Age (-8.1 to -5.4 m), Middle Iron Age (-5.3 to -2.4 m) and Late Iron Age (-2.4 m to surface). Radiocarbon dates obtained from the Early Iron Age layers, suggest an occupation beginning slightly after the abandonment of Oursi. Archaeological and archaeobotanical finds were studied for square L6 only, and for the faunal remains, both squares (L6 and M6) where the maximal depth was reached were analysed, besides the bones retrieved from archaeobotanical samples in square L6. A glass pearl found in Middle Iron Age layers possibly points to trade contacts (Höhn *et al.*, 2004). In the course of the site's occupation the relative amounts of collected fruits and seeds decrease, probably because more and more space was taken for the cultivation of pearl millet (Kahlheber, 2003, 223).

Fig. 6: View of Oursi hu-beero (BF 97/30) during excavations (Picture SFB 268)

Oursi hu-beero (BF 97/30) (Hallier and Petit, 2001)

Oursi hu-beero is situated on top of a mound that is part of the same group of mounds as Oursi village. As mentioned in 2.2.1.1., this is one of the rare settlement mounds of northern Burkina Faso where house structures have been found, thanks to a fire that "baked" their mud brick walls. Mud brick debris sealed off the building and preserved it in the exact state it was left in during the Late Iron Age, around AD 1000. Because of the special nature of the site, excavation techniques were different than at the mound sites. Archaeological features were respected while excavating, and sieving was limited to special contexts. Except for some bone remains from a test-sieved sample and from archaeobotanical sampling, all animal remains from the site were hand-collected. All finds are considered to belong to the same occupation phase, but some parts of the site are disturbed by erosion, which makes their dating less secure (Petit, pers. comm.). An iron object retrieved from Oursi hu-beero has been interpreted as a slave-chain, which indicates that the inhabitants of the site may have taken part in the slave-trade economy across the Sahara (Petit and Hallier, pers. comm.). Inside the house, the remains of two adult humans and one child, probably murdered, have also been recovered (von Czerniewicz *et al.*, 2004).

Sites near Kissi

A second group of sites, consisting of settlement mounds, cemeteries and stone structures, is situated near Kissi, an Iklan dominated village (Krings, 1980, Map 3), in the dune field at the northern edge of the medium-sized lake of Kissi (Magnavita *et al.*, 2002). The most important of these sites is the cemetery of Kissi 3 where burials with rich grave goods have been found, dating to the fifth-seventh century AD. The cemetery is situated at the western end of a settlement mound, Kissi 3b, and is surrounded by many more. The chronology of the sites near Kissi covers almost the entire Iron Age. The fauna from Kissi 3b was not available for faunal analysis, but material from two other settlement mounds from the same group was studied. Sieves used at the Kissi excavations had smaller meshes than usual (2.5 and 1 mm) because there were always a lot of small beads in the archaeological layers.

Kissi 22 (BF 96/22) (Magnavita *et al.*, 2002; S. Magnavita, pers. comm.)

The settlement mound Kissi 22 is located about 100 m south of Kissi 3 and 600 m north of the lake. A small test pit was dug at its top, from where a radiocarbon date, falling in the Early Iron Age, was obtained. Very little faunal material was recovered from the test pit, and as this was collected with finer sampling techniques than that of the other Kissi sites, it has not been included in this study. The actual excavation at Kissi 22 (Kissi 22B), a trench measuring 2 by 4 m at the surface, was done at the foot of the mound at a location where a stone structure was visible. A layer of about 10 cm thick, eroding from its surface, has been interpreted as the clay bottom of a house. Unlike at the other excavations in northern Burkina Faso, the trench was excavated in 5 cm spits and over the complete surface, without a subdivision in square metres. Below the floor, several pits and hearths were discovered up to a depth of about 4.0 m, where the excavations were stopped. Most of the animal bones have been found inside these features. In the lowest layer a human burial was unearthed, dated to the very beginning of the Iron Age. Radiocarbon dates for the layers point to occupation in the Middle Iron Age.

Kissi 40 (BF 97/31) (Magnavita *et al.*, 2002; S. Magnavita, pers. comm.)

Kissi 40 is a large complex composed of two settlement mounds, situated about 500 m east of Kissi 3 and Kissi 22. Excavations in a 2 by 2 m trench between the two mounds yielded clay house floors with pits and hearths, dated to the ninth to twelfth centuries AD, thus falling in the Late Iron Age. At about -2.5 m, the clay layers stopped and from that point downwards a sandy cultural sediment dominated. The pits, on the contrary, continued down to about 5.0 m deep, where the excavations were ended.

Sites near Saouga

Situated 2 km east of the village of Saouga, with a predominantly Songhay population, there is another group of 14 settlement mounds (von Czerniewicz, 2004, 72). About 2 km north of this group runs the river Gorouol, which today only carries water during the rainy season. Two trenches were excavated on the largest mound of the groups, BF 94/120 on its top and BF 95/7 at its foot.

Saouga 94/120 (BF 94/120) (Vogelsang *et al.*, 1999; Vogelsang, 2000; von Czerniewicz, 2004, 72-98)

The trench of Saouga 94/120 was 3 by 3 m in size and although cultural layers continued further down, the excavations had to be stopped at a depth of 5.7 m for safety reasons. The excavated surface was gradually reduced with increasing depth. Radiocarbon dates indicate that the six-metre stratigraphy of the site built up in less than 400 years. Archaeologists and archaeobotanists concentrated on the finds of square G5, while all squares were considered for the faunal analysis.

Saouga 95/7 (BF 95/7) (Vogelsang *et al.*, 1999; Vogelsang, 2000; von Czerniewicz, 2004, 72-98)

A second trench, 2 by 3 m, was opened at the foot of the same mound to collect more information on formation processes of the mounds and their horizontal extent, but deposited layers were interpreted as mainly erosion material from surrounding mounds. Cultural deposits at Saouga 95/7 still reached almost 3.0 m deep. Depth labelling started at -1.8 m, and not 0, in order to make depths correspond to those of Saouga 94/120. The fauna of all squares was studied, detailed archaeological and archaeobotanical analyses, on the other hand, were limited to square G9.

Fig. 7: View of Saouga 94/120 (BF 94/120) during excavations (Picture SFB 268)

2.2.3. Nigeria

2.2.3.1. General

Earliest archaeological evidence

The earliest traces of Holocene occupation in north-eastern Nigeria were found at Dufuna, in the Yobe state, bordering the research area in the Borno State, where a dugout of about 8.5 m long was excavated, dating to around 6000 BC (Breunig, 1996). It is the oldest known African boat, and its date implies a possible association with the Mega-Chad period. The Dufuna boat testifies to an aquatic adaptation of at least part of the hunter-gatherer groups around that time (Breunig and Neumann, 2004). At a site near Konduga, on the Bama Ridge, a few potsherds dating to the sixth millennium BC were found (Thiemeyer, 1992; Breunig *et al.*, 1996). They are decorated in a style and technique known from contemporary sites in the Sahara. Broadly contemporary with Konduga is the site of Blabli (David and Sterner, 1987, 1989), also close to the Bama Ridge. However, there seems to be a problem with the radiocarbon dates obtained for the site, because they turned out to be significantly older than the dates estimated from the cultural material, between 2500 BC and the beginning of our era (David and Sterner, 1989). After Konduga and Blabli, there is a hiatus of nearly 3000 years in the archaeology of the southern Lake Chad area.

Bama Deltaic Complex (or Gajiganna area)

With increasing aridity after the mid-Holocene, Lake Mega-Chad gradually disappeared, opening up a large area for human occupation. The archaeological complex, representing the first colonists of the south-western Lake Chad Basin, is called the Gajiganna Complex, after the village where the first discoveries were made (Breunig *et al.*, 1992). The origins of the Gajiganna Culture should probably be looked for in the Central Sahara, somewhere between the Ennedi in the east, the Hoggar in the north and the river Niger in the west (Breunig and Neumann, 2002a). The Gajiganna Complex is dated to between approximately 1800 and 400 BC (Breunig *et al.*, 2006) and about 120 Gajiganna sites have been located in the study area (Breunig and Neumann, 2002a). There is a dense concentration of sites in the north-west (Gajiganna group), where the core area of the complex might have been situated. Another cluster is visible at the south-eastern fringes of the Gajiganna Complex (Walasa group) and a third concentration was found on the Bama Ridge and its vicinity to the north (Bama-Konduga group). The Bama Ridge also seems to be the western fringe of the Gajiganna sites. The sites are mainly settlement mounds with diameters of 150 m and more; some are quite flat while others have cultural deposits with a maximum accumulation of 2 to 4 m. Apart from the sites in the vicinity of the Bama Ridge, most sites are situated on slightly elevated sand areas, surrounded by seasonally inundated clay depressions. There are several indications, however, that the relief in the area was slightly more pronounced at the time of occupation of the sites than it is today (Breunig, 1995). The central Gajiganna area was then situated near the shores of a lagoon from Lake Chad reaching as far as Maiduguri (*Ibid.*).

Based on a stylistic and technological analysis of pottery from the Gajiganna group, the Gajiganna Complex was initially divided into two phases: phase I, dated to about 1800 to 1400 BC and phase II to about 1500 to 800 BC, further subdivided into phase IIa (1500-1200 BC), phase IIb (1200-1000 BC) and phase IIc (1000-800 BC) (Wendt, 2005). Phases IIa and IIb are usually taken together because the distinction is hard to make, unless an in-depth statistical analysis of the pottery is made (Breunig, pers. comm.). Recently, a third phase, dating to between approximately 600 and 400 BC, has been added to the sequence (Magnavita, 2003). In the Bama-Konduga area no Gajiganna sites younger than phase IIb have been recognised.

Archaeobotanical work for the oldest two Gajiganna phases concentrated on impressions in ceramics, since carbonised plant remains are very poorly preserved (Klee and Zach, 1999; Klee *et al.*, 2004). For phases I only wild grasses, rice (*Oryza* sp.) and Panicae, have been found, which were thus probably the main sources of carbohydrates of the Gajiganna people at that time. During phase IIa, around 1200 BC, the first impressions of domesticated pearl millet appear, but they do not make up more than 10 % of the identifiable plant remains. By phase IIc, 200 years later, their portion has risen to more than 50 %. Changing percentages do not necessarily reflect agricultural intensification, but may also be a result of evolving pottery techniques. The available evidence points to an introduction of pearl millet from outside, rather than to a local domestication.

Throughout the first two Gajiganna phases, settlement types vary (Breunig and Neumann, 2002a). Phase I sites are usually very flat with a poor accumulation of cultural material, indicating mobility. Ceramic finds are nevertheless more numerous than would be expected from highly mobile nomads (Breunig, 2001). Settlement structures were probably light and simple constructions made from organic material and some sites have been interpreted as short-term camps. Phase IIa and IIb sites on the contrary are large mounds, which were probably formed by decomposed wattle and daub house constructions, pointing to more sedentary and village-organized communities, while sites from phase IIc are again flat, pointing to renewed mobility. From the described indications on settlement strategies, archaeobotanical data, and results from previous archaeozoological analysis (Breunig *et al.*, 1996) phase I is defined as pastoral, while phase II is considered to be agropastoral (Breunig and Neumann, 2002a).

Despite changes in the pottery, the archaeological material shows continuity between the pastoral and the agropastoral phase (Breunig and Neumann, 2002a). The main stone artefacts during both phases are grinding stones, polished axes and flakes broken from these axes, besides bifacially retouched arrowpoints of Saharan types. As mentioned in the description of the present Lake Chad Basin (2.1.2.), there are no sources of raw stone material in the area. The Gajiganna people had to import their stones from at least 70 km away (Rupp, 2004). It seems that manufactured tools were imported rather than the raw material, as no knapping debris has been found on any of the Gajiganna sites (Breunig and Neumann, 2002a). Breunig (2004) proposes that this demand for raw stone material may have stimulated trade and exchange contacts. To compensate for the lithic shortage, bone tools (harpoons, chisels and bone points or gorge-hooks) were widely used (Kottusch, 1999; Breunig and Neumann, 2002a). A last find category is the figurines of untempered, baked clay, in anthropomorphic or zoomorphic shapes (Breunig, 1994). Cattle seem to be the predominantly depicted animals among the figurines. Mostly cone shaped pieces, interpreted as fragments of cattle horns, are found, but there is also, for example, a cattle body with an udder. The function ascribed to this kind of figurine is simply as a children's toy (Gronenborn, 2000, 243-244).

Radical changes appear around the middle of the first millennium BC, with the start of what has been labelled Gajiganna phase III. A key-site of the final phase of the Gajiganna Culture, dated between 600 and 400 BC, is the site of Zilum (NA 97/37). The apparent chronological gap with the previous Gajiganna phase is due to a plateau in the calibration curves around that time (Reimer *et al.*, 2004). Zilum shows early urban features and seems to have functioned as a focal hub (Magnavita and Magnavita, 2001). Magnavita *et al.* (2004) summarise as main characteristics for the period "a near doubling of the size of the largest settlements, the apparent emergence of site hierarchies, a change in the social organisation that included intensive pit digging for house construction and for storage, and the appearance of a peripheral ditch-and-rampart system in the largest settlement". The appearance of new crops, such as cowpeas (*Vigna unguiculata*), new farming systems based on intercropping and the apparently increased necessity of storing food surpluses, can probably be seen as responses to the unstable environment that existed in the Chad Basin during the first millennium BC. Iron has not been found at Gajiganna phase III sites, although it was known in other parts of the Chad Basin by that time (e.g., MacEachern, 1996).

High aridity has been recorded for the entire first millennium BC, but Gajiganna phase III settlements appear to have profited from a period of temporarily better circumstances (Breunig *et al.*, in press). Once these ended, the Gajiganna Culture seems to have come to an end. One of the consequences of the increasing aridity appears to have been a higher mobility, archaeologically documented by the replacement of the mounds through small accumulations of potsherds (Breunig and Neumann, 2002b).

The period between the abandonment of the Gajiganna phase III sites and renewed settlement of the area by Iron Age peoples, at the beginning of the first millennium AD or slightly earlier, still remains archaeologically undocumented (C. Magnavita, pers. comm.). As in Burkina Faso, the Iron Age in the Bama Deltaic Complex starts relatively late compared to other parts of West Africa, where iron smelting has been attested from 500 BC onwards (Jemkur, 2004). However, new evidence from the Gajiganna area probably pushes the start of the Iron Age in the region a few hundred years back in time (Breunig, pers. comm.). Magnavita (2003, 178-179) connects the beginning of the Iron Age in the area with better ecological circumstances.

All Iron Age sites in the Bama Deltaic Complex consist of relatively low settlement mounds, either isolated or in small groups (Magnavita, 2002). Some of the mound clusters reach a surface of up to 5 ha, while isolated mounds are in general only 1-2 ha large. There is no evidence that the Iron Age ceramic in the area is a continuation of the Gajiganna tradition, but it would rather be introduced from the *firgi* (Magnavita, 2003, 179). As the name of the period indicates, iron slag and iron objects have been commonly found at Iron Age sites. Evidence for domesticated sorghum is present from about the third century AD onwards (Magnavita, 2002). It is the dominant crop in the Bama Deltaic Complex from then onwards (Kahlheber, submitted).

Since the Frankfurt project concentrated mainly on the beginning of food production and the transition between the Late Stone Age and the Iron Age, the period between the Early Iron Age and present in the Bama Deltaic Complex remains largely un-investigated. With the exception of one subrecent site, Galaga (NA 92/2C), no fauna of this period has been analysed. Unlike northern Burkina Faso, however, the Bama Deltaic Complex shows traces of human occupation for the entire chronological sequence after the beginning of our era (Breunig, pers. comm.).

Firgi area (or Chad Lagoonal Complex)

Around the time when the Gajiganna Culture ended, occupation became possible in the *firgi* area. The area was probably opened up during the same arid phase as had caused the collapse of the Gajiganna Culture. It has been proposed that the Gajiganna people followed retreating waters into the *firgi* area (Breunig and Neumann, 2002b), although the archaeological evidence for this hypothesis is still scanty. Wiesmüller (2001, 218) has for example concluded from her pottery analysis that emigration from the Gajiganna to the *firgi* area is improbable. The earliest occupation of the *firgi* area is documented at Kursakata (NA 93/46), where Late Stone Age layers date to between 1000 and 800 BC (Gronenborn, 1998). Around the same time other large, West African floodplains, like the Inland Niger Delta and the Senegal River Valley, which had been inaccessible before, also became available for human habitation (McIntosh, 1999).

As said earlier, permanent habitation in the *firgi* is only possible on sand dunes that protrude through the clay plains, since the plains themselves are inundated during several months of the year. Because of this particular geographic situation large settlement mounds were formed of sizes comparable to Near Eastern tells (Breunig, 1995). The mounds can be several hundred metres large and up to 10 m high. The Yedseram River approximately marks the western border of their distribution area.

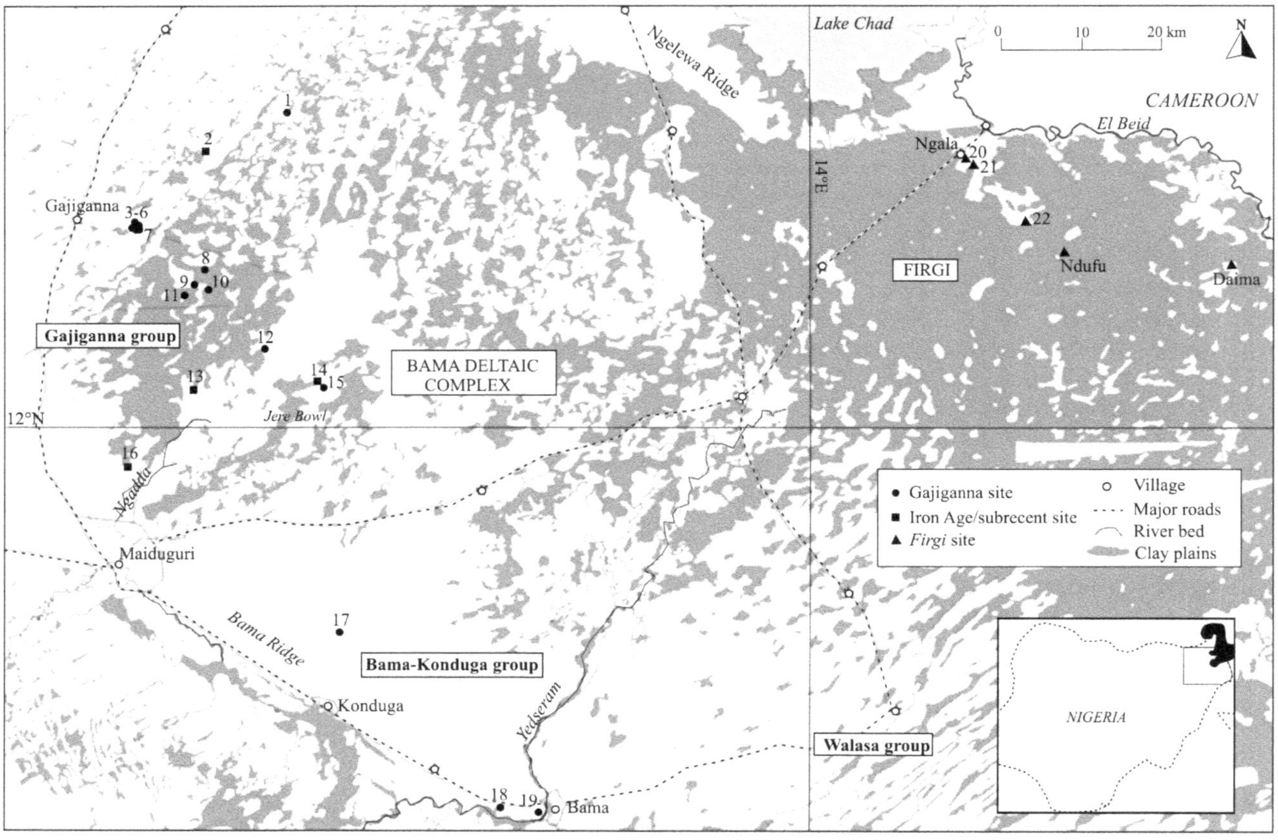

Fig. 8: Studied sites in north-eastern Nigeria

1: Zilum (NA 97/37), 2: Galaga (NA 92/2C), 3: Gajiganna A (NA 90/5A), 4: Gajiganna BI and BII (NA 90/BI-II), 5: Gajiganna C (NA 90/5C), 6: Gajiganna D (NA 90/5D), 7: Gilgila (NA 99/65), 8: NA 93/36, 9: NA 93/10, 10: NA 93/42, 11: NA 91/1A, 12: Dumge A (NA 97/18), 13: Elkido North (NA 99/75), 14: Labe (NA 97/24), 15: Labe Kanuri (NA 97/26), 16: Dorota (NA 97/13), 17: Bukarkurari (NA 97/33), 18: Kariari C (NA 95/1), 19: Kelumeri (NA 96/45), 20: Ngala (NA 93/45), 21: Kursakata (NA 93/46), 22: Mege (NA 94/7)

Much of the archaeological research G. Connah (1981) carried out in the 1960s in northern Nigeria concentrated on the *firgi* type settlement mounds. His most extensive excavation was that of Daima, where he recognised three occupation phases (Daima I-III), which served as a basic chronological framework for the area. The earliest occupation of the *firgi* is attributed to the Late Stone Age and Wiesmüller's (2001, 236) analysis of the ceramic typology has shown that the area was inhabited without long interruptions up to the present-day. The phases she defined for the *firgi* are Late Stone Age (1300/800 – 800/500 BC), Early Iron Age (800/500 BC – AD 400/600) and Late Iron Age (AD 400/600-1600), besides a historical (sixteenth-nineteenth century AD) and subrecent period (nineteenth-twentieth century AD) (Wiesmüller, 2001, 235). The most important ceramic change Wiesmüller has observed is between Late Stone Age and Iron Age, around 500 BC, which she has tentatively attributed to renewed influences from the West- or Central-Sahara. The stratigraphic sequence of all sites in the area additionally showed a discontinuity between Late Stone Age and Iron Age layers (Gronenborn, 1998). There appears to have been a shift from semi-permanent to permanent settlement between the two periods and the beginning of large-scale farming also seems to be situated at the Late Stone Age-Iron Age transition (Klee *et al.*, 2000). Nevertheless, house structures were recorded in Late Stone Age layers at Ndufu (NA 93/47) (Gronenborn, 1998). The onset of more arid ecological conditions has also been recorded for the period in which the cultural changes took place (Klee *et al.*, 2000).

Archaeobotanical studies have shown that pearl millet was known in the *firgi* from the initial Late Stone Age occupation onwards, although the crop did not gain major economic importance before the beginning of the Iron Age (Klee *et al.*, 2000). The oldest archaeobotanical evidence for sorghum in the area does not date to before 800 AD (Connah, 1981, 188-189; Zach *et al.*, 1996; Klee and Zach, 1999). Nonetheless, Magnavita (2002) hypothesises that sorghum was cultivated in the area at the latest by the first half of the first millennium AD, in view of the older finds in the Gajiganna area. Judging from historical evidence, the special *masakwa* technique for sorghum cultivation in the *firgi* (see 2.1.2.) was known since at least the sixteenth century AD (Gronenborn, 2001). In contrast to the predominantly sandy areas of northern Burkina Faso, the presence of both sand and clay soils in the southern Lake Chad area must have allowed the cultivation of crops during both the rainy season and the dry season (Breunig and Neumann, 2004).

From their earliest occupation, the *firgi* sites have also yielded clay figurines that bear some resemblance to the Gajiganna finds (Breunig, 1994; Gronenborn, 1998). Animal depictions are predominant, but the represented taxa (cattle, sheep, goat, wild animals) seem to vary through time (Connah, 1981, 135-136, 156, 182-183).

Contacts with northern Africa may have initiated much later in the *firgi*, and the southern Lake Chad area in general, than in Burkina Faso. Gronenborn (2000, 263) argues that the area did not become part of the international trade- and exchange networks before the thirteenth/fourteenth-sixteenth century AD. Connah (1981, 164) on the other hand, dated contacts back to around AD 800, which seems to coincide with the penetration of Arabs into the Lake Chad Basin from the ninth century AD onwards, described by Braukämper (2004). In the sixteenth century AD, after almost two centuries of resistance, the southern Lake Chad area came under the control of the Kanem-Borno Empire[7] (Gronenborn, 1998). As a consequence of the pressure exerted by the Empire, local principalities were formed in the southern Lake Chad area from about the fourteenth century AD onwards. The described political changes did not cause any marked changes or interruptions in the archaeological record and the Iron Age in the *firgi* area can be broadly considered as a period of cultural continuity (Breunig and Neumann, 2004), with the exception perhaps of changes recorded around the seventh/eighth century AD (Gronenborn, 2000, 350, 361-362).

2.2.3.2. Description of the studied sites

The main attention of the archaeologists working in the Nigerian southern Lake Chad area, as in Burkina Faso, was focused on the establishment of a good chronological framework for the region and trenches were therefore also relatively small. Excavations were conducted in spits of 10 cm and sediment was systematically sieved on 5 mm meshes. The sites from the Cameroonian part of the southern Lake Chad area are also discussed here. Radiocarbon dates for the sites are given in Appendix A. Faunal weights by site can be found in Appendix B, but for more details on amounts of bone material recovered by locality the reader is referred to 5.6.

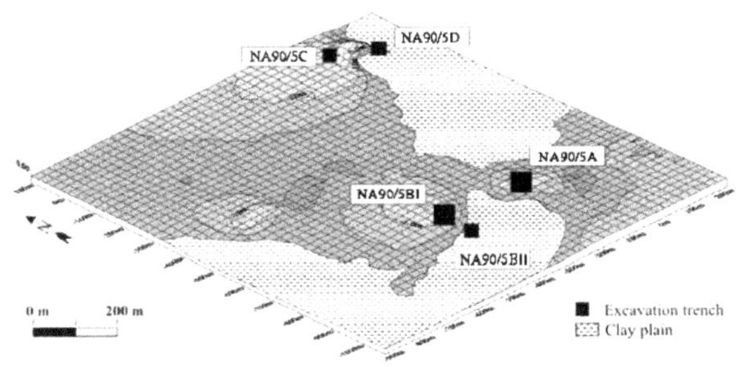

Fig. 9: Location of the Gajiganna A-D sites (Wendt, 2005, Fig. 6)

Gajiganna A-D sites

Present Gajiganna that gave its name to the archaeological complex, is a fairly recently founded Kanuri village. A group of four settlement mounds (NA 90/5A-D) is situated about 12 km to its east. Near the settlement mounds are some pools, surrounded by a few trees, which can hold water until late in the dry season. The pools are formed on clay patches, which are believed to be the remnants of a former lagoon that still existed at the time of the sites' occupation (Breunig *et al.*, 1996).

Gajiganna D (NA 90/5D) (Wendt, 2005, 24, 33-34; Breunig *et al.*, in prep.)

The oldest mound of the group near Gajiganna lies directly at the fringe of a clay plain. The trench that was excavated on its top measured 3 by 3 m and yielded a stratigraphy of about 1.8 m. Two cultural layers were recognised. The basal one was attributed to phase I of the Gajiganna Culture, while the upper layer (surface to about -0.9 m) appeared to be mixed with post-Gajiganna material. Only 225 g of faunal material was recovered from the site, all stemming from the mixed levels, and with very few identifiable remains. Although it was studied the faunal assemblage from Gajiganna D is, therefore, not further discussed.

Gajiganna C (NA 90/5C) (Wendt, 2005, 24, 33; Breunig *et al.*, in prep.)

A 3 by 3 metre square was excavated at about 20 m northwest of the highest point of this second mound of the same group, situated about 300 m west of previously described Gajiganna D. Seven layers were recorded in its

[7] The Kanem-Borno Empire is among the oldest sub-Saharan African empires, with one of the longest durations (Gronenborn, 2000, 57-65). Kanem is the name for its part north of Lake Chad, while Borno points to its territories south of the lake.

2.5 m thick stratigraphical sequence, three of which were anthropogenic (-0.2 to -0.6 m, -0.6 to -1.2 m, -1.2 to -1.4 m). They were all attributed to Gajiganna phase IIa. From the second cultural layer, at a level of -0.8 m, a pit appeared to be dug down, which at the time of its construction must have been about 1.7 m deep. The pit was bell-shaped and round in cross section, with a diameter of roughly 1.3 m at the bottom. It was interpreted as a storage facility for wild or early domestic grasses (Gronenborn, 1997).

Gajiganna A (NA 90/5A) (Breunig *et al.*, 1996; Wendt, 2005, 23-24, 28-30; Breunig *et al.*, in prep.)

Gajiganna A is the most south-eastern of the four mounds at Gajiganna. Its highest point reaches 1.8 m above the neighbouring clay plains to the south and north. A 5 by 5 m trench was excavated on the top of the mound and yielded a stratigraphic sequence of about 2.2 m thick. Two cultural layers were recognised (upper cultural layer: -0.3 to -1.3 m, lower cultural layer: -1.6 to -1.8 m), separated by an almost sterile sand layer. Both cultural layers were attributed to phase IIb of the Gajiganna Culture (Breunig and Neumann, 2002a). Four inhumated individuals were discovered in the upper cultural layer. About twenty clay figurines, mainly zoomorphic but also including five anthropomorphic specimens, were collected during the excavation at Gajiganna A, but no clear concentration was noticed (Breunig, 1994).

Gajiganna B (NA 90/5BI and NA 90/5BII) (Breunig *et al.*, 1996; Wendt, 2005, 24, 30-32)

Gajiganna B is situated 200 m west of the previous mound and has about the same height. A first 5 by 5 m trench (NA 90/5BI) was dug on the south-eastern part of its summit and later a second trench (3 by 3 m) (NA 90/5BII) was excavated 100 m south of the first, in the seasonally inundated plain (Fig. 9). The latter part of the site must have been inhabited during a drier phase, when the depression did not carry any water. Archaeological deposits at NA 90/5I were discovered to 4.5 m below the present surface, with altogether about 2.6 m of cultural material. Two cultural layers could be recognised, divided into an upper (-0.8 to -1.3 m) and a lower one (-1.3 and -3.8 m). At the transition of the upper and lower cultural layer a cluster of clay figurines was found (Breunig, 1994). From the lower cultural layer, at approximately -2.0 m, a pit appeared to be dug down to nearly 4.5 m below the surface. The function of the pit remains unknown. Besides many large fragments of ceramic vessels, a large amount of animal remains were recovered from it. The upper cultural layer from NA 90/5BI contained very little faunal remains. It was also recognised in trench NA 90/5BII, where it was covered with more than 1 m of sandy clay deposits. NA 90/5BII was excavated down to a depth of about 3.6 m, from -2.2 m onwards in spits of 20 instead of 10 cm. Most material from NA 90/5BI is dated to Gajiganna phase IIa/b, while radiocarbon dates place the upper cultural layer in Gajiganna phase IIc. NA 90/5BII, attributed to phase IIc, belongs to the youngest representatives of the second phase of the Gajiganna Culture.

Other sites of the Gajiganna group

NA 91/1A (Wendt, 2005, 24, 34)

The site NA 91/1A is located about 18 km south-east of the village of Gajiganna, on the east side of a large sandy area, and reaches approximately 1 m above the clay plain. The excavated surface measured 1.8 by 1.8 m and was not subdivided into smaller squares. It was dug down according to the five layers visible in the stratigraphy, two of which anthropogenic (0 to -0.2 m and -0.2 to -0.5 m). Maximal depth was reached at -1.2 m. Insufficient charcoal was available from the site for radiocarbon dating, but ceramic decoration allowed an attribution to Gajiganna phase I.

NA 93/42 (Wendt, 2005, 27, 39)

The site NA 93/42 is situated on the north-eastern corner of a 2 km long, triangular, sand island. The largest height difference with the clay plains is about 1.5 m. A surface of 1.2 by 2.2 m, without subdivision in one-metre squares, was excavated in the transition zone between the hilltop and its north-eastern slope. Depth labelling at the site only started underneath a sterile aeolian sand layer. Below this, the site yielded two cultural layers (0 to -0.1 m and -0.1 m to -0.6 m), together about 0.6 m thick. A 1.6 m deep pit appeared to be dug from the second cultural layer to about -2.2 m. The site has the oldest date of all Gajiganna sites. The faunal remains stem from the second cultural layer and from the bottom of the pit.

Dumge A (NA 97/18) (Breunig *et al.*, in prep.)

Dumge A is a small mound, with a diameter of about 40 m and a maximum height of 1 m, situated about 30 km northeast of Maiduguri on the eastern border of a vast area with several clay plains. A shallow depression north of the site, where water is stored longer than in the surrounding plains, might have favoured human habitation. On top of the mound a 2 by 2 m trench has been excavated down to 1.5 m below the surface. Its upper 0.7 m was composed of sterile aeolian sand. The lower cultural section was build up of dark brown, solid, clay where a very small quantity of cultural remains was found belonging to Gajiganna phase I. This is in contrast to the eroded mound slopes covered all over by potsherds of both phase I and phase II. Apparently the settlement was founded on lagoonal clay deposits, only suitable for habitation during the dry season. Dumge A was therefore most probably a dry season camp of the pastoralists of phase I of the Gajiganna Culture. The finds of the later phase II either point to a mixture with later settlement remains or to a transitional phase, with elements of phase I and II. One small bag with faunal remains (41 g), collected from surface deposits all over the site but still in situ, was available for analysis.

Labe (NA 97/24) (Breunig *et al.*, in prep.)

This shallow settlement mound (height 2 m, diameter about 150 m) is situated at the north-eastern border of an extensive clay plain, about 30 km northeast of Maiduguri. The clay depressions around the site today carry water during at least part of the dry season. The site is one of the rare Gajiganna I settlement mounds, and therefore a 2 by 2 m excavation (no subdivision in square metres) was carried out on its top. Archaeological deposits were unearthed up to about 2.2 m below the surface. Cultural finds were scarce, as was expected from the relative paucity of cultural material on the surface. A seasonal and low-density occupation was therefore assumed for the site.

NA 93/36 (Wendt, 2005, 24, 36-38)

The settlement mound at this site has a circumference of about 200 m and protrudes 2.5 m above the clay plains that surround it. An L-shaped pond is situated at its south-western slope. The excavated surface measured 4.5 by 1.5 m and was dug down without further subdivision into square metres. The 2.5 m stratigraphy was complicated, with about 15 sand and clay layers that often could not clearly be identified as natural or anthropogenic. The entire occupation of the site was attributed to Gajiganna phase IIa (Breunig and Neumann, 2002a).

Gilgila (NA 99/65) (Magnavita, 2003, 26-29; Breunig *et al.*, in prep.)

The settlement mound of Gilgila is situated only 1.5 km south-west of Gajiganna A-D. With a height of about 1.5 m it is rather shallow and it has a surface of about 0.5 ha. A 2 by 3 trench was dug at its highest point. Archaeological deposits were about 1.7 m thick and unlike many others of the Gajiganna sites, they were not covered by sterile aeolian sand layers. Three phases have been recognised: Gajiganna IIa (-1.7 to -1.4 m), Gajiganna IIb (-1.3 to -0.8 m) and Gajiganna IIc (-0.7 m to surface). Finds consisted of a few bone artefacts and a fragmented ceramic cattle figurine, apart from some ceramics and stone tools.

NA 93/10 (Wendt, 2005, 24, 35-36)

Of the two trenches dug at this site, NA 93/10C and NA 93/10ST1, only the latter has yielded faunal remains. The trench was excavated to investigate the nature of a rectangular pit. It was lowered according to stratigraphic units, five altogether. The pit measured about 2.2 by 2.2 m, was 0.4 m deep and had a flat bottom. It is without archaeological parallels in the area so far. NA 93/10 has been assigned to Gajiganna phase IIc (Breunig and Neumann, 2002a).

Zilum (NA 97/37) (Magnavita and Magnavita, 2001; Magnavita, 2003, 32-52; Breunig *et al.*, in prep.; C. Magnavita, pers. comm.)

The site of Zilum is located about 20 km northeast of the Gajiganna A-D sites in an area that is much sandier. Several present-day villages are situated on a sand ridge west of the site, among which the Kanuri village Zilum. The site is located at the edge of a clay depression that is seasonally flooded. With its surface of 12 to 13 ha, Zilum is the second largest site in the Bama Deltaic Complex. Enormous quantities of eroded cultural material are exposed on its surface but the shallow cultural layers suggest an occupation of relatively short duration. The site has become the type-site of phase III of the Gajiganna Culture (see 2.2.3.1.). Among the cultural finds at the site are large amounts of ceramics, pits for grain storage and bone points. Excavations at Zilum are still going on in the framework of the project „Ecological and cultural change in West and Central Africa“ (see 1.1.). The fauna retained for this study was collected during the first field season at the site (winter 2000/2001). Material from later excavations will be dealt with in future publications.

During the first season six test excavations were conducted in different areas of the site. The excavations were executed in 10 cm spits, as usual, but without the one square metre grid system. Zilum 1 was a small, 1 m^2 test pit, where sterile layers were hit at 1.8 m below the surface. Zilum 2 measured 1 by 2 m and the depth of the deposits was approximately 1.0 m, but no fauna has been recovered from them. A third, 3.5 m^2 pit, Zilum 3, was dug to investigate the nature of round features visible at the site's surface and which appeared to be basins built of clay. The basin in Zilum 3 reached about 0.7 m deep, but the excavation itself continued to -1.4 m. The test excavation at Zilum 4 measured 6 m^2 at the surface. A layer of sterile sands, 0.7 m thick, covered the cultural deposits. Below this layer the excavated surface was narrowed down to 4 m^2 and maximal depth reached was at -2.6 m. Zilum 5 was started as a 2 m^2 trench, but was later enlarged with an extra 7 m^2. The excavation was stopped at -1.3 m. Zilum 5 was dug to verify if the hard clay elevations on the sites surface were house floors. No house structures were recovered but indirect evidence, including the presence of two storage or refuse pits and two human burials, seems to confirm the floor hypothesis. Zilum 6 was the largest of the excavated trenches, measuring 17 m^2 and intended to investigate a wall like structure visible at the surface. It is the only of the described test excavations were a grid system was applied. At a depth of 0.3 m, underneath the supposed wall, a human burial was found. Five pits were recognised during the excavation; all eventually used for refuse dumping. Pit 1 clearly yielded the richest organic finds: charred pearl millet grains and several turtle carapaces (see 4.4.1.). Excavations at Zilum 6 stopped at -1.4 m, but pit 1 was dug further down to -1.8 m.

Sites of the Bama-Konduga group

The Nigerian sites described above are all part of the Gajiganna group of the Gajiganna Culture. Sites of the Walasa group, archaeologically investigated by M. Hambolu (2000), University of Maiduguri (Nigeria), yielded only small faunal assemblages that were not sent to the RMCA for analysis. The sites discussed in what follows belong to the Bama-Konduga group. Characteristic for sites of this group, compared to those from both other groups mentioned, is their setting in a different environment, with the rivers Yedseram and Ngadda that replace the clay plains as a source of water (Breunig and Neumann, 2002a). No archaeobotanical data are available for the sites of the Bama-Konduga group.

Bukarkurari (NA 97/33) (Breunig *et al.*, in prep.; Breunig, pers. comm.)

With a height of about 4 m, Bukarkurari is one of the largest mounds recorded for the Gajiganna Culture so far. It is the only site where both phases of the Gajiganna Culture were recognised. The excavated trench on the top of the mound was 9 m^2 large, but was reduced to 4 m^2 from around -1.0 m. Based on the pottery analysis, levels from -2.1 m down to the basal layer at -3.7 m, were attributed to Gajiganna phase I, whereas the higher levels belong to phase II. The levels of phase I correspond to the levels with high quantities of cultural material, mostly ceramics, indicating some kind of sedentism of the pastoral communities. Worked bones and stone artefacts, of the usual types, are not very numerous.

Kelumeri (NA 96/45) (Breunig *et al.*, in prep.; Breunig, pers. comm.)

Kelumeri is situated a few kilometres north-west of the town of Bama. The site is a medium-large settlement mound located on the Bama Ride. It was first tested by G. Connah (1981, 85-91) in the 1960s, who labelled it "Borno 38". The trench excavated by the team from Frankfurt measured 3 by 10 m and was dug to a depth of about 3.2 m. A set of radiocarbon dates points to occupation during a later phase of the Gajiganna Culture. The site is characterised by a pottery different to the one found in the area further north-west. The fauna of one square (A7) was sent to the RMCA for analysis.

Kariari C (NA 95/1) (Breunig *et al.*, in prep.; Breunig, pers. comm.)

Kariari is a settlement mound on the Bama Ridge, about 80 km south-east of Maiduguri. The ridge is almost completely eroded in its surroundings, probably as a result of a former course of the river Yedseram, which is still very close today. An area of 3 by 10 m was cut close to the mounds highest point, where sterile ground was reached at a depth of roughly 2.0 m. The archaeological inventory, not analysed in detail, is very similar to the one excavated at Kelumeri. The site was attributed to phase II of the Gajiganna Culture (Breunig and Neumann, 2002a).

Early Iron Age and subrecent sites in the Bama Deltaic Complex

Labe Kanuri (NA 97/26) (Magnavita, 2002, 2003, 54-55; Breunig *et al.*, in prep.; C. Magnavita, pers. comm.)

With a surface of about 1 ha, the settlement mound of Labe Kanuri has an unusually large extent, while its height (ca. 1.5 m) is within normal range. It is situated about 2 km north-west of the Gajiganna site Labe (NA 97/24) and is part of a group of equally large mounds that is spread over an area of about 1 km^2. Labe Kanuri lies in the middle of an area locally called "Jere Bowl", which is the inland delta of the river Ngadda. The site lies close to the flooding area of the delta, which is under water during the rainy season until about the end of November. A 2 m^2 trench was excavated near the highest spot of the mound and dug down to -2.3 m. No pits or other features were recorded in it. The site of Labe Kanuri appeared to be the earliest of the investigated Iron Age sites in the area, probably occupied between the first and fifth century AD.

Elkido North (NA 99/75) (Magnavita 2003, 55-58; Breunig *et al.*, in prep.)

Elkido North consists of four neighbouring settlement mounds and stretches over a surface of about 4 ha. The highest of the mounds reaches 2 m above the surrounding plains, which include many clay depressions that fill up during the rainy seasons. Excavations were limited to a 2 m^2 trench on top of the highest of the mounds and reached a depth of about 3.3 m. Three cultural layers were recognised (surface to -1.1 m, -1.2 to -1.6 m, -1.7/1.8 to -2.7 m). Besides a relatively large amount of cultural materials, a structure was found in the upper part of layer 2, between about -1.2 and -1.4 m, interpreted as a smith's hearth or forge. At different depths in layers 2 and 3 at least four human burials were present, indicating that this part of the site was used as a cemetery. Apart from potsherds, stone tools and pieces of iron slag, the excavated deposits yielded beads made out of molluscs and ostrich eggshell, re-used sherds, an anthropomorphic ceramic figurine and a few iron artefacts. Judging from the radiocarbon dates, Elkido North was inhabited between the second and fifth/sixth century AD and thus at least partly contemporaneous with the previous site, Labe Kanuri.

Dorota (NA 97/13) (Magnavita, 2002, 2003, 58-63; Breunig *et al.*, in prep.)

The settlement mound of Dorota is low (max. 1.5 m), but with ca. 25 ha the site covers the largest surface of all known sites in the Bama Deltaic Complex. Dorota is situated at the southern edge of a sandy area, in the middle of a clay depression. Between the second and fourth century AD, the site would have already extended over a 12 ha surface, while between the fifth and seventh century AD, the whole visible range of the site was occupied. Three relatively small excavation units were investigated, one in the north (2 by 2 m), one in the centre (1 by 1 m) and one in the south of the site (1 by 1 m). Only the fauna from the northern unit, Dorota 1, was brought to the RMCA for analysis. Dorota 1 was excavated near the highest part of the mound to a depth of 2.3 m. From about 0.8 m below the surface a pit like feature was visible in squares A1 and B1. Its shape and dimensions (1.4 m deep; 1.0 m wide) are reminiscent of present-day storage pits in the area. The pit was probably re-used later on as a refuse pit and contained well-preserved archaeological and archaeobotanical finds such as charred sorghum grains, iron slag, vitrified sand, some iron objects and two figurines probably representing a sheep and a cow. Most of the animal bone remains of the location also stem from the pit.

Galaga (NA 92/2C) (Breunig *et al.*, in prep.; Breunig, pers. comm., Gronenborn, pers. comm.)

Galaga is a Kanuri toponym, but the village, situated about 20 km north-east of Gajiganna, has now been completely abandoned. With 3 large mounds the archaeological site is rather prominent in the landscape. Its environmental situation is similar to the Gajiganna A-D sites, being located in the westernmost clay plains of the Nigerian Chad Basin. The village is mentioned in the travel accounts of Rohlfs (1872), who apparently spent a night there. From his text, Galaga should not date much earlier than the middle of the nineteenth century AD, which is in accordance with the pottery and radiocarbon dates. Excavation was carried out in spits of 20 cm, instead of the usual 10 cm. At the start, the excavated surface measured 10 by 10 m, but was quickly reduced to 50 m^2, and further down even to 25 m^2. The fauna of only 2 squares was studied (A1 and A4). Accumulation probably happened quickly and the different layers are therefore probably of similar age. Cultural deposits reached about 2.4 m deep.

Firgi sites

Kursakata (NA 93/46) (Gronenborn, 1998; Wiesmüller, 2001, 28-31)

The settlement mound of Kursakata was first tested by Connah (1981, 91-98). It is 7 m high and located very close to an extensive *firgi* plain to the west and on the edge of what are believed to be the eroded remains of the Ngelewa Beach Ridge. The site might have been right at the shore of a larger lagoon. Kursakata was first occupied around 1000 BC and abandoned in the first centuries AD, thus covering the Late Stone Age and Early Iron Age. The excavated trench was dug at about 60 m south-east of the mound top. It measured 1 by 2 m at the surface and was gradually narrowed down from about -2.0 m. Finally, at the bottom (-7.0 m), the trench measured about 1 by 1.5 m. Between -3.5 and -4.0 m there was a discontinuity in the stratigraphy, probably corresponding with the Late Stone Age-Iron Age transition. In the excavation reports, the layers around -1.8 m are described as a fish landing place because they were packed with fish bones. Kursakata yielded what are among the oldest remains of pearl millet in West Africa (Klee and Zach, 1999). With the transition between Late Stone Age and Iron Age, the importance of the crop rises, but gradually decreases again in the upper layers and it is completely absent from the top metre. A shift from seasonal to permanent occupation has also been assumed, from the archaeobotanical remains, for the transition between the Late Stone Age and the Iron Age at the site. Besides the bones recovered from sampling during the excavation, animal remains collected from archaeobotanical sampling were also studied.

Mege (NA 94/7) (Gronenborn, 1998, 2000, 247-252; Wiesmüller, 2001, 31-33; Breunig *et al.*, in prep.)

Mege is located about 10 km south of Kursakata, on a slightly undulating sand plain stretching over a few kilometres and surrounded by clay plains. A small stream runs not far south-east of the site but carries hardly any water during the dry season. Mege is a large mound with a diameter of about 0.5 km and a height of around 8 m. Its occupation started around 800 BC and continued up to 1983, albeit with several hiatuses. Part of the site is covered with the remains of a subrecent Kanuri and Shuwa Arab settlement, depopulated around 1980 after repeated attacks by robbers. At present, pearl millet is still farmed within the limits of the abandoned village. The excavated square, dug near the mound summit, measured 3 by 3 m and cultural layers continued down to a depth of about 6.8 m. Mege has the most complete archaeological sequence of all *firgi* sites, from Late Stone Age, through all Iron Age phases, to the historical and subrecent phase (Table 1). Similar to Kursakata, the lowest cultural layer is situated directly on top of a thin clay band. This seems to indicate that initial settlement occurred on a dry spot close to the flooded area, or immediately after floodwaters had retreated from the surroundings. Gronenborn (1998) interpreted a layer of decomposed cow dung in Late Stone Age layers (-5.0 m) as the remains of a kraal, i.e. an enclosure made up of poles and branches and used as a sleeping place for cattle. Judging from the radiocarbon dates, the 3.2 m of Late Stone Age layers were deposited in only 300 years, while it took 2500 years to build up the 3.6 m of all younger layers. Wiesmüller (2001, 37) suggests that this might be a consequence of a smaller habitation surface during the former period. Archaeobotanical analysis has indicated that plant food at the site was mainly composed of wild grasses, while the amount of grains of domesticated pearl millet is small throughout the whole sequence and sorghum appears only very late in the cultural layers (Klee and Zach, 1999). Out of the nine excavated squares, only one (B2) was studied archaeozoologically.

phase	depth	approximate date
IV	-1.3 m to surface	AD 1150-1983
III	-2.4 to -1.3 m	Daima III (AD 700-1150)
II	-3.5 to -2.4 m	Daima II (AD 50-700)
I	below -3.5 m	Daima I (550 BC-AD 50)

Table 1: Phases at Mege (NA 94/7) with corresponding depths and dates

Ngala (NA 93/45) (Gronenbron, 1998, 2000, 217-247)

The present village of Ngala is situated only about 1.5 km from Kursakata, close to the Cameroonian border. On top of the settlement mound of Ngala there is still a palace of a local ruler, *Mai* Ibrahim Laminu *Ngalama*, but which is now largely in ruins. The extent of the archaeological site cannot be established anymore because of the growth of the traditional town and considerable reconstruction in the 1960s and 1970s. An archaeological excavation was executed in the courtyard of the palace, until recently used as kitchen. Nearly complete storage vessels and two traditional clay ovens[8] are still present as remnants of the former kitchen. The excavated trench measured 3 by 3 m and cultural layers ended at a depth of slightly over 5.0 m. First occupation of the site probably should be situated at around the second half of the first millennium AD. Five phases have been defined for the total period of habitation, based on the study of the ceramics (Table 2). A palace was presumably already at this location from about -1.0 m upward, which corresponds with the start of phase IIIb. Below the cultural layers, the burial of an adult human was found, dating to the eighth century BC. In the northern profile, at a depth between 3.5 and 4.0 m a second burial was discovered, but not further excavated.

phase	depth	approximate date
IV	-0.6 m to surface	19-20th cent. AD
IIIb	-0.6 m to -1.0 m	14/16-18th cent. AD
IIIa	-1.0 to -1.9 m	10-12th cent. AD
II	-1.9 to -3.6 m	8-10/11th cent. AD
I	below -3.6 m	Mixed/?-7th cent. AD

Table 2: Phases at Ngala (NA 93/45) with corresponding depths and dates

The Blé Mound Complex

The Blé Mound Complex (Holl, 2002, 73-130) is situated in the Cameroonian part of the southern Lake Chad area, in the bend of the Abani fossil channel that used to evacuate excess water from the River Logone. It lies just east of the *yaéré* clay plains, on land above flood level, and has thus a different environmental setting than the Nigerian *firgi* sites. The Blé Mound Complex consists of five settlement mounds, Blé Mound A to E, each extending over a surface between 0.5 and 12 ha. All of the mounds are characterised by evidence for intensive fishing activities in the form of features for smoking fish. One fish-smoking feature at the northern part of Mound A was tested. The pit appeared to be elongated and measured 0.8 m N-S, 0.6 m E-W and was 0.8 m deep. Its bottom and sides consisted of fire-hardened clay walls of 5 to 8 cm thick.

[8] The traditional ovens are called *denderu* in Kanuri. The walls of these ovens are heated by inserting burning wood. When temperatures are high enough, the wood is removed and meat is inserted for roasting. Meat prepared in *denderus* is associated with feasts of high status individuals (*Mai* Ibrahim Laminu *Ngalama*, pers. comm.).

Large parts of what seem to have been above-ground clay walls were found in the upper part of the pit fill. Between 29 and 445 of such features were counted on the surface of each mound.

At all mounds an excavation area of 12 m^2 was opened. The excavations were executed in 20 cm spits, but adjusted to account for successive exposed living surfaces or house floors, labelled "Occupation Horizons" (OH). Find sampling was done without sieving. The period of occupation of the mounds ranges between about the end of the first millennium BC and the sixteenth century AD. The archaeological evidence suggests a permanent sedentary settlement for the earliest phases, shifting to seasonal occupation, by fishing parties from the surrounding area, at the end of the sequence, and ending with evidence of warfare. Judging from the ceramic finds there does not seem to be a close cultural relationship between the Blé sites and the *firgi* sites (C. Magnavita, pers. comm.). Plant remains were extremely rare in the archaeological deposits of the Blé sites and several attempts at flotation failed to produce archaeobotanical evidence (Holl, 2001, 238). An initial analysis of the fauna from the Blé Mound Complex was done by A. Holl (2002, 73-137), after which part of the animal remains from Blé Mound A, B, C and E were sent to the RMCA for further study.

Blé Mound A (Holl, 2002, 73-91)

This is the largest mound of the complex (maximum height 5 m), accumulated above a sand and silt island situated in an ancient Logone delta. The high concentration of headrests on its surface suggests that the locality may have been a production centre for this piece of domestic equipment. Blé Mound A also yielded hundreds of zoomorphic and anthropomorphic figurines. The site would have been settled at the end of the first millennium AD or the very beginning of the second millennium AD and was inhabited permanently until its abandonment in the sixteenth century AD. Archaeological deposits were about 4.5 m thick and 16 occupation horizons have been recognised. Three shallow pits in Occupation Horizon III, filled with livestock dung, were interpreted as drinking troughs for small livestock. Two human burials were excavated; both are probably associated with Occupation Horizon VII.

Blé Mound B (Holl, 2002, 92-101)

Mound B is the smallest of the complex and is situated about 20 m south of Mound A, on the same extensive sand island. Blé Mound B was settled by the middle of the first millennium AD and was inhabited up to about the fifteenth century AD. Twelve occupation horizons were recorded over a depth of about 3.4 m.

Blé Mound C (Holl, 2002, 101-111)

Occupation at Blé Mound C developed above a former channel of the ancient Logone Delta, situated in the east of the Blé Mound A-B sand island. The archaeological layers at the site appear to have accumulated between AD 900 and AD 1200. The site yielded 8 occupation horizons over a depth of about 2.8 m. An adult burial was found in the upper Occupation Horizon VIII.

Blé Mound E (Holl, 2002, 120-130)

Mound E, occupied between the tenth and thirteenth century AD, was identified as a special purpose site, devoted to fish processing and fish smoking activities. Archaeological deposits were 2.8 m thick and 5 occupation horizons were recognised. Fish concentrations were recorded to be exceptionally high in Occupation Horizon III. In the same horizon, a human burial was found.

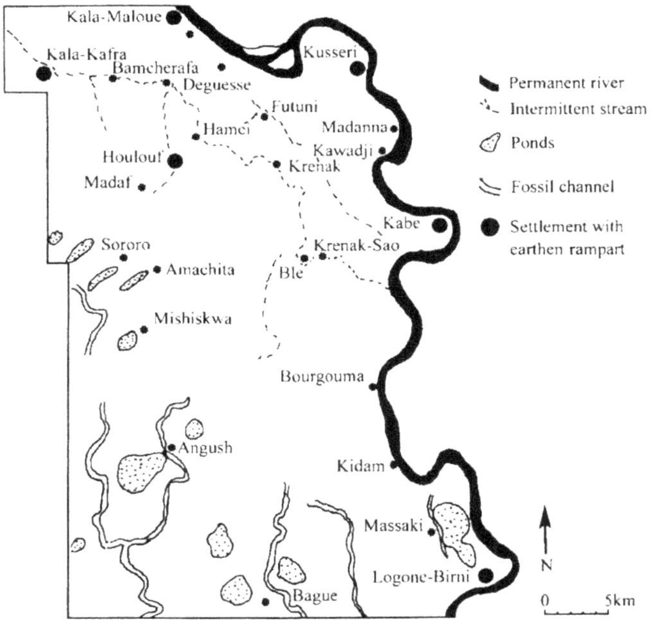

Fig. 10: The Houlouf area in northern Cameroon with archaeological sites indicated (Holl, 2002, Fig. 10)

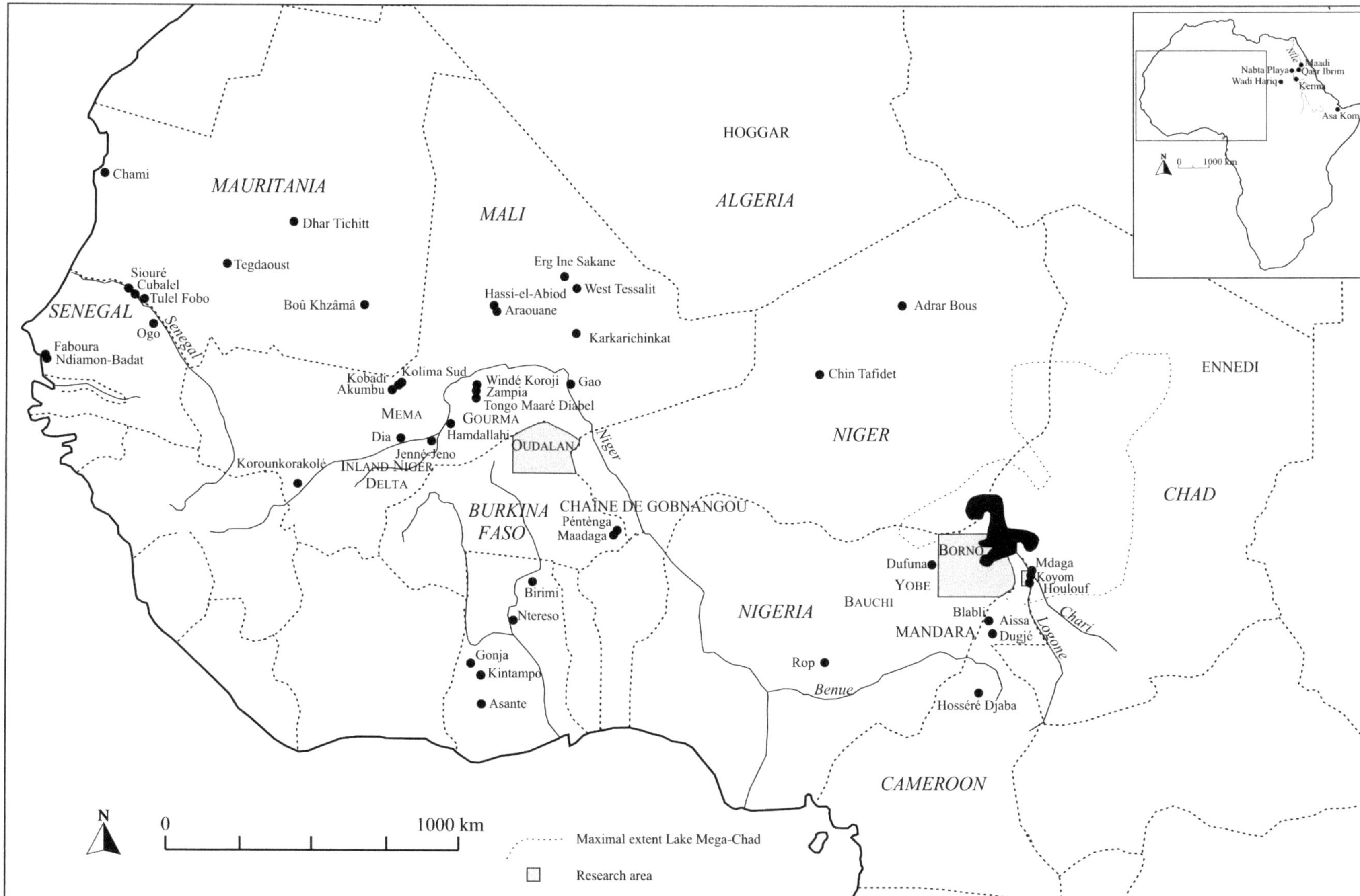

Fig. 11: Localities mentioned in Chapters 2-7
For details on the research area see Fig. 5, Fig. 8 and Fig. 10

Chapter 3. Material and methods

This chapter deals mainly with the methods applied during excavation and study of the faunal remains, and is intended both as a description and as a justification of the methodology chosen. Some technical terms used in the following chapter are also clarified. In the taphonomical analysis in Chapter 5, some specific consequences of the methodological choices on the faunal data and their interpretation are discussed. Although every effort was made to ensure consistency in order to allow an easy and straightforward comparison and interpretation of results from all sites, this was not always entirely the case due to the involvement of several people, with slightly different methodological approaches, during the excavations as well as the study of the fauna.[9]

3.1. Excavation and sampling

In the previous chapter, the excavation and sampling methods were described for each site. With the exception of the site of Oursi hu-beero (BF 97/30), no faunal specialist was present in the field. As explained earlier, the excavations were usually conducted in grids of 1 m^2 and by 10 cm spit. At some sites several archaeological phases appeared to be represented. The phases were usually not separated by clear-cut borders in the stratigraphy or abrupt changes in the material culture (e.g., Gronenborn, 2000, 215). The phase attributions for the spits should therefore be treated with caution. Mixture between layers of different phases, even when separated by a long period of abandonment, has also occasionally been noticed. In general few archaeological features, e.g., pits or settlement structures, were found at the sites.

Information on the sampling techniques used for animal bone recovery is principally based on oral information obtained from the excavating archaeologists. At the Blé Mound Complex only hand collecting was carried out. All the Frankfurt archaeologists applied dry sieving during the excavations: in Burkina Faso sieving was usually done on 10 mm meshes and in Nigeria on 5 mm meshes. . The high clay content of the sediments, causing the soil to stick together in large lumps, hampered bone recovery at the Nigerian sites. Finer methods were applied for archaeobotanical samples, which were, as said, sieved on 2.5, 1 and 0.5 mm meshes and fractions then flotated. Information on volumes of the sieved archaeobotanical samples have not systematically been transmitted, but changed with mesh size. Good sampling methods are also essential for archaeozoological studies. Small animals and small bones of larger taxa will be underrepresented when sieving is not carried out, or only on large mesh sizes (e.g., Payne, 1972; Garson, 1980; Meadow, 1980). To guarantee a good recovery of animal bones from a given site, screening of the sediment through 2 or 1 mm mesh sizes is generally recommended. It is thus self-evident that sampling methods have to be kept in mind while interpreting results from faunal identifications. When only part of the sediment is sieved, volumes have to be indicated in order to extrapolate quantifications to the total site. Incomplete information on the volumes screened for archaeobotanical sampling at the studied localities excludes quantitative interpretation of the bone remains recovered from them, although a qualitative evaluation still remains possible.

Fig. 12: Sieving at an excavation in north-eastern Nigeria (Picture SFB 268)

For all studied localities the fauna was packed in the field by archaeological unit (i.e., by square and by spit), whereby hand collected and sieved material was lumped together. The faunal assemblages of the majority of the sites therefore consisted of many small bags of bones, each corresponding with one archaeological unit. Bones recovered from archaeobotanical sampling were kept separate, but without differentiation between material from different mesh sizes and flotated fractions. Such bones were available for a few excavations in Burkina Faso, Oursi (BF 94/45), Oursi village (BF 97/13) and Oursi hu-beero (BF 97/30), as well as Kursakata (NA 93/46) in Nigeria. All material was shipped to Europe and eventually transported to the RMCA. For Kelumeri (NA 96/45) the entire faunal assemblage was not brought to Belgium, but just that of one square. Equally for the Blé sites, only subsamples of the excavated faunal material

[9] The majority of the faunal remains discussed were analysed by the present author, but this study also incorporates results from archaeozoological work carried out by W. Van Neer (Breunig *et al.*, 1993a, 1993b; Gronenborn *et al.*, 1995; Breunig *et al.*, 1996; unpublished data) and S. Lambrecht (1997; unpublished data) at the RMCA. Two biology students from the University of Leuven also contributed to the faunal analysis (Loonbeek, 2001; De Cock, 2002).

were submitted for faunal analysis, but it is not clear what portion they represent of the total sample.

3.2. Weights and selection of studied contexts

Before any study was conducted on the animal bones, they were weighed by archaeological unit, without cleaning them or removing incorrectly sorted finds. This procedure had been applied since the beginning for both animal bone and ceramics in order to help define occupation phases, and to document the vertical and horizontal distribution of finds. Animal bones were not weighed for the sites of Corcoba (BF 97/5), Oursi hu-beero (BF 97/30), Gajiganna A (NA 90/5A), Gajiganna BI (NA 90/5BI) and the Blé Mound Complex. Bone weights are summarised by locality in Appendix B, but Tin Akof (BF 94/133) and Dori (BF 94/40 and BF 94/96) are not included. Because the localities yielded very little faunal remains, only their total bone weight is mentioned in the site descriptions (see 2.2.1.2.). Dumge A (NA 97/18) is not included either as its faunal sample consists entirely of surface collected material. In Appendix B, weights of bones from archaeobotanical samples are also given, but are mainly intended as an overview of the units for which these were available for analysis.

Usually all animal remains available for a site were studied, but when assemblages were large subsampling was sometimes practiced. For Mege (NA 94/7) and Galaga (NA 92/2C), respectively one (B2) and two squares (A1 and A4) were chosen that, judging from the bone weights, yielded the most material and for which the total sequence was then analysed. For Ngala (NA 93/45) material from all squares was studied, but for the fish remains only two squares were included (A3 and C1) because they were very abundant in comparison to remains from other taxa. Squares were again chosen taking into account bone weights. Estimation of the total number of fish bones for the complete site was done by considering the relative weight of bones from squares A3 and C1 compared to that of the total sample of the site. Finally, also for Oursi village (BF 97/13) only two squares were studied (L6 and M6) but this was more a practical decision, since these were the only two where maximal depth was reached.

3.3. Material and preservation

Bones and teeth form the vast majority of the studied faunal material, and the term "animal bones" is therefore sometimes used as a synonym for the animal remains in general. Besides bones, some mollusc shells, bird eggshells and a few animal coprolites were also recovered. Bone preservation at the studied sites is usually rather poor, but varies depending on the locality and date. An important group of destructive processes takes place prior to deposition. Faunal remains from sites in Burkina Faso are generally more cracked and more commonly crumble down into splinters than those from Nigeria, which may mean that for some reason they had been lying on the surface longer before deposition, thus being exposed to the destructive effects of the elements. Damage caused to animal remains by carnivore gnawing is limited at all sites, judging from the low number of bones recorded with traces of gnawing or regurgitation (see 3.9. and 4.6.5.2.). Trampling of bones by people or animals must have been substantial, since they were apparently lying on the living surfaces, as could be seen at Oursi hu-beero (BF 97/30) (Linseele, in press). Rapidly buried bones are usually better preserved, since they have been less influenced by predepositional processes. These include most bones found in pit structures or other features. At the Late Stone Age site of Tin Akof (BF 94/133), organic remains were indeed only found when preserved in such pits.

Fig. 13: Faunal material from one archaeological unit at Oursi village (BF 97/13) (Scale = 1 cm)

An important postdepositional factor is soil type. At the sites in Burkina Faso soils are usually sandy, while sediments are more clayey near Lake Chad. In the West African Sudan and Sahel zone, faunal remains generally do not preserve well, which is in the first place related to their climatic regime with yearly alternations of very dry periods with periods of heavy rainfall. This is especially important on clay soils, since it causes alternations of swelling and shrinking and cracking of the sediment. However, preservation in arid parts of West Africa is much better than in forested zones, where bones are destroyed because of the acidic soils. Material from the Late Stone Age sites in Burkina Faso is poorly preserved, probably because the archaeological material is deposited in loose dune sand where it can easily be moved through trampling or aeolian processes (Kahlheber *et al.*, 2001). The majority of the studied sites are settlement mounds, where sediment deposition seems to have happened quickly, favouring preservation. At some of the older studied sites in the Gajiganna area, e.g., in the lower layers of NA 93/42 and at Gajiganna BII (NA 90/5BII), a calcareous matrix covered bone surfaces.

Besides the conservation factors that are at work prior to the archaeological investigations, excavation and subsequent transport can, judging from the presence of fresh breaks, also cause considerable additional fragmentation of the bones.

A standard system to describe the state of preservation of bone remains has been designed by Behrensmeyer (1978), who uses so-called weathering stages, ranging from 0 to 5, between very good and very bad preservation. Bones from the studied sites generally had a weathering stage 3 or 4, at Tin Akof (BF 94/133) and Dori (BF 94/40 and BF 94/96) even 5. In 5.6., the identification rate, the species and types of bones present will be used as parameters to describe preservation conditions at the sites more precisely. The identification rate gives indications on preservation, as more bones will be identifiable the better they are preserved. Some species and skeletal elements preserve better than others (differential preservation, see 5.3.1.), and identified species and skeletal distribution can thus also be indicative.

3.4. Cleaning and consolidation

Although washing bones in water can make their osteological details clearer, as well as traces on them (e.g., cut-marks, see 3.9.), the studied faunal remains were not washed. This was too time-consuming because of the many small archaeological units from which the material had to be kept separate. It was also feared that water could cause further damage to the brittle bone assemblages. The remains were, however, cleaned with a dry toothbrush when necessary, and in a few case a little water was additionally used. Finally, pieces of broken bone that matched were glued together when this allowed the taking of extra measurements (see 3.8.).

3.5. Identification

Before identification was begun, a list of animal species that could be expected in the research area was made. The basis for this list is data on the animals that can be found in the area presently. It is important to recognise, however, that the composition of animal populations has changed through time, due to climatological changes and anthropogenic influences. Although this is probably to simplistic, climatic changes have, broadly speaking, caused north-south movements of ecological zones (see 2.2.1.) and, therefore, species of more southerly and northerly regions were also considered. The most drastic human impact on animal populations has occurred since the colonial period, through the introduction of firearms for example, and can therefore be relatively well assessed through historical sources.

Detailed explanations of identification procedures are given under the description of the respective species (see Chapter 4). Bones were identified by comparing them to reference skeletons of recent animals and using identification manuals. Not all possible species were represented in the reference collection at the RMCA, although this is one of the largest and most complete collections in the world for the research area. However, related species, with similar skeletons, were usually available. In addition to the collection from the RMCA, the "Staatssammlung für Anthropologie und Palaeoanatomie", Munich, was also consulted. The most important criteria used for identification were morphology and size. However, the latter may be variable within a certain species, according to ecological circumstances, or sex, for example (Klein and Cruz-Uribe, 1984, 92-98; Davis, 1987, 68-72). In particular, domestic animals are known to display considerable size variations, mainly due to human breeding strategies (Gautier, 1990, 45). It was outside the scope of this study to develop identification keys for the separation of related taxa, where these do not yet exist. As a consequence, some identifications stranded above species level. Nonetheless, this did usually not affect the interpretative value of the faunal data.

A bone was considered as identified when the taxon could be determined beyond class level (mollusc, fish, amphibian, reptile, bird, and mammal). In the vast majority of the cases, this implies that the skeletal element was also named. There are some exceptions, however, where only the animal taxon could be determined. Identification in such cases was based mainly on the bone structure of the taxon, e.g., amphibian long bones, which are typically very light, hollow and having thin walls. Some authors (e.g., Uerpmann, 1973) divide the remaining unidentified bones in size categories, such as small, medium-sized and large mammal, but for many of the small bone fragments from the studied localities even the size category of the animal they belong to was undeterminable. In this study, unidentified bones were therefore not further subdivided by size class. It is also believed that more subdivisions would cause confusion with other categories used, such as "large bovid" and "medium-sized carnivore". An important assumption for the interpretation is that the animal taxa represented among the unidentified remains qualitatively and quantitatively do not differ markedly from those identified.

3.6. Quantification

Quantification is perhaps the most debated subject in archaeozoological methodology (Reitz and Wing, 1999,

191-202; O'Connor, 2000, 54-67). Numerous methods have been described and used, but NISP (Numbers of Identified Specimens) and MNI (Minimum Number of Individuals) remain the two main methods. For NISP each identified bone or specimen is counted as one, while for MNI the minimum number of individuals necessary to explain the bones present is calculated. The basic principles seem simple, but in reality several different ways are used to make the calculations, particularly for MNI estimations. Since neither MNI nor NISP are standardised methods, each faunal study using either one of them, should contain a complete description of how they were calculated.

The usefulness of calculating MNI estimates has been questioned by Gautier (1984a), mainly because of the small chances of interdependence, i.e. that several bones of one individual will be recovered and identified in an archaeological context. Nevertheless, at several occasions during the analysis of the animal remains from the sites under study here, a number of elements that seemed to be from the same individual have been found. Another major criticism of MNI, formulated by Poplin (1976), is that the reliability of the calculations decreases when studied samples become larger. Moreover, for the studied sites MNI estimates are extremely difficult to calculate. This is mainly due to the fact that they are mounds of which the maximal extent is not known and which usually did not yield settlement structures or distinguishable habitation levels. This renders it unclear how to define the units for which the MNI calculations should be done. Because of the excavation methods (see 3.1.), material that could possibly belong to one individual may also be spread over several find bags, causing other practical problems. Nevertheless, it should be mentioned that MacDonald (e.g., 1995) did calculate MNI, besides NISP, at similar mound sites in Mali where he considered the material by trench and by recognised phase. Because of the formulated drawbacks, calculations of MNI in the present study were only done when there were indications for the disposal of (parts of) complete carcasses.

Quantification serves in the first place for making intersite comparisons and NISP data have proven to give the best results, using simple calculations (Gautier, 1984a). It is therefore the principal method chosen here, which means that each identified bone was counted as one. However, fragments that clearly belonged to the same bone, or bones that articulated, were considered as one specimen. A problem here is, again, that these may be spread over several find bags and that therefore it may have gone unnoticed that they belong together. Tooth fragments were not counted separately either, but the minimum number of teeth (Minimum Number of Elements, MNE, cf. Lyman, 1994, 102-104) they represented was estimated by archaeological unit. Ordinary fragment counts would have led to an overestimation of their relative importance, because, when preservation conditions are poor, teeth usually fall apart into many small pieces. Estimates were made similarly for molluscs and bird eggshells. For bivalves, numbers of umbo fragments were counted by archaeological unit and each valve was then listed as one. All other fragments were counted together as one as well, in order to ensure that units with bivalve fragments, but no pieces of the umbo, would also appear on the faunal lists. For gastropods numbers of apertural fragments and opercula were counted, while all other fragments within a unit were considered as one. The bird eggshell fragments in one archaeological unit were always counted together, unless it was obvious from their thickness that several specimens must have been present.

3.7. Ageing and sexing

Data on age and sex are an important source of information on domestic herd management on the one hand and hunting and fishing strategies on the other. Fish and reptiles keep on growing throughout their entire life at documented speeds and their size (see 3.8.) can therefore also be an indication of their age. The most important terms used in the text of the following chapter, when referring to the age at death of mammals are, in order of increasing age, foetus, neonate, juvenile, subadult and adult. Bones were attributed to foetuses, when they were clearly smaller than those of newborn animals of the corresponding taxon in the reference collection at the RMCA. When in about the same size range, they were called "neonate". Bones of both groups have a very light and porous aspect and distinction between "foetus" and "neonate" was often not possible on fragmentary remains. For foetuses of domestic species, using Habermehl (1975), a more precise estimate of their age at death could be made from the length of long bones without articulations. The label "juvenile" has been used for bones that have clearly reached larger sizes than those of newborn specimens but are still not well-ossified. Subadult is used for well-ossified bones, with deciduous teeth for cranial parts and with unfused articulations for bones of the axial or postcranial skeleton. In all other cases, i.e. when permanent teeth are in and when all articulations are fused, bones were called adult.

Fusion of bird bones happens very early in life and it is not possible to estimate the age of the animals macroscopically once they have reached osteological maturity. For domestic mammals and some wild species, on the other hand, the fusion state can be used for age determinations because the age at which the epiphyses of each element fuse to the diaphysis is known (Silver, 1963; Habermehl, 1975, 1985). Tooth eruption and replacement also happens at known moments in life and their wear is an indication for age at death (see mandibular wear stages of Grant, 1982). For domestics, the described methods have not been designed using types from sub-Saharan Africa, however. Since the growth rate of animals as well as their tooth replacement and wear depends on the living conditions, it is unclear how accurate these methods are when applied on assemblages from that region. However, as reconstruction of relative age distribution is important for archaeozoological interpretation, rather than absolute ages at death, this should not be a problem. The relatively poor preservation of the studied material, with (nearly) complete mandibles usually missing, excluded the use of tooth eruption or mandibular wear stages for age estimates. Instead the fusion states of long bones were mainly used. Age estimates for taxa, not included in the cited studies on fusion ages, were made by comparison with related species of similar size. Fusion states of vertebrae were not included, as it was experienced that they are not reliable, with for example one end of a vertebra fused while its other end is still unfused.

Data on the sex of the animals found were even more difficult to obtain than data on their age. Again poor preservation was a limiting factor, which virtually excluded the use of metrical data, females being usually smaller and more slender than males, as a character for differentiation. In a few isolated cases horncore shape and size and the shape of the pelvic bone could be used to sex bovids. Finally, in domestic fowl (*Gallus gallus* f. domestica) spurred tarsometatarsals were used as an indication for the presence of male individuals.

3.8. Measurements and size reconstructions

When preservation allowed, mammal and bird bones were measured according to the standard methods of A. von den Driesch (1976), with a precision of 1 or 0.1 mm. Measurements taken are indicated by the abbreviations and numbers used by this author. Although she advises limiting measurements to adult bones because only these have definitely reached maximal size (von den Driesch, 1976, 4), bones with unfused articular ends were also measured when well-ossified. This was decided because the state of preservation already severely limited measurable elements and because measurements on unfused elements did not seem to deviate from their fused counterparts. On top of the measurements described in von den Driesch two additional measurements were taken on bovid third phalanges, the proximal width (Bp) and the height (H) or maximum distance between the sole and the processus extensorius (Van Neer, 1989a, 7). Measurements served principally as an argument for identification. For domestic animals they were also used to determine possible types present since size appeared to be an important discriminating factor (see 6.7.3.1.). Chronological or geographic changes in the dimensions moreover shed light on variation in animal sizes, which may in turn give indications of breeding strategies or ecological circumstances.

Reconstructions of withers heights of domestic species were made by multiplying lengths of long bones by factors summarised in von den Driesch and Boessneck (1974). A problem for the studied sites and (West) African archaeological contexts in general, is that these dimensions are rarely available because long bones are usually found in a fragmentary state. Small, compact, elements like phalanges, on the other hand, are more often preserved completely but their influence on withers height is small (von den Driesch and Boessneck, 1974). They were therefore not used here for size estimates, although this technique was employed by MacDonald and MacDonald (2000) for West African locations. In addition, for the archaeological assemblages studied here, there are rarely enough measurable elements to compare dimensions on the same skeletal parts. It was therefore decided to calculate Logarithmic Size Indices (LSI) *sensu* Meadow (1999), going back to the Standard Index (SI) of Ducos (1968, 37), because they allow the combination of measurements on different skeletal elements on the same graph. Meadow (1999) has warned, however, not to mix cranial and postcranial measurements, or length and breadth measurements. During this study, the original SI's of Ducos proved to be easier to calculate than LSI's, yielding similar but more comprehensible results, which directly reflect the relative size of the archaeological specimen compared to the reference specimen. Nevertheless, LSI was chosen because it is the most widely used method and in order to allow comparison of the results to LSI graphs made by other authors. LSI equals log x/y, with "x" a measurement of an archaeological specimen and "y" the same measurement on a reference animal. Values are negative when the archaeological specimen is smaller than the reference animal, positive when it is larger and 0 when it has the same size. The dimensions used to calculate LSI's were breadth measurements on postcranial elements, which are chiefly connected to the animals' robustness. More information on the reference animals used and the dimensions taken on them are mentioned under the relevant species in the following chapter. An important weakness of LSI's and related methods, and in extension also the methods for directly calculating withers heights, is that they use the underlying assumption that proportions within the skeleton are the same between the reference animal and the archaeological specimens. However, proportions may be variable, especially between the different breeds of domestic animals (Gautier, 1990, 48).

A few methods are available for the reconstruction of fish sizes from isolated bones (Wheeler and Jones, 1989, 139-148). Fish sizes are usually expressed as standard lengths (SL), i.e. the distance from the tip of the snout to the beginning of the tail. Accurate results for fish size reconstructions are, for example, obtained through the calculation of correlations between given bone measurements and the standard length of the fish from a large number of comparative specimens. Van Neer and Lesur (2004) calculated such equation curves for clariids (Clariidae) and tilapia (tribe Tilapiini). As the curves became available only near the end of the practical work for this study and since no equations are available for the other fish taxa found, standard lengths were estimated for the fish remains by comparing them directly with skeletons from modern fish of known size. Although not as accurate as the other method described, this allowed a quick and simple subdivision of the fish specimens in size classes of 10 cm. Size reconstructions were only possible when elements were well preserved and are more accurate the better the reference specimens cover all size intervals. In addition to indications of the age of the fish, sizes can also be used to reconstruct where fish were caught and to trace possible overfishing, among others.

3.9. Marks and pathologies

Marks or traces were only recorded for identified faunal remains. Since their presence may render bones unidentifiable, this methodological approach possibly lead to an underestimation of their abundance. Because the studied bones were not washed, marks on them may have also occasionally been overlooked. Furthermore, they can also have disappeared because of preservation circumstances altering bone surfaces. In what follows the different types of marks that have been attested in the faunal assemblages are explained.

The first type that was discerned is burning or charring of the bones. This causes alterations in their colour, which is indicative of the temperature they were exposed to, with brownish (charred) at the lower end and grey-white (calcined) at the upper end of the scale (see summary in Lyman, 1994, Fig. 9.9). At the studied localities it was often not clear, however, if (dark) brown colours were the consequence of exposure to fire or rather of diagenetic processes. Whenever bone is burned, but especially under very hot circumstances, it is also reported to shrink (Reitz and Wing, 1999, 133), but no obvious examples for such shrinking were seen in the investigated samples. Burning can happen either intentionally or accidentally. Although distinction is rarely straightforward, mostly the former is relevant for archaeological interpretation because it is usually related to food preparation. Meat adhering parts are only burned when roasted to beyond the point where the flesh is still edible, while exposed bone ends are usually burned during roasting. Examples of the latter were not noticed at the studied localities. Another possibility for intentional bone burning may be the burning of disposed refuse when quantities become too large, but this seems to be mainly a recent practice in the research area (Jones, 1997). Although during faunal analysis it was recorded whether bones were clearly burned or not, the data cannot be used because of the difficulties in distinguishing between burning and other colour alterations on the one hand and between naturally and accidentally burned specimens on the other. In Linseele (in press) bone burning at the site of Oursi hu-beero (BF 97/30), which was destroyed by a fire, is discussed.

An important group of marks, certainly from an interpretative point of view, are those left by tools. Shallow marks have been named cut-marks, deeper ones as chop-marks and bones can also be chopped right through. The former mainly result from cutting away meat or removing skin with knife-like implements, while the latter are usually formed while dividing up carcasses with tools like axes or cleavers (O'Connor, 2000, 45-46). A final type that was recorded is that of shaving marks, probably resulting from filleting (*Ibid.*). Bones were labelled "worked" when altered by humans to serve as the raw material for the manufacture of any kind of bone object.

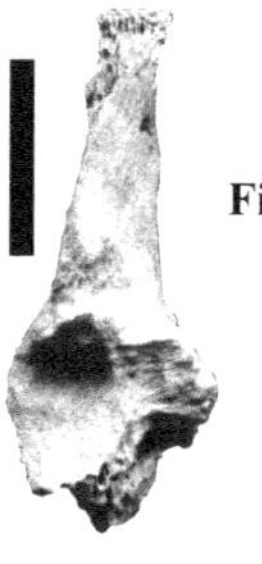

Fig. 14: Etched bone from Kissi 22 (BF 96/22) (Scale = 1 cm)

Another type of mark that was recorded is that of animal gnawing. Several animal taxa are known to gnaw bones for various reasons, for example bovids to obtain minerals or rodents to sharpen their teeth. Besides chewing and breaking bones, domestic dogs and other carnivores can also swallow small bone fragments, which they vomit out or pass out with their faeces. When attacked by stomach juices such fragments may completely disappear, or show varying degrees of etching (Payne and Munson, 1985). In relation to the studied localities, only gnawing by rodents and domestic dogs appeared to be relevant. Only for the latter were quantifications made, since the interpretative value of small rodent gnawing is not clear in the studied contexts. Besides gnawing, other natural marks were rarely seen, but some isolated bones may have been attacked by plant roots.

Several types of pathologies have been seen on bones from the studied faunal assemblages. A first type is called exostosis, i.e. the formation of new, abnormal tissue on the outside of a bone (Baker and Brothwell, 1980, 225). A second type is lipping or "the extension of the articular surface by excess bone formation" (Bartosiewicz *et al.*, 1997, 33). Fusion of adjacent bones has also been recorded, as well as other types of pathologies such as abnormal growth due to malnutrition, healed fractures and unusual tooth wear patterns. Finally, on turtle carapaces holes caused during life by fungi have been attested.

Chapter 4. Description of the faunal remains

This chapter contains a description of the identified animal remains by taxon. In Appendix C, for each of the taxonomic groups, a list is given of the recognised elements by site, or where several archaeological phases were discerned, by site and phase. No distinction by occupation horizon is made for the Blé sites. Remains from archaeobotanical samples, indicated by the abbreviation AB, are listed separately. Detailed faunal lists by location can be found in Appendix D. All animal descriptions have the following pattern: an overview of the identification criteria used, data on age and sex, measurements, size reconstructions, presence of traces and pathologies, and a summary of the identified animal's present zoogeographic distribution, habitat requirements, behaviour, etc. Because more publications are available for northern Nigeria than for northern Burkina Faso, the present status of both wild and domestic fauna of the former region is usually described in more detail. Where relevant, data from other (West) African archaeological sites are included, as well as available historical and ethnographical information. For domestic animals a short summary of the *status quaestionis* on the history of their spread and introduction into West Africa is given in addition. The actual discussion of this subject follows in 6.7.

4.1. Molluscs

Because molluscs are particularly fragile, they were often found in a heavily fragmented state. In many cases identification therefore remained limited to a classification as either gastropod or bivalve. The well-preserved specimens were identified with the aid of a small comparative collection of the most common African freshwater and terrestrial molluscs held at the RMCA. Part of the material was sent to Dr. D. Van Damme, Ghent University, who provided identifications. Table C.1 gives an overview of the identified mollusc taxa for each locality and phase.

4.1.1. Marine gastropods

Cowry (*Cypraea moneta/annulus*)

Cowries are marine gastropods of the family of the Cypraeidae of which a few species, *Cypraea spurca*, *C. pyrum*, *C. lurida* and *C. achatidae*, occur along the West African coast (Lepetit, 1989). The main species found at archaeological sites inland, however, *C. moneta* and *C. annulus*, are imports from the Indian Ocean. They seem to have appeared in West Africa around the end of the first, or the beginning of the second, millennium AD, and usually served as ornaments or, more importantly, as a form of currency (Johnson, 1970). During the excavations of the studied archaeological sites, cowry shells were normally treated as artefacts and therefore not sorted with the animal remains transmitted for study. The specimens identified in the faunal sample from both Saouga sites (BF 94/120 and BF 95/7) may not have been recognised in the field. Since the inventory is incomplete, cowry shells will not be discussed further in this study.

4.1.2. Freshwater gastropods

Pila wernei

Nearly all the studied sites in the southern Lake Chad area have yielded remains of large ampullarid snails (Ampullaridae) of the genus *Pila*. Some juvenile specimens were recovered from Labe Kanuri (NA 97/26) and phase I at Ngala (NA 93/45). Many of the unidentified fragments of gastropod shell from the region may be *Pila* as well. The species that have to be taken into account, *P. ovata* and *P. wernei*, can be distinguished by the shape of their operculum (Van Damme, 1984, 8-11). Only *P. wernei* opercula have been recognised for the studied sites and all fragments of *Pila* shell therefore presumably also belong to this species. No sites in Burkina Faso yielded specimens of *Pila* that were well enough preserved for identification, but it may be represented among the unidentified gastropod fragments. *P. wernei* and *P. ovata* have similar habitat requirements, i.e. river connected swamps, floodplains and other stagnant waters (*Ibid.*). Aestivating *Pila* can easily be collected from swamps in the dry season (Gautier, 1983). According to Arkell (1949, 28-29), Nilotic ethnic groups on the Upper Nile do not eat snails, but do use their meat as fish bait and their shells as cups or spoons. On the other hand, *Pila* shells are collected in large quantities for food in north-eastern Congo, but they are taboo to the younger people (Pilsbry and Bequaert, 1927). Connah and McMillan (1995) mention *Pila wernei* on sale in the Maiduguri market, as recently as 1966, but do not comment on their use.

Lanistes varicus

Like *Pila*, the genus *Lanistes* belongs to the family of the Ampullaridae. In Africa west of Nigeria it is represented only by the species *Lanistes varicus* (Van Damme, 1984, 11-13) and five *Lanistes* shell fragments from the site Saouga 94/120 (BF 94/120) in Burkina Faso have therefore been attributed to this species. It can often be found in shallow pools with abundant aquatic vegetation (Brown, 1980, 50).

Cleopatra bulimoides

This freshwater gastropod was identified from Elkido North (NA 99/75) only. It lives in stagnant or slowly flowing waters with sandy or muddy substrata (Lévêque, 1972).

Lymnaea natalensis

A gastropod of the genus *Lymnaea* was identified at the site Labe Kanuri (NA 97/26) in Nigeria. Seven species of this genus occur in Africa, but only *L. natalensis* is ubiquitous (Van Damme, 1984, 25-31) and the specimen has therefore been tentatively attributed to this species. *L. natalensis* occurs in all kinds of stagnant or slow flowing waters, but is rare in seasonal pools.

4.1.3. Freshwater bivalves

Chambardia sp./*Spathopsis* sp.

The studied contexts in northern Burkina Faso have mainly yielded *Chambardia/Spathopsis* shells. Both of these large bivalve genera belong to the family of the Mutelidae. Distinction is difficult and could not be made for the specimens studied, as these were usually incomplete. *Chambardia/Spathopsis* was the most common bivalve type present, and, therefore, most of the large bivalve fragments that were too small for identification probably also belong to these genera. At Gajiganna A (NA 90/5A) and Galaga (NA 92/2C) a few records have been made of shells with polished edges. *Chambardia* prefers fluviatile environments, in contrast to *Spathopsis* that can mainly be found in lakes (Van Damme, 1984, 62).

Mutela dubia

Two species of the genus *Mutela* can be found in the Lake Chad area, *M. dubia* and *M. rostrata* (Van Damme, 1984, 70-75). *M. dubia* can be differentiated from *M. rostrata* by its larger dimensions, its less elongated shape and its thicker and heavier shell. These features allowed identification of a *Mutela* shell from Blé Mound B as *M. dubia*. The recent distribution of the species includes large rivers, streams and lakes in the Senegal, Niger, Congo, Lake Chad and Nile basins.

Nile oyster (*Etheria elliptica*)

Etheria elliptica is a very characteristic bivalve that resembles a large oyster. It was found at both excavations at Saouga (BF 94/120 and BF 95/7) and at Blé Mound C and E in Cameroon. It can occur in permanent rivers or in lakes, and is easiest to collect when water levels are at their lowest (Van Damme, 1984, 75). In lakes individuals often remain solitary and can develop spines. The specimens found in the studied archaeological assemblages lack spines and seem to be rather of the river form, which corresponds well with the geographic location of the sites, which are in the vicinity of a river. No *Etheria elliptica* shells were retrieved from sites in the Nigerian part of the southern Lake Chad area, but Connah and McMillan (1995) have identified the species from Daima III levels in Daima, where it was associated with an infant burial. They argue that it is not local but probably brought from the Mandara Mountains, at least 130 km away. However, Lake Chad is mentioned by Van Damme (1984, 75) in the list of rivers and lakes where the species is presently distributed and there is consequently no need to propose import. Many people in present northern Nigeria eat the flesh of *Etheria elliptica* and during the dry season smoked ones can regularly be seen at riverside market places (Reed *et al.*, 1967, 121). Near old village sites large heaps of empty shells could frequently be found.

Fig. 15: Nile oyster from Saouga 94/120 (BF 94/120) (Scale = 1 cm)

Eupera ferruginea

One shell of a small bivalve from Saouga 94/120 (BF 94/120) was attributed to *Eupera ferruginea*. It can be found in rivers and lakes of Tropical Africa (Van Damme, 1984, 80).

4.1.4. Terrestrial gastropods

Limicolaria sp.

Terrestrial gastropods of the genus *Limicolaria*, from the family of the Achatinidae, were only found at sites in the southern Lake Chad area, where they sometimes occurred in relatively large numbers. The genus comprises at least three species in this region, *L. kambeul*, *L. aurora* and *L. flammea* (Crowley and Pain, 1970), but identification to species level was impossible because of insufficient reference material. *Limicolaria* favours cultivated land

and can often be found on the outskirts of settlements and farms (Crowley and Pain, 1970, 1-2). It is known to aestivate at appreciable depths in places with loose soils, e.g., rubbish and compost heaps, and prefers damp and shady places. Achatinidae are edible snails, but mainly the large representatives, belonging to the genera *Archachtina* and *Achatina* appear to be exploited. In more humid regions of West Africa these are well known to be collected from the bush along with other edible molluscs (Blench, 2000b; Van Damme, pers. comm.).

Pseudoglessula sp.

Pseudoglessula is part of the family of the Subulinidae. A specimen of this genus of relatively small land snails was recovered at Saouga 95/7 (BF 95/7), but a more precise identification was not possible. *Pseudoglessula* snails are burrowing taxa and occur in arid environments (Van Damme, pers. comm.).

4.2. Fish

Information on the fish taxa that could be expected in the studied faunal assemblages and the maximal sizes they can reach was taken from Paugy *et al.* (2003), unless otherwise specified in the descriptions below. The number of species that had to be taken into account while identifying fish remains from both parts of the research area was relatively limited. Present Lake Chad is characterised by a rather low species richness: around ninety fish species can be found in it (Hopson, 1967; Paugy *et al.*, 2003). This low number may be due to the uniformity of habitats available in the lake (Hopson, 1967; Beadle, 1981, 217). However, possible diachronic changes in the fish composition of Lake Chad, related to the changes in its size and level, must be taken into account. In periods of lake shrinkage fish species could survive by retreating to areas where enough water was still available, for example to the lake's large tributaries, such as the Chari, and the effects of fluctuating lake sizes must therefore have been relatively limited (Beadle, 1981, 230). A large number of species was not expected for the sites in Burkina Faso either, because the area only has small bodies of water, which are not connected to large river systems (Lévêque and Quensière, 1988). Identification of fish remains was facilitated by the presence of an extensive reference collection of recent skeletons of African freshwater fish at the RMCA. In addition descriptions of characteristics to distinguish between some species that closely resemble each other are available in the literature (see species descriptions). A further advantage when working with archaeological fish bones is that they are usually easier to identify from small remains than those of other taxa. Wheeler and Jones (1989, Table 7.1) and Brinkhuizen (1989, Table 3.2) were consulted for the correct (Latin) names of fish skeletal elements and the order in which to present them in the tables.

4.2.1. Lepidosireniformes

Lungfish (*Protopterus annectens*)

Four lungfish species of the genus *Protopterus*, from the family of the Protopteridae, are known from present-day Africa, but *P. annectens* is the only one that occurs in West Africa. Lungfish form a separate evolutionary branch within the bony fish. Their bones have a unique morphology and are therefore relatively easy to recognise. Naming lungfish skeletal elements, on the other hand, can be problematic, because homologies with elements of other bony fish are often unclear. In this study the nomenclature by De Beer (1937, Plate 24) was followed. Only a few lungfish skeletal elements are well ossified (the jaws, the cranial roof, and part of the hyoid) and can be preserved in archaeological contexts. Nevertheless, bones of lungfish were found rather ubiquitously at the studied sites (Table C.2a). Lungfish can attain lengths up to about 80 cm and among the archaeological remains specimens from nearly all size classes have been recorded (Table C.2b). A lungfish cranial rib (SL: 60-70 cm) bearing cut-marks was found at Oursi hu-beero (BF 97/30), while a lungfish ceratohyal (SL: 50-60 cm) from Kissi 40 (BF 97/31) seemed to have been broken during the animal's life and then healed later on.

Fig. 16: Lungfish upper jaw from Early Iron Age Oursi (BF 94/95) (Scale =1 cm)

Lungfish are abundant in inundation plains and temporary lakes, where they lead an aquatic life style during high waters. When water levels drop they can survive until the next floods thanks to their ability to breathe atmospheric oxygen and to bury themselves in the humid ground, where they form a cocoon that prevents desiccation (Lévêque and Paugy, 2003). They can easily be caught during this period of aestivation by digging them out of their burrows, although they also seem to be frequently captured from their shallow water habitats. Reed *et al.* (1967) mention that lungfish, although not uncommon, do not form a significant part of the commercial fish catch in present northern Nigeria. Their flesh is said to be tasty, but taboos are attached to its consumption among many ethnic groups in the area. Lebeuf *et al.* (1980, 103-106) have reported clay figurines depicting lungfish from Mdaga in southern Chad, dating to middle of the nineteenth century AD. Similar figurines are worn as talismans nowadays by young boys and girls who fall ill after eating lungfish, which is normally taboo for them.

4.2.2. Polypteriformes

Polypterus sp.

Fishes of the family of the Polypteridae can be distinguished from other African freshwater fishes by their elongated body, covered by very strong, rhombic, ganoid scales. *Polypterus* is the only genus of the family that has to be considered, judging from its present zoogeographic distribution. Bones of *Polypterus* are relatively easy to identify. Most frequently found elements are scales, vertebrae, skull bones and dorsal spines (Table C.3a). Their diagnostic morphological criteria are described in Van Neer (1995) and in Van Neer and Bocoum (1991). In West Africa, the genus *Polypterus* includes four species, but no criteria are known to distinguish them osteomorphologically. Only two species, *P. bichir* and *P. endlicheri*, can attain lengths of over 50 cm and at least one of those thus has to be represented at the studied sites judging from the size estimates (Table C.3b). A pathology was observed in phase II at Ngala (NA 93/45) where two fused vertebrae occurred.

Like lungfishes, Polypteridae can breathe atmospheric oxygen, which makes them capable of surviving out of water for many hours, and all species of the family are bottom dwellers (Poll, 1939). *Polypterus* prefers shallow, slow running or standing water with sandy bottoms and abundant vegetation (Welcomme and de Merona, 1988). The species can be found on the floodplain or in marginal areas of the main water body. In present northern Nigeria, *Polypterus* forms a significant part of the commercial fish catch, especially in places where fishing is confined to swamps (Reed *et al.*, 1967, 10). The fish seems most easy to catch when swamps are starting to dry up by blocking off creeks that lead to the main river (Budgett, 1900). In the river itself, on the other hand, *Polypterus* appears to be one of the most difficult fish to catch. Its flesh is lean and is known for its excellent taste (Reed *et al.*, 1967, 10). When boiled or roasted, the scales of Polypteridae can easily be flaked off and their flesh removed. Among the Kotoko, consumption of *Polypterus (senegalus)* is very popular and the fish is omnipresent in their symbolic world (Lebeuf, 1976, 42).

4.2.3. Osteoglossiformes

Heterotis niloticus

Heterotis niloticus is the only West African representative of the family of the Arapaimidae. Remains attributed to the species after comparison with recent skeletons from the reference collection at the RMCA are listed in Table C.4a. Small fragments of dermal bones of *Heterotis niloticus* may have occasionally been confused with cranial roof fragments of silurids (Siluriformes) or with carapace fragments of softshell turtles (Trionychidae), since these all have a similar outer ornamentation. For recent *Heterotis niloticus* specimens sizes up to nearly one metre have been recorded, but the largest individual in the reference collection at the RMCA has a standard length of 66 cm. *Heterotis niloticus* individuals from the studied sites do not seem to have reached much larger dimensions either and very small individuals are also underrepresented (Table C.4b).

Heterotis niloticus is very common in places with muddy bottoms and abundant aquatic vegetation (Daget, 1954, 60-62). It can survive in deoxygenated waters thanks to its auxiliary branchial air breathing organs (Welcomme, 1988, 125). The species spawns on the flood plain in more-or-less circular nests made from plant chunks, measuring 1-1.5 m in diameter (Daget, 1954, 60-62). One of the parents guards the nest after the eggs have hatched, and *Heterotis niloticus* is therefore very vulnerable to human predation during this period. Reed *et al.* (1967, 15-16) report that the species can be caught throughout the year, but particularly in rising water levels. In present Lake Chad, *Heterotis niloticus* is one of the most commercially important fish (Beadle, 1981, 224). The species has rather tough and dry flesh, with low oil content, and can therefore be more easily kept in fresh condition than many other species (Reed *et al.*, 1967, 15-16). The flesh tastes strong and not very pleasant, and it is therefore only moderately appreciated by local people.

4.2.4. Mormyriformes

Hyperopisus bebe

Hyperopisus is a genus within the mormyrid family (Mormyridae), comprising several fish taxa with a characteristic elongated snout. The genus' only West African representative is *H. bebe*. The skeleton of the species does not seem to preserve well in archaeological contexts. Only three *H. bebe* bones, all from the site Ngala (NA 93/45), could be identified: a parasphenoid from phase I layers, and a hyomandibular and parasphenoid from phase II layers. With estimated standard lengths of 40-50 cm, the bones moreover appear to have belonged to individuals in the upper size range of the species. Reed *et al.* (1967, 18-19) report that *H. bebe* inhabits both rivers and swamps in present North Nigeria and that it is one of the most common mormyrids in the commercial catches of the area. Its flesh has a fine, delicate flavour, but is rather oily. It is not suited for smoking and is therefore chiefly sold in fresh condition.

Mormyrid (Mormyridae)

Besides the *Hyperopisus bebe* bones described above, a few sites have yielded mormyrid vertebrae, identifiable from the external lateral morphology of the centra (von den Driesch and Boessneck, 1985, Fig. 48b), and a mormyrid quadrate was found at Blé Mound C (Table C.5a). The bones could not be identified beyond family level due to a lack of sufficient comparative material. Their estimated standard length is usually over 30 cm, but individuals recovered from archaeobotanical sampling are smaller (Table C.5b). Mormyrids are typical bottom dwellers, feeding on insects, worms and other small invertebrates (Welcomme and de Merona, 1988). The preferred habitat of many species seems to be either deep, rocky pools, or deep-water around fallen trees, where the current is not too strong (Reed *et al.*, 1967, 17-18). They are rather difficult to catch and are therefore of negligible importance in commercial fish catches in present northern Nigeria. Most of them are moreover not considered first-class fish.

Gymnarchus niloticus

Bones of *Gymnarchus niloticus*, an eel-like fish of the Gymnarchidae family closely related to the mormyrids, are not difficult to recognise. The morphology of the species' vertebrae and cranial elements is unique (Van Neer and Gayet, 1988) and vertebrae were mainly represented among the identified *Gymnarchus niloticus* remains (Table C.6a). The fish can reach very large sizes, up to 1.5 m or more. The majority of the archaeological remains seem to be from rather large specimens, with estimated standard lengths predominantly ranging between about 1 and 1.5 m (Table C.6b). Some of the *Gymnarchus* vertebrae look like they have been bored through in the middle, but perforation probably rather happened naturally, due to the lighter structure of the bone at this spot (Fig. 17).

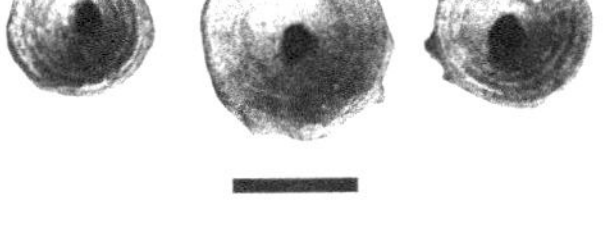

Fig. 17: Naturally bored vertebrae of *Gymnarchus niloticus* from Early Iron Age Kursakata (NA 93/46) (Scale = 1 cm)

Gymnarchus niloticus lives mainly in swamps where it builds very conspicuous floating nests at the beginning of the flood season. Made from grass stalks, these nests can measure about 1.5 by 0.8 m (Daget, 1954, 104; Reed *et al.*, 1967, 32). In present northern Nigeria, the species is caught throughout the year, but is most abundant in commercial fish catches during the wet season, when it can easily be speared while swimming around its nest (Reed *et al.*, 1967, 32). Flesh of *Gymnarchus niloticus* is very oily and has a strong, rich, flavour, which is greatly esteemed by most Africans. In some regions, however, people refuse to eat it (Copley, 1952, 194).

4.2.5. Characiformes

Tigerfish (*Hydrocynus* sp.)

When dealing with Alestidae remains of the genus *Hydrocynus* collected at the studied sites, three species, *H. brevis, H. vittatus* and *H. forskahlii*, have to be considered, but these cannot be differentiated osteologically. A *Hydrocynus* premaxilla and tooth have been recovered from Early Iron Age levels at Kursakata (NA 93/46), both from fish with an estimated standard length of 40-50 cm. In the same archaeological unit two Alestidae caudal vertebrae (SL: 30-40 cm) were found. Because of the associated *Hydrocynus* elements they were also attributed to the genus, although this could not be ascertained from their osteomorphology. *Hydrocynus* occasionally occurs in swamps but can predominantly be found in the main river (Reed *et al.*, 1967, 33-34). Tigerfish can attain sizes of about 80 cm. They can be captured throughout the year, but are not of large commercial importance (*Ibid.*).

No remains of other Alestidae have been identified at the studied sites. This is surprising since, with the exception of the genus *Hydrocynus*, most taxa of the family are numerous in present northern Nigeria, where they make up an important part of commercial catches (Reed *et al.*, 1967, 33). One of the most abundant and commercially important fish in the Chad Basin nowadays is *Alestes baremoze* (Neiland, 1992). Flesh of Alestidae is white and tasty, but the fish are difficult to eat because of their innumerable fine and sharp bones (Reed *et al.*, 1967, 33). Therefore they are often prepared in brine (Van Neer, pers. comm.).

Distichodontidae or Citharinidae

A few Nigerian sites have yielded bone remains, mainly vertebrae, belonging either to Distichodontidae or Citharinidae. Fish of both families are osteomorphologically similar and identifications to a lower taxonomic level were therefore not possible. Maximal size recorded for species of the families is about

80 cm standard length and among the archaeological remains, all size classes up to 70 cm standard length are represented (Table C.7b). A cluster of bones at Blé Mound E of specimens with a standard length of 50-60 cm may represent one individual. Distichodontidae and Citharinidae mainly occur in swamps (Reed *et al.*, 1967, 46, 50). Although their flesh is generally of mediocre quality, they are commercially important fish in the present Lake Chad area (Beadle, 1981, 224).

4.2.6. Cypriniformes

Cyprinid (Cyprinidae)

Several cyprinid genera occur in West Africa, but since the remains found at the studied sites are from individuals with a standard length of over 30 cm (Table C.8b), only two of them, *Labeo* and *Barbus*, have to be considered. Both genera are very rich in species but distinction between them was not possible from the skeletal elements retrieved (Table C.8a). Cyprinids can be found both on the floodplain and in the main river (Welcomme and de Merona, 1988). Larger specimens, with a standard length 50 cm or more, are more likely to come from the main river. Flesh of *Labeo* is apparently not very tasty and also contains many fine bones (Reed *et al.*, 1967, 58).

4.2.7. Siluriformes

Catfish 1 (*Bagrus bajad*)

Of the three West African species of the genus *Bagrus*, only two can be found in the Chad Basin, *B. bajad* and *B. docmak*, where the former is much more common than the latter (Daget, 1954, 161). A few elements allow osteological distinction between the two species (Boessneck and von den Driesch, 1982, Plate 8). A hyomandibular from Early Iron Age levels in Kursakata (NA 93/46) showed *B. bajad* morphological traits, but the other *Bagrus* remains found could not be identified to species level (Table C.9a). Most individuals have an estimated standard length of under 60 cm (Table C.9b). This corresponds with the maximum size that can be reached by *B. bajad*, while *B. docmak* can grow to more than one metre standard length. Therefore, and because of the zoogeographical and osteomorphological reasons mentioned, all *Bagrus* remains were tentatively attributed to *B. bajad*. Only at Blé Mound C could we be dealing with *B. docmak*.

Fish of the genus *Bagrus* can be found in open, deep, waters (Welcomme and de Merona, 1988), but they enter the floodplain for spawning and can also occur in swamps (Reed *et al.*, 1967, 66). In northern Nigeria today, *Bagrus* is captured throughout the year and makes up a small but significant part of commercial catches (Reed *et al.*, 1967, 66). It is one of the finest tasting fish of the world and is highly esteemed by Africans and Westerners alike.

Catfish 2 (*Auchenoglanis* sp.)

Of all studied contexts, Late Stone Age layers at Corcoba (BF 97/5) yielded most *Auchenoglanis* bones. The catfish was not found at any other site in Burkina Faso, while in the southern Lake Chad area it was confined to the *firgi* and the Blé sites. Two catfish species of the genus *Auchenoglanis* can be found in West Africa today, *A. biscutatus* and *A. occidentalis*, but the *Auchenoglanis* remains found at the sites under investigation could not be identified to species level (Table C.10a). Remains from Corcoba are of considerably smaller individuals than those from the other sites (Table C.10b).

A. biscutatus and *A. occidentalis* have similar habitat requirements. They can be found in inundation zones and in parts of the main water body with muddy bottoms (Daget, 1954, 171). *Auchenoglanis* is fairly common in northern Nigeria, especially in swamps, and its flesh is of reasonable quality (Reed *et al.*, 1967, 73).

Catfish 3 (*Schilbe* sp.)

Schilbe is the only genus of the family of the Schilbeidae with fish reaching a standard length of more than 20 cm and this was used as an argument for the identification of a cleithrum found at Blé Mound E (SL: 20-30 cm). Distinction between the possible species, *S. intermedius, S. uranoscopus* and *S. mystus*, was not possible. In the Chad Basin, *Schilbe* seems to occur in very varied hydrological conditions (Daget, 1954, 182). Its flesh is said to be very tasty (Reed *et al.*, 1967, 78).

Catfish 4, clariid catfish (Clariidae)

Remains of clariid catfish are by far the most numerous among the fish bones recovered from the studied sites. Fragments of the cranial roof are especially frequent. It is possible that fragments of the cranial roof of other silurids, dermal bones of *Heterotis niloticus* and carapace fragments of softshell turtles have occasionally been confused with them, especially when only small pieces were available. Nevertheless, even when a few of these bones may have been misclassified, the predominance of clariids remains clear from the large amount of other types of skeletal elements attributed to them (Table C.11a).

Two genera, *Heterobranchus* and *Clarias*, have to be considered when dealing with clariid remains from West Africa. Osteological distinction can most easily be made from the articular facet of the pectoral spine (von den

Driesch, 1983; Gayet and Van Neer, 1990). The numbers of pectoral spines attributed to both genera are included in Table C.11a. Only five out of more than 600 pectoral spines, or less than one percent, could be identified as *Heterobranchus*, all others are from *Clarias*. It can therefore safely be assumed that also the other clariid remains are mainly from the genus *Clarias*. Three species of *Clarias* can be found in western Africa, *C. gariepinus*, *C. anguillaris* and *C. albopunctatus*. While the latter can often be excluded because of its small size, standard length not over 20 cm, the first two can be distinguished from the shape of their vomerine toothplate (Gautier and Van Neer, 1989). In general, *C. gariepinus* appeared to be about four times as common as *C. anguillaris*, only at Gajiganna A (NA 90/5A) *C. anguillaris* was more numerous.

Clariid catfishes can attain sizes of up to one and a half metre, but specimens from the studied sites rarely reached sizes of over one metre (Table C.11b). Most of the clariids caught seem to have had a standard length between 30 and 50 cm. Among the thousands of clariid remains, only one was recovered bearing traces of butchery: a Weberian apparatus with cut-marks (SL: 40-50 cm) from Early Iron Age levels at Kursakata (NA 93/46). A branchiostegal from Phase III layers at Mege (NA 94/7) shows evidence for a healed fracture and from phase I at the same site a few fused precaudal vertebrae were retrieved. Among the clariid remains from Kariari C (NA 95/1), a pectoral spine (SL: 40-50 cm), identified as *Clarias* sp., has the second pectoral ray fused to it.

All clariids have an accessory breathing organ, enabling them to use oxygen from the atmosphere (Teugels, 2003). This ability and their high tolerance to elevated temperatures allow them to survive in adverse conditions. *Clarias* can withstand temporary habitat desiccation by seeking out, or creating, temporary burrows in areas that have at least a limited supply of both air and water (Bruton, 1979a). The taxon is most vulnerable to human predation at the very beginning of the flood season, in the first few days after its spawning run, when it reproduces in shallow, marginal, areas of the floodplain, and, later on during dropping water levels, when it can be found concentrated in residual pools (Bruton, 1979b). Small specimens are mostly likely to have been captured in such residual pools, while larger animals are probably caught while spawning (Van Neer, 2004). First maturity in African *Clarias* populations is observed at median standard lengths of 22-65 cm (Bruton, 1975b). In northern Nigeria, *Clarias* is of primary importance in commercial fish catches at the beginning of the dry season (Reed *et al.*, 1967, 79). The oily flesh of *Clarias* is very popular among local people, either fresh or smoked. *Heterobranchus* lives in swamps and rivers, but prefers the former habitat (Reed *et al.*, 1967, 81). The genus is reported to be common in commercial fish catches in present northern Nigeria, particularly in areas with many swamps. Flesh of *Heterobranchus* is less oily than that of *Clarias*, but it is also highly prized.

Catfish 5 (*Synodontis* sp.)

Remains of *Synodontis* catfish have commonly been identified from the studied sites, but, except at the Blé Mound Complex, usually in low numbers (Table C.12a). In West Africa, the genus comprises more than 30 species. No osteological differentiation could be made, although the shape and ornamentation of the cleithrum is characteristic for many species (Poll, 1971, Plates 10-12). *Synodontis* can reach a maximum standard length of about 40 cm, and all size classes are represented at the studied sites (Table C.12b). Specimens from Mege (NA 94/7) and the Blé sites seem to be larger than those from other localities. *Synodontis* catfish occur in a variety of habitats according to the species (Welcomme and de Merona, 1988). In present-day northern Nigeria, fish of the genus are commonly caught throughout the year and form a very important part of commercial fish catches (Reed *et al.*, 1967, 86). Flesh of all species is of similar quality, white in colour, slightly soft, and of excellent flavour.

Silurid (Siluriformes)

At most sites, the vast majority of silurid remains that were identifiable to a lower taxonomic level pertained to clariid catfish. At these locations, fragments of silurid bones that could not be identified more precisely on an osteological basis, mainly parts of the cranial roof and vertebrae, were also attributed to clariids. When a possible confusion with other catfish species was suspected, however, they were labelled "Siluriformes" (Table C.13). This category is especially large for Late Stone Age levels at Corcoba (BF 97/5), where preservation was not so good (see 5.6.1.) and where the presence of many *Synodontis* catfish remains had to be taken into account.

4.2.8. Perciformes

Parachanna obscura

This only representative of the Channidae in African freshwater has not been very frequently recorded in the studied faunal assemblages. Only from Early Iron Age levels at Kursakata (NA 93/46) were considerable numbers of bones retrieved, including a ceratohyal (SL: 20-30 cm) with cut-marks (Table C.14a). Most of the identified specimens have an estimated standard length of 30-40 cm (Table C.14b). Some are larger, approaching the maximal size of 50 cm standard length recorded for the species in present-day Nigeria (Olaosebikan and Raji,

1998, 71). *Parachanna obscura* has accessory breathing organs enabling it to survive in deoxygenated conditions (Daget, 1954, 366-377). The species is found in vegetation on floodplains but is not very common. It can be easily caught because it spends long periods of time almost immobile close to the surface.

Nile perch (*Lates niloticus*)

Remains of Nile perch, the only West African freshwater species of the family of the Latidae, can often be separated from those of other Perciformes, like tilapia (tribe Tilapiini), by their large size. Morphology also allows the distinction of smaller pieces (Gautier and Van Neer, 1989). Because of their large size and sturdiness, Nile perch bones preserve well in archaeological contexts. The species was regularly encountered at the studied sites, although in very variable numbers (Table C15.a). An exoccipitale with cut-marks was retrieved from Early Iron Age layers at Kursakata (NA 93/46) and a chopped cleithrum (SL: 60-70 cm) was recovered from phase II layers at Ngala (NA 93/45).

Nile perch is presently extremely common in the Chad Basin. The Hausa people in the area call it "giwan ruwa" or "elephant of the water" (Reed *et al.*, 1967, 111). This name is very appropriate in view of the large sizes, up to two metres, that it can reach. The Nile perch at the studied localities was usually not larger than one metre, however (Table C.15b). Large adults are mainly confined to deep-water, while young individuals, of sizes up to about 35 cm standard length, can also be found on the floodplain (Daget, 1954, 359; Blache *et al.*, 1964, 228). For spawning, Nile perch follows the floodwaters of the main waterbodies and lays its eggs more-or-less hidden in herbage (Copley, 1952, 229). A good way to catch Nile perch is with baited hooks. After initial resistance, the fish will succumb relatively rapidly because it is very sensitive to a lack of oxygen (Daget and Iltis, 1965, 222). It is one of the most important commercial fish of Lake Chad, where research has shown that the average size of specimens caught rises with the distance from the shores (Beadle, 1981, 224; Tobor, 1974, 10). The flesh of Nile perch is the most prized of all African freshwater fish (Reed *et al.*, 1967, 111).

Tilapia (tribe Tilapiini)

Bones of tilapia are very well represented among the studied fish remains (Table C.16a). Fish of the Tilapiini tribe are subdivided into three genera, *Tilapia*, *Oreochromis* and *Sarotherodon*, but distinction is not possible osteologically. According to Reed *et al.* (1967, 135) *Tilapia* and *Oreochromis* are the only ones whose members grow large enough to be important as food fish. Only *Oreochromis* can reach a standard length of well over 30 cm, its maximal size is about 40 cm, and the genus must thus be present among the archaeological remains (Table C.16b). A tilapia anal spine (SL: 30-40 cm) from Early Iron Age levels at Kursakata (NA 93/46) showed cut-marks.

Fish of the tribe of the Tilapiini are resistant to adverse hydrological conditions with low oxygen contents and high salinity (Phillipart and Ruwet, 1982). They do not have accessory breathing organs, but the haemoglobin of most species has a very high affinity for oxygen (Fish, 1956; Dusart, 1963). Tilapia prefers shallow water and is therefore easy prey for human predation. It can be caught in large numbers near the shores of large water bodies, but is most vulnerable at the beginning of the floods when it migrates to the floodplain for spawning. Tilapia can be easily traced while breeding in its circular nests in the soft sediment of lake and river margins and since the fish is a repetitive breeder, it can be exploited over a longer period of time (Bruton and Bollt, 1975).

4.3. Amphibians

Frog or toad (Anura)

Altogether, nearly 1000 frog or toad bones have been found (Table C.17). In general, they appeared to be more common in the studied faunal assemblages from Burkina Faso than in those from the Lake Chad area. Identification of frog or toad remains to a lower taxonomic level was usually not possible, because of a lack of comparative material for African species. Moreover, no comparative osteological study seems to exist for frogs and toads of the research area. In addition, no specialist in amphibian remains could be found that was able to help with the identifications. Nevertheless, one large frog species, bullfrog (*Pyxicephalus edulis*), could be recognised because of its cranial ornamentation, which is missing in other frog and toad taxa (Sheil, 1999). A well-preserved skull was recovered from Kissi 22 (BF 96/22) (Fig. 18) in association with disarticulated cranial and postcranial remains of a minimum of four individuals. The bones from Kissi 22 were then used as comparative material for the study of frog and toad bones from other sites. Judging from their size and morphology, most of these are also bullfrog, although other species are probably represented as well. One frog or toad bone with traces of butchery, a coracoid from Kissi 22 (BF 96/22), was recovered.

Rödel (1996, 67-70) does not mention Burkina Faso among the countries were bullfrog is presently distributed. In fact, the species' geographical range within West Africa ought to be limited to the easternmost countries, Cameroon and Nigeria. Bullfrog mainly lives in arid savannahs, and can be found on sandy as well as clayey substrates. It spends most parts of the year burrowed in the ground in cocoons and only comes out during the rainy season, when it can be found in large numbers in marshy areas (Lamotte and Xavier, 1981). Several recipes testify that bullfrog is a popular food item today (Channing *et al.*, 1994). Even its scientific species name, *Pyxicephalus edulis*, refers to the fact that it is an edible species. When asked whether they eat frogs, people from Oursi replied that they themselves do not eat them, but that inhabitants of nearby villages do. From Houlouf, in northern Cameroon, a clay figurine in the shape of a frog or toad is mentioned (Holl, 1988, Table 60, 195). No illustration is provided for the figurine, which dates from a recent occupation phase at the site.

Fig. 18: Bullfrog skull from Kissi 22 (BF 96/22) (Scale = 1 cm)

4.4. Reptiles

The ends of reptilian long bones remain cartilaginous, allowing continued growth. This allowed easy differentiation from mammal and bird bones where growth stops with the fusion of the epiphyses to the diaphyses. It was also noticed that long bones of mammals and birds usually have larger medullary cavities than those of reptiles. Numbers of unidentified reptile remains are very low, but is not entirely excluded that some fragmentary bones were put with the unidentified mammals.

4.4.1. Turtles

The number of turtle taxa that could be expected at the studied sites was limited, judging from the species distribution maps of Iverson (1992). Osteological criteria to distinguish between the possible taxa were provided by Dr. F. de Lapparent-de Broin, Muséum National d'Histoire Naturelle, Paris, who also assisted with the identification of the turtle remains. Except when heavily fragmented, skeletal remains could easily be subdivided according to family: tortoises (Testudinidae), softshell turtles (Trionychidae) and terrapins (Pelomedusidae). Besides bones, a clay figurine depicting a turtle was also recovered from one of the studied sites, more precisely from Kursakata (NA 93/46) (Gronenborn, 1996). A similar example was also found in subrecent layers at Houlouf (Holl, 1988, Table 60, 195).

Sahelian giant tortoise (*Geocheolone sulcata*)

At only one of the studied sites, Oursi hu-beero (BF 97/30), were tortoise remains found. Eleven carapace fragments were collected from adjacent units at the site. Since some of the pieces fit together, they probably all belong to one individual. In addition two fragments of burnt long bones were tentatively attributed to the same tortoise specimen. The carapace must have been about 40 cm long and this large size allowed identification as *Geochelone sulcata*, the largest African tortoise (Villiers, 1958, 129). Other tortoise species that occur at present in the research area, all of the genus *Kinixys*, have not been attested. At first sight, their bones may seem similar to those of terrapins, but in fact, their shape and general appearance allows relatively easy distinction (see below).

Geochelone sulcata occurs in the Sahel and adjacent parts of the Sudan zone, but is most frequent in very arid areas, where it buries itself during the dry season (Villiers, 1958, 129). Al-Bakri, writing in the eleventh century AD, describes how a giant tortoise can be dragged out of its burrow, by attaching ropes to it and then pulling it with many men (Levtzion and Hopkins, 1981, 83). In the Kitab al-Istibsar (AD 1191) there is a reference to the use of tortoises for food in western Africa (Levtzion and Hopkins, 1981, 149). Blench (2000b) mentions that very large tortoises are presently kept in some courts of the Muslim emirs and that the Dogon in Mali raise giant tortoises for their meat.

Softshell turtle (Trionychidae)

Long bones of softshell turtles were identified, as well as pieces of the carapace and plastron, recognisable from their outer ornamentation. The species that had to be taken into account were African softshell turtle (*Trionyx triunguis*), Senegal flapshell turtle (*Cyclanorbis senegalensis*) and Aubrey's flapshell turtle (*Cycloderma aubryi*). The shape of the elements of the carapace and plastron and the nature of their ornamentation, allowed positive identification of both African softshell turtle and Senegal flapshell turtle (Table C.18, Fig. 19). A softshell turtle plastron fragment from Early Iron Age levels at Kursakata (NA 93/46) was worked into what looks like a pendant.

Softshell turtles are all predominantly freshwater animals and African softshell turtle is mainly confined to large waterbodies, especially large rivers (Villiers, 1958, 197-219). Softshell turtles only come out of the water to lay their eggs. When water levels drop during the dry season, they bury themselves in the mud until the next rains. The animals are generally very aggressive and they can bite ferociously. African softshell turtle can be especially dangerous, while Senegal flapshell turtle is much less aggressive. Meat of softshell turtles has a very fine taste and is much appreciated in many parts of Africa.

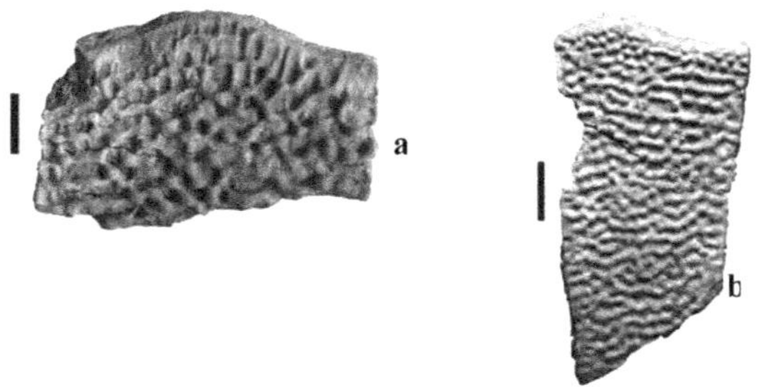

Fig. 19: Carapace of African softshell turtle (a) and of Senegal flapshell turtle (b) from phase II at Ngala (NA 93/45) (Scale = 1 cm)

Adanson's mud turtle (*Pelusios adansonii*)

Terrapins are by far the most common turtle family recorded at the studied sites. Fragmentary turtle bones that could not be identified more precisely on a purely osteomorphological basis were, therefore, also attributed to terrapin. Two genera, *Pelomedusa* and *Pelusios*, had to be taken into account while dealing with remains of terrapins. Typical osteological features of *Pelusios* that are missing in *Pelomedusa* are its hinged plastron, without fontanelle in adult specimens, its higher carapace, and the fusion of its pelvic girdle to both carapace and plastron. On pieces of the carapace or plastron, discrimination between *Pelomedusa* and *Pelusios* was possible from the shape of the sutures and sulci. Not all skeletal elements, or parts of skeletal elements, allowed the separation of *Pelusios* and *Pelomedusa*, but as no positive identification of *Pelomedusa* could be made, all Pelomedusidae remains were attributed to *Pelusios*. The species that could possibly be present are *P. adansonii* and *P. castaneus*. The distinction on both carapace and plastron was, again, made on the basis of the shape of the sulci relative to the sutures. In addition, the carapace of *P. adansonii* has a very typical smooth surface. In contrast to *P. adansonii*, no bones of *P. castaneus* could be recognised and it was therefore assumed that mainly the former species must be present. An overview of the *Pelusios* remains is given in Table C.20. Pit 1 in trench 6 at Zilum (NA 97/37) yielded the remains of at least five *Pelusios* specimens, with all parts of the skeleton represented and with two specimens that had large parts of the carapace preserved in anatomic position (Fig. 20). At a few locations, in Gajiganna BI (NA 90/5BI), Gajiganna phase II layers at Bukarkurari (NA 97/33), Kariari C (NA 95/1), phase IIIb and IV layers at Ngala (NA 93/45), Blé Mound B and Blé Mound E, peripheral parts of *Pelusios* carapaces with polished or worn edges were found.

Terrapins of the genus *Pelusios* have semi-aquatic habits. When the pools in which they live start to dry out, they burrow themselves in the mud and remain there until the next rains (Villiers, 1958, 223). Historical sources indicate that freshwater turtles played "some part" in the food of the people of West Africa during the Middle Ages (Lewicki, 1974, 97). Many present African ethnic groups eat terrapins, in spite of their revolting smell (Villiers, 1958, 71). Blench (1995, 2000b) reports that, in north-eastern Nigeria and neighbouring areas, tortoises and terrapins are often captured for food. The latter can also often be found in waterpots where their main function is to eat mosquito larvae and other possible worm infestations, but they are also reared for human consumption.

Fig. 20: Carapace of Adanson's mud turtle from Zilum (NA 97/37) (Scale = 1 cm)

4.4.2. Lizards

Agama (*Agama* sp.)

Most Iron Age contexts from Burkina Faso and a few Nigerian sites have yielded remains of a small lizard (Table C.19). The morphology of the dentaries and maxillas corresponds well with *Agama* and all the remains have therefore been attributed to the genus. A more precise identification was not possible because the reference collection at the RMCA is insufficient. According to Böhme *et al.* (1996), *A. agama* is the most widespread and abundant species of the genus in present-day Burkina Faso and Mali. They observed the animal in rocky habitats but also often found it in and around human settlements.

Monitor lizard (*Varanus* sp.)

Two monitor lizard species can be found in sub-Saharan Africa, *Varanus niloticus* and *V. exanthematicus*. The species can be distinguished osteologically from the shape of the parietal bone (Mertens, 1942, 131). Only one well-preserved parietal bone, from Early Iron Age Oursi (BF 94/45), was available and seemed to show *V. exanthematicus* characteristics. Altogether, four *Varanus* bones with cut-marks have been recovered: a metapodial from Saouga 94/120 (BF 94/120), an ilium from Saouga 95/7 (BF 95/7), a cervical vertebra from phase III at Mege (NA 94/7) and a vertebra from Blé Mound E.

V. niloticus is mainly confined to areas with a rich plant cover and close to water, while *V. exanthematicus* is a species of open and dry areas (Mertens, 1942, 36-40). Records of the former in present-day Burkina Faso are extremely rare (Böhme *et al.*, 1996). The typical habitat of *V. niloticus* in the Lake Chad area is shallow water with abundant vegetation, often some distance away from firm ground or in the midst of the numerous islands that line the lake's shores (de Bufffénil, 1993, 76-78). Specialised hunters from the region mainly catch the animal during high waters by posting baited hooks. Similar techniques are used in aquatic areas in other parts of West Africa, but on firm ground hunting strategies are more varied (de Bufffénil, 1993, 113). In Mali, for example, monitor lizards are killed with a blow of a spade when encountered coincidentally. During more planned hunting operations, traps are laid near their dens and sometimes they are pursued with the aid of hunting dogs.

Monitor lizard is eaten with relish all over Africa (Mertens, 1942, 50), although several reports exist of villages or areas, for example in Mali (de Bufffénil, 1993, 112), where there is a taboo against hunting the animal. Leo Africanus, writing in the sixteenth century AD, refers to the eating of meat of monitor lizard by Saharan Arabs, who cut off the animal's head and tail because they consider those parts as poisonous (Lewicki, 1974, 96). Besides the animal's meat, its eggs are also eaten (de Bufffénil, 1983, 79-80, 114). Monitor lizard is at present also greatly sought after because of its skin, which is locally used, for example for the making of drums, but which is also an important export product. It seems, however, that both for the skin and meat, *V. niloticus* is preferred over *V. exanthematicus* (de Bufffénil, 1993, 56). In Mali, dried monitor lizard heads are apparently also kept for medicinal reasons (de Bufffénil, 1993, 114) and from central Nigeria there are records of a few villages where monitor lizards are caught in the wild and then fattened in pots or basins before being sold for food (Blench *et al.*, 1992, 310).

4.4.3. Snakes

Many of the studied sites have yielded snake remains, almost exclusively vertebrae (Table C.22). A few dozens of species can be found in the research area (Chippaux, 2001) and a more precise identification was therefore not possible. Judging from the size of the vertebrae, however, it is mainly large species such as python (*Python* sp.) that are present. During the cool periods of the year, snakes hibernate in protected places, e.g., holes dug in the ground by other animals and crevices in rocks (Angel, 1950, 65). The Arab author Al-Idrisi, writing in the twelfth century AD, reports that West Africans eat snakes, usually after cutting off the head (Levtzion and Hopkins, 1981, 118). Small snake vertebrae found in burials at the site Akumbu in Mali, dated to AD 600-1000, were thought to have a possible symbolic significance (MacDonald and Van Neer, 1993). According to one of the branches within Islam, the Ibadite branch, snakes are forbidden food (Benkheira, 2000, 128).

4.4.4. Crocodiles

Crocodile (*Crocodylus* sp.)

Two crocodile genera occur in sub-Saharan Africa, *Crocodylus* and *Osteolaemus*. As the latter is characteristic for Guinean forests (Villiers, 1958, 324) it is unlikely to be present in archaeozoological assemblages from arid West Africa. The genus *Crocodylus* is represented by two species, *C. niloticus* and *C. cataphractus* (Villiers, 1958, 312-314). *C. niloticus* is spread all over Africa. It can live in large waterbodies, small streams and temporary lakes. The species seems most typical in the Sudan and Sahel zone. *C. cataphractus* can also be found in savannahs, but is most common in forest zones. It occurs mainly in large lakes or rivers. From their zoogeographical distribution we would expect to find *C. niloticus*, although the presence of *C. cataphractus* cannot be excluded.

Crocodiles have now disappeared from much of their original distribution zone because of overhunting (Villiers, 1958, 289-293). The animals are exploited for their meat, which is eaten in many parts of Africa and is apparently very tasty, even to European standards. More importantly, however, they are hunted for their skin, which is an important export product. Crocodiles are part of West African mythology and are often depicted in art objects (*Ibid.*). One of the local workmen at the excavation of Oursi hu-beero (BF 97/30) and other sites in northern Burkina Faso claimed that his grandmother used to guard the crocodiles that lived in the lake of Oursi. Although she had strongly forbidden it, a hunter went to the lake and killed the largest of the crocodiles with his spear. Shortly afterwards all crocodiles were said to have left the lake and were never seen again. From many parts of Africa examples are known of animals kept in captivity (Blench, 2000b). In many cases this concerns young crocodiles that are kept in hand-dug ponds and eaten for medicinal reasons. Historical sources on medieval West Africa are silent on the eating of crocodiles (Lewicki, 1974, 96). From nineteenth century Borno it is known, through Barth and Nachtigal, that Muslim inhabitants of the area considered crocodile as unclean, while others were very fond of it (*Ibid.*).

4.5. Birds

Identification of bird remains was in general relatively difficult, first of all because of the large number of species within this class in both Burkina Faso (453 species) and Nigeria (862 species) (Dowsett and Dowsett-Lemaire, 1993, 69, 91), and secondly because no reference skeletons were available for many of the species. Bird remains are also fragile and in the studied archaeological assemblages they were therefore usually found fragmented, which hampered identification even more. Many of the bird bones thus remain unidentified. The unidentified bones mainly consist of shaft fragments and pieces of less diagnostic elements (e.g., ribs, vertebrae), but some articular ends and occasionally even complete bones could not be identified either. Among the unidentified bird bone shafts there are some worked specimens. Most of the identified bird remains are from adult animals. Remains from juvenile and subadult birds are uncommon. When found, they were difficult to identify because osteomorphological features are not yet well developed in young specimens.

4.5.1. Pelicaniformes

Long-tailed cormorant (*Phalacrocorax africanus*)

The only one of the studied bird bones attributed to long-tailed cormorant is a coracoid from Early Iron Age layers at Kursakata (NA 93/46). The other species of cormorant that occurs in the Lake Chad area, the great cormorant (*P. carbo*), could be excluded because of its larger size (Brown *et al.*, 1982, 108-110). Long-tailed cormorant is common on Nigerian inland freshwaters (Elgood, 1982, 53). It is resident throughout, but densities may change locally with fluctuating water levels. This species mainly feeds on fish (Brown *et al.*, 1982, 118).

4.5.2. Ciconiiformes

Cattle egret *(Bubulcus ibis)*

A tarsometatarsal (Bd: 9.2 mm) from Saouga 94/120 (BF 94/120) corresponded well in morphology and size with cattle egret. The species is widespread and common throughout West Africa, except in thick forest (Serle *et al.*, 1977, 20). It is gregarious and rather tame, and often follows cattle or other large ungulates for the insects they disturb. Cattle egrets are most common on dry, often cultivated, grounds but can occasionally also be seen in swamps or rice fields. According to the local people from Oursi, there is an Islamic prohibition against the eating of the bird.

Egret (*Egretta* sp.)

The morphology and size of an ulna from Zilum (NA 97/37) fitted well with the genus *Egretta*. However, only one, incomplete, reference specimen of this genus, more precisely a skeleton of little egret (*E. garzetta*), was available. Therefore, a first phalanx (GL: 29.9 mm, L: 28.7 mm) of the second anterior digit, retrieved from Gajiganna phase IIb levels at Gilgila (NA 99/65), could not be compared to the same element of an *Egretta* reference skeleton. However, because of its Ardeidae morphology that did not match other genera from the family and because of its size, it was also attributed to *Egretta*. Four species of the genus can be found in the Lake Chad area, one of which, great white egret (*E. alba*), could be eliminated because of its larger size (Brown *et al.*, 1982, 151-159), but distinction between the three other species, black egret (*E. ardesiaca*), little egret (*E. garzetta*) and yellow-billed egret (*E. intermedia*), was not possible. All Nigerian egret species are resident, only little egret has a partly Palearctic migrant population. Egrets can mostly be found on inland or coastal water.

Goliath heron (*Ardea goliath*)

Because of its size, a proximal part of an Ardeidae radius from Saouga 95/7 (BF 95/7) could be attributed to the largest representative of this family in Burkina Faso, the goliath heron. The species is resident and widely spread throughout West Africa, although never very common (Serle *et al.*, 1977, 23). It is attracted by swamps, the sandbanks of large rivers, estuaries and mangrove creeks.

Heron *(Ardea* sp.)

A scapula from Early Iron Age levels at Oursi (BF 94/45), a tibiotarsal and a quadrate from Gajiganna A (NA 90/5A), a humerus from phase IV at Ngala (NA 93/45) and a mandible from Blé Mound E were attributed to *Ardea*. In both Burkina Faso and Nigeria this genus comprises four species, purple heron (*A. purperea*), grey heron (*A. cinerea*), black-headed heron (*A. melanocephala*) and goliath heron (*A. goliath*) (Brown *et al.*, 1982, 160-168). The latter could be excluded because of its larger size, but no distinction could be made between the other species. Confusion with great white egret, in the same size range as *Ardea*, is moreover not excluded. The *Ardea* species from Burkina Faso and Nigeria are all resident, but parts of the grey heron population are Palearctic migrants. They all usually occur near water.

Stork (Ciconiidae)

Three Nigerian sites have yielded bones from birds of the stork family (Table C.25). The seven species, belonging to four genera, that occur in the area today are given in Table C.24 (Brown *et al.*, 1982, 172-190). Only for three of them was reference material available and the identification manual that exists for (Egyptian) Ciconiidae (Gruber, 1990) concerns only the same three species. This hampered the determination of the stork bones to species level, but their size has nonetheless permitted some tentative species identifications. The largest stork bones found probably pertain to white stork (*Ciconia ciconia*) although marabou (*Leptoptilus crumeniferus*), a slightly larger species, cannot be ruled out. The smallest stork remains are possibly of Abdim's stork (*Ciconia abdimii*). A third species of intermediate size, in the range of yellow-billed stork (*Mycteria ibis*), also seems present. The three size classes are also reflected in the measurements (Table C.26).

Ciconiidae are a family of large to very large long-legged water birds. Most feed on fish or other aquatic organisms but some, like Abdim's stork, are mainly terrestrial, feeding on insects (Brown *et al.*, 1982, 172-173). Marabou lives largely on carrion and is also an active predator. Some of the Ciconiidae in Burkina Faso and Nigeria are resident (e.g., marabou), while others are intra-African migrants (e.g., Abdim's stork) (Brown *et al.*, 1982, 172-190). White stork is a Palearctic migrant and can be quite common on seasonally flooded areas in the Sudan and Sahel zones (Elgood, 1982, 58).

Sacred ibis *(Threskiornis aethiopica)*

A Threskiornithidae humerus was identified at Galaga (NA 92/2C). Judging from its size, three species living in the Lake Chad area today have to be taken into account, sacred ibis (*Threskiornis aethiopica*), Eurasian spoonbill (*Platalea leucorodia*) and African spoonbill (*P. alba*) (Brown *et al.*, 1982, 200-209). When the distal breadth of the humerus is compared to the same measurement on recent skeletons of the species in the RMCA, it corresponds best with sacred ibis (Table C.27). From Phase II layers at Ngala (NA 93/45) another Threskiornithidae bone, a scapula fragment with cut-marks, was recovered. This bone could not be measured but seems to pertain to a bird of the same size class as the humerus from Galaga. Sacred ibis, African and Eurasian spoonbill are all resident in West Africa. European spoonbill additionally has a partly Palearctic migrant population. The three species mainly occur near water.

4.5.3. Anseriformes

In the study area the order of the Anseriformes, all waterbirds, contains only the family of the ducks and geese (Anatidae). The species that have to be taken into account when identifying Anatidae remains from Burkina Faso and Nigeria are listed in Table C.28. The list is quite extensive and only a minority of the species are represented in the available reference collection, or treated in the comparative osteological study of (European) Anatidae by Woelfle (1967). This rendered the identification of the Anatidae remains particularly problematic. They were therefore divided into size classes, although even this was not always straightforward because the duck and goose species of the area cover a continuous size range between very small and very large.

4.5.3.1. Small and medium-sized ducks and geese (Anatidae)

Anatidae remains of about the size as common teal (*A. crecca*) and garganey (*A. querquedula*) from the reference collection at the RMCA were classified as "small" (Table C.29). Bones in the size range of northern pintail (*A. acuta*) and white-faced whistling duck (*Dendrocygna viduata*) were labelled "medium-sized" (Table C.30). The twenty medium-sized duck or goose remains from Galaga (NA 92/2C) are in fact larger than pintail from the reference collection and even larger than mallard (*A. platyrhynchos*), a bird with an average total length of 58 cm (Heinzel *et al.*, 1982, 52). They are, however, still smaller than Egyptian goose (*Alopochen aegyptiacus*), as is also clear from the measurements taken on them (Table C.32), and it was therefore decided to classify them among the medium-sized ducks or geese. As Galaga is a site from the nineteenth century AD, the possible presence of muscovy duck (*Cairina moschata*) also has to be taken into account. For this species no reference skeletons were available, but its bones would be comparable in size to those of Egyptian goose (Hamilton-Dyer, pers. comm.). Muscovy duck is a rather large duck species, with a total length between 66 and 84 cm (del Hoyo *et al.*, 1992, 594), and was introduced into Africa during the sixteenth century AD by the Portuguese, who brought it from South America (Blench, 2000b). The species has seen a rapid spread to the West African inland, both from the coast and across the Sahara with the Arabs, despite the fact that large parts of the Muslim population regarded ducks as unclean (Blench, 2000b, submitted).

4.5.3.2. Large ducks and geese (Anatidae)

The bones attributed to large ducks and geese are given in Table C.31. Several of the remains bear cut or chop-marks: a proximal humerus, a proximal radius and a distal tibiotarsal from phase II at Ngala (NA 93/45) and a proximal scapula from phase IIIb at the same site, in addition there is a distal humerus from Blé Mound B and an ulna from Blé Mound C. An ulna shaft fragment from Early Iron Age levels at Kursakata (NA 93/46) seems to have been worked. Three species have to be considered for the size class of large ducks and geese, Egyptian goose, spur-winged goose (*Plectropterus gambensis*) and knob-billed duck (*Sarkidiornis melanotos*). Only the first species is represented in the reference collection of the RMCA, but there are some archaeological remains of spur-winged goose from the subrecent site Koyom (Rivallain and Van Neer, 1983, 1984) in southern Chad, which can serve as comparative material. Measurements taken on the large duck and goose bones are listed in Table C.32 together with measurements on the Egyptian goose specimens and on the available spur-winged goose bones from the reference collection. Even when not measurable, remains of spur-winged goose could often be distinguished from those of the two other species because of their considerably larger size. Bones in about the same size range as Egyptian goose are either of Egyptian goose or knob-billed duck, but the morphology of a carpometacarpal from Gajiganna A (NA 90/5A) allowed the exclusion of Egyptian goose.

At the localities in the southern Lake Chad area, spur-winged goose seems to have been caught especially. The species is an intra-African migrant and can be found in the area between February and May (Fry *et al.*, 1982, 246). Spur-winged goose, kept as a backyard species, has been recorded in present-day Mali and north-eastern Nigeria (Blench, 2000b). Already in the fourteenth century AD, the Arab author Al-Umari mentions that the people of Mali keep geese (Levtzion and Hopkins, 1981, 267). In communities at the end of trans-Saharan trade routes these may have been greylag geese (*Anser anser*), instead of local species, although it is generally assumed that geese did not cross the Sahara (Blench, 2000b).

4.5.4. Falconiformes

Accipitridae

Three sites have yielded phalanges of specimens from this rich family of birds of prey, Oursi (BF 94/45) (n=1) and Saouga 94/120 (BF 94/120) (n=2) in Burkina Faso and Gajiganna A (NA 90/5A) (n=1) in Nigeria. Because the elements are not very diagnostic and because of a lack of sufficient comparative material, the identifications could not be carried out to a lower taxonomic level. Judging from the size of the phalanges, at least two species are represented, however, one of about the size of black kite (*Milvus migrans*) (BF 94/120) and another nearly as large as Rueppell's vulture (*Gyps rueppellii*) (BF 94/45 and NA 90/5A). According to the workmen at the excavation of Oursi hu-beero (BF 97/30), vultures are not eaten.

4.5.5. Galliforms

4.5.5.1. Small galliforms

Quail (*Coturnix coturnix*)

A small Phasianidae tibiotarsal (Bd: 4.5 mm) from Early Iron Age layers in Oursi (BF 94/45) belongs to quail. Judging from the geographic distribution of the species of the genus *Coturnix* (Urban *et al.*, 1986, 13-19), it is probably of common quail (*Coturnix coturnix*). Common quail is a Palearctic winter visitor in Burkina Faso. It occurs in open grass fields and is particularly attracted to agricultural lands (growing crops, fallow fields, pastures) and grasslands regenerating after burning (Urban *et al.*, 1986, 15).

4.5.5.2. Large galliforms

Both for Nigeria and Burkina Faso, several taxa have to be considered when dealing with remains of large galliforms. First of all the presence of large francolins is possible, including double-spurred francolin (*Francolinus bicalcaratus*), scaly francolin (*F. squamatus*) and Clapperton's francolin (*F. clappertonii*) (not in Burkina Faso), besides that of another, exotic Phasianidae, the domestic fowl (*Gallus gallus* f. domestica). Finally, helmeted guineafowl (*Numida meleagris*) can also occur. The other West African guineafowl species were not taken into account because their distribution is limited to forested areas (Serle *et al.*, 1977, 58-59). Osteological distinction between the large francolins, domestic fowl and helmeted guineafowl proved to be difficult, in spite of the existence of an osteomorphological and osteometric study for the differentiation of their remains by MacDonald (1992). A problem with the latter study is that only a few elements (cranium, sternum, coracoid, scapula, radius, carpometacarpal, pelvis and tarsometatarsal) are suitable for morphological distinction and that they also need to be fairly complete. Francolin and guineafowl appear to be relatively easily distinguishable from their size, however, as most bones of the former seem to fall below the size range of the latter. The ranges given by MacDonald for measurements on ten modern domestic fowl specimens, of which neither type nor locality are mentioned but which are presumably West African, show considerable overlap with those for measurements on both helmeted guineafowl and francolin skeletons.

Osteometry thus appears to be useless, once a possible presence of domestic fowl has to be assumed.

Francolin (*Francolinus* sp.)

A tarsometatarsal from a large galliform found at Gajiganna A (NA 90/5A) is too small for helmeted guineafowl. The presence of domestic fowl at this site can be excluded in view of its occupation date (see below) and the bone was therefore attributed to francolin. A second francolin tarsometatarsal was retrieved from Galaga (NA 92/2C) (Fig. 21). There is no doubt about the identification of this bone, since it has a pronounced hypotarsal ridge and two spurs (MacDonald, 1992). A coracoid from the same site also showed francolin morphological traits, although it is larger than the size range given for the bird, and its attribution to francolin therefore remains uncertain. Identifications to species level were not possible for the francolin bones. Measurements taken on them are given in Table C.36, together with measurements on the other large galliform remains. Francolins occur in a wide range of vegetation types and are extremely sedentary (Urban *et al.*, 1986, 24). They are at present extensively hunted, both for sport and for food.

Fig. 21: Francolin tarsometatarsal from Galaga (NA 92/2C) (Scale = 1 cm)

Domestic fowl (*Gallus gallus* f. domestica)

Table C.33 summarises the bone remains that could be attributed to domestic fowl, using the morphological features described in MacDonald (1992). Measurements are given in Table C.36. One of the bones, a coracoid from Saouga 95/7 (BF 95/7), shows cut-marks. All tarsometatarsals classified as domestic fowl are spurred. This is not due to a predominance of male specimens, but because the distinction between fowl and guineafowl cannot be made for unspurred tarsometatarsals.

Domestic fowl is not indigenous to Africa and is generally considered to have evolved from jungle fowl (*Gallus gallus*) some time before the sixth millennium BC, in the area where jungle fowl can presently be found, between eastern India and Java (Blench and MacDonald, 2000). The history of domestic fowl on the African continent has been described by MacDonald (1992), who mentions that the earliest known skeletal and iconographic evidence in Africa is from the Egyptian mid-second millennium BC. The oldest remains of domestic fowl in West Africa were found at Jenné-Jeno in Mali, in layers dating to AD 500-850 (MacDonald 1992, 1995; MacDonald and MacDonald 2000). The early occurrence at this locality would probably be related to its function as a trade centre. The oldest written evidence for domestic fowl in West Africa dates from the fourteenth century AD and is by the hand of Ibn Battuta (Levtzion and Hopkins, 1981, 300).

Domestic fowl is a very important livestock animal in West Africa today, and it is kept by both sedentary and mobile groups. In Nigeria it outnumbers the other livestock species and can be found all throughout the country, where it is typically kept under low input, free-range, systems of management (Bourn *et al.*, 1994). However, it should be stressed that the present situation in Nigeria is somewhat distorted, because of recent efforts by the government to encourage the keeping of domestic fowl.

Helmeted guineafowl (*Numida meleagris*)

The presence of domestic fowl at sites of the Gajiganna Culture is unlikely because they date from a time before which an introduction of the bird into West Africa seems possible (see above). Large galliform remains from these sites that fall in the size range of helmeted guineafowl have therefore been attributed to this species, but they were checked against the characters of MacDonald (1992) where possible. For more recent sites, identifications of guineafowl were only carried out on bones where the mentioned osteomorphological criteria could be used. The guineafowl bones identified are listed in Table C.34 and measurements taken on them in Table C.36. Two coracoids, one from Gilgila (NA 99/65) and one from Galaga (NA 92/2C), are probably from juvenile specimens, since they are not well ossified.

The wild helmeted guineafowl is widely distributed over Africa. It is a resident species from Senegal to Ethiopia and Somalia, and south to Namibia and South Africa (Urban *et al.*, 1986, 8-10). A very small relic population has survived in northern Africa, more precisely in north-western Morocco, although this could instead represent escaped domestic birds (Blench, 2000b). Helmeted guineafowl can be found in all open-country vegetation from forest edges to arid savannahs, from sea level to above 3000 metre (Urban *et al.*, 1986, 8-10). It can locally be common to very abundant, and is especially attracted by savannahs mixed with cultivation. The availability of drinking water and elevated roosts are the habitat features that determine its occurrence. During the dry season guineafowl flocks of up to 300 birds have been recorded and there is possibly a direct relationship between aridity and infrequent water places on the one hand and flock size on the other (Donkin, 1991, 5). Guineafowl breeds mainly in, or just after, the rainy season (Urban *et al.*, 1986, 8-10). The bird suffers from heavy hunting pressure and egg collecting, but when there is sufficient habitat it is rarely threatened by extinction.

Within the African continent guineafowl is also reared. At present, this is chiefly done in the semi-arid west, although the species cannot be considered as fully domesticated (Blench, 2000b). Birds kept are usually not provided with housing and roost in nearby trees at night, but rarely fly away. They are left to find their own food and, like their relatives in the wild, they lay their eggs scattered in the bush, rather than in a single place. Since the living conditions of the domestic (or tamed) form of guineafowl are very close those of the wild form, distinction between the two on a purely osteological basis may be impossible, but unfortunately this never seems to have been investigated in a formal study. Although the two forms are very close, the meat of wild guineafowl is said to be much tastier than that of its relatives kept in captivity (Pelzer, pers. comm.). In this study guineafowl was included in the wild birds, although it may concern tamed animals.

Before the present study, positive identifications of guineafowl in West African archaeological contexts were limited, with the oldest ones from early second millennium BC rockshelter K6 in Kintampo (Stahl, 1985). Other remains were found at sites in the Middle Senegal Valley (MacDonald and MacDonald, 2000; Van Neer and Bocoum, 1991), dating to the first millennium AD, at Akumbu (AD 600-1200) (MacDonald and Van Neer, 1993), in Daima I layers at Daima (Connah, 1981, 137) and at the subrecent site Koyom (Rivallain and Van Neer, 1983, 1984). There is also a historical mention of guineafowl in medieval Borno (Lewicki, 1974, 91).

Unidentified large galliforms and metrical analysis of all large galliform remains

Table C.35 lists all large galliform remains, which could not be identified to species level. The numbers of francolin, domestic fowl and guineafowl bones found at each locality are also mentioned, making it clear that species identifications were rather rare. Among the large galliform bones, cut-marks occurred mostly on distal tibiotarsals, as attested at Oursi hu-beero (BF 97/30), twice at Saouga 94/120 (BF 94/120), in phase IV layers at Ngala (NA 93/45), in phase IV layers at Mege (NA 94/7) and at Blé Mound E. Other bones with cut-marks were a distal humerus from Saouga 94/120, a femur shaft fragment from phase II and a scapula from phase I at Ngala. The ossification state of five bones from Galaga (NA 92/2C), besides one from phase II and two from phase IIIb at Ngala (NA 93/45), points to juvenile specimens. A large galliform tibiotarsal from Phase II at Ngala (NA 93/45) and a femur from Galaga (NA 92/2C) both showed evidence of a healed fracture. In addition, two thoracic vertebrae, pathologically fused together, were recorded at Saouga 95/7 (BF 95/7).

Using the measurements taken on them (Table C.37), an attempt was made to determine the species present among the unidentified large galliform remains. To this end, Logarithmic Size Indices (LSI) were calculated (see 3.8.). The reference specimen chosen was a skeleton from a recent (domestic) guineafowl of unknown sex, collected in northern Burkina Faso, and measurements on this skeleton, used for the calculations, are given in Table C.37.

To begin with, LSI values for large galliform remains identified to species level were plotted by species (Fig. 22). From this graph, it is clear that the francolin remains show large size variability, even when the coracoid from Galaga (NA 92/2C) with an uncertain identification would be left out. This may be due to the presence of more than one species, presumably double-spurred francolin and a smaller species, although size differences between the francolin species are rather small. If the identification of the coracoid is correct, then a small size overlap of francolin and guineafowl bones should apparently also be taken into account. The graph furthermore shows that size ranges are limited for guineafowl and that its values all cluster nicely around the reference guineafowl specimen. Domestic fowl remains, on the other hand, seem to exhibit much variability in size. The graph confirms what has been formulated above, that guineafowl and francolin, perhaps with consideration of a small overlap, can be separated based on size, but that domestic fowl largely overlaps with these two species.

Subsequently, on the same graph, LSI values for large galliform remains were plotted by site, with indication of the values for bones identified to species level. The size ranges recorded for the sites correspond well with the species that were identified morphologically from them. Unfortunately, due to the aforementioned size overlap between the large galliform species, no additional species identifications could be made based on the LSI values or the original measurements. The LSI values for Early Iron Age levels at Oursi (BF 94/45) exclude the presence of guineafowl at the site, however. It could represent remains of (double-spurred) francolin, since more

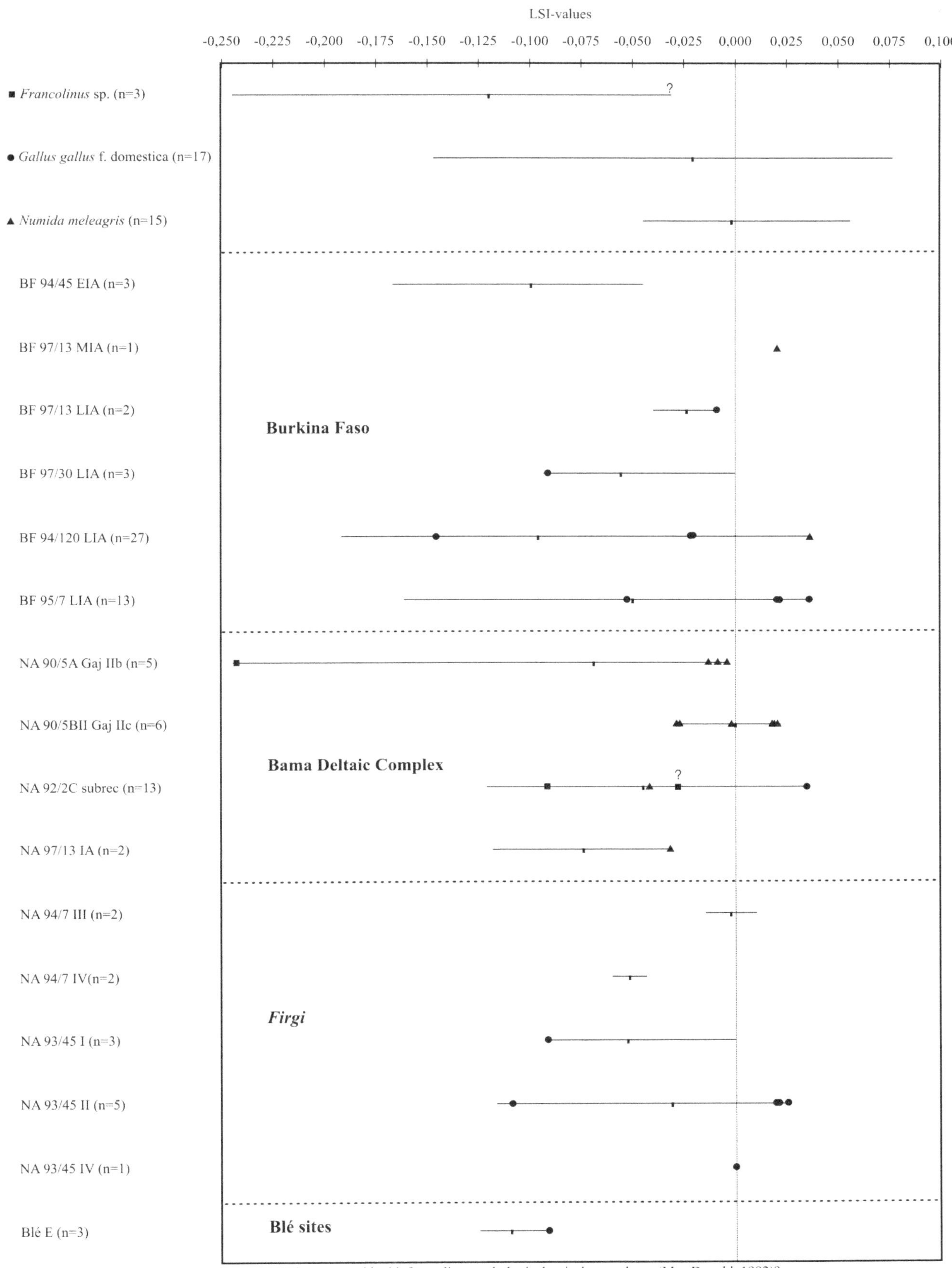

? LSI-value from coracoid with francolin morphological traits but too large (Mac Donald, 1992)?

Fig. 22: LSI analysis of large galliforms by site and phase

dispersed values would perhaps be expected if domestic fowl was present.

4.5.6. Gruiformes

Bustard (Otididae)

Bustards are terrestrial birds that inhabit semi-desert, grassland, savannah, open patches in scrub, light woodland and cultivated areas (Urban *et al.*, 1986, 148). Two bustard bones have been found at Nigerian sites. An ulna (Did: 22.2 mm) from Gajiganna BI (NA 90/5BI) pertains to one of the largest two representatives of the Otididae in Nigeria, Arabian bustard (*Ardeotis arabs*) and Denham's bustard (*Neotis denhami*). Reference specimens were available only for the former, and identification to species level was therefore impossible. Arabian bustard is partly resident, and partly an intra-African migrant (Urban *et al.*, 1986, 158-159). It is a common species but has suffered much from persecution in the past few decades. Denham's bustard is an endemic resident and intra-African migrant and is, like Arabian bustard, a popular prey (Urban *et al.*, 1986, 151). In Nigeria it is reported to be uncommon in the open savannahs of the north (Elgood, 1982, 80). A smaller bustard species was present at Galaga (NA 92/2C), where a distal tibiotarsal bearing cut-marks was found, approximately in the size range of white-bellied bustard (*Eupodotis senegalensis*) and black-bellied bustard (*E. melanogaster*). Both species are resident in Nigeria, where the former appears to be more common than the latter (Elgood, 1982, 80).

4.5.7. Charadriiformes

A scapula and a humerus from Early Iron Age Kursakata (NA 93/46) and a humerus from phase II at Ngala (NA 93/45) were attributed to the order of the Charadriiformes. The bones could not be identified more precisely because of a lack of sufficient comparative material. The Charadriiformes comprise many species and are a diverse group of mainly wading birds, occurring in marshes, freshwater and marine habitats, besides open fields, grasslands and mudflats (Urban *et al.*, 1986, 180).

4.5.8. Columbiformes

Pigeon and dove (Columbidae)

A few of the studied sites in Burkina Faso have yielded remains of pigeons or doves (Table C.38). More than ten species of the Columbidae family occur in Burkina Faso today (Urban *et al.*, 1986, 442-497). Sizes of most species are similar and for only three of them were reference specimens available. Identifications could therefore not be carried out beyond family level. Columbidae are medium-sized arboreal and terrestrial birds that eat grain, fruit and other plant material (Urban *et al.*, 1986, 442). Al-Umari, writing in the fourteenth century AD, reports that the peoples of West Africa kept domestic pigeons (Levtzion and Hopkins, 1981, 267). On linguistic grounds, Blench (submitted) argues for a trans-Saharan introduction of the culture of pigeon-keeping, probably in the early Middle Ages. Nevertheless, the bird has been put with the wild species in the faunal lists of Appendix D.

4.5.9. Cuculiformes

Large cuckoo (Cuculidae)

A humerus of a large cuckoo was found at Kariari C (NA 95/1). Its size corresponds well with an African cuckoo (*Cuculus gularis*) skeleton from the reference collection at the RMCA. A few other Nigerian cuckoo species fall in the same size range, however. No comparative skeletons were available for these species and identification could therefore not be carried out beyond family level. Most African cuckoos are migratory and the yearly arrival of many of them in the drier savannahs and arid habitats during the rainy season seems to depend on the availability of butterfly caterpillars, their preferred food (Fry *et al.*, 1988, 58-59).

4.5.10. Coraciiformes

A radius from Kissi 22 (BF 96/22) belongs to a bird of the order of the Coraciiformes, but the element could not be identified more precisely. The Coraciiformes are a heterogeneous assemblage of perching birds and its families are adapted to very varied ways of life (Fry *et al.*, 1988, 255).

4.5.11. Passeriformes

The songbirds (Passeriformes) are by far the largest order of African birds and identification of their remains, even to family level, is therefore generally impossible. A bill fragment and ulna from Labe Kanuri (NA 97/26) were attributed to a small specimen of this order.

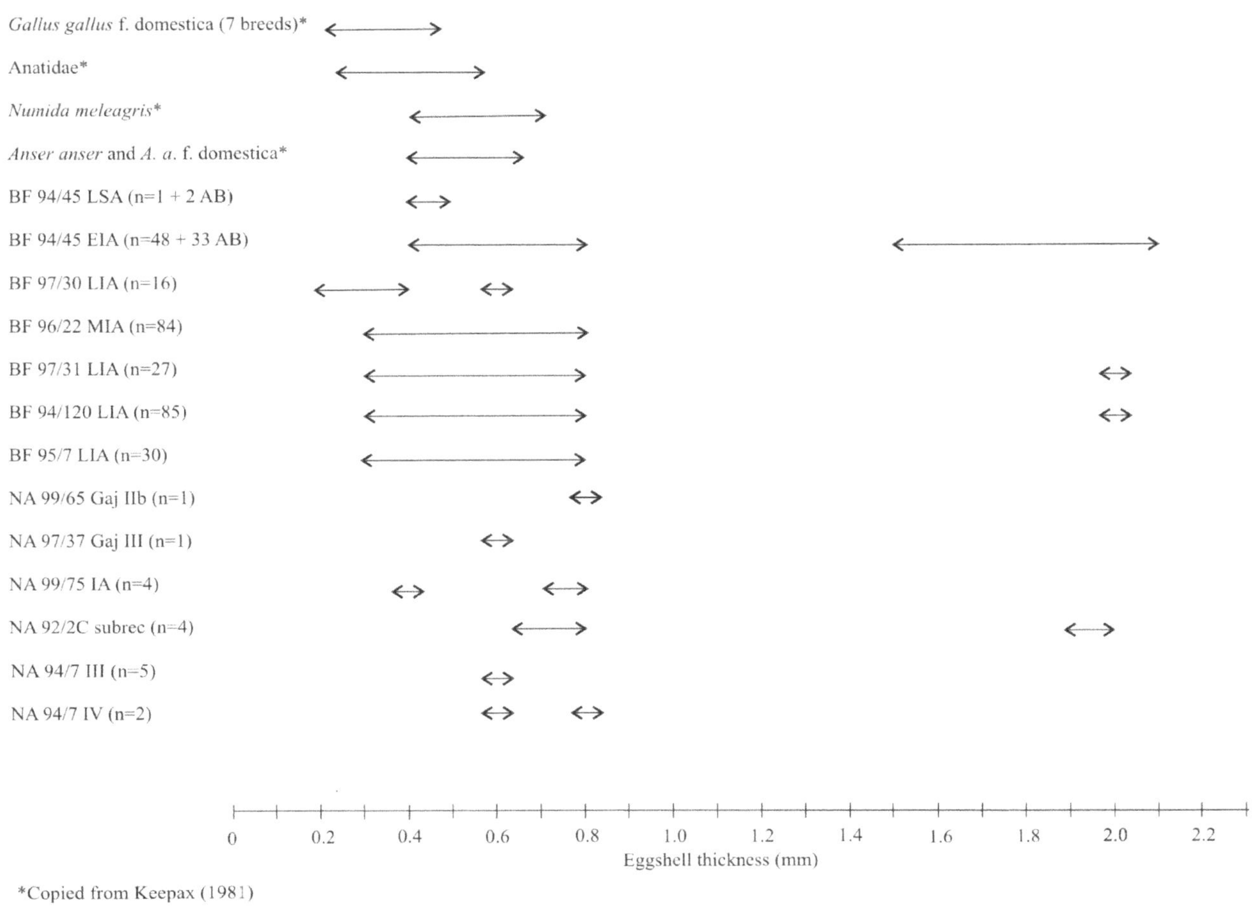

Fig. 23: Bird eggshell thickness by site and phase

4.5.12. Eggshells

At several sites bird eggshell fragments were found. The thickness of the fragments was measured since a study by Keepax (1981) has shown that, besides analysis of microscopical features, eggshell thickness can also give an indication of the species present.

Ostrich (*Struthio camelus*)

Ostrich eggshell fragments were very easy to separate from eggshells from other bird species because they are much thicker, around 2.0 mm. They were most numerous in the Early Iron Age levels at Oursi (BF 94/45), from where eight fragments with polished edges were also retrieved. They are probably waste from ostrich eggshell bead production at the site, of which the end product has also been recovered (von Czerniewicz, 2004, 45). No ostrich eggshells were found among the faunal remains from the studied sites in the southern Lake Chad area, except at the subrecent site Galaga (NA 92/2C). At some localities, however, they appear to have been grouped with the cultural finds. This was the case at Labe Kanuri (NA 97/26) and Elkido North (NA 99/75) (Breunig *et al.*, in prep.). Connah (1981, 175-178, 188) has also found ostrich eggshells in Daima III contexts at Daima, mostly as grave goods but also in occupation layers. Preferred habitats of ostrich are open short-grass plains and semi-desert (Brown *et al.*, 1982, 32-34). Judging from the present geographic distribution of the species, the eggs could probably have been collected not very far from the sites. The birds themselves do not seem to have been hunted, as their bones are missing from the archaeological assemblages. Apparently, no known archaeological site in West Africa has yielded ostrich bones, although there is historical evidence for ostrich hunting on horseback during the medieval period (Lewicki, 1974, 94). At present, ostrich eggs are widely used ornaments in Islamic buildings (Blench, 2000b).

Unidentified bird

Species identification was almost impossible for eggshell fragments of birds other than ostriches. On Fig. 23 their thickness is compared to those measured for eggshells of a few European bird types, which broadly seem to fall into two groups, a thicker group (0.50-0.85 mm) comparable with goose or guineafowl, and a thinner group (0.20-0.50 mm), matching medium to large modern domestic fowl eggs (Keepax, 1981). Eggshell fragments from Late Stone Age and Early Iron Age Oursi (BF 94/45) have about the thickness of large ducks or geese eggs of which bone remains have also been found at the site (see 4.5.3.2.). Late Iron Age contexts from Burkina Faso all yielded some egg fragments thinner than those from Oursi, which could be related to the presence of domestic fowl, attested from that period onwards (see 4.5.5.2.). The presence of thin shells at Middle Iron Age Kissi 22 (BF 96/22) might mean that domestic fowl was introduced earlier in northern Burkina Faso than could be attested from the bone remains. The Nigerian sites studied have yielded many fewer eggshell remains than those in Burkina Faso, although at Mege (NA 94/7) there were a few units, in phase III and IV layers, with concentrations of eggshell fragments. Because of the quantification method used (all shells of same thickness in one unit lumped together, see 3.6.), however, the concentrations do not appear in the faunal list of the site (Table D.32). The thickness of eggshell fragments from Nigerian sites usually falls in the upper range of those from sites in Burkina Faso and seems to be too large for domestic fowl. Some eggs are even too thick for guineafowl or geese. A few other West African archaeological sites also yielded pieces of bird eggshell. Possible guineafowl egg fragments were present at Cubalel and Siouré (AD 0-1400) in the Middle Senegal Valley (MacDonald and MacDonald, 2000) and domestic fowl eggs were identified at Tegdaoust (AD 600-1200) in Mauritania (Bouchud, 1993), while thin, brown, eggshell, in the size range of domestic fowl or guineafowl eggs, were recorded in several contexts at Akumbu (AD 400-1400) in the Malian Méma (McDonald and Van Neer, 1993).

4.6. Mammals

A relatively large number of species had to be taken into account while identifying mammalian remains, but their study was facilitated by the presence of an extensive reference collection at the RMCA and by the availability of identification manuals for several taxa. The manuals used will be mentioned under the relevant species, but in the initial stage Walker (1985) was sometimes consulted as an aid to determine the family or large taxonomic group in which bones should be placed. The best pieces for identification were cranial parts, especially with teeth preserved, articulations of long bones and small compact bones like phalanges, while ribs and vertebrae were less diagnostic. (Small) fragments of long bone shafts were also problematic because of the absence of identification criteria. Their presence is mostly responsible for the low identification rate of mammalian bones at sites with poor preservation (see 5.6.).

4.6.1. Primates

Olive baboon (*Papio anubis*)

After comparison with recent reference skeletons, a distal fibula and a humerus shaft fragment from Gajiganna phase IIc layers at Gilgila (NA 99/65), were identified as olive baboon, a species of the family of the Cercopithecidae and the largest monkey of the West African savannahs. Olive baboon is the most extensively distributed of all baboons (Kingdon, 1997, 34-35). In Nigeria it is common in protected areas, but is absent from the Lake Chad area (Happold, 1987, 90). It can be found in wooded savannahs and on rocky inselbergs. Baboons live within reach of water and need to drink daily (Kingdon, 1971, 182). They can sometimes cause serious damage to cultivated crops, and despite intensive human persecution, they are still very widespread and common (Kingdon, 1971, 189; Happold, 1987, 90). One of the animal figurines from Daima III layers at the site of Daima was thought to represent baboon (Connah, 1981, 184-185).

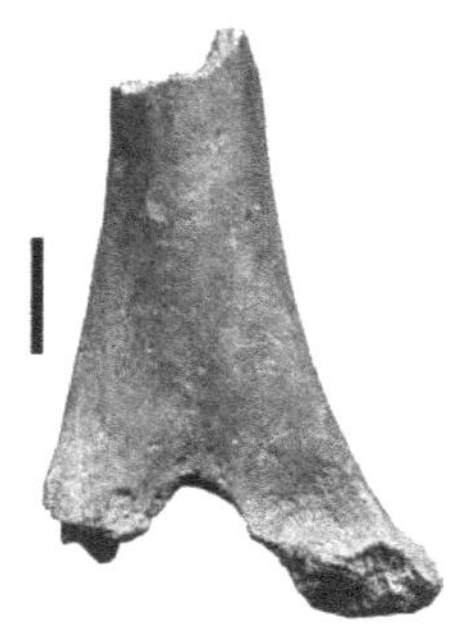

Fig. 24: Anubis baboon humerus shaft from Gajiganna IIc layers at Gilgila (NA 99/65) (Scale = 1 cm)

Savannah or patas monkey (*Cercopithecus aethiops/ patas*)

Some Cercopithecidae remains from two Gajiganna sites were too small for olive baboon and could belong either to savannah monkey or to patas monkey. Osteological distinction was not possible. A scapula, an ulna, a first and a fifth metatarsal were found at Gajiganna A (NA 90/5A). Gajiganna phase IIb layers at Gilgila (NA 99/65) yielded a fragment of a mandible and a loose canine, while from Gajiganna phase IIc layers at the same site a loose premolar, two loose lower molars and a distal humerus were retrieved. All *Cercopithecus* remains from Gilgila were found in adjacent archaeological units and it is therefore suspected that they represent one individual.

The fact that they are spread over two different phases should not be an objection as divisions between phases are usually artificial and also that postdepositional processes could have caused mixing (see 3.1).

C. aethiops is by far the most widespread monkey in the whole of Africa, but it does not occur in the two most extreme environments, forests and deserts (Kingdon, 1997, 58-63). Typically living near watercourses and in riverine forests the species is now disappearing from its northernmost distribution zone as a consequence of tree-clearance in the Sahel. This has also caused a population decline of patas monkeys, accelerated by intensified cultivation and hunting (Kingdon, 1997, 57-58). Patas monkeys can be found in vegetation types ranging from open grassland to dry woodland, but are most common in thinly bushed acacia-woodland grassland. At the beginning of the twentieth century AD the German geographer Schultze (1968, 114-115) mentions that monkeys of the genus *Cercopithecus* are widely distributed over Borno. Both savannah and patas monkeys still occur in the area (Happold, 1987, 94-95) and the animals are notorious for their raids on domestic crops (Kingdon, 1971, 216, 269). In present south-eastern Nigeria, they are traded in markets, either as bushmeat, or as pets (Angelici *et al.*, 1999).

Human (*Homo sapiens sapiens*)

In the faunal assemblages from the studied localities, human remains have been encountered relatively frequently (Table C.39) and some presumably belong to human burials that were recognised during the sites' excavation. Except perhaps for loose teeth, it can be assumed that the others are also from buried individuals, as inhumations of the deceased within the settlement of the living seem to have been common practice in the research area (see 2.2.2. and 2.2.3.). No detailed age determination of the recovered human remains has been carried out, but babies as well as older children and adults seem to be present. Except for the buried individuals at Gajiganna A (NA 90/5A) (Breunig *et al.*, 1993a), no anthropological studies have been conducted on human remains found at the sites included in this work.

4.6.2. Insectivores

When teeth were preserved, cranial remains of insectivores could readily be identified because of the typical insectivore tooth pattern. Fragmentary postcranial bones, on the other hand, could perhaps have been confused with small rodent remains. Two genera of insectivores had to be taken into account for both parts of the research area, *Atelerix* and *Crocidura*, which were easy to separate since individuals of the former are considerably larger than those of the latter.

African hedgehog (*Atelerix albiventris*)

This sole representative of the genus *Atelerix* was attested at localities in Burkina Faso only, from where not more than a handful remains have been recovered altogether (Table C.40). They are predominantly mandible fragments, but in archaeobotanical samples from Early Iron Age layers at Oursi (BF 94/45) a humerus and tibia were also found. The only measurement that could be taken on *Atelerix* bones was the toothrow length (18.1 mm) of a mandible from Kissi 22 (BF 96/22). *Atelerix* has a marked preference for relatively open, dry, or seasonal habitats with sparse or patchy grass cover, especially in overgrazed regions with dense ungulate populations (Kingdon, 1997, 141). The animal is very easy to catch since it moves very slowly (Moreno-García, 2003). From the Nigerian Bauchi State, there is an isolated record of hedgehogs being caught and then fattened for consumption (Blench *et al.*, 1992, 309). In Eastern Africa, many superstitions exist about the animals (Kingdon, 1974, 36).

White-toothed shrew (*Crocidura* sp.)

The genus *Crocidura* is extremely rich in species, over one hundred of them can be found on the African continent alone (Kingdon, 1997, 145-146), and identification to species level from bone remains is therefore not possible. Altogether only four remains of white-toothed shrew were identified. Late Stone Age layers at Oursi (BF 94/45), Oursi hu-beero (BF 97/30) and Saouga 95/7 (BF 95/7) all yielded a mandible. A skull fragment was recovered from phase I at Ngala (NA 93/45). Only the mandible from Oursi was well enough preserved to allow measurement of the toothrow length (6.7 mm). Shrews can be found in all vegetation zones, but since they need a continuous supply of insects and small invertebrates for food, they are confined to moist and protected habitats (Happold, 1987, 24). Some shrew species are commensal.

4.6.3. Lagomorphs

True hare (*Lepus capensis/saxatilis*)

Hare remains were mainly found at Iron Age localities in Burkina Faso (Table C.41). The two species that occur in West Africa today, *Lepus capensis* and *L. saxatilis*, can be differentiated from the shape of their upper incisors (Petter, 1959, Fig. 2). Since none of the studied sites has provided such elements, this character could not be used. Measurements on hare bones from the studied sites do not point to large size variation (Table C.42), but the bones are consistently smaller than those from reference skeletons of European *L. capensis* at the RMCA. There were, however, no *L. saxatilis* reference specimens available for comparison. The upper limit of the body and head length of *L. capensis* is said to be about 20 % higher than that of *L. saxatilis* but the size of all true hare species is strongly variable according to local resources and climate (Harrison and Bates, 1991, 214; Kingdon, 1997, 153-154) and (small) size can therefore not be used as an identification criterion. Testifying to the size variation within the species *L. capensis*, is the skeleton of a recent Egyptian individual in the reference collection at the RMCA, which is much smaller than that of the European specimens mentioned above.

Most of the identified hare bones are from adult animals, but a few elements with unfused diaphysis and epiphysis have been found. An unfused distal tibia was recovered from Early Iron Age Oursi (BF 94/45) and Middle Iron Age Oursi village (BF 97/13), besides an unfused third metatarsal from Kissi 22 (BF 96/22). In European hares both elements fuse before the age of five to six months (Habermehl, 1985, 109). The other elements of subadult animals are a distal femur from Kissi 40 (BF 97/31), a proximal humerus and two distal femurs from Saouga 94/120 (BF 94/120); elements which all fuse some time after the age of five to six months in European hares. Some time before that age, at four to five months, young hares have already reached adult weight.

Hares are typical savannah species. *L. capensis* lives in completely open grasslands, steppes and subdeserts while *L. saxatilis* prefers greener grasslands (Kingdon, 1997, 154-155). The latter has also been seen in man-made habitats, such as old farmlands, and it is the only hare species living at the Nigerian side of Lake Chad today (Happold, 1987, 104-106). Hares are sometimes persecuted because they cause damage to young crops, but they are generally hunted for their meat and the excitement of the chase (Kingdon, 1974, 351). Historical sources on medieval West Africa also mention that hares were hunted (Lewicki, 1974, 93). In the late nineteenth century AD, a new lagomorph species, the domestic rabbit (*Oryctolagus cuniculus* f. domestica), was introduced into Africa through Christian missions and the colonial agricultural service (Blench, 2000b), although Dutch seafarers had probably already brought it to South Africa about two centuries earlier (Plug and Badenhorst, 2001, 219). The species is now well established over many parts of Africa. At the beginning of the 1990s, 1.7 million rabbits would have been present in Nigeria (Bourn *et al.*, 1994). Unlike hares, rabbits burrow (Kingdon, 1997, 154-156).

4.6.4. Rodents

4.6.4.1. Small rodents

In the present study, all rodents smaller than giant pouched rat (*Cricetomys gambianus*) have been classified as "small rodent". Well-preserved cranial parts were identified to species level with the help of W. Wendelen, RMCA. Because of the large species variety and because most fragmentary cranial remains and postcranial bones do not have diagnostic features, these were simply listed as "small rodent". There are no indications that other species are present among these bones other than those identified from the well-preserved cranial remains. Judging from the fusion state of the long bones, adult individuals are represented as well as subadult ones. That small rodents were living at the sites could also be deduced indirectly from gnaw-marks found on bones from other taxa. The marks can be distinguished from gnawing of other animal taxa by their shallow, parallel, grooves. Rodents gnaw at dry bone to obtain minerals and to sharpen and shorten their permanently growing incisors. Most small rodent species are commensal.

Striped ground squirrel (*Euxerus erythropus*)

This species was attested at only four of the studied sites (Table C.43). According to Rosevear (1969, 131) it can be found in all vegetation belts from high forest to sub-desert, but it occurs only in cleared parts of forested areas. Ground squirrels are the only squirrels that spend their life on the ground and are unable to climb trees (Happold, 1987, 115). The animals live in rock and tree-root crevices, in termitaria and self-dug burrows (Kingdon, 1997, 161). Kingdon (1974, 441) has observed in Uganda that striped ground squirrel is characteristically found in cultivated land. It appears to be a nuisance to farmers and foresters, but is hunted and eaten in some East-African areas.

Gerbil (*Gerbillus* sp.)

The only evidence for gerbils (Gerbillinae) of the genus *Gerbillus* is a mandible retrieved from Early Iron Age levels at Oursi (BF 94/45). No more precise identification was possible, since the genus counts over 50 African representatives. *Gerbillus* ranges throughout the drier parts of the African continent (Rosevear, 1969, 181).

Tatera gerbil (*Tatera* sp.)

The molar tooth pattern of two mandibles found at Kissi 22 (BF 96/22) corresponded well with the subfamily of the Gerbillinae and its size best fitted individuals of the genus *Tatera*. It is not clear how many species the genus comprises and identification to species level could therefore not be carried out. The geographical distribution of the genera *Gerbillus* and *Tatera* largely coincides, but *Tatera* seems to occupy more humid ecological niches (Rosevear, 1969, 191), although data in Happold (1987, 119) is contradictory.

Dwarf gerbil (*Desmodilliscus braueri*)

A skull and two maxilla fragments from Saouga 94/120 (BF 94/120), found in the same archaeological unit and thus most probably pertaining to one individual, were attributed to dwarf gerbil. The species is not represented in the modern skull collection at the RMCA, but the identification was based on the small size of the maxillas and on their morphology, which was clearly Gerbillinae but did not fit with *Gerbillus*. Dwarf gerbils are tiny, relatively uncommon, gerbils inhabiting the Sahel and Sudan woodlands from Senegal to Sudan (Rosevear, 1969, 208).

Multimammate rat (*Mastomys natalensis*)

Jaws and teeth of the two indigenous murid (Muridae) genera, *Arvicanthis* and *Mastomys*, that could be expected at the studied sites are morphologically close. The modern skulls from the reference collection at the RMCA show that maxillary and mandibular toothrows of more than 6 mm long are usually *Arvicanthis*, while those under 6 mm are usually *Mastomys*. Tooth row measurements on the archaeological specimens are not given here because the state of preservation usually allowed estimates only. Of all studied cranial remains of small rodents, most appeared to be *Mastomys* (Table C.43). The genus comprises only one West African species, *M. natalensis*, which seems to be the most common and most widespread indigenous rodent of Tropical Africa (Rosevear, 1969, 413). Although it can be found at some distance from habitation, multimammate rat is chiefly associated with man, his houses, food stores, farms and other clearings (Rosevear, 1969, 417). When very numerous, it can become a plague, especially in millet (*Pennisetum americanum*) and maize (*Zea mays*) crops (Happold, 1987, 144).

Recently, multimammate rat has faced competition from two newly introduced murid species, black rat (*Rattus rattus*) and brown rat (*R. norvegicus*). Black rat was presumably first brought to West Africa by Portuguese explorers, during the fifteenth century AD or shortly after, while brown rat did not appear before the nineteenth century AD (Rosevear, 1969, 267-284). However, finds from the later first-early second millennium AD in southern Africa, point to a pre-European spread of the black rat, probably through an east-west route from the Arabian Peninsula (Ervynck, 1989, 119). After its introduction to the West African coast by the Portuguese, the black rat has progressively spread inland and is now not uncommon in towns and villages throughout the forest belt (Rosevear, 1969, 267-284). In Nigeria, the animal has also been observed outside forest areas and the northern border of its distribution coincides approximately with 10°N (Happold, 1987, 128). Brown rat, on the other hand, does not appear to have penetrated beyond the ports (Rosevear, 1969, 285). Although Westerners do not consider rats as palatable food, they are frequently eaten in non-Muslim parts of West Africa (Blench, 2000b).

Unstriped grass rat (*Arvicanthis niloticus*)

Arvicanthis is the second most common small rodent that was recorded at the studied sites (Table C.43). It is one of the most widespread and most abundant genera of Afrotropical rats of which *A. niloticus* is the only West African representative (Rosevear, 1969, 285, 295). Unstriped grass rat is known in all vegetation belts from sub-desert to seashore, but it is more common in dry than in moist ecological conditions, and seems to be most abundant in open grassland (Rosevear, 1969, 286). According to Happold (1987, 130), unstriped grass rats are widely spread throughout Nigeria, but are very localised in distribution because of their large water requirements. The species is extremely common near farms and villages and can also be seen inside houses (Rosevear, 1969, 288). From the north of Nigeria there are records of unstriped grass rats becoming true pests when very numerous in crops like sorghum (*Sorghum bicolor*) and millet (Happold, 1987, 130).

4.6.4.2. Large rodents

Savannah cane rat (*Thryonomys swinderianus*)

The only studied context that has yielded bones of this large rodent is Late Stone Age Corcoba (BF 97/5), where a mandible, two loose molars, a talus, a calcaneus and two metapodials have been found. Cane rats live in dense grasses, in seasonally waterlogged valleys throughout the moister parts of Africa, but are absent from present northern Burkina Faso (Barral, 1977, 20-22; Kingdon, 1997, 189). They are hunted and trapped in many parts of the continent as their meat is considered a delicacy. The animals are commonly traded and can be seen in large numbers in bushmeat markets in south-eastern Nigeria (Angelici *et al.*, 1999). Experiments with the breeding of cane rats in captivity have recently been conducted in Nigeria, Ghana and Congo (Tewe *et al.*, 1984; Van De Velde, 1991).

Giant pouched rat (*Cricetomys gambianus*)

Two *Cricetomys* species can be found in western Africa, *C. gambianus* and *C. emini* (Kingdon, 1997, 199-200). Since *C. emini* is mainly confined to rainforests, it has not been taken into account. *C. gambianus* on the other hand, can be found in a variety of habitats, excepting very arid areas. It is one of the most common and frequently seen rodents of Nigeria (Happold, 1987, 125). Very few

	small carnivores	SH (cm)	medium-sized carnivores	SH (cm)	large carnivores	SH (cm)
Canidae			*Vulpes pallida*	+/-25		
			Canis adustus	40-45		
			Canis aureus	45-50		
			Canis lupus f. familiaris	35-50		
Herpestidae/Viverridae	all species	10-25	except *Civettictis civetta*	35-40		
Hyaenidae					*Hyaena hyaena*	65-80
					Crocuta crocuta	70-90
Felidae			*Felis silvestris* (f. catus)	30-36		
			Felis caracal	40-50		
			Felis serval	45-55		

SH (Shoulder Height) from Haltenorth *et al.* (1979) and Epstein (1971, 35-51)

Table 3: Size classes defined for carnivores and taxa identified by class

remains of the species have been found at the studied sites (Table C.44), especially considering that it is a very much appreciated food animal in Africa today (Dorst and Dandelot, 1970, 30; Happold, 1987, 125). The majority of the bones seem to be of adult animals. Only one out of the total number of recorded *Cricetomys* bones, an unfused proximal tibia from Gajiganna A (NA 90/5A), was of a subadult specimen. Measurements that could be taken on the giant pouched rat bones are given in Table C.45.

In some parts of Nigeria, giant pouched rats have been captive bred since the 1960s or perhaps even earlier (Tewe *et al.*, 1984; Blench *et al.*, 1992, 301; Blench, 2000b). Their raising and consumption is mainly associated with ethnic groups from the south of the country, while in the Muslim states of the north there is a religious objection against the rearing of animals that look like house-rats (Blench *et al.*, 1992, 301). In West African towns giant pouched rats are known to invade sewers and in rural areas they sometimes cause damage to domestic crops (Kingdon, 1974, 554).

4.6.5. Carnivores

The carnivores that can occur in the research area have been divided into arbitrary size classes, small, medium-sized and large (Table 3), and are discussed in systematic order by size class. This division was in the first place meant as an aid during the identification of carnivore remains, but it also allowed a more detailed description of badly preserved, or undiagnostic, carnivore bones.

4.6.5.1. Small carnivores

Mongoose (Herpestidae), genet or civet (Viverridae)

No well-preserved cranial or dental remains that allowed a specific or generic identification were available for this group; a precise identification of the postcranial archaeological remains also appeared to be impossible without an in-depth osteomorphological study of recent mongoose, genet and civet skeletons. The species that have to be taken into account for both parts of the research area are given in Table C.46. The presence of African civet (*Civettictis civetta*) could be excluded because it is considerably larger than the other species. In fact it falls more neatly in the size range of what was defined as medium-sized carnivores. Judging from their size variation, several other species must be represented among the recovered mongoose, genet or civet bones, listed in Table C.47. There were no indications of subadult animals, except for an unfused proximal humerus from Oursi hu-beero (BF 97/30). There are no data on the age at which skeletal elements of Herpestidae and Viverridae fuse, but when compared to domestic cats (*Felis silvestris* f. catus) proximal humeri are among the last elements to fuse, around the age of one year (Habermehl, 1975, 177). Seven bones of Herpestidae or Viverridae, all burned, were found in roof material from room 9 at Oursi hu-beero. Because they were found in the same context, it is assumed that these are all from one individual.

Mongooses are the most common African carnivores (Kingdon, 1997, 238). They are ground-dwelling predators of invertebrates and small vertebrates. Most species are water-dependant inhabitants of forests, woodlands, savannahs and marshes. The range of genets and civets seems to cover all major habitats with the exception of true desert (Kingdon, 1997, 266-275). Historically, civets have been used in African commerce, both for their skins and for their glands from which perfumes are made (Haltenorth *et al.*, 1979, 177). At the beginning of the twentieth century, the inhabitants of Borno were reported to keep African civet for perfume production, and also the smaller genet and civet species to exterminate mice and rats (Schultze, 1968, 120).

4.6.5.2. Medium-sized carnivores

Domestic dog (*Canis lupus* f. familiaris) and jackal (*Canis aureus/adustus*)

Size allowed separation of the bones of domestic dogs and jackals, including common jackal (*C. aureus*) and side-striped jackal (*C. adustus*), from those of other species of the canid family (Canidae) that could be expected at the sites, the smaller sandfox (*Vulpes pallida*) and the larger African wild dog (*Lycaon pictus*). An overview of the recovered remains, particularly abundant at Middle and Late Iron Age sites in Burkina Faso, is given in Table C.48. Besides bone remains, some indirect evidence for medium-sized canids has been found at the studied sites as well. A few locations yielded coprolites of the animals and also traces of etching and gnawing, recognisable from the typical pits left by their teeth, could be recorded (Table C.49).

Domestic dogs are not considered to be native to Africa, but were probably introduced from the Middle East in the course of the fifth millennium BC, most likely as shepherd dogs together with small livestock (Gallant, 2002, 51; Gautier, 2002). They were already present in Saharan West Africa by the second millennium BC, as apparent from the find of three buried individuals at Chin Tafidet in Niger, dated to about 2600-1300 BC (Paris 1984, 66, Figs 52 and 53; Paris, 2000). South of the present Sahara, the oldest indications for domestic dogs are from phase I/II levels (250 BC-AD 400) at Jenné-Jeno (MacDonald, 1995). Of the three major domestic dog types of present-day Africa, the pariah, the greyhound and the africanis, only the first occurs in West Africa (Epstein, 1971, 28-82; Gallant, 2002). Several definitions or meanings for the term "pariah" seem to exist. In this study it is used for dogs of a primitive generalised racial type that are not attached to human masters, and that can be considered as descendants of domestic dogs that have turned semi-feral (Epstein, 1971, 28). Both regional and individual variation can be seen in pariah dogs, which may partly be a consequence of differences in ancestral domestic stocks (Epstein, 1971, 28-58).

Distinction between domestic dogs and jackal species is particularly difficult from osteological remains. Moreover, a possible occurrence of hybrids between both taxa has to be considered. Recent examples of interbreeding are known, although this is probably a rare event (Kingdon, 1977, 14). Osborn and Helmy (1980, 365-366) described osteomorphological traits for the separation of cranial bones of domestic dogs and jackals. The cranial remains found at the studied sites are usually not well enough preserved to use the characters, but a mandible from Saouga 95/7 (BF 95/7) clearly corresponds with domestic dog (Fig. 25). The criteria used by other researchers working on medium-sized canid remains from African archaeological sites are mainly osteometrical. MacDonald and MacDonald (2000) have, for example, plotted mandibular molar length and mandibular height of archaeological specimens from the sites Cubabel and Siouré (AD 0-1400) in the Middle Senegal Valley against the same measurements on reference skeletons of (Indian!) pariah dogs and jackals to make the differentiation. When measurements taken on a mandible from Oursi hu-beero (BF 97/30) and one from Saouga 94/120 (BF 94/120) are added on the graph, they seem to fall within the range for pariah dogs, while another mandible from Oursi hu-beero and a mandible from Saouga 95/7, different from the one mentioned above, are in the range of jackals (Fig. 26). However, the mandible from Saouga 95/7 is more curved than any of the jackal mandibles in the reference collection at the RMCA and rather points to domestic dog (Osborn and Helmy, 1980, 365). A second osteometric character that can allow identification is the length of the second lower molar (MacDonald, 1995). The specimens from the studied sites all fall in the range for African pariah dogs, although the larger ones reach the lower size limits for common jackal (Fig. 27). The breadth of the first lower molar is also a distinguishing trait between dogs and jackals (MacDonald and MacDonald, 2000). This measurement could be taken only once, on a tooth from Saouga 95/7 (GB: 7.4 mm), confirming the earlier identification of the mandible it was in as domestic dog. Guérin and Faure (1996) have compared measurements on postcranial medium-sized canid bones from the site of Asa Koma in Djibouti (ca. 2000-1500 BC) against measurements on recent jackal specimens and on pariah dogs from burials at Kerma, Sudan (ca. 2400-1500 BC). They identified the remains as jackal and the canid remains of the studied sites fit well with the size ranges they give for common jackal (Table C.49). Measurements on bones from the analysed assemblages also compare well with those on a recent side-striped jackal skeleton, given in Van Neer (1989a, Table 28), although the metapodials are systematically smaller.

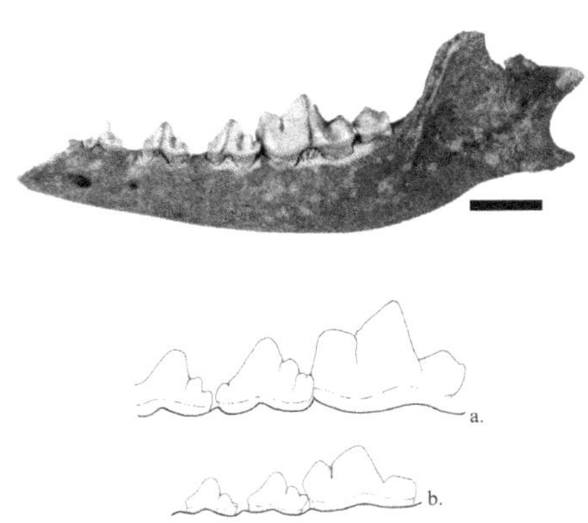

Fig. 25: Domestic dog mandible from Saouga 95/7 (BF 95/7) (Scale = 1 cm) compared against jackal (a) and domestic dog (b) mandibular teeth (not to scale) (Osborn and Helmy, 1980, Fig. 111)

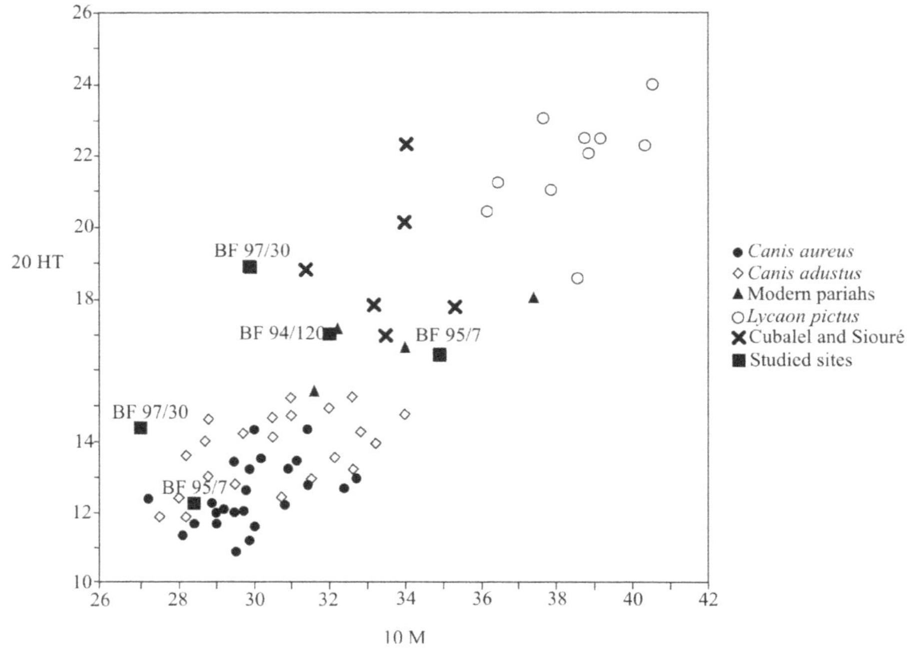

Fig. 26: Mandibular height (20 HT) (mm) and mandibular molar row length (10 M) (mm) of domestic dogs and jackals (adapted from MacDonald and MacDonald, 2000)

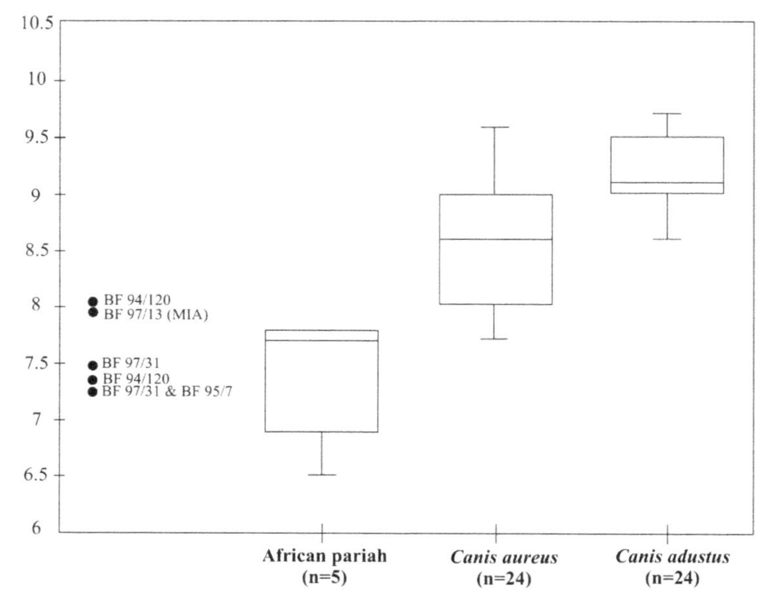

Fig. 27: Length of lower second molar (mm) of domestic dogs and jackals (adapted from MacDonald, 1995)

From the previous paragraph, it is clear that existing studies on jackal and domestic dog skeletons are insufficient to allow for secure species identifications on archaeological remains. Nevertheless, most of the bone remains from this study seem to point to domestic dog. It is possible that the osteometric studies mentioned do not take size variation of pariahs sufficiently into account and this may be the reason why some of the measurements obtained on canid bones through this study fall outside the range they give for domestic dogs. Supporting the identification of domestic dogs at the studied sites is the presence of coprolites, which indicates that the animals may have been living in the settlement. Strangely, however, these are most common in the Nigerian sites where very little bone remains of domestic dogs have been found. Large numbers of hunted jackals do not make sense for the Iron Age sites in Burkina Faso, because among the other mammal remains, domestic species largely predominate and specialised jackal hunting does not seem to be known ethnographically. Nevertheless, the mentioned jackal identifications by Guérin and Faure (1996) were from a context dominated by domestic cattle (*Bos primigenius* f. taurus). To sum up, it is believed that most of the medium-sized canid bones from the studied sites are domestic dog; although it is not entirely excluded that a few jackal remains may be present as well.

With the exception of two unfused bones from Saouga 95/7 (BF 95/7), all recovered medium-sized canid remains from the studied sites in Burkina Faso seem to be of adult animals (Table C.50). In contrast, all bones from Nigerian sites of which the fusion state could be determined, show unfused epiphyses. A few canid bones were found bearing traces of butchery, while others look as if they are worked (Table C.52) (see Linseele, 2003).

Common jackal occurs patchily across the Sahara and the Horn of Africa (Kingdon, 1997, 218). The species can be found in dry, open, country and is especially common around villages and small towns. Side-striped jackal ranges in a broad belt both north and south of the rainforest, but is absent from the driest parts of the African continent (Kingdon, 1997, 219). In East Africa jackals appear to be the subject of much superstition (Kingdon, 1977, 28). The animals are persecuted because they represent a threat to livestock (Moreno-García, 2003). The most common use of domestic dogs in sub-Saharan Africa would be for food, but the animals also have a wide variety of other functions (Boettger, 1958, 13-19). In present north-eastern Nigeria they are kept for hunting and as guards, but they are not eaten (Blench, 1995). In the northernmost part, herders do not use them, although more to the south, some pastoral groups do keep them to guard their herds. In the central zone of the country dogs are widely sold and traded and their meat is considered a delicacy (Blench *et al.*, 1992, 292). The animals are primarily raised for other purposes, however, and then sold off at the end of their useful life, while feral dogs provide additional supplies. Lewicki (1974, 89-90) mentions that dogs were also eaten in medieval West Africa and that the animals were particularly common in agricultural areas. A zoomorphic figurine from Houlouf, from a layer with an estimated date of AD 1275-1400, is thought to represent a canid (Holl, 2002, 163, 165).

Sandfox (*Vulpes pallida*)

This is the only species of the genus *Vulpes* that can be found in the research area. A tarsal from Middle Iron Age levels at Oursi village (BF 97/13), a calcaneus (GB: 7.5 mm) from Saouga 94/120 (BF 94/120) and a first phalanx and a talus from Saouga 95/7 (BF 95/7), all of adult individuals, were attributed to it. Sandfox occurs in the southern Sahara and the Sahel from Mauritania to the Red Sea (Kingdon, 1997, 223-224). It is one of the most common carnivores in the sub-Saharan region and most typically occurs in very dry zones (Rosevear, 1974, 71). According to Barral (1977, 21) it is at present extremely common in northern Burkina Faso. The species digs extensive burrows and can tolerate heat very well, but it cannot survive in completely waterless conditions (Kingdon, 1997, 223-224). Foxes seem to favour the proximity of human habitations and cultivated fields because prey is more readily available there than in natural habitats (Rosevear, 1974, 65).

African wild or domestic cat (*Felis silvestris* or *F. s.* f. catus)

Except perhaps at the oldest sites (see below), *Felis silvestris* remains found in the research area (Table C.53) could pertain to either the wild or the domestic form of this felid (Felidae). Osteological distinction between the two is almost impossible, because domestic cats are usually semi-feral and interbreed with their relatives in the wild (Kingdon, 1977, 313). Even for living animals the distinction between wild and domestic forms is hard to make. Measurements that could be taken on the elements found are given in Table C.54. With the exception of an unfused proximal humeral epiphysis from Early Iron Age Oursi (BF 94/45) and an axis of a juvenile specimen from Galaga (NA 92/2C), all bones seem to be of adult individuals. In domestic cats proximal humeri are among the elements that fuse latest in life, around the age of one year (Habermehl, 1975, 177).

African wild cat has a wide ecological and geographic range; it can be found in woodlands, grasslands and steppes (Kingdon, 1997, 276-277). The species is the wild ancestor of the domestic cat and the domestication centre is traditionally placed in Ancient Egypt (Gautier, 1990, 4-5; Clutton-Brock, 1993). Although much earlier evidence for possible taming is known from predynastic Egypt (ca. 5000-3000 BC) (Van Neer *et al.*, 2004b) and from around 7500 BC in Neolithic Cyprus (Vigne *et al.*, 2004), domestic cats only appear in Egyptian art from the Middle Kingdom (from 1991 BC) onwards (Clutton-Brock, 1993). The domestication process is believed to have been gradual, however, reaching completion by about 1000 BC (Blench, 2000b). The history of the domestic cat on the African continent, outside Egypt, remains largely undocumented because of the difficulties in distinguishing it from its wild predecessor (Clutton-Brock, 1993).

Wild cats are often attracted by the rats and insects that can be found near villages, where they are sometimes killed by dogs or trapped and shot because they steal domestic fowl (Kingdon, 1977, 316.). However, they are usually ignored in countries in eastern parts of the African continent, where they are not killed because of a widespread superstition (Kingdon, 1977, 316). Domestic cats can be found all over present Africa. During a survey in Nigeria in the early 1990s, for example, their numbers were estimated at not less than three million (Bourn *et al.*, 1994). Domestic cats are used to hunt vermin or for medicinal and magical purposes, in which contexts they are sometimes eaten, although the animals are usually considered as forbidden food (Blench, 2000b).

Serval cat or caracal (*Felis serval/caracal*)

These two felids cannot be separated from their postcranial bones. Remains have been found at two Late Iron Age sites in Burkina Faso and at a few sites in the Lake Chad area (Table C.55). Two subadult bones have been recovered: a distal radius from Gajiganna A (NA 90/5A) with unfused epiphyses, and a distally unfused second metatarsal from Gajiganna C (NA 90/5C). Measurements that could be taken on the serval cat or caracal bones are given in Table C.56. All serval cat or caracal remains from Oursi hu-beero (BF 97/30) were found close to each other and are probably from one individual. Three out of the four first phalanges from the site show cut-marks all around their distal end. The distal end of a metapodial found at Gajiganna C (NA 90/5C) and a canine from phase II at Ngala (NA 93/45) are bored through.

Serval cat can be found in almost all West African vegetation zones, but as it needs dense cover and the proximity of water it avoids deserts (Rosevear, 1974, 413). Caracal is distributed over large parts of the African continent, except in dense forests and in the central Sahara, and lives mainly in arid bush country (Rosevear, 1974, 401-411; Estes, 1991, 364). It has a reputation of raiding domestic stock and poultry (Estes, 1991, 364). While serval cat skin is highly valued in East Africa, caracal fur is of little commercial value (Kingdon, 1977, 329, 339). Skins and claws of cats traditionally have a symbolic meaning in Africa (Van Neer, 1989a, 53).

Unidentified medium-sized carnivore

Badly preserved bones of medium-sized carnivores, mainly fragments of teeth and metapodials, could often not be identified more precisely. They are listed in Table C.57, together with the number of specimens identified for the medium-sized carnivore taxa described above. Since domestic dog is the most abundant of these taxa, the majority of the unidentified carnivore bones of the same size class probably also belong to dog. The majority of the unidentified bones are of adult animals, as has been observed for the identified remains (Table C.58). Mustelids (Mustelidae) are the only family of medium-sized carnivores of which no remains at all have been attested at the studied sites. The animals can mainly be found in cooler habitats and are most common in wet upland or in temperate conditions (Kingdon, 1997, 228).

4.6.5.3. Large carnivores

Striped (*Hyaena hyaena*) or spotted hyena (*Crocuta crocuta*)

The only family of large carnivores represented at the studied sites is that of the hyaenids (Hyaenidae), although three large felids, leopard (*Panthera pardus*), lion (*Panthera leo*) and cheetah (*Acinonyx jubatus*), and one large canid, African wild dog, can be found in the area today. At Saouga 94/120 (BF 94/120) four hyaena bones were identified, two first phalanges, one metapodial and a lower premolar, while a canine was retrieved from Gajiganna phase IIc layers at Gilgila (NA 99/65). Osteological distinction between the two species of hyaena that can be found in the research area, striped and spotted hyaena, was not possible. Both species occur in open habitats, although striped hyaena seems to be mostly confined to arid areas (Kingdon, 1997). Striped hyaena is omnivorous and opportunistic but primarily adapted to eating bones and carrion. Spotted hyaena is an opportunistic carnivore and scavenger, which today largely lives on household or town rubbish.

4.6.6. Tubulidentates

Aardvark (*Orycteropus afer*)

This is the only African species within the order of the Tubulidentates. The morphology of its postcranial bones, determined by its peculiar way of life, can hardly be confused with any other taxon. Two metapodials, three first phalanges and one second phalanx were found at Oursi hu-beero (BF 97/30), probably pertaining to one individual. One of the first phalanges has cut-marks. At Kariari C (NA 95/1) a second phalanx of aardvark has been identified. The animal can be found over large parts of sub-Saharan Africa, but is seldom seen (Kingdon, 1997, 294-295). Its occurrence is in the first place related to a year-round abundance of ants, termites and beetle larvae. It avoids very hard soils and regularly flooded areas to dig its deep and complex system of burrows, where it passes most of its time. The species is hunted nowadays mainly because the holes are inconvenient or dangerous for humans and their stock. Its flesh is greatly relished by some East African ethnic groups and parts of the animal are used as charms (Kingdon, 1971, 385). From South African Iron Age sites there are indications that aardvark may have been of some ritual importance (Plug and Badenhorst, 2001, 111).

4.6.7. Proboscids

African elephant (*Loxodonta africana*)

Two subspecies of elephant can be found on the African continent, bush elephant (*Loxodonta africana cyclotis*) and savannah elephant (*L. a. africana*). Taking their present zoogeographic distribution into account, only the presence of the latter seems possible at the studied sites. A burnt elephant caudal vertebra with fused epiphyses was found at Oursi hu-beero (BF 97/30). From Saouga 94/120 (BF 94/120), a badly preserved long bone was retrieved that, judging from its size, could only pertain to either elephant or giraffe (*Giraffa camelopardalis*) (see 4.6.9.). Phase IIIa layers at Ngala (NA 93/45) yielded a fragmented piece of ivory that is either from elephant or hippo. At nearby Daima, elephant bones were reported from Daima I and Daima III layers (Connah, 1981, Table 8.3). Finally at Blé Mound E, a shaft fragment of a poorly preserved elephant tibia was found. Beside bone remains, there is also mention of a clay figurine from one of the studied sites, Kursakata (NA 93/46), that possibly depicts an elephant (Gronenborn, 1996).

Elephants used to be ubiquitous south of the Sahara wherever water and trees occurred (Estes, 1991, 259-260). As their names suggest, bush elephants are typical for rainforests and adjacent parts of the savannah, while savannah elephants are distributed over the remaining parts of the continent. At present, the African savannah elephant has disappeared from much of its original distribution zone, but a relic population has remained in the Malian Gourma and during the rainy season its range extends as far south as Burkina Faso (Roth and Douglas-Hamilton, 1991). Elephants can still be found in the Lake Chad area as well, but as elsewhere, their populations had greatly decreased during the previous century (Happold, 1987, 188-191; Roth and Douglas-Hamilton, 1991). At the end of the nineteenth century AD, Rohlfs (1868, 61) already mentions that their numbers in the region had dropped because of the demand for elephant ivory. Nevertheless, the main reason for the disappearance of the elephant in the Sahel zone seems to be competition for water with men and livestock, while hunting for ivory in early colonial times seems to be the most important cause in the Sudan zone (Roth and Douglas-Hamilton, 1991). Where present, elephants have an enormous impact on the environment (Estes, 1991, 259-260). They can, for example, cause large-scale destruction of trees.

An account by Leo Africanus from the sixteenth century AD reports on elephant hunts using traps, bows and arrows (Lewicki, 1974, 97). Elephant ivory was one of the important commodities in trans-Saharan trade, but medieval Arabic sources contain very little information on it. The situation is furthermore complicated by the fact that no distinction is made in Arabic between elephant or hippo ivory (Insoll, 1995). However, the use of feathers, horns, hoofs or tusks derived from animals that were not ritually slaughtered appears to have been much disapproved of by the Muslim population of medieval West Africa (Levtzion and Hopkins, 1981, 55).

4.6.8. Perissodactyls

Horse (*Equus ferus* f. caballus) and donkey (*Equus africanus* f. asinus)

The Holocene range of the African wild equids, wild donkey (*Equus africanus*) and three species of zebra (*E. quagga, E. zebra* and *E. greyvi*), apparently did not include West Africa (Kingdon, 1997, 309-317). Wild donkeys seem to have been confined to northern and eastern Africa, while zebras were limited to the south and the east of the continent. Therefore, when equid remains were found at the studied sites, they were assumed to be of domestic animals, i.e. horse, donkey or perhaps crosses between the two. Three horse types can at present be found in the study area: two larger ones, Barb and Dongolawi, around 140-150 cm tall, and the smaller pony, with a withers height of 100-120 cm (Epstein, 1971, 436-480).

For donkey there is now genetic evidence that domestication took place solely in Africa (Beja-Pereira *et al.*, 2004; Vilà et al., 2006), while previously the Near East was also named as a possible domestication centre (Gautier, 1990, 4-5; Uerpmann, 1991). The oldest archaeozoological evidence for the animal has been found at the predynastic Egyptian site Maadi (3400-3100 BC) (Boessneck *et al.*, 1989). Until recently, the origins of the domestic horse were placed in the present Ukraine, around 3500 BC (Gautier, 1990, 4-5). However, genetic research has now indicated the existence of several independent domestication centres over Eurasia (Jansen *et al.*, 2002), with the earliest undisputed archaeological evidence not older than 2000 BC. The horse found at the site of Tell Heboua in the Sinai in Egypt, dated to the second intermediate period (1786-1552 BC), is the earliest indication for this species on the African continent (Chaix, 2000). In sub-Saharan West Africa, the earliest donkey identification was on equid remains from contexts at Siouré in the Middle Senegal Valley, dated to AD 0-250 (MacDonald and MacDonald, 2000). The oldest evidence for horse is a buried animal from Aissa Dugjé, in northern Cameroon, dated to around the 8th century AD (MacEachern *et al.*, 2001), although remains from Akumbu in Mali could be as old, or even a little older (MacDonald and Van Neer, 1993). The latter were found in layers dated to AD 600-1000. Holl (2002, 149, 206) has also identified horse at Houlouf, dating between AD 0-500, but because some of his faunal determinations seem questionable (see 1.2.), it may be unwise to take this horse as evidence for the animal's presence in West Africa before the middle of the first millennium AD.

Aissa Dugjé has yielded four more horse skeletons, besides the described animal, and additionally some clay figurines were found, tentatively identified as horse but not associated with the buried animals (MacEachern *et al.*, 2001). Rock art depictions of horses and riders, probably dated to the first millennium AD, have been recorded at Aìré Soroba, in the Inland Niger Delta of Niger (Marchi, 1998), and one of the oldest equestrian figurines of West Africa comes from the site of Tongo Maaré Diabel in Mali dating to the end of the first millennium AD (MacDonald and MacDonald, 2000). For the described archaeological and iconographic evidence, the possible type of horse is not indicated, but there are a few historical references to the occurrence of small animals in medieval West Africa (Levtzion and Hopkins, 1981, 81, 185, 263).

Fig. 28: Horse lower molar from Saouga 94/120 (BF 94/120) (Scale = 1 cm)

The study area in northern Burkina Faso yielded bone remains of domestic equids, be it very scanty. The oldest possible find for the region is a distal scapula fragment (coracoid process) from Middle Iron Age layers at Oursi village (BF 97/13), tentatively attributed to domestic equid. From Late Iron Age layers at the same site an equid proximal metatarsal and an equid tooth were unearthed that could not be identified beyond family level. An equid femur shaft and tooth from Oursi hu-beero (BF 97/30) were also impossible to identify more precisely. An equid coprolite from the same site was tentatively identified as donkey because it seemed to be too small for horse (Linseele, in press). A lower molar from Saouga 94/120 (BF 94/120) (Fig. 28) could be attributed to horse using the morphological traits described in Armitage and Chapman (1979). An equid upper molar was also attributed to domestic horse because it came from the same archaeological unit. Finally, an equid tooth and a sesamoid bone from a small horse or a donkey were found at Saouga 95/7 (BF 95/7). From about the same period onwards as horses appear to have been introduced in Burkina Faso, Kahlheber (2003, 226) found higher amounts of leguminosa in archaeobotanical samples. Leguminosa are very rich in proteins and minerals and she therefore interpreted them as fodder plants brought to the site for the animals. No remains of equids have been identified at all from the studied sites in the southern Lake Chad area, but there is some indirect evidence for the animals' presence. Phase III at Mege (NA 94/7) has yielded a clay figurine that was interpreted as a horse (Gronenborn, 1996). Another example from phase II at Ngala (NA 93/45) was thought to represent a zebra (Gronenborn, 2000, 243). However, as explained above, this is improbable on zoogeographical grounds and it may thus also be a domestic equid. There are other sites in the Chad Basin, which have also yielded clay figurines of horses, e.g., Houlouf (Holl, 2002, 163, 165) and Mdaga (Lebeuf *et al.*, 1980, 67, 80, 86, 89), apparently all dating to the second millennium AD.

Donkeys are generally the most widely used animals in the study area today for riding and load-carrying, except in the northernmost areas where they give way to one-humped camels (*Camelus thomasi* f. dromedarius) (Krings, 1980, 46; Connah, 1981, 43). Less common is their use in traction, for example for pulling ploughs or carts, which are both post-European introductions (Starkey, 2000). Donkeys can stay healthy on very varied and often poor-quality diets and need no substantial management input (Blench, 2000c). They are very hardy but are susceptible to parasites and trypanosomiasis. Reproduction is therefore problematic in more humid areas and donkeys are mainly bred in Sahelian countries from where they are exported. Donkey milk is rarely consumed in Africa and is of negligible economic importance, but it is used in magical remedies in parts of West Africa (*Ibid.*). There is an Islamic prohibition against donkey meat but, in non-Muslim areas, donkeys are widely eaten when too old or too sick for other uses (*Ibid.*). People from the present village of Oursi deny eating donkeys themselves, but say they know the habit from the area near Ouagadougou. The eating of donkey meat already seems to have been in use in medieval West Africa (Lewicki, 1974, 89).

Horses are not commonly eaten in present West Africa (Law, 1980, 169-171, 1995). They are mainly kept by wealthier people for personal transport and as a sign of prestige, but local conditions pose major logistic problems to their maintenance (Connah, 1981, 43; Geis-Tronich, 1991, 438-452). Moreover, in seasonally inundated areas, like the Nigerian *firgi*, horses are useless during high water and suffer greatly from insect bites (Connah, 1981, 227). Horses are very susceptible to humidity related diseases like trypanosomiasis, and within western Africa their state of health progressively diminishes the further one moves to the south (Law, 1980, 76-82). The animals do not tolerate heat very well either and can therefore only be used during the cooler parts of the day (Geis-Tronich, 1991, 438-452).

However, the pony breeds are better adapted to local West African circumstances and they are trypano-tolerant (Blench, 1993). In Islamic societies in pre-colonial West Africa there was a close association between the possession of horses, elite social status and political power (Law, 1980, 176-196, 1995). The symbolic value attached to horses grew historically because of their association with warfare and raids. Whoever possessed them had the significant strategic advantage of fast movement and high manoeuvrability. Since horses could not be bred in the southern parts because of the diseases inherent to those areas, they were imported from the north, thus creating one of the major sectors of long distance trade in pre-colonial West Africa (Law, 1980, 54-58, 1995).

Mules, produced from a male donkey with a female horse, are presently used throughout North Africa and are very much associated with Arab culture (Blench, 2000c). They have also been recorded in parts of West Africa, where they were apparently imports from the north (Doutresolle, 1947, 264). In other areas, however, like Niger and Nigeria, mules are not bred because the intentional breeding of an unfertile animal is considered unnatural (Blench, 2000c). Al-Umari, writing in the fourteenth century AD, mentions mules among the domestic livestock species known at that time in western Africa (Levtzion and Hopkins, 1981, 267). Hinnies, crossbreds between a male horse and a female donkey, are biologically difficult to produce and are as good as unknown from the African continent (Blench, 2000c).

4.6.9. Artiodactyls, except bovids

Hippopotamus (*Hippopotamus amphibius*)

From Early Iron Age levels at Kursakata (NA 93/46) and from phase IIIa at Ngala (NA 93/45) some jugal tooth fragments of this animal were recovered. As mentioned earlier, the latter also yielded a piece of ivory that may be of hippo. Hippos are enormous semi-aquatic mammals that need permanent water and open grazing (Kingdon, 1997, 325-326). They still occur in the Lake Chad area today but are in general not very common (Happold, 1987, 205). At present, they are mainly hunted for their meat, which is said to be very palatable, and because of the damage they cause to crops (Kingdon, 1997, 325-326). There is a record on hippo hunting in the Senegal Valley during the eleventh century AD by the hand of Al-Bakri, who mentions that the animals' meat is eaten and that whips are made from their skin (Levtzion and Hopkins, 1981, 78, 456). In the fourteenth century AD, Ibn Battuta writes that, along the Niger hippos are hunted for their meat with harpoons, but according to other Arabic sources, the animal is not hunted near Lake Chad (Levtzion and Hopkins, 1981, 188, 297). At least two millennia ago, the animal may already have been hunted in sub-Saharan West Africa, as can be deduced from finds at Kobadi, dated to the second-first millennium BC (Raimbault *et al.*, 1987; Jousse and Chenal-Vélardé, 2001-2002). Hippopotami are predominantly depicted among the zoomorphic clay figurines at Houlouf in northern Cameroon, occupied between about the beginning of our era and AD 1400 (Holl, 2002, 139-188). Some dangers and difficulties seem to be attached to the hunt of hippos (Plug and Badenhorst, 2001, 120). Nevertheless, Insoll (1995) has argued that it involves fewer risks than elephant hunting and that this may be an important reason for the preference of hippo ivory to elephant ivory as a trans-Saharan trade commodity. At the site of Gao, a terminus for trans-Saharan trade, a cache of hippo ivory was found, in a layer dating to the tenth-eleventh century AD (*Ibid.*).

Common warthog (*Phacochoerus africanus*)

Not all warthog remains found could be distinguished from bush pig (*Potamochoerus porcus*), using the osteomorphological traits described in Van Neer (1989a, 56-59). However, the presence of bush pig at the studied sites can probably be excluded on zoogeographical grounds since the animal rarely occurs outside rainforests (Kingdon, 1997, 331). The presence of a third species, domestic pig (*Sus scrofa* f. domestica), also has to be considered, at least for the younger sites (see below). The low number of suid (Suidae) remains seems to be in contradiction with a possible rearing of this livestock animal, however (Table C.59). Part of the bones retrieved from the studied sites points to subadult animals. An unfused distal metapodial was found in Early Iron Age levels at Oursi (BF 94/45). In domestic pigs metacarpals fuse at the age of about two years, while metatarsals fuse around three months later in life (Silver, 1963). In Gajiganna phase I contexts at Bukarkurari (NA 97/33), as well as at Galaga (NA 92/2C), an unfused distal humerus has been recorded, which points to the presence of individuals of under one year old. Only three pig skeletal elements could be measured (Table C.60).

Although its wild ancestor, the wild boar (*Sus scrofa*), can be found in northern Africa, nothing points to a local African domestication of the domestic pig. The domestication centre of the animal is placed in the Near East, where the oldest evidence dates to the seventh millennium BC (Peters *et al.*, 1999), and it is widespread in Egypt by the predynastic period (e.g., von den Driesch and Boessneck, 1985, 23-29). The date of introduction into West Africa is highly controversial, but there is no archaeozoological evidence for an arrival of the animal in the area before the coming of the Europeans. Pigs are perhaps the most classic of the taboo animals in Islam. They are at present prohibited in the Muslim northern and central part of Borno (Blench, 1995), but more than three million domestic pigs have been counted in the non-Muslim parts of Nigeria (Bourn *et al.*, 1994). They are kept all over the forest-savannah belt of West Africa, usually by settled farmers (Blench, 2000d).

Warthogs can survive in arid and open areas, which is unusual for suids, but are most common on alluvial soils in lightly wooded country (Kingdon, 1997, 335). They normally stay within walking distance of water but can subsist on succulents and other water-conserving plants for a while. The Islamic prohibition on pig meat also concerns the wild forms and warthogs are therefore not hunted in parts of northern Nigeria and consequently they are relatively numerous in the area (Happold, 1987, 202). Intensive stock rearing or agriculture can also cause a decline in warthog populations (Kingdon, 1974, 249).

There does not seem to be any reference in Arabic historical sources to the keeping or hunting of pigs in medieval West Africa. Some sites in the southern Lake Chad area have yielded possible iconographic evidence for the animals. Holl (2002, 163, 165) mentions a wild boar clay figurine from layers at Houlouf dating to AD 1275-1400 and a clay figurine from Daima III levels at Daima has a pig-like appearance but bears spiky protuberances (Connah, 1981, 184-185).

Giraffe (*Giraffa camelopardalis*)

The only possible indication for this species is a fragment of a badly preserved long bone from Saouga 94/120 (BF 94/120), which could equally be from elephant, however (see 4.6.7.). Giraffes were formerly widespread throughout the drier savannahs of Africa (Kingdon, 1997, 343). They have now been eliminated from large parts of their former West African range (Estes, 1991, 202), but small numbers of giraffes can still be found near the Nigerian shores of Lake Chad (Happold, 1987, 208). Giraffes live in savannahs, open woodlands and seasonal floodplains, and are especially common in areas with a scattered low and medium-high woody growth (Kingdon, 1997, 343). During the wet season, they are dispersed because vegetation is abundant, while in the dry season they concentrate in areas where evergreens survive. Although giraffes drink irregularly, their distribution is determined by the availability of water (Happold, 1987, 208). Giraffe is one of the most frequently depicted animals in African rock art and was probably one of the favourite prey animals of Saharan hunters (Kingdon, 1979, 337). Nachtigal wrote that meat of young giraffes was very much sought after by gourmets in Borno and Barth considered the meat of giraffe to be the best of all foods (Lewicki, 1974, 93-94). Both Europeans travelled to West Africa during the nineteenth century AD.

4.6.10. Bovids

Bovids generally predominate among the mammalian bone remains from the studied sites. The large numbers of species of the family in Africa hampers identification, but a few osteomorphological and osteometric studies are available for the distinction between the possible taxa (see species descriptions below). There is moreover a good reference collection for the group at the RMCA. When pieces were too fragmented or not diagnostic enough to allow more precise identification, they were subdivided into size classes. Brain (1974), working in southern Africa, was the first to propose such a system and later authors (e.g., Peters, 1986c; Van Neer, 1989a; MacDonald, 1995) have adopted it, but usually in a slightly modified form to fit better with the faunal assemblages studied. Table 4 contains the size classes that have been defined here, together with the species that are certainly present for each of them. The term "antelope" has been used to designate all wild bovids (cf. Peters, 1986c, 67).

A few of the studied localities, like Ngala (NA 93/45), and other West African archaeological sites such as Daima (Connah, 1981, 184-185), have yielded clay figurines depicting antelopes, but attribution to a more precise taxon was not possible. Blench *et al.* (1992, 309) mention that throughout Nigeria various types of antelopes, especially duikers, are kept as pets and that there have also been attempts, albeit unsuccessful, to raise antelopes for their meat.

4.6.10.1. Small bovids

Forest duiker (*Cephalophus* cf. *rufilatus*)

Three localities in the southern Lake Chad area have yielded remains of a very small antelope of the genus *Cephalophus*. Two loose upper molars and one loose lower molar have been found at Gajiganna BI (NA 90/5BI). There was also a first phalanx of the genus from Gajiganna phase IIb at Gilgila (NA 99/65) and a first, second and third phalanx of presumably one animal at Elkido North (NA 99/75). In the southern Lake Chad area no forest duikers can be found at present, although Schultze (1968, 123) mentions the animals among the fauna of early twentieth century AD Borno. From the present distribution of *Cephalophus* species, red-flanked duiker (*C. rufilatus*) seems most likely (Happold, 1987, 217-225; Kingdon, 1997, 369-382). All elements are fused, except the proximal epiphysis of the second phalanx from Elkido North, which is still fusing. In domestic sheep (*Ovis ammon* f. aries) fusion of this element happens rather early in life, shortly after it reaches an age of one year (Silver, 1963). Measurements on the *Cephalophus* elements from Elkido North (Table C.61) are greater than the ranges given by

	wild	SH (cm)	domestic	SH (cm)
small bovids	**very small antelopes**			
	Red-flanked duiker (*Cephalophus rufilatus*)	30-38		
	small antelopes			
	Bush duiker (*Sylvicapra grimmia*)	45-65		
	Oribi (*Ourebia ourebi*)	50-67		
	medium-sized antelopes		Sheep (*Ovis ammon* f. aries)	45-100
	Bushbuck (*Tragelaphus scriptus*)	65-100	Goat (*Capra aegagrus* f. hircus)	
	Bohor reedbuck (*Redunca redunca*)	65-90		
	Kob (*Kobus kob*)	70-105		
	Red-fronted gazelle (*Gazella rufifrons*)	65-82		
large bovids	**large antelopes**			
	Roan antelope (*Hippotragus equinus*)	140-160		
	Tiang (*Damaliscus korrigum*)	100-130		
	Hartebeest (*Alcelaphus buselaphus*)	120-145	Cattle (*Bos primigenius* f. taurus)	90-150
	very large antelopes			
	Savannah buffalo (*Syncerus caffer*)	150-165		

SH (Shoulder Height) from Haltenorth *et al.* (1979, 32-96) and Epstein (1971)

Table 4: Size classes defined for bovids and taxa identified by class

Van Neer (1989a, Table 43) for *C. rufilatus*. Robustness of the second phalanx and the morphology of the third phalanx excluded identification as bush duiker (*Sylvicapra grimmia*) or oribi (*Ourebia ourebi*) (Van Neer, 1989a, 65-67), however, with which *Cephalophus* may sometimes be confused. The identification as *Cephalophus* is therefore retained. As their name suggests, forest duikers are typically confined to forests and red-flanked duiker needs at least relic forest close to water (Happold, 1987, 222). The species used to be fairly common but its numbers have seriously declined in recent years due to overhunting and habitat changes.

Bush duiker (*Sylvicapra grimmia*) and oribi (*Ourebia ourebi*)

Bone remains of these two small antelope species resemble each other closely but can be separated using the criteria described in Van Neer (1989a, 63-67). Both have been recognised in the studied assemblages, although not in large numbers (Tables C.62 and C.63). For about twenty bones, distinction between bush duiker and oribi was not possible. The remaining small antelope bones could not be identified to a lower taxonomic level on a purely osteological basis. It is assumed that they all belong to either bush duiker or oribi as well, since no other small antelope species have been positively identified. The remains have therefore been lumped with the bush duiker/oribi bones (Table C.64). Most bones are of adult animals; the others are listed in Table C.65 together with their fusing ages in domestic sheep (Silver, 1963). Some appear to be of quite young animals.

Bush duikers do not need water and are distributed over the savannahs and woodlands of sub-Saharan Africa, but are absent from extreme habitats, such as forests, open plains, deserts and subdeserts (Kingdon, 1997, 368). These animals can survive in areas with a fairly dense human population and thanks to their ability to breed twice a year, can also withstand heavy exploitation (Spinage, 1986, 178-179). They are hunted for their meat but also because they cause damage to vegetable gardens (Kingdon, 1982, 326). Among the Kotoko, meat of bush duikers is forbidden food (Lebeuf, 1976, 16), while in Eastern Africa horns of the species are frequently used as a pendant or as charms against evil spirits (Kingdon, 1982, 326). Oribis can predominantly be found on open grassland maintained by fires or heavy grazing (Kingdon, 1997, 389) and do not seem to be water dependent (Estes, 1991, 58). The animals are absent from present northern Burkina Faso (Barral, 1977, 20-22). They occasionally cause damage to field crops (Kingdon, 1982, 222).

Bushbuck (*Tragelaphus scriptus*)

A first phalanx from Gajiganna BII (NA 90/5BII) was recognised as bushbuck using the osteometric and osteomorphological characters described in Peters (1986b). Bushbucks need thick vegetation cover and are confined to habitats close to water (Happold, 1987, 214-216; Kingdon, 1997, 352). In some regions they congregate during the dry season in places where green grasses can still be found, for example in dry river beds (Happold, 1987, 214-216). Bushbucks are nocturnal and solitary, which allows them to maintain population numbers despite heavy hunting pressure.

Bohor reedbuck (*Redunca redunca*)

This species has a typical Reduncini tooth pattern (Gentry, 1978). Its smaller size allows distinction from kob (*Kobus kob*), the other representative of the tribe that can be expected in the research area. Identification criteria for postcranial skeletal elements are described in Peters (1986b) and in Van Neer (1989a, 68-72). The authors also provide measurements on recent skeletons to which measurements obtained from the studied assemblages were compared (Table C.68). Bohor reedbuck has been attested at several sites and the animal appeared to be especially common at Gajiganna A (NA 90/5A) (Table C.67). At localities where bohor reedbuck was positively identified, medium-sized antelope bones in the same size range were attributed to the species as well. The bohor reedbuck bones are generally of adult animals, the subadult ones are listed in Table C.69, together with the fusing ages of the same bones in domestic sheep (Silver, 1963). Apart from a juvenile calcaneus from Blé Mound B, nothing points to the presence of very young animals. A second phalanx from Early Iron Age layers at Oursi (BF 94/45) bears cut-marks and is also bored through. A distal metacarpal from Blé Mound B was noticed to have cut and chop-marks and from the same site there was also a radius diaphysis with cut-marks.

Bohor reedbucks typically live in highly unstable, large grasslands with extensive annual flooding, droughts and fires (Kingdon, 1997, 401). During dry-season droughts, they can sometimes be found in concentrations near remaining waterholes and in unburned grassland (Estes, 1991, 95). Kingdon (1982, 358) writes that in present Uganda, bohor reedbucks can be caught in large numbers during the dry season using hunting dogs and nets.

Kob (*Kobus kob*)

This is the most frequently encountered antelope species at the *firgi* sites. It was especially abundant in phase I layers at Mege (NA 94/7). Identification was done with the aid of the osteomorphological criteria and size ranges given in Van Neer (1989a, 68-72, Table 50). At localities where kob was positively identified, remains of medium-sized antelopes in the same size range were also attributed to it. Associated with the eighth century BC human burial at Ngala (NA 93/45), three pendants made of bovid second phalanges were found. They have not been sent to the RMCA for analysis, but judging from the (unscaled) drawings in Gronenborn (2000, Fig. 2.35), they may be of kob. Most of the kob bones are from adult animals. Elements of subadult individuals are indicated in Table C.71 and compared against the fusing ages of the same elements in domestic sheep (Silver, 1963). At Blé Mound B a few elements of juvenile specimens were found: a second phalanx, a metacarpal, a tibia and a metatarsal. Blé Mound E yielded a juvenile mandible. Table C.72 summarises the measurements that could be taken on kob bones when preservation allowed. A few kob remains with cut-marks were recorded: a second phalanx from Blé Mound B, besides the juvenile mandible, a distal humerus and a proximal tibia from Blé Mound E. A pathological bone was recovered in phase I layers at Mege, where a proximal fragment of a first phalanx with exostosis was found.

Kob is the most common antelope of western Africa (Spinage, 1986, 182), but seems to be absent from present northern Burkina Faso (Barral, 1977, 20-21). It can mainly be found in areas without seasonal extremes, on plains close to permanent water (Kingdon, 1997, 403). The species is totally dependent on regular drinking and is therefore tied to habitats within a short distance of water. It is very gregarious, sometimes occurring in dense concentrations numbering several thousands (Spinage, 1986, 182). Kobs are highly territorial animals that keep on living in an area even after human settlements have been founded there (Kingdon, 1982, 381). They can very easily be caught using hunting dogs, nets, bows and arrows and young can be speared without many problems. Because of the ease with which they can be killed, kob populations rapidly decline near villages.

Red-fronted gazelle (*Gazella rufifrons*)

Gazelles have an Antilopini tooth pattern (Gentry, 1978) that may sometimes be confused with that of domestic ovicaprines, especially when teeth are fragmentary or deciduous. Three species of gazelle can be found in the research area, dorcas gazelle (*Gazella dorcas*), red-fronted gazelle and dama gazelle (*G. dama*), which can easily be separated from their size. Postcranial remains were identified with the aid of the criteria and size ranges given in Peters (1986b). All gazelle remains found have been attributed to red-fronted gazelle, which seems to be the most common antelope species at the studied sites in Burkina Faso (Table C.73). The red-fronted gazelle bones mostly belong to adult specimens. Only a first phalanx from Early Iron Age layers at Oursi (BF 94/45), a calcaneus from Kissi 22 (BF 96/22) and an ulna from Kissi 40 (BF 97/31) are of subadult individuals. In domestic sheep, first phalanges fuse early in life (13-16 months), while both calcanei and ulnae fuse considerably later (2.5-3 years) (Silver, 1963). A mandible from Galaga (NA 92/2C) had heavily worn teeth and thus probably belongs to a very old individual. Measurements taken on red-fronted gazelle remains when the state of preservation allowed it are given in Table C.74. A few gazelle phalanges with cut-marks were noticed: two second phalanges from Early Iron Age Oursi and a first phalanx from Galaga.

Red-fronted gazelles thrive in dry grasslands and shrubby steppes of the Sahel (Kingdon, 1997, 412). They prefer heavily grazed, trampled or burnt grasslands or naturally open steppe. They are known to stay on pastures long deserted by larger herbivores as long as some sparse growth remains. However, red-fronted gazelles are less adapted to dry conditions than their relatives, dorcas gazelles and dama gazelles, and are found in more southerly regions where the dry season is shorter (Happold, 1987, 242).

Medium-sized antelope

Out of about 450 medium-sized antelope bones, around 200 could not be identified more precisely, mostly small fragments or undiagnostic elements. Presumably, the species described above are represented among them in about the same proportions as among the identified medium-sized antelope bones. A few of the retrieved elements point to the presence of subadult animals. They are listed in Table C.76 and compared against the fusing data of the same elements for domestic sheep (Silver, 1963). Medium-sized antelope bones with traces of butchery have only been recovered from the Blé sites: a proximal femur and a second phalanx with cut-marks were found at Blé Mound B, and three ribs with cut or chop-marks were recorded at Blé Mound E. A second phalanx from phase II at Ngala (NA 93/45) showed proximal exostosis.

Sheep (*Ovis ammon* f. aries) and goat (*Capra aegagrus* f. hircus)

Looking at their shape and general sturdiness, well-preserved bone remains of domestic ovicaprines could readily be differentiated from bones of small and medium-sized antelopes. Separation was not attempted on ribs and vertebrae however, except on the axis, atlas and sacrum, because these bones are of little diagnostic value. Osteological remains of sheep and goat resemble each other very closely and several publications exist to distinguish between them, including Boessneck *et al.* (1964), Schramm (1967), Payne (1969, 1985), Prummel and Frisch (1986), Helmer (2000), Halstead *et al.* (2002) and Balasse and Ambrose (2005). Boessneck *et al.* (1964) was principally used here, although only some of the distinguishing traits were applied. The traits that were selected all appear to work well according to the experience of Dr. B. De Cupere, Royal Belgian Institute for Natural Sciences, Brussels, who has studied large samples of ovicaprines from Anatolian sites. Although never systematically investigated, the validity for African types of at least part of the characters described in Boessneck *et al.* (1964), and other studies, has been questioned (Welbourne, 1975; Marshall, 1990; MacDonald, 1995; Badenhorst and Plug, 2003). The identifications from this study should perhaps be considered with some reservation, therefore. Altogether 2705 ovicaprine bones were identified, of which 229 were attributed to sheep and 423 to goat, for all the others distinction could not be made (Tables C.77-C.79). Finds of ovicaprine faeces in archaeological layers at Saouga 94/120 (BF 94/120) and a concentration of them in one room at Oursi hu-beero (BF 97/30), also point indirectly to the presence of the animals.

A local African domestication of sheep and goat is excluded a priori, because their wild ancestors never seem to have inhabited the continent. They were introduced from the Near East, where available evidence indicates that they were domesticated by the eighth millennium BC (Peters *et al.*, 1999). Introduction probably happened through one or more routes, by way of the Sinai (Smith, 1984b; Close, 2002) or, as Gifford-Gonzalez (2005) has suggested, perhaps by boat from the Arabian Peninsula, based on early dates (around 6000 BC) for ovicaprines, including certain goat, from Sodmein in the Egyptian Eastern Desert (Vermeersch *et al.*, 1994, 1996). In West Africa south of the Sahara, the oldest identifications for domestic ovicaprines are from Windé Koroji Ouest in Mali, not so far north of the research area in Burkina Faso, and dating to around 2200-950 BC (MacDonald, 1996; MacDonald and MacDonald, 2000). An older date for ovicaprines was obtained from Blabli, in Cameroon, where radiocarbon dates give a minimal date of 2400 BC (David and Sterner, 1989). However, as mentioned in 2.2.3.1., the site suffered from dating problems (David and Sterner, 1987, 1989). Multiple introductions of ovicaprines into western Africa must have taken place in the course of time, as suggested by the various types of both sheep and goats that can be found in the area today (see below). Little is known on the early history of the types, however (Clutton-Brock, 1993). A basic problem is that often no distinction between sheep and goat is made in archaeozoological reports.

For African sheep in general, Epstein (1971, 21-191) uses a subdivision into four major groups: hairy thin-tailed sheep, woolly thin-tailed sheep, fat-tailed sheep and fat-rumped sheep. The first type, which is well adapted to arid circumstances, is the most common and is assumed to have been introduced from Asia through the Horn of Africa (Blench, 1995). The type was recognised at the Sudanian site Kerma (ca. 2400-1500 BC), where some complete and well-preserved individuals were found (Chaix and Grant, 1987). Hairy thin-tailed sheep is the only group that can be found in present Northeast Nigeria, where three important breeds occur: the Balami (syn. Balani, Bornu, Fellata, White Bororo) and Uda (syn. Bororo, Bali-Bali), sometimes collectively called Fulani sheep and dominant among mobile pastoral groups, besides the Yankasa, usually associated with agriculturalists (Blench, 1995). All three have a tall stature, exceeding heights at the withers of

65 cm, and Uda sheep can even grow to 70-100 cm tall (Epstein, 1971, 21, 41)! Because of their exceptionally long legs, they have been named long-legged sheep. The only other type of sheep that seems to be present in arid parts of West Africa is the Macina sheep, a woolly thin-tailed animal that occurs in the central Niger Delta and is assumed to be related to North African sheep populations (Epstein, 1971, 89). It is probably a marginal and relatively late introduction (Blench, submitted). Additionally, in forested zones, a dwarf form of hairy thin-tailed sheep, i.e. with withers heights under 50 cm, can be found (Epstein, 1971, 53). The animal is kept in small numbers by sedentary people. It is extremely hardy and has a high resistance to trypanosomiasis. Hairy sheep are also mentioned in Arabic historical sources on medieval West Africa (Levtzion and Hopkins, 1981, 76).

For African goats, Epstein (1971, 210-309) again makes a subdivision into four broad groups: dwarf goats, savannah goats, Nubian goats and Maltese goats. Only the first two have to be considered when working in western Africa. Savannah goats are long-legged and, as their name suggests, they occur along a broad belt in the savannah zones south of the Sahara (Epstein, 1971, 240). Like long-legged sheep, they are tolerant of arid conditions but are susceptible to trypanosomiasis (Blench, 1993). Two breeds can be found in North Nigeria, first of all the Sahel goats (syn. Balami), restricted to a strip along the frontier with Niger and kept in mobile pastoral systems, and secondly the Sokoto Red (syn. Maradi), raised in villages throughout the area (Bourn *et al.*, 1994). The latter stand approximately 65 cm at the shoulder and are famous for their skin. In more tropical zones, approximately to the south of 15° N latitude, but also in northern Nigeria, dwarf types, with shoulder heights between, on average, 40 and 50 cm, can be found as well (Epstein, 1971, 210-211, 240; Blench, 1995). Dwarf goats are very well adapted to local circumstances and are associated with sedentary groups.

Except for dwarf goats (Chang and Landauer, 1950), no studies are available on the osteology of the described recent West African ovicaprine types and it is therefore difficult to determine to which types the ovicaprines from the studied sites may correspond. However, sizes could be useful, since both sheep and goat have larger West African forms, which can be found in the arid zones, and dwarf forms, which occur in the more humid areas (see above). Measurements taken on sheep and goat bones (Tables C.83 and C.84) have therefore been analysed and results compared to other West African localities. Only measurements on postcranial skeletal elements have been used since the connection between skull dimensions and body size is not sufficiently known. Results are discussed in 6.7.3.1, together with their possible economic and ecological implications.

Fusion dates of the ovicaprine long bones are summarised in Tables C.80-C.82. Animals of all age classes seem to be present, including foetuses. In Central and Western Europe ovicaprines attain their maximal meat weight at the age of about 1-2 years (Uerpmann, 1973). According to Dahl and Hjort (1976, 94) African sheep can reach ages of about 5 years, while at 5.5-6 years, the life expectancy of goats is slightly higher. At the studied sites, clusters of foetal remains were often found in the same, or in adjacent, archaeological units. Some of the foetal bones were well enough preserved to allow a more precise age estimate based on their length. A radius from Gajiganna A (NA 90/5A) has a shaft length of about 52 mm, a metatarsal from Gajiganna BII (NA 90/5BII) of 41 mm and a radius of Labe Kanuri (NA 97/26) of about 38 mm, while a metatarsal from phase IV at Ngala (NA 93/45) measures 48 mm. All these data point to animals that died about three or four months after conception (Habermehl, 1975, Tables 11 and 12). This was also estimated for several foetal bones of one individual at Oursi hu-beero (BF 97/30), on which no measurements could be taken. Habermehl (1975, 110) indicates that sheep carry their young for about 5 months before giving birth. Sex determination of ovicaprine remains was only possible in two exceptional cases. Part of a goat skull with large horns from Early Iron Age Oursi (BF 94/45) was attributed to a male, while a sheep pelvic bone from Gajiganna IIc layers at Gilgila (NA 99/65) probably stems from a female.

Heights at the withers of sheep and goat could be estimated from the lengths of completely preserved long bones, using the factors summarised in von den Driesch and Boessneck (1974). Proportions within the skeleton can be different according to the breed, however. Chaix and Grant (1987) have, for example, remarked that the tall stature of long-legged sheep is mainly due to the particular development of the radius and metapodials. Nevertheless, they used the factors of Teichert (von den Driesch and Boessneck, 1974), obtained from pre- and early historical sheep, to calculate withers heights. These were also used in this study for withers height calculations of sheep, besides those of Schramm for goat, mentioned in von den Driesch and Boessneck, because there does not seem to be a suitable alternative. A sheep metatarsal from Galaga (NA 92/2C) yielded an estimated withers height of 71 cm, a sheep radius from Blé Mound B yielded one of 54 cm. A goat metacarpal from phase II at Ngala (NA 93/45) belonged to an animal that was approximately 53 cm tall. The length could be measured of a radius from Saouga 95/7 (BF 95/7), but it was not identified to species level. Withers height estimation was therefore done with the factors for both sheep and for goat, giving a result of 58-59 cm.

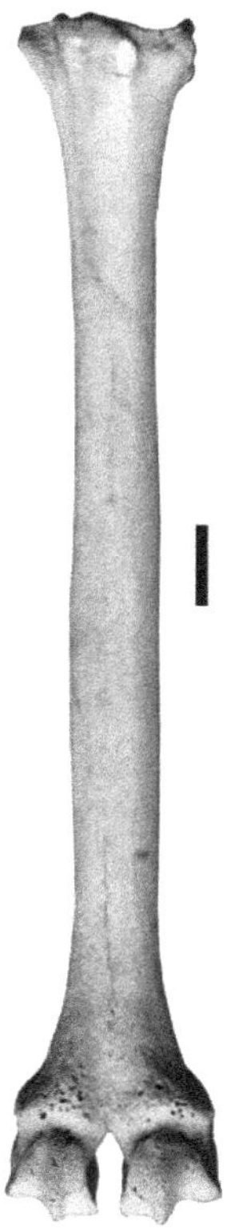

Fig. 29: Sheep metatarsal from Galaga (NA 92/2C) (Scale = 1 cm)

Of all recovered ovicaprine bones, 65 were found bearing traces of butchery (Table C.86). There were also a few worked remains: a distal tibia from Early Iron Age Oursi hu-beero (BF 97/30), a distal metacarpal and a distal phalanx from Kissi 22 (BF 96/22), and finally a first phalanx from phase II at Ngala (NA 93/45). Two first phalanges from the latter site, one from phase II and one from phase IIIa, were bored through. A handful of ovicaprine bones showed pathological features. An ovicaprine distal tibia and malleolar bone from Early Iron Age Oursi (BF 94/45) were grown together, a sheep first phalanx from Oursi hu-beero had distal exostosis and an ovicaprine second phalanx from Kissi 22 (BF 96/22) had proximal and distal exostosis and proximal lipping. From phase III at Mege an ovicaprine mandible with malocclusion of the teeth was recovered. However, the most striking example was a very long and slender sheep metatarsal from Galaga (NA 92/2C) (Fig. 29), possibly because of distorted growth of the animal due to malnutrition. However, since no reference material of recent long-legged sheep or descriptions from the literature are available, it cannot entirely be ruled out that this is the normal appearance of long-legged sheep metatarsals.

In the early 1990s, numbers of sheep in Nigeria were estimated at 22 million, while about 34.5 million goats had been counted (Bourn *et al.*, 1994). Larger sheep and goat types are usually associated with nomadic herders, while smaller types are kept as village livestock. Nevertheless in present north-eastern Nigeria, long-legged sheep are also kept as scavengers around towns and villages (de Leeuw *et al.*, 1972, 54). Dwarf ovicaprines are not able to travel more than a few consecutive days and must have water each day, in contrast to the savannah types which may travel for months and drink only about every three or four days (MacDonald, 1995). Nonetheless, pastoral animals are said to be more productive than village ones. In mobile pastoral systems sheep are the second most important species after cattle, but in village livestock small ruminants predominate (Bourn *et al.*, 1994). In general, ovicaprines have to be watered more frequently than cattle, but their intake is smaller (Dyson-Hudson, 1966, 45; Dahl and Hjort, 1976, 249). Sheep are grazers, concentrating on grass species, but goats are browsers and can cause considerable damage to vegetation because they uproot or strip any plants they can reach (FAO, 2001).

Small bovid

For more than half of the about 7000 small bovid remains, a more precise identification was not possible. A large part of them are undiagnostic bones, mainly ribs and vertebrae, but the principal reason for the low number of identified remains is high fragmentation at the sites. It is assumed that wild and domestic small bovids are represented in similar proportions among the unidentified as among the identified bones, which implies a predominance of domestic ovicaprines for most localities. The fusion state of the long bones is summarised in Table C.88. Animals of all age groups seem to be present, although bones that fuse after the age of two years in domestic sheep are usually not fused in this material (Silver, 1963). A summary of the traces of butchery that have been found on the unidentified small bovid bones is given in Table C.89. There are additionally also a few worked remains: a second phalanx from Gajiganna A (NA 90/5A) and two metapodials from Gajiganna BII (NA 90/5BII). Pathologies included two fused carpal bones from Late Iron Age levels at Oursi village (BF 97/13) and two fused thoracic vertebrae from the neighbouring site of Oursi hu-beero (BF 97/30). The same site yielded two fused lumbar vertebrae, which also have cut-marks all around.

4.6.10.2. Large bovids

Roan antelope (*Hippotragus equinus*)

Altogether, only four bones have been recognised of this species, which is the largest non-bovine antelope that can at present be found in the study area. Two third phalanges and a distal fragment of a humerus were found at Saouga 94/120 (BF 94/120), while from Gajiganna BII (NA 90/5BII) a first phalanx was retrieved. The elements were identified with the aid of the comparative traits and measurements given in Van Neer (1989a, 72-77, Table 59). All recovered roan antelope bones are from adult individuals. Measurements that could be taken on them are given in Table C.90.

Roan antelope is one of the most common West African antelopes, favouring the higher rainfall Sudan and Guinean zones (Spinage, 1986, 183). It used to be widespread in northern savannahs and woodlands. In Sahelian regions it only occurs within short distances from water and roan antelope populations are known to concentrate near water points during the dry season (Kingdon, 1997, 436-437). The animal generally prefers habitats where there is little competition with other herbivores and where there are few carnivores.

Hartebeest (*Alcelaphus buselaphus*) or topi (*Damaliscus lunatus*)

A few localities in Burkina Faso and Nigeria have yielded together eleven bones of adult specimens of hartebeest or topi (Table C.91). The bones were recognised using the traits described in Van Neer (1989a, 72-77, Tables 54 and 55), but distinction between the two species was not possible from the available osteological remains. Measurements on well-preserved bones are given in Table C.92.

Hartebeests are grazers that live in open grass plains, often on boundaries with parkland, woodland or scrub (Kingdon, 1997, 427-428). Where water is available, they drink regularly, but they can also meet their needs by eating water-holding plants (Estes, 1991, 139). Hartebeest populations decline when there is much competition with domestic cattle (Kingdon, 1997, 427-428). The animals are easy to hunt and are known for their tasty meat. Topi is common on floodplains in otherwise relatively dry regions south of the Sahara (Kingdon, 1997, 427-428). The species lives in seasonally inundated grasslands where it follows receding water during the dry season and retreats onto higher ground with rising water levels. The densest populations occur where green pastures persist throughout the dry season, notably on plains subject to seasonal flooding bordering lakes and rivers (Estes, 1991, 142-143).

Large antelope

As for the medium-sized antelopes, some of the large antelope bones could not be identified to a lower taxonomic level (Table C.93). It is assumed that the large antelope taxa described above are represented among them in similar proportions as among the identified remains. Two bones of subadult specimens were recorded: an unfused distal metatarsal from Gajiganna BII (NA 90/5BII) and an unfused first phalanx from Kelumeri (NA 96/45), while all other bones are from adult animals. In domestic cattle phalanges fuse at the age of about 18 months, while distal metatarsals fuse considerably later, around 27-36 months (Silver, 1963). A pelvic bone from Gajiganna BII (NA 90/5BII) was chopped ventrally on the ilium.

African buffalo (*Syncerus caffer*)

To separate bone remains of the bovines African buffalo and domestic cattle, criteria described by Gentry and Gentry (1978, 316) and by Peters (1986a, 1988) were used. The African buffalo has two subspecies, savannah buffalo (*Syncerus caffer caffer*) and forest buffalo (*S. c. nanus*). From their present geographic distribution, we would expect to find the savannah rather than the forest form, which is the larger of the two (Kingdon, 1997, 348-349). Buffalo has been recognised at three of the studied localities. A loose lower fourth premolar was found at Tin Akof (BF 94/133) and at Blé Mound E the same tooth was recorded in association with other bovine lower teeth and mandible fragments (Fig. 30). Blé Mound E also yielded a buffalo anterior first phalanx and a talus. A remarkable find was an almost complete bovine skeleton from Gajiganna BII (NA 90/5BII), with some bones that seemed to show morphological features of buffalo (distal radio-ulna, proximal metacarpal, distal tibia, talus, calcaneus, naviculo-cuboid, posterior first and second phalanx) while others showed characteristics more like cattle (fibula, one anterior first and two anterior second phalanges) (Peters, 1986a, 1988). The remains were therefore sent to Prof. Dr. J. Peters, Institut für Paläoanatomie und Geschichte der Tiermedizin, Ludwig-Maximilians-University, Munich, who confirmed the identification as buffalo. A list of all recovered elements is given in Table C.94. All long bones of the animal have fused epiphyses, which has happened by the age of 4 years in domestic cattle (Silver, 1963). A humerus and a femur diaphysis of this buffalo from Gajiganna BII appeared to be chopped. Measurements on the postcranial buffalo skeletal elements found at Blé Mound E and Gajiganna BII are given in Table C.95. They all clearly fall outside the size ranges of cattle remains that have been found in the research area (see below). Only a distal tibia from Oursi hu-beero (BF 97/30), identified as cattle on morphological grounds, reaches a size as large as that recorded for buffalo (Table C.99) (Peters, 1986a, 1988). This suggests that there may be some limited size overlap between cattle and buffalo in arid West Africa.

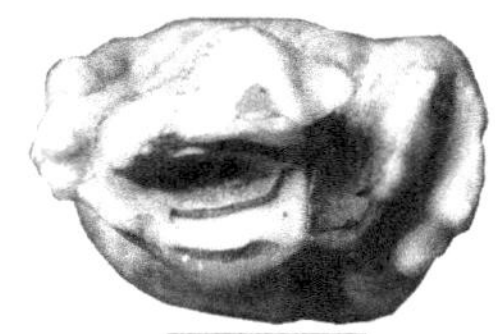

Fig. 30: Buffalo lower fourth premolar from Blé E (Scale = 1 cm)

Savannah buffaloes do not occur more than 20 km away from water since they need to drink daily (Happold, 1987, 197, 211-213; Kingdon, 1997, 348-349). Combined with their need for dense cover and grass, this makes them favour mosaics and savannahs with patches of thicket, reeds or forest. Buffaloes are known to aggregate in herds (Estes, 1991, 196-197). Over the past two centuries, their numbers have greatly decreased due to overhunting, habitat loss, and especially the rinderpest epidemic of the 1890s (Kingdon, 1997, 348-349). In Nigeria today they can no longer be found north of 12°N (Happold, 1987, 211). They are also absent from the present Oudalan province in Burkina Faso (Barral, 1977, 20-22). Al-

Umari, writing in the fourteenth century AD, describes a pursuit of buffaloes with poisoned arrows (Levtzion and Hopkins, 1981, 264), but the animals are usually considered dangerous to hunt (Plug, 1997).

Domestic cattle (*Bos primigenius* f. taurus)

Robustness and size of the bones usually allowed the separation of bovines from non-bovine large antelopes, although the largest of these, roan antelope, showed some size overlap. Remains for which the separation could not be made, mostly ribs, vertebrae and badly fragmented pieces of skulls or long bones, were put in the general category "large bovids". As explained earlier, the criteria described in Peters (1986a, 1988) and Gentry and Gentry (1978, 316) were used to distinguish between domestic cattle and buffalo. Since large antelopes are not well represented (see above), in contrast to domestic cattle, it was decided to simplify the tables by lumping all the unidentified large bovid bones together with cattle. It is not excluded that some buffalo or other large antelope remains are present among these bones, but it is believed that their presence would not greatly affect the general proportion of the species and that they would certainly be noticed among the measurable remains. Altogether, more than 5000 bones of this category have been found (Table C.96).

Three main cattle types can presently be found in West Africa.[10] Taurine or unhumped cattle (*taurus* type) and zebu or humped cattle (*indicus* type) are the most common ones, together with a whole range of crossbreeds. The first cattle were domesticated around the eighth millennium BC, the zebu on the Indian subcontinent (Meadow, 1993) and taurine cattle in the Near East (Peters *et al.*, 1999). In addition, Africa is named as a third domestication centre for cattle, by both geneticists (Bradley *et al.,* 1996; Bradley and Loftus, 2000; Troy *et al.*, 2000; Hanotte *et al.*, 2002) and archaeozoologists (e.g., Gautier, 1984b, 2002), although this is still a matter of debate. Relic populations of these putative indigenous African cattle (*africanus* type), collectively called Sanga, are sometimes considered as a third domestic taxon (Grigson, 1991). Sanga cattle are usually unhumped, but in males a small, muscular hump is possible in a cervico-thoracic position, as opposed to the usual thoracic position of the hump in zebu cattle. Some authors do not accept the Sanga cattle as being the indigenous African forms, but consider them as part of the many crossbreeds between zebu and taurine cattle (MacDonald, 1995; MacHugh *et al.*, 1997).

Osteomorphologically, there are a few features that can be used to distinguish between zebu and taurine cattle. There is first of all a list of cranial characteristics typical of zebu, including the long, narrow skull, the shape of the ridge between the horns, the convexity of the forehead, the concavity of the occipital region, the less prominent orbits, flat orbital rims and the diagonally upward direction of the horns at the base (Grigson, 1980). A criterion of the list that has been used in archaeological contexts is the presence of flattened jugal rims. However, this trait cannot be taken as good proof for zebu cattle unless it can be shown that the rims come from elderly animals (Grigson, 1991, 2000), which often appears to be problematic. Grigson (2000), for example, points to an incorrect identification of zebu cattle in Kenyan contexts from the beginning of our era (Marshall, 1990), where it was not proven that the orbital rims were from old animals. The shape of the thoracic vertebrae is another method of making the distinction between zebus and taurines, since the dorsal spines on some of the posterior ones (6th-9th caudally) are bifurcated in zebu, but usually not in taurine cattle (Epstein, 1971, 388). The trait is difficult to use in archaeological contexts because the spines do not seem to preserve well. Crosses between zebus and taurines may also have bifurcated dorsal spines on their thoracic vertebrae, but bifurcation usually not reaches as deep as in pure zebu cattle and is in some cases hardly perceptible (Epstein, 1971, 538). C. Magnavita, for example, brought spinal columns of three recent individuals of humped cattle (Shuwa breed) from northern Nigeria to Europe, but no bifurcated spines were visible on any of the columns. A third possibility for differentiation between zebu and taurine cattle osteologically would be the relative slenderness of the limb bones of the former (Grigson, 2000), but this character is also hard to use in archaeological contexts because complete long bones are only very exceptionally available. Very little is known on the osteology of *africanus* type cattle. Characters listed by Grigson (1991) are long faces, straight intercornual ridges, flat foreheads, prominent orbits and long legs. Furthermore, their horns are "usually directed diagonally outwards and upwards at the base, and tend to be crescentic in males and lyre-shaped in cows and castrates".

Because of the difficulties in determining cattle types archaeologically, their propagation over the African continent remains poorly documented. Several models have been proposed for the propagation of taurine/Sanga cattle over the African continent, all of which have north-eastern Africa as the focal point for the initial dispersion (e.g., Payne, 1964; Shaw, 1981; Gautier, 1987a; Krzyzaniak, 1992; Breunig *et al.*, 1996; Hassan, 2000,

[10] There is as yet no consensus among (archaeo) zoologists on the use of a uniform nomenclature for domestic animals (Gautier, 1993; Gentry *et al.*, 2004). In this study, all domestic cattle types are referred to as *Bos primigenius* f. taurus, in accordance with the system of scientific names by Bohlken (1961), used for the other domestic species (cf. Gautier, 2002). Alternative names often encountered in the literature are *Bos taurus* (taurine cattle), *Bos indicus* (zebu) and *Bos africanus* (Sanga); a system that has for example been defended by Gentry *et al.* (2004). Uerpmann (1993) has tried to solve the confusion in nomenclature of domestic animals and proposed to name taurine cattle "*BOS (TAURIBOS)*" and zebu cattle "*BOS (INDIBOS)*", but his system has never been adopted in practice.

2002). In that area the oldest domestic cattle may date as far back as the ninth millennium BC (Gautier, 1984b, 2002) although more secure evidence is at least a millennium younger (Chenal-Vélardé, 1997). The cattle remains from Windé Koroji Ouest, dating to 2200-950 BC, possibly represent the earliest evidence for sub-Saharan West Africa (MacDonald, 1996; MacDonald and MacDonald, 2000). After that date cattle rapidly appear at other sites throughout West Africa, and also at relatively southern localities like rockshelter K6 at Kintampo, dating to around the beginning of the second millennium BC (Carter and Flight, 1972). According to Payne (1964), the earliest cattle type propagated was a shorthorn that entered West Africa by way of the Sahara, which was followed by a second wave of introduction, in this case of longhorn cattle,[11] along the Atlantic coast. This does not seem to corroborate with data from Saharan rock art, however, where shorthorns only appear after 1000 BC, i.e. a few millennia later than longhorns (Muzzolini, 2000). Neither does it fit with osteological evidence, with all early Saharan cattle being longhorn (Gautier, 1987a; Chenal-Vélardé, 1997). Gautier (1987a) therefore proposes that the diverse shorthorn breeds may have developed out of longhorns. This also fits better with the view of Grigson (1991) who has argued that the longhorns are the indigenous African cattle. Archaeozoological evidence from Morocco and the West African Atlantic coast is as yet insufficient to support or refute Payne's (1964) route along the western Atlantic coast (Jousse, 2004b).

Zebus seem to have entered the African continent in considerable numbers only from the fourth century AD onwards (Epstein, 1971, 519), although, judging from depictions in tombs and temples, they were known in New Kingdom Egypt (1569-1076 BC) through feudal tributes and spoils of war (Nicolotti and Guérin, 1992). The actual introduction of zebu cattle was probably from Asia by way of the Arabian Sea, entering the African continent at the Horn (Hanotte *et al.*, 2002). Between AD 700 and 1400 the animal must have dispersed rapidly over the African continent (Grigson, 1991; Blench, 1993; Meghen *et al.*, 1994). So far, no West African archaeological site has yielded good osteological evidence for zebu cattle. They may be present in phase IV at Jenné-Jeno and in phase IV/V at Siouré, both dated to around AD 800-1400, based on metrical data and the presence of flattened jugal rims (MacDonald, 1995; MacDonald and MacDonald, 2000). Iconographic evidence in West Africa for zebu, or better, humped cattle, is available in the form of a clay figurine from the upper spits at Daima, dated to Daima III, a phase which ends around 1150 AD (Connah, 1981, Fig. 8.9, 183). However, disturbance of these spits indicates that material from as late as the seventeenth century AD may be present (Magnavita, in press). Another zebu figurine was retrieved from Mdaga (425 BC-AD 1780), but has not been precisely dated (Lebeuf *et al.*, 1980, 160).

At the end of the nineteenth century AD, taurine/Sanga cattle were eradicated from much of their original distribution zone by the rinderpest pandemic. Remaining populations occur in West and West-Central Africa, where three different types can be found: West African Dwarf Shorthorn (WAD) (syn. Muturu) and two humpless longhorns, N'Dama (syn. Gambia Longhorn, Futa, Futa Longhorn, Malinke, Mandingo) and Kuri (syn. Buduma cattle). Grigson (1991) thinks the latter two are Sanga breeds. WAD and N'Dama are both trypano-tolerant. WAD's occur in a broad belt in southern West Africa, between Liberia in the West and Cameroon in the East (Epstein, 1971, Fig. 295). According to Epstein (1971, 201), N'Dama cattle originate from the Fouta Djallon plateau in Guinea, and can now also be found in Gambia, Sierra Leone, Liberia and western Mali. Kuri are confined to the area around Lake Chad and its eastern shores, but there is mention of a relic population in Ethiopia as well, although it is not entirely clear if these are really of the same type (Alberro and Haile-Mariam, 1982). Kuri cattle are very tall animals with long horns that emphasize their size even more. Some have very thick and bulbous horns, conical in shape, while in others horn thickness is more proportionate to horn length (Epstein, 1971, 208). Microsatellite analysis showed that Kuri cattle are genetically not markedly different from N'Dama cattle, despite their different appearance (Meghen *et al.*, 2000) and their origins are still unknown. They live outside the main tsetse-infested areas, but they are said to have a high tolerance for biting insects (Blench *et al.*, 1992, 144). Nevertheless, Quéval *et al.* (1971) recorded high mortality rates among Kuri cattle due to trypanosomiasis. The animals spend much of their time withers-deep in water and are excellent swimmers (Blench *et al.*, 1992, 143-149). Kuri cattle have a preference for fresh grass, which determines their geographic distribution. Apparently there is no medieval Arabic account of them, the traveller Barth first recorded them in 1851 (Barth, 1857-8, 200).

Zebus and zebu-taurine/Sanga crossbreeds are the most common and the most widespread types of cattle in Africa today. As mentioned in 2.1.1., the present-day cattle in northern Burkina Faso also belong to the zebu type. During the rinderpest pandemic of the 1890s, zebus replaced taurine/Sanga populations over the continent because they are more resistant to the disease (Payne, 1964). Zebus are moreover better adapted to living in semi-arid environments and they are more productive (Porter, 1991, 209). Finally, they are also said to be better suited for long-distance migrations (FAO, 2001). They are, on the other hand, much more susceptible to trypanosomiasis than taurine/Sanga cattle and therefore

[11] Following Gautier (1987a), in this study animals with horncores up to 20 cm long are considered as shorthorn, while those with horncores longer than 25 cm are longhorn. However, all longhorns go through a shorthorn phase in the course of their life.

do not thrive in tsetse-infected regions (Porter, 1991, 209). Zebu crossbreeds are widely distributed because they combine the favourable characteristics of the two. In fact, all African cattle types seem to have taurine/Sanga origins, while zeboid characters, both morphological and genetic, appear to be the result of varying degrees of zebu introgression (Loftus and Cunningham, 2000). The zebu introgression in African cattle declines from east to west and from north to south and is believed to correspond with the direction along which the zebu was spread over the African continent (Meghen *et al.*, 1994; MacHugh *et al.*, 1997; Hanotte *et al.*, 2000; Hanotte *et al.*, 2002).

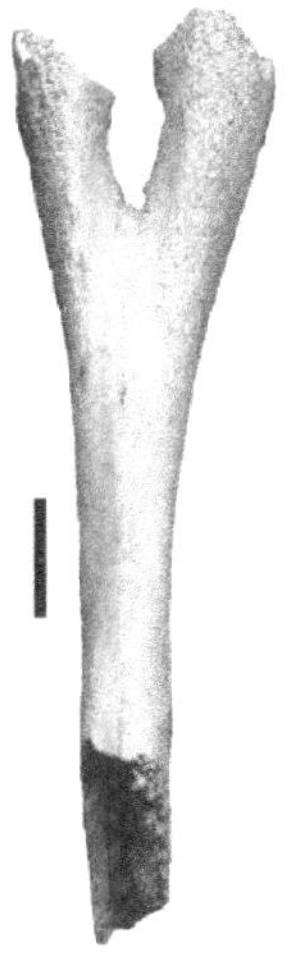

Fig. 31: Bifid neural spine of thoracic vertebra, probably of zebu cattle, from Galaga (NA 92/2C) (Scale = 1 cm)

No osteomorphological indications for a possible presence of zebu cattle were found at the studied sites, except for a bifurcated dorsal spine of a thoracic vertebra found at the subrecent site Galaga (NA 92/2C) (Fig. 31). A cattle orbital rim from Gajiganna BI (NA 90/5BI) was sharp, as usual in taurine cattle. The state of preservation of the studied faunal material did not allow the gathering of data on the shape and size of cattle horncores, which excluded another possible way to get information on cattle types. However, cattle figurines from the Gajiganna Culture seem to depict shorthorn animals (Breunig, 1994), while a figurine from Iron Age Dorota (NA 97/13) is probably of a longhorn (Breunig *et al.*, in prep.). Other sub-Saharan West African sites did yield cattle horns. At Ntereso, dating to the early second millennium BC, longhorn cattle were probably present (Gautier and Van Neer, 2005), while shorthorn cattle were identified at Kobadi in a second-first millennium BC context (Jousse and Chenal-Vélardé, 2001-2002).

Grigson (1991) has stated that size is not an adequate criterion for the distinction of the three types of cattle. Nonetheless, for most of the studied contexts size is the only criterion available and it is believed that it is possible to more-or-less define metrical characteristics of the present-day types to which the archaeological cattle remains can be compared. This has already been carried out for West African archaeological sites by MacDonald and MacDonald (2000), who divided present West African cattle types into three broad size categories. The WAD is the smallest type, with a height at the withers of 90-110 cm (Epstein, 1971, 276), whereas the N'Dama is slightly larger (between 95 and 120 cm tall) (Epstein, 1971, 202). Zebu and Kuri both fall in the large size category, the former measuring 115-140 cm at the withers (Epstein, 1971, 397, 400-408) and the latter 130-150 cm (Epstein, 1971, 209). An analysis of the available metrical data on cattle bones (Table C.99), with hypotheses on possible types present, is given in 6.7.3.1. As for the ovicaprines, only measurements on postcranial skeletal elements were included since the relations between skull dimensions and body sizes are not clear.

Fusion dates of the cattle long bones found are listed in Table C.97. Animals of all age groups, including foetuses, seem to be present. Cattle usually reach mature size at an age of between 4 and 5 years (Dahl and Hjort, 1976, 167). Almost 80 cattle bones showed traces of butchery (Table C.98). A proximal part of a metacarpal from Oursi hu-beero (BF 97/30) resembled waste of a bone working activity. From Early Iron Age Oursi (BF 94/45) a perforated incisor and a perforated second phalanx were retrieved, while a worked proximal radius and scapula were found at Gajiganna BI (NA 90/5BI). Finally, at Kariari C (NA 95/1) a polished scapula was recorded. A rib from Galaga (NA 92/2C) had a pathological, longitudinal, depression at its dorsal side. No pathologies were found on phalanges and metapodials that might have indicated draught cattle (Bartosiewicz *et al.*, 1997). Measurements of cattle bones are listed in Table C.99. A complete metacarpal from Gajiganna A (NA 90/5A), with a length of 179 mm, allowed withers height estimation using the factors summarised in von den Driesch and Boessneck (1974). Depending on the sex of the animal it must have measured between 107 and 113 cm at the withers.

Data on the ecological requirements of cattle have already been mentioned above while describing the different types, but in general, good conditions, quality pasture and enough drinking water, are needed to sustain large cattle herds. Zebu cattle are especially demanding, requiring a particular pasture composition, while taurines/Sanga eat almost any type of vegetation, e.g., grass, herbs or woody vegetation (Blench, 1999, 40). The WAD is known in particular for the high proportion of browse it can digest (FAO, 2001). Cattle are usually kept within an 8 km radius of water because they have to be watered frequently (Dahl and Hjort, 1976, 238). They are the predominant species in pastoral nomadic systems (Bourn *et al.*, 1994). There are also records of cattle kept in surprisingly adverse circumstances. In the Mandara Mountains, for example, bulls are stall-fed for a year or more, after which they are released as part of a festival (David and Sterner, 1989).

Cattle (*Bos primigenius* f. taurus) or buffalo (*Syncerus caffer*), large bovid

At the studied Late Stone Age sites in Burkina Faso, buffalo is the only large bovid that has been positively identified. Faunal samples from the period and region are too limited to have a good idea of the complete species spectrum possible. Therefore, identifications that were made on a pure osteological basis are given for this site (Tables C.100 and C.101), as opposed to what has been done for the other localities.

Chapter 5. Taphonomical analysis

Taphonomy is a term borrowed from classical palaeontology and is used to refer to the processes that an organism is subjected to between its death and its recovery as a fossil (Efremov, 1940). From the model in Fig. 32, it is clear that the faunal assemblage that is eventually recovered archaeologically is a greatly reduced sample of the initial living animal population. There are several factors that cause bone loss, which we need to understand before starting to interpret the faunal data described in the previous chapter. In what follows these factors are discussed: cultural practices of the sites' inhabitants, type of site and disposal, differential preservation and tertiary deposition, area chosen for excavation and the sampling techniques used during the excavation. Animals that went through the same history between their death and their archaeological recovery are part of the same taphonomic group (*sensu* Gautier, 1987b). It is mainly the cultural practices of the human groups that appear to determine the attribution of remains to a certain taphonomic group, and the taphonomic groups are therefore discussed under "cultural practices". The final subjects of this chapter are the faunal assemblages of each of the studied sites. The sizes of the bone samples are looked at, as well as their state of preservation. How these assemblages were influenced by the different taphonomic factors is also examined.

5.1. Cultural practices (taphonomic groups)

Which of the animals in an original living population that end up in the bone assemblage of a given site depends mainly on the cultural practices of its human inhabitants. Only intrusives can become part of the assemblage independent from human action. The animal remains of the studied sites are divided into taphonomic groups, intrusives, food refuse, artisanal refuse and carcasses, according to the reasons why they may have become part of the bone assemblage. With the exception of the intrusives, this is mainly related to the purpose they have served the humans. A summary of the taphonomic groups assigned to each taxon is given in Table 5. Data mentioned in the previous chapters, on ecology, ethnography, sites specifics, etc., are used to this end and references to the data can be found there. Some taxa have been put in several groups, since animals may have served multiple purposes.

5.1.1. Intrusives

The taphonomic group of the intrusives includes the remains of all animals that were not brought to the site intentionally by people, or by other living agents such as predators. It can be subdivided into penecontemporaneous and late intrusives. The former arrived at the site during, or shortly before or after, its human occupation, and the latter long after its abandonment, although it is often not possible to make the distinction between the two. Typical for intrusive animals is that their skeletons are usually recovered more-or-less completely and that traces of butchery or skinning are lacking on their remains. Late intrusives can sometimes be discerned from their better state of preservation compared to the rest of the faunal sample. Penecontemporaneous intrusives are potentially important ecological indicators, since they date from around the time of the site's occupation but have not undergone a cultural filter. Another type of intrusives that may be present are reworked or geological intrusives. Gautier (1987b) defined them as "fossils from deposits underlying the site reworked in the deposits by human activities, or brought in by other agents from existing deposits, by fluviatile erosion and transport or other geological processes."

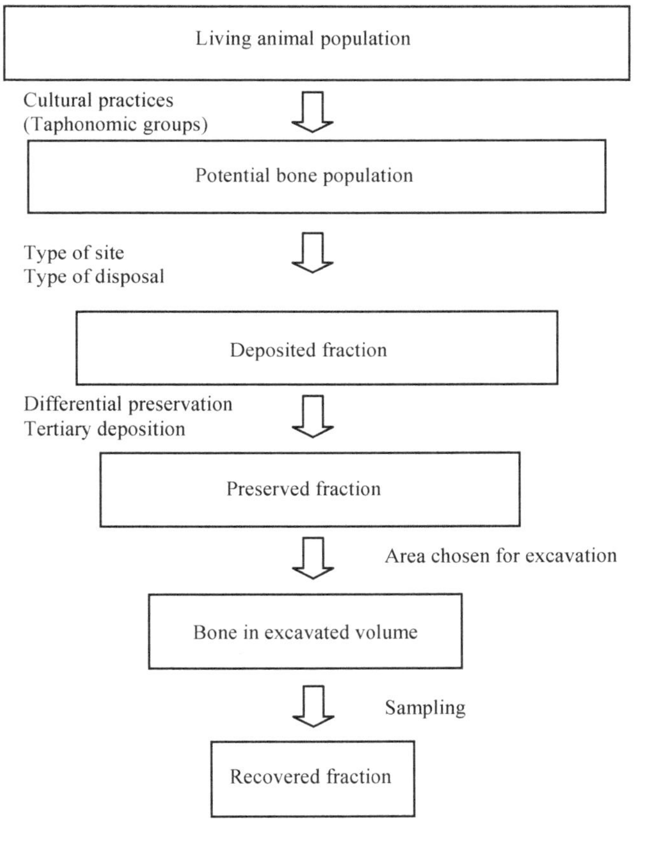

Fig. 32: Theoretical scheme of taphonomic history (After Meadow, 1980)

Commensal animals are usually counted among the penecontemporaneous intrusives. At the studied sites these probably include white-toothed shrew (*Crocidura* sp.) and most small rodent species. Food habits of recent African populations can alert us to the possibility, however, that even small rodents, such as ground squirrels (*Euxerus erythropus*) or rats (*Mastomys natalensis*, *Arvicanthis niloticus*), may have been consumed. Small rodents were especially common at Oursi hu-beero (BF 97/30) where they were often found in crevices and cavities in the walls (Linseele, in press). Small rodents also left traces at the studied sites in the form of gnawing marks on bones of other taxa. Other small animals that may have lived and died at the sites, unnoticed or undisturbed by humans, are the small mollusc taxa (*Lymnaea natalensis* and *Eupera ferruginea*), the agama (*Agama* sp.) and the songbirds (Passeriformes). Some of the recovered eggshells may represent hatched eggs of domestic birds that lived at the site, or of wild birds that made their nests at the site after its abandonment.

Burrowing taxa are also often intrusive, either contemporaneous or late, and their presence may cause disturbance of archaeological deposits. Disturbance of the sites, either by burrowing animals or due to other processes, may also have caused the intrusion of recent animal remains into the archaeological deposits. The fresh appearance of the medium-sized carnivore vertebra found at the Late Stone Age site Dori 94/40 (BF 94/40), indicates that its presence should probably be explained in this way. The terrestrial snail *Limicoloria* is often considered as a penecontemporaneous intrusive when present in archaeological layers, because it is known to aestivate in places with loose soils (e.g., Gautier, 1983; Van Neer and Bocoum, 1991). This is also proposed for the sites under study here, although some *Limicolaria* shells seem to have been collected as a raw material and perhaps also for food. The small snail *Pseudoglessula* is probably also an intrusive that burrowed into the archaeological deposits. The taphonomic status of the amphibians at the studied sites, mainly bullfrog (*Pyxicephalus edulis*), is not entirely clear. Good preservation of most remains and finds of more-or-less complete skeletons seem to indicate that they are intrusives that died naturally in their dry-season burrows. The animals may have been attracted by clayey spots that retained water some time after the rainy season, created by abandoned and collapsed mud-brick house structures. It seems therefore that they can be placed in the category of the penecontemporaneous intrusives, since they probably arrived at the site not long after its human occupation. The proposed scenario also offers an explanation for why there were so little frog remains at Oursi hu-beero (BF 97/30). At this location mud-brick constructions did not collapse under the influence of rains because they were accidentally fired (but see 5.5.). Although most evidence thus points to an intrusive nature of amphibians, a coracoid with cut-marks suggests that they may occasionally have been consumed as well. This bone could have been hit while cutting open an animal's abdomen. As mentioned in 4.3., consumption of bullfrog is also well-documented ethnographically. Besides the amphibian bones, most snake remains are probably also of animals that buried themselves inside the deposits where they died. Some small rodent taxa can apparently burrow to appreciable depths as well. This was particularly clear from the settlement mound of Oursi village (BF 97/13) where a concentration of small rodent bones was found in the levels below -6.0 m. Other taxa, like sandfox (*Vulpes pallida*), also have burrowing habits but there are no indications in the studied assemblages for such animals that died naturally in their burrows.

No geological or reworked intrusives have been found. It has been considered that some of the fish remains may perhaps have arrived at the sites embedded in the clay used for house construction, but there were no direct indications to assume such a scenario. Nevertheless, in recent West African houses, bones can be observed mixed in the mud used as a building material (Vélardé, 1995).

5.1.2. Food refuse

The majority of the recovered bones are from animals that were used for human consumption. They are typically found scattered, are usually fragmentary and traces of butchery are often recorded on them. From ethnographical data mentioned in the previous chapter, it is clear that probably more taxa appeared on the menu than those considered as classical food animals by Westerners. Many Africans also seem to be aware of the fact that Westerners find some of their food habits unpalatable and may therefore not answer truthfully when interviewed (Simoons, 1981). However, it should also be taken into account that exploitation patterns of wild species have been changing recently (see 6.6.2.). It is also apparent from the previous chapter, that which species are allowed to be eaten and which are not is often culturally determined. There seem to be very little general trends in traditional African societies, and customs may differ between neighbouring groups without any clear reason. Islam is particularly well-known for its food restrictions, although these also differ depending on the branch. Animals that are forbidden food according to the religion can be broadly divided into three groups: carnivores, scatophague animals and vermin (Benkheira, 2000).

The *Pila* shells found at the studied sites are probably the remains of snails collected for food, although in some parts of present Africa they are only used as a raw material. Other freshwater molluscs, except for the small taxa, have also been put among the food debris, but some seem to have served more than just culinary purposes and, as said, at least part of the *Limicolaria* shells may be intrusives instead. Nonetheless, Connah and McMillan (1995), studying material from the Nigerian part of the southern Lake Chad area, have argued that molluscan shells are not a very palatable source of food and that they are only eaten in times of famine.

All fish are thought to have been eaten, even when only a few remains show traces of butchery. This may be related to the way in which they have been prepared (see 6.5.4.). In several occasions fish bones have been found in anatomical connection but this is not in contradiction with consumption, since articulating parts of the same specimen may have been thrown away. It could, however, also be indicative of entire fish that were discarded because they had, for example, gone bad and perhaps some are (forgotten or spoiled) supplies of dried or smoked fish.

Even though only bones of monitor lizards (*Varanus* sp.) showed traces of butchery, most reptile taxa, with the possible exception of the very small agama, may have been eaten. As said earlier, however, at least part of the snake remains are probably intrusives. The five more-or-less complete skeletons of Adanson's mud turtle (*Pelusios adansonii*) from pit 1 in trench 6 at Zilum (NA 97/37) were at first not counted with the food refuse either. It was thought that the pit was perhaps a water reservoir and that the turtles lived inside but died when it dried out. However, the archaeologist responsible for the excavations at Zilum thinks this scenario is highly unlikely because the potsherds and ash in the pit clearly pointed to its use for refuse disposal (C. Magnavita, pers. comm.). Therefore, it seems that the turtles may have been eaten after all, or that complete animals were discarded. Concentrations of bones from several individuals of one taxon have also been noticed at other locations within Zilum, e.g., in pit 2 at Zilum 5 where a concentration of clariid catfish (Clariidae) bones was found.

Apart from the small songbirds (Passeriformes), all bird taxa, both domestic and wild, are considered to have been killed for consumption. This is corroborated by the presence of butchery marks. However, birds of the Accipitridae family, mainly vultures, may not have been eaten, as many human groups, including the present inhabitants of Oursi, do not like their taste. Besides remains of hatched eggs, eggshell fragments of unidentified birds may also represent waste of consumed eggs, collected from domestic or wild birds' nests. Traditionally, Africans rarely eat eggs, however, and many groups have food prohibitions concerning eggs (Lagercrantz, 1950, 39-44). Historical sources are also silent on the use of eggs as food in West Africa during the medieval period (Lewicki, 1974, 112). Nevertheless, at present, both raided eggs of wild birds and domestic fowl eggs are part of the normal diet (Bourn *et al.*, 1994), although this may largely be a consequence of government politics and campaigns by development organisations (FAO, 2004).

Bones of most mammalian species, olive baboon (*Papio anubis*), savannah or patas monkey (*Cercopithecus aethiops/patas*), African hedgehog (*Atelerix albiventris*), hare (*Lepus capensis/saxatilis*), large rodents, aardvark (*Orycteropus afer*), warthog (*Phacochoerus africanus*) and wild and domestic bovids, have also been put among the food refuse. Modern parallels show that meat of wild carnivores was probably eaten as well, although in the Western view they are only useful for their fur. At the Iron Age sites, especially the Middle and Late Iron Age sites in Burkina Faso, meat of domestic dogs (*Canis lupus* f. familiaris) also seems to have contributed significantly to the amount of animal protein consumed, although it was probably not seen as an ordinary source of food (Linseele, 2003). Elephant (*Loxodonta africana*) meat may have been part of the diet as well. Nonetheless, the tail vertebra found at Oursi hu-beero (BF 97/30), was tentatively interpreted as part of a tail that was taken as a trophy from an animal that was hunted or found dead (Linseele, in press). Domestic horse (*Equus ferus* f. caballus) and donkey (*Equus africanus* f. asinus), although in the first place kept for transport, may have been slaughtered and eaten at the end of their useful life. No evidence, in the form of butchery marks on the bones, is available, however, to support this. Only jugal teeth fragments have been found of hippo (*Hippopotamus amphibius*) and evidence of its consumption is therefore lacking, although the animal's meat is much appreciated today.

5.1.3. Artisanal refuse

This category unites remains of animals that served (also) other than culinary purposes after their death. First of all bones and shells are included that have been modified into tools or other objects, or the waste of their production. They can unmistakably be recognised when cut or sawn, polished, bored, etc. When considering worked material, attributing a bone to a certain taxon is usually rendered difficult or impossible and artisanal refuse may therefore often end up among the unidentified remains. The use of bones as a raw material can also influence skeletal distributions, leading to an over- or underrepresentation of skeletal elements that are best suited for the fabrication of objects.

Polished edges recorded on *Chambardia/Spathopsis* shells may be due to their use as a tool to smooth pottery surfaces (Van Neer, 2002b). Although examples are missing from the studied sites, the gastropods can also be used for adornment by simply removing their apex and then threading them on a wire (Crowley and Pain, 1989).

Mutela dubia and Nile oyster (*Etheria elliptica*) may also have been used as raw material, although there are no direct indications for it from the analysed sites. The latter bivalve is of poor quality because it exfoliates easily. Another possible use for large bivalves is as a receptacle or spoon. Finds of worked shells that were not sent to the RMCA for faunal analysis include beads from Dorota (NA 97/13) and Elkido North (NA 99/75), as well as a denticulated shell from Kelumeri (NA 96/45) that probably served as a tool to decorate pottery surfaces (Breunig *et al.*, in prep). Records of polished edges of peripheral pieces of *Pelusios* carapaces possibly means that these were used as some kind of scraping or digging tool. On one occasion, in Early Iron Age levels at Kursakata (NA 93/46), indications for the manufacture of what looks like a pendant from a piece of softshell turtle (Trionychidae) plastron have been found. Worked bird long bones, mainly of unidentified taxa but including one spur-winged goose (*Plectropterus gambensis*) ulna, were also regularly encountered. Apparently bird bones, being hollow and having thin walls, were popular as a raw material. The purpose of the objects made from them remains unclear and most of the available material may represent waste of artisanal activities. Ostrich (*Struthio camelus*) eggs were probably brought to the site intentionally as a raw material, mainly for bead making. At Early Iron Age Oursi (BF 94/45) in particular, ostrich eggshells are probably the refuse of local ostrich bead production, as pieces of shell with polished edges as well as end products have been recovered (von Czerniewicz, 2004, 45). Among the mammalian bones, worked specimens seem to be mainly of domestic species, cattle (*Bos primigenius* f. taurus), ovicaprines (*Ovis ammon* f. aries and *Capra aegagrus* f. hircus) and domestic dogs. From the site of Mdaga in southern Chad (Lebeuf *et al.*, 1980, 30), not so far from the Nigerian part of the southern Lake Chad area, horse bone is also reported to have been used for tool making (no date given). In Kottusch (1999) more details can be found on bone artefacts from the Gajiganna sites, where, as said in previous chapters, the use of bone was a way of overcoming the local lack of stone raw material. Here also, bones of domestic ruminants appear to have been predominantly used (Kottusch, 1999, 17-18), probably because they were most easily available. A few records have been made of (small) bones of mammalian taxa, including a caracal/serval (*Felis caracal/serval*) metapodial and canine, a bohor reedbuck (*Redunca redunca*) second phalanx, possibly three kob (*Kobus kob*) second phalanges, two ovicaprine first phalanges and finally a cattle incisor and second phalanx, that are bored and may thus have served as pendants. The retrieved elephant or hippo ivory was probably used as raw material as well, although the piece does not look worked. Why the hippo jugal teeth ended up in the studied faunal samples is not clear, as only the animal's canines and incisors yield ivory.

Besides worked specimens, the artisanal refuse comprises a second important group of bone remains, those of animals of which the skin or fur has been removed for use. This can be documented by the find of traces on skeletal parts where the skin lies close to the bone, e.g., metapodials or the skull. Some bones (phalanges, metapodials) typically stay attached to the fur or hide when it is removed, and therefore the skeletal distribution may also be indicative of skinning. Van Neer (1989a, 87) considered the serval and leopard size carnivores he found as animals hunted for their fur, because they were only represented by skull fragments and foot elements. However, such a skeletal distribution may also be an artefact of differential preservation (see 5.3.), combined with possibilities of identification. Looking at ethnographic parallels, Africans traditionally do not kill an animal solely for its fur or skin, even not carnivores.

Fig. 33: Crocodile skin drying at Kano (Nigeria)

Although there are no clear indications recorded for the studied sites, skins of monitor lizards and crocodiles were probably used, judging from their present popularity in the traditional artisanal sector. The same is suspected for the fur of wild carnivores, for which there are two good examples from Oursi hu-beero (BF 97/30). At the site, where finds are in situ, metapodials and phalanges of one caracal/serval individual, some with cut-marks, were found. Both the traces and the skeletal distribution suggest we may be dealing with the remains of a caracal/serval skin (Linseele, in press). The seven metapodials and phalanges of Herpestidae or Viverridae from a single context at the site may also be the remains of a skin. Skeletal distribution of aardvark at the same site and cut-marks recorded on its bones, could point to the presence of a hide of this animal as well. However, aardvark skin is thick and difficult to work, and the bones should therefore perhaps be interpreted differently, for example as a piece of leg kept for ritual reasons.

	intrusive			food	carcass	artisanal	
	penec.	late	rework.			raw mat.	skin & derivates
Molluscs							
Pila wernei	-	-	-	x	-	x?	-
Lanistes varicus	-	-	-	x	-	-	-
Cleopatra bulimoides	-	-	-	x	-	-	-
Lymnaea natalensis	x	-	-	-	-	-	-
Chambardia sp./*Spathopsis* sp.	-	-	-	x	-	x	-
Mutela dubia	-	-	-	x	-	x?	-
Etheria elliptica	-	-	-	x	-	x?	-
Eupera ferruginea	x	-	-	-	-	-	-
Limicolaria sp.	x	-	-	x?	-	x	-
Pseudoglessula sp.	x	-	-	-	-	-	-
Fish	-	-	x?	x	-	-	-
Amphibians	x	x	-	x?	-	-	-
Reptiles							
Turtle	-	-	-	x	-	x	-
Agama sp.	x	-	-	-	-	-	-
Varanus sp.	-	-	-	x	-	-	x
Snake	x	x	-	x?	-	-	-
Crocodylus sp.	-	-	-	x	-	-	x
Birds	-	-	-	x	-	-	-
except Passeriformes	x	-	-	x?	-	-	-
Eggshell ostrich (*Struthio camelus*)	-	-	-	x?	-	x	-
Eggshell unidentified bird	x	x	-	x?	-	-	-
Mammals							
Papio anubis	-	-	-	x	-	-	x?
Cercopithecus aethiops/patas	-	-	-	x	-	-	x?
Atelerix albiventris	-	-	-	x	-	-	x?
Crocidura sp.	x	-	-	-	-	-	-
Lepus capensis/saxatilis	-	-	-	x	-	-	x?
Small rodent	x	x	-	-	-	-	-
Large rodent	-	-	-	x	-	-	x?
Herpestidae and Viverridae	-	-	-	x	-	-	x
Canis lupus f. familiaris	-	-	-	x	x	x	x
Vulpes pallida	-	-	-	x	-	-	x
Felis silvestris or *F. s.* f. catus	-	-	-	x	x	-	x
Felis caracal/serval	-	-	-	x	-	x	x
Hyaena hyaena/Crocuta crocuta	-	-	-	x	-	-	x
Orycteropus afer	-	-	-	x	-	-	x
Loxodonta africana	-	-	-	x	-	x	-
Domestic equid	-	-	-	x?	x	-	x?
Hippopotamus amphibius	-	-	-	x?	-	x?	x?
Phacochoerus africanus	-	-	-	x	-	-	x?
Giraffa camelopardalis	-	-	-	x	-	-	x?
Wild and domestic bovid	-	-	-	x	-	x	x?
Foetuses and juveniles domestic bovid	-	-	-	-	x	-	-

Table 5: Taxa attributed to each taphonomic group defined

Generally, the skin of bovids, and especially of domestic species, was probably also removed for use, which can be seen from cut-marks on bones like phalanges and metapodials. However, when an animal is killed only for the meat, its skin may have been removed during butchery, thus leaving skinning marks. Other taxa of which the skin may have been used but for which good indications are missing are hedgehog (including its spines), baboon, savannah or patas monkey, hare, large rodent, domestic equids and common warthog.

5.1.4. Carcasses

Carcasses include the remains of animals that were discarded more-or-less complete after their death. They are usually found with their skeletal elements in anatomical position; at least when covered rapidly after they died. In the most classical sense it concerns domestic species, like horses, donkeys and dogs, which died naturally and which were not eaten after their life as a transport or companion animal but their dead bodies simply disposed of. The identified domestic equids and the domestic dogs from the Nigerian sites could belong to this category, although their remains were found scattered and a domestic dog femur from Zilum (NA 97/37) is possibly chopped. Finds of scattered bones of animals that were not eaten do not seem to be unusual, however. Holl (2003, 379-382) has, for example, described that although Shuwa Arabs do not eat equids, remains of the animals can be found within their settlements. Especially when the dead animals were not buried, their remains may have easily become spread over a large surface, for example by dogs or other scavengers. Nevertheless, the inhabitants of the studied sites may have killed their equids at the end of their useful life and then still have eaten them, as is done in some non-Muslim West African areas today. Foetal, neonate and also juvenile bones of cattle and ovicaprines, often found in concentrations, have been considered as carcasses as well. Presumably the presence of foetuses has to be explained by their removal from dead pregnant animals or by spontaneous abortion. Meat of very young individuals, and especially of foetuses, was probably not eaten because it is watery and sometimes considered tasteless. Although it may seem unpalatable to Western eyes, non-Islamic Africans do not refrain from eating animals that have deceased from a natural death. Medieval Arab travellers have also reported this habit from western Africa at that time, which they considered revolting (Lewicki, 1974, 79).

In the second instance carcasses also include wild animals that have been killed because they were dangerous or a nuisance, and then have not been consumed afterwards. In the studied faunal assemblages, nothing clearly points to specimens of this category. Animals were probably eaten even if this may not have been the reason why they were initially killed. The buffalo (*Syncerus caffer*) skeleton from the site Gajiganna BII (NA 90/5BII) is somewhat puzzling. The find of almost complete skeletons can be indicative for carcasses, but then there should be no butchery marks on the bones. As mentioned in the description in 4.6.10.2., a humerus and femur of the animal were chopped and its bones were also not found in anatomical position. This could mean we are rather dealing with an animal that was killed and butchered at the site, with all or most of its skeletal parts disposed of together.

5.2. Type of site and disposal

The type or function of a site will greatly determine the kind of taxa, the number of remains and the skeletal parts that can be found at it. All studied localities are considered as habitation sites by the archaeologists. The mound sites are probably not large refuse heaps, neither in Burkina Faso nor in Nigeria. Rather, they build up due to the construction and collapse of mud-brick house structures and because of rubbish disposal during habitation and after settlement abandonment (Jones, 1997; Gronenborn, 2000, 211-216). A second factor causing the building up of deposits in the southern Lake Chad area is the influence of phases of inundation of the sites, although their importance has recently been minimised (Brunk and Gronenborn, 2004). Intriguingly, in Burkina Faso house structures have been found both on top of a mound, at Oursi hu-beero (BF 97/30), and between mounds, at some of the Kissi sites (BF 96/22 and BF 97/31) (see 2.2.2.2.). Because of the latter examples S. Magnavita (pers. comm.), who excavated those sites, thinks the mounds do represent refuse heaps after all. Her viewpoint thus deviates from the more generally defended theory in West African archaeology. Anyhow, the evidence is there, although it is not clear to what extent the situation at Kissi can be generalised.

Occupation floors have usually not been recognised during the excavations at the settlement mounds, except at the Blé sites. Most material thus represents a mixture of longer periods of occupation, rendering the recognition of separate activities almost impossible. There are some mounds with indications for special activities, however (see 2.2.2. and 2.2.3.). High concentrations of fish bones, for example, noticed around a depth of 1.8 m during excavations at Kursakata (NA 93/46), have been interpreted as the remnants of a fish landing place. Such accumulations build up today at locations were fish is landed from Lake Chad and smoked before transportation to the surrounding markets (Gronenborn, 1998). At the studied sites, most bones found in pits seem to have been deliberately dumped there and, therefore, they probably represent a relatively short time span. The special nature of the site of Oursi hu-beero (BF 97/30), abandoned suddenly after a fire, offers the potential of recognising activity zones. Bone remains from the site were probably found at the place were they were thrown immediately after use, but there are indications that the floors of certain rooms of the house were cleaned regularly (Linseele, in press). The site can serve as a good example to explain the difference between primary and secondary disposal. "Primary disposal" involves the discard of rubbish at the locus of creation (waste on the floors at Ouri hu-beero), while "secondary disposal", implies the removal of rubbish from its locus of origin to dumping areas (cleaning of the floors) (Meadow, 1980). Gajiganna BII (NA 90/5BII) may have been such a dumping place outside the actual living area, or rather a butchery area as could be guessed from the find of the buffalo skeleton (see 5.1.4.). Zilum (NA 97/37), as a flat site showing early urban features, is a completely different type of site than the settlement mounds. Spatial differentiation of activities, including refuse disposal, was visible at the locality. During the study of the fauna from the site, it was observed, for example, that faunal remains of the same or related taxa were sometimes concentrated, with the find of five turtle individuals in one pit as the most extreme case (see 5.1.2.). It therefore seems that the refuse at Zilum is much less a mixture of several activities, than at the settlement mounds. This will be studied in more detail in the future when larger faunal samples from the site have been analysed.

Disposal of carcasses outside habitation areas may be the reason for the low numbers of bones attributable to this taphonomic group and it can also explain why certain domestic species, like domestic equids and dromedary (*Camelus thomasi* f. dromedarius), are rare or even completely missing from the assemblages analysed. When very large mammals are killed during hunting events, large bones are often left behind while only the meat, with smaller bones or bone fragments adhering, is removed and taken to the habitation site. This process has been named the "*Schlepp Effekt*" by Klein and Cruz-Uribe (1984, 64-65), although Baker and Gautier (1997) think its influence is much overrated. The "*Schlepp Effekt*" may account for the low number of recovered bones of very large wild mammals like elephant, giraffe (*Giraffa camelopardalis*) and hippopotamus (*Hippopotamus amphibius*) at the studied sites. However, taphonomic reasons alone are not sufficient to explain the near absence of such species, because they have been found at other African archaeological locations (e.g., Peters, 1995).

Sites may have been inhabited temporarily/seasonally or year round. This will have repercussions on the deposited fauna, since available resources are variable according to the time of year. Summarising data from the first chapter, the Late Stone Age in northern Burkina Faso was characterised by high mobility, but there was a shift to full sedentism once the Iron Age began. In the Bama Deltaic Complex, Iron Age and later sites are also thought to have belonged to fully sedentised communities. The situation at the time of the Gajiganna Culture in the area seems to have been more complex, with relatively high mobility during phase I and IIc, and permanent, or semi-permanent, habitation during the intermediate phases IIa and IIb, while the Gajiganna phase III site Zilum (NA 97/37) shows urban features. Nonetheless, even within one phase of the Gajiganna Culture, sites seem to show considerable variation. For Gajiganna phase I at Bukarkurari (NA 97/33) some kind of sedentism has been assumed, for example. At the *firgi* sites in Nigeria a change from seasonal to permanent habitation has been proposed for the transition between Late Stone Age and Iron Age. An inverse trend has been mentioned for the Blé sites, with first permanent and then seasonal settlement towards the end of their occupation. Among the studied sites, localities that were either inhabited permanently or repeatedly for a shorter period thus prevail. This is probably a consequence of their greater archaeological visibility and the nomadic component in the landscape may thus be largely missing.

Judging from the apparent uniformity of archaeological material by phase, none of the studied localities showed indications for use by different cultural groups during the same period. However, at the site Akumbu in the Malian Méma, where archaeozoological evidence pointed to the presence of two economically different human groups for the period AD 600-1000, this is not reflected in a difference in the material culture (MacDonald and Van Neer, 1994).

5.3. Differential preservation and tertiary deposition

5.3.1. Differential preservation

The preservation conditions that determine what is left of the originally deposited bone assemblage by the time of its excavation have already been described for the studied assemblages in general terms (see 3.3.). As the conditions can be rather variable depending on the site, they are discussed by locality and in more detail in 5.6. Equally, or even more, important is that bones of some taxa and some skeletal parts are more affected by destructive processes than others. This is difficult to evaluate but a few trends have been noticed during the faunal analysis. In the discussion on differential preservation, it should also be remembered that not only are postdepositional processes responsible for bone destruction and fragmentation, but that butchery and food preparation practices are equally important factors (see 6.5.4, 6.6.4 and 6.7.9.).

Most striking at the site Mege (NA 94/7) was the absence of molluscan shells and bird eggshells in the lower layers, while they were common in the higher levels of the stratigraphy. At Gajiganna BI (NA 90/5BI) a reverse trend has been observed, with more molluscan remains in the lower layers. A possible explanation is differential destruction of shells in places where soils are more acidic (Connah and McMillan, 1995), although this cannot be ascertained at the studied sites because the pH was not measured. Nevertheless, acidity should equally have affected bone remains, which was not noticed.

Large differences in preservation potential seem to exist for fish taxa (Wheeler and Jones, 1989, 62-63). Some, like clariid catfish (Clariidae) and Nile perch (*Lates niloticus*), have sturdy bones and preserve well in archaeological contexts, while bones of other species, such as *Hyperopisus bebe*, are much more fragile and are thus more easily destroyed. Not all bones of clariids seem to preserve equally well, however. Vertebrae are more easily destroyed than other elements (Van Neer, 2004) and it was observed that the coracoid also has a relatively low chance of preservation. Pieces of the cranial roof, on the other hand, preserve particularly well. Differential destruction may also explain the very low number of Alestidae remains recovered from the studied sites, seemingly in contrast with the family's present economic importance. However, teeth, the only skeletal elements from fish of this family that do preserve well, were also rare. As mentioned in 4.2.1., in lungfish (*Protopterus annectens*) only skull bones are well ossified and they were therefore the sole elements of this species found in the investigated archaeological contexts. Altogether, at least 19 fish taxa were present in the faunal assemblages from the studied sites, but many more can nowadays be fished today from the water bodies in their vicinity, at least from Lake Chad and its tributaries, and thus some fish taxa probably do not preserve at all.

Preservation conditions often caused heavy fragmentation of mammalian bones. Consequently, neither length measurements on long bones nor measurements on cranial parts could frequently be taken. Another consequence was the absence of enough mandibular tooth rows to use eruption or wear stages for age determinations (see 3.3.). Parts of mammals that were apparently less liable to destruction are small compact bones, like petrous bones, phalanges and metapodials. Bones of larger mammal species, like domestic cattle, must have preserved better than those of less heavily built animals, such as domestic ovicaprines. In extension, differential preservation could also have distorted proportions between animal classes towards those with the largest taxa, for example mammals compared to fish. Differential destruction may also have affected bones of young animals, which are more fragile because they are not well ossified. The observed foetal bones possibly stayed preserved because they were buried intentionally.

It is believe that, in general, skeletal distributions of the animals recorded at the studied sites mainly reflect differential preservation. Bones that preserve well are overrepresented, phalanges, carpals and tarsals of bovids for example (e.g., Table C.96), and the cranial parts of clariid catfish (Table C.11a). For the former animals this is clearly reflected in the elements that could be measured (e.g., Table C.99). In addition, parts that are recognisable from small fragments, like teeth, have been identified in high numbers.

5.3.2. Tertiary deposition

Besides primary and secondary deposition, explained in 5.2., the term tertiary deposition is used to refer to situations where, after initial deposition, sediment is moved and redeposited elsewhere (Meadow, 1980). None of the studied sites seems to have shown evidence of serious displacement, but disturbance has been recorded, for example through finds of disarticulated human remains that must have been part of buried individuals or through mixture of material of layers from different phases. Factors causing displacement are burrowing animals, for example, and also, more seriously, erosion. From excavations at Saouga, where a pit was dug at the top (BF 94/120) and at the foot of a mound (BF 95/7), it appeared that the youngest occupation layers of a mound might have eroded away. Erosion of the top layers probably has to be assumed for all settlement mounds in Burkina Faso.

5.4. Area chosen for excavation

The area that is chosen for excavation can also have great influence on the fauna that is recovered, because waste disposal usually does not happen randomly across a site. At Jenné-Jeno, in Mali (250 BC-AD 1400), for example, several contemporary, but spatially separated, excavation units yielded faunal samples with differing composition (MacDonald, 1995). During excavations around Aissa Hardè in Cameroon, differentiation of activities across the site's surface was also recorded (Bourges *et al.*, 1999). For the studied sites, where usually one relatively small trench was dug, differentiation across the site is not possible to trace. As said earlier, only for Oursi hu-beero (BF 97/30) (Linseele, in press) and Zilum (NA 97/37) can spatial analysis be done. For the latter site this will be carried out in the future, when more fauna has been studied. For the studied settlement mounds, differences in faunal composition can be expected between excavations on top, where the habitation area probably was (but see 5.2.), and excavations between, or at the foot, of mounds. Shifts of activity zones in the course of a site's occupation, like the one recorded for the fish landing place in Early Iron layers at Kursakata (NA 93/46), should also probably have detectable repercussions on the fauna.

5.5. Sampling

The importance for archaeozoological analysis of good sampling techniques during excavation has been stressed in 3.1. As explained, sieving at the studied sites was efficient and satisfactory, but not fine enough to retrieve all bones from the sediment and, therefore, some taxa or type of elements will be underrepresented or even missing. Bones retrieved from archaeobotanical sampling allow an evaluation of the impact of a lack of fine sieving, when compared with the "ordinary" faunal sample from the same site. Oursi (BF 94/45) appeared to be the most useful site for such an evaluation (see Table D.5). Some animals found in the finely sieved samples (finest mesh 0.5 mm) are completely lacking in the others (10 mm), notably two fish genera, *Polypterus* sp. and *Hydrocynus* sp., and also striped ground squirrel. The abundance of other taxa is underestimated when sieving with small meshes has not been carried out. This seems to be the case for tilapia, agama, amphibians, snakes, African hedgehog and small rodents. It also appears that the degree of this underestimation can be very serious. As expected, the taxa missing or underrepresented when mesh sizes are not fine enough are all relatively small animals. For other taxa, 10 mm meshes are apparently sufficient to retrieve the large majority of their remains. Identifiable elements of these found in small fraction residues are typically recognisable from small fragments, for example clariid cranial roof parts or bovid teeth. For the fish remains, sampling techniques also have repercussions on the distribution of reconstructed sizes. Small individuals appear to be underrepresented when fine sieving is not carried out (e.g., Table C.16b).

Although sampling bias may have resulted in the underrepresentation of certain taxa, it should be emphasised, that inter-site and intra-site comparisons are meaningful because the strategies were identical in each region. Because sampling was finer at Nigerian sites (5 mm meshes) than at those in Burkina Faso (10 mm meshes) or in the Blé Mound Complex (hand), higher proportions of smaller animals might be expected at the former. In reality, this does not seem to be the case, however. On the contrary, sites in Burkina Faso often yielded smaller remains, e.g., bones of amphibians or small mammals and bird eggshells. This might be related to the clay sediment in Nigeria, hampering find recovery compared to the more sandy sediments at the sites in Burkina Faso, especially since water could not be used during sieving. During the excavations at Kissi (BF 96/22 and BF 97/31) sieving was finer than at the other sites in Burkina Faso. This seems to have had as a most striking consequence, the retrieval of a very large amount of amphibian bones. At Oursi hu-beero (BF 97/30), on the other hand, the only site in Burkina Faso where sieves were not used, amphibians were almost absent. The only frog or toad bone that was retrieved there came from a test-sieved sample. However, as proposed earlier, sampling may not have been the only reason for a paucity of amphibian bones at the site (see 5.1.1.). Finally, the absence of small rodent bones at the Blé sites may also be indicative of the lack of sieving.

5.6. Faunal assemblages by site. Productivity, identification rates and preservation

With the different taphonomic factors in mind, a discussion of faunal assemblages by site, trench and phase follows. To this end, some figures concerning (the analysed squares of) each of the localities have been put in Table 6: total bone weight, number of bones, excavated volume, bone weight by excavated volume, number of bones by excavated volume, and finally the identification rate of both fish and mammal remains. Bones recovered from archaeobotanical sampling have not been taken into account in this table. In Fig. 34 bone weights and numbers of bones by volume have been plotted.

Calculating number of bones by excavated volume has been carried out previously by Gautier (1983), who used it as a measure to express "archaeozoological productivity". One of the main ideas behind it is that numbers will be variable according to occupation intensity, e.g., temporary versus permanent. However, it can also be influenced by other factors, like preservation and fragmentation of the bones, sedimentation rate or sampling techniques. In addition, bone weight by excavated volume was calculated, because weight will only be influenced slightly by the sampling techniques, extra bones collected with fine techniques being mostly small and thus light (see 5.5.). Indications from archaeozoological productivity on occupation intensity will be compared to archaeofaunal data concerning the season of habitation further on.

One of the reasons for looking closer at identification rates by locality is that they are related to the state of preservation of the material (see 3.3.). However, the relationship is not always straightforward. It will become clearer in the discussion below, but poor preservation can lead to high identification rates, when it is only the robust bones of larger animals that have not been destroyed. Good preservation, on the other hand, can result in low identification rates, because the assemblage will contain more types of bones and of more diverse taxa, which may hamper identification. Only identification rates for fish and mammals have been selected, because these classes are usually best represented in the faunal samples. As already indicated in the previous chapter, identification rates for fish are usually much higher than those for mammals since they are easier to identify from small remains. The figures for mammals therefore proved to be more useful parameters for preservation. In addition, within animal classes, the ease of identification depends much on the taxon.

5.6.1. Burkina Faso

Faunal remains from the Late Stone Age sites of Tin Akof (BF 94/133) and Dori (BF 94/40 and BF 94/96) are scarce compared to their excavated volume. This is clearer from the bone weights than from the numbers of bones because of the highly fragmented nature of the remains. The only available bone from Dori 94/40 (BF 94/40) is moreover from an intrusive animal. Small sample size is probably due to high mobility of the human groups and to poor preservation. Because the samples are so small, identification rates are not meaningful. Larger Late Stone Age samples are available from the site of Corcoba (BF 97/5), thanks to the special preservation conditions in the pit found at the site. From Corcoba, Early Iron Age material has been studied as well. At the site identification rates of both the fish and mammal remains are lower for the Late Stone Age than for the Early Iron Age contexts. This is undoubtedly related to poorer preservation of the older material, which was already observed during identification of the fauna. In contrast to Corcoba, a great difference in the amount of

			Total bone weight (g)	Number of bones	Excavated volume (m^3)	Bone weight/volume (g/m^3)	Number of bones/volume	% fish identified	% mammals identified
1	BF 94/133	LSA	350	803	6,8	51	118	-	1,2
2	BF 94/40	LSA	1	1	2,4	0	0	-	100,0
3	BF 94/96	LSA	14	8	2,4	6	3	-	62,5
4	BF 97/5-II	LSA	-	136	1,4	-	97	69,6	3,5
5	BF 97/5-III	LSA	-	115	1,4	-	82	80,0	1,0
6	BF 97/5-IV	LSA	-	90	1,4	-	64	66,7	3,5
7	BF 97/5-V	LSA	-	2188	1,4	-	1563	68,9	3,5
8	BF 97/5-VI	LSA	-	514	1,4	-	367	45,6	3,5
9	BF 97/5-VIII	EIA	-	973	0,6	-	1622	95,3	15,0
10	BF 94/45	LSA	202	363	17,9	11	20	83,3	4,0
11	BF 94/45	EIA	10761	10435	38,1	282	274	93,4	6,0
12	BF 97/13	EIA	563	609	7,8	72	78	96,3	14,7
13	BF 97/13	MIA	11682	4260	13,4	872	318	91,1	7,0
14	BF 97/13	LIA	27797	8389	19,8	1404	424	96,2	4,5
15	BF 97/30	LIA	-	7573	-	-	-	98,7	15,5
16	BF 96/22	MIA	2126	3566	32,0	66	111	100,0	12,6
17	BF 97/31	LIA	2751	4524	20,0	138	226	98,4	9,3
18	BF 94/120	LIA	17407	27739	37,2	468	746	96,4	5,6
19	BF 95/7	LIA	5659	7849	18,0	314	436	98,6	11,3
20	NA 90/5C	Gaj IIa	-	190	22,5	-	8	-	11,8
21	NA 90/5A	Gaj IIb	-	12278	50,5	-	243	100,0	19,7
22	NA 90/5BI	Gaj IIa/b	-	5512	112,5	-	49	99,8	10,5
23	NA 90/5BII	Gaj IIc	12605	2007	32,4	389	62	100,0	19,7
24	NA 91/1A	Gaj I	221	7	3,9	57	2	-	85,7
25	NA 93/42	Gaj I	972	718	6,1	159	118	100,0	21,7
26	NA 97/18	Gaj I	41	53	6,0	7	9	100,0	4,8
27	NA 97/24	Gaj I	371	16	8,8	42	2	-	93,3
28	NA 93/36	Gaj IIa	418	315	16,9	25	19	85,7	4,0
29	NA 99/65	Gaj IIa	381	26	2,4	159	11	-	4,0
30	NA 99/65	Gaj IIb	2679	1425	3,6	744	396	96,7	5,0
31	NA 99/65	Gaj IIc	2312	1417	4,2	550	337	100,0	5,3
32	NA 93/10	Gaj IIc	925	252	1,9	478	130	-	11,6
33	NA 97/37-1	Gaj III	4	1	1,8	2	1	-	-
34	NA 97/37-3	Gaj III	73	7	2,0	37	4	100,0	50,0
35	NA 97/37-4	Gaj III	156	73	4,9	32	15	100,0	6,3
36	NA 97/37-5	Gaj III	2435	567	7,2	338	79	90,0	11,6
37	NA 97/37-6	Gaj III	3060	1084	23,8	129	46	90,8	16,5
38	NA 97/33	Gaj I	10601	2545	9,8	1082	260	100,0	25,0
39	NA 97/33	Gaj IIa/b	5222	1265	14,0	373	90	100,0	21,9
40	NA 96/45	Gaj IIa/b	5392	1034	3,2	1685	323	100,0	34,5
41	NA 95/1	Gaj IIa/b	8738	5134	63,0	139	81	97,2	12,2
42	NA 97/26	IA	949	971	4,6	206	211	100,0	13,3
43	NA 99/75	IA	1064	649	6,6	161	98	83,0	28,0
44	NA 97/13	IA	1035	947	9,2	113	103	-	12,5
45	NA 92/2C	subrec	9927	3939	5,2	1909	758	91,1	26,7
46	NA 93/46	LSA	1509	1683	5,1	296	330	68,9	4,8
47	NA 93/46	EIA	19535	37859	7,8	2504	4854	69,8	3,6
48	NA 94/7	I	19091	10409	3,3	5785	3154	79,1	9,7
49	NA 94/7	II	2713	1824	1,1	2466	1658	68,0	7,6
50	NA 94/7	III	1939	1496	1,1	1763	1360	58,9	10,3
51	NA 94/7	IV	811	1131	1,3	624	870	70,0	6,4
52	NA 93/45	I	5244	2532	15,3	343	165	74,3	56,9
53	NA 93/45	II	7463	5212	15,3	488	341	79,2	27,9
54	NA 93/45	IIIa	3620	2721	8,1	447	336	78,3	9,7
55	NA 93/45	IIIb	4184	2005	3,6	1162	557	89,0	6,6
56	NA 93/45	IV	3073	2281	5,4	569	422	87,2	10,9
57	Blé A		-	420	?	-	?	83,5	18,6
58	Blé B		-	1075	?	-	?	87,0	10,5
59	Blé C		-	883	?	-	?	76,4	9,2
60	Blé E		-	1534	?	-	?	78,3	13,4

Table 6: Archaeozoological productivity and identification rates by site, trench and phase. Legend to Fig. 34

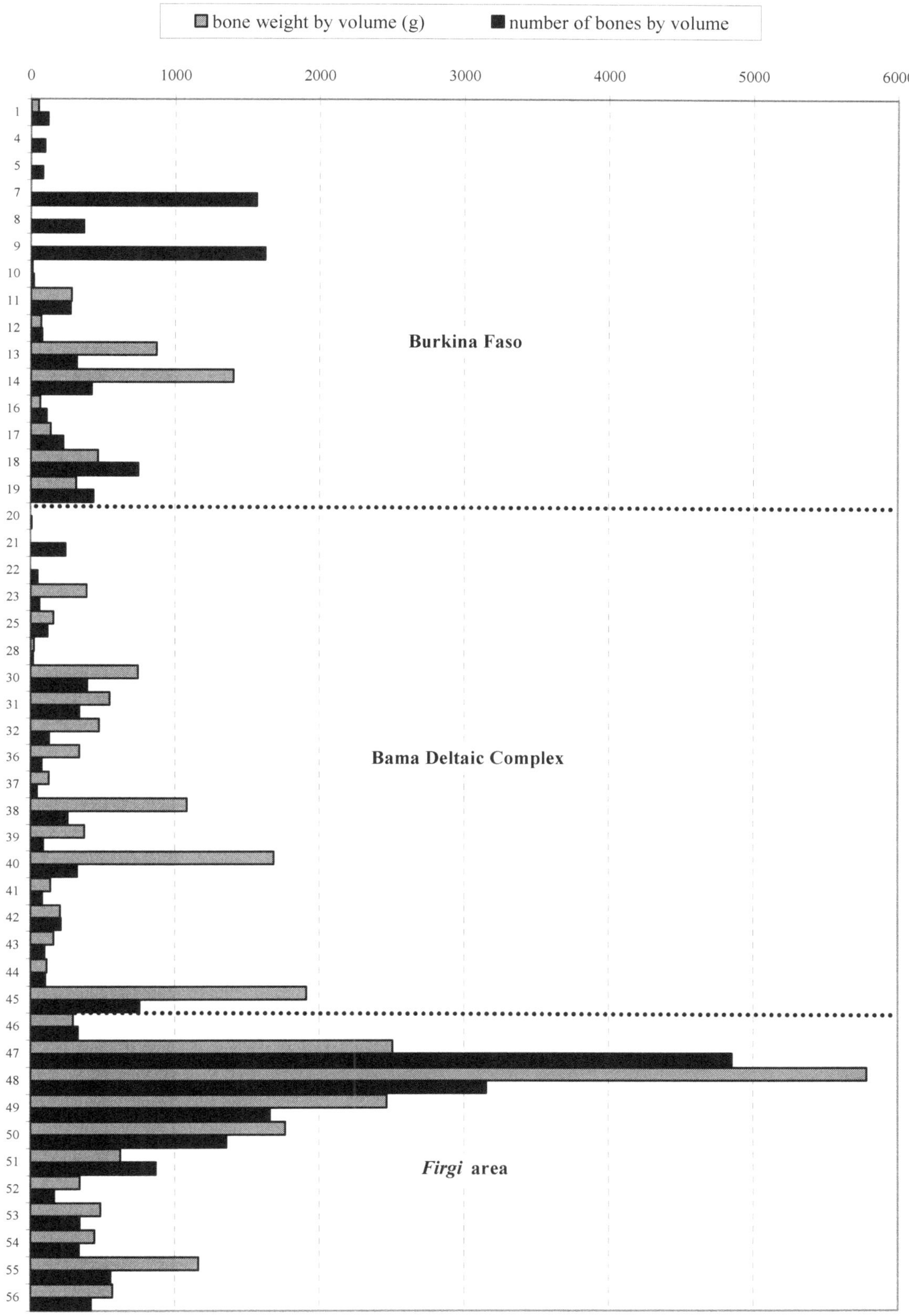

Fig. 34: Archaeozoological productivity by site, trench and phase (Only samples with more than 100 specimens)

material between Late Stone Age and Early Iron Age layers is visible at Oursi (BF 94/45). Productivity in the latter is clearly higher. Again, identification rates are lower for the older contexts. At Oursi village (BF 97/13), bone productivity rises when occupation phases are younger. Better preservation is a plausible explanation, but it is not clear to what extent more intense occupation could have played a role as well. Identification rates of mammalian bones at the site are higher the older the occupation phases, which is additionally emphasised by a relative large number of (well-preserved) bones of intrusives in its lower layers. Of all Iron Age sites in Burkina Faso, the highest identification rate for mammalian bones, a little over 15 %, was obtained for Oursi hu-beero (BF 97/30), presumably because the site was abandoned and covered suddenly. Nevertheless, in terms of absolute numbers, the percentage is still low, probably because the material was trampled while lying on the habitation surface and because the site's inhabitants may have regularly removed the largest bones from the floor (Linseele, in press). Of the two Kissi sites studied, the youngest, Kissi 40 (BF 97/31), had the highest bone productivity, again presumably related to better preservation and perhaps also to higher occupation intensity. However, compared to the other Late Iron Age sites in Burkina Faso, Kissi 40 has a relatively low bone yield. This may be due to the location of the excavation trench between two mounds instead of on top of a mound as at the other sites. Despite its older age, Kissi 22 (BF 96/22) has higher identification rates than Kissi 40. Looking at numbers of bones by volume, Saouga 94/120 (BF 94/120) has the highest bone productivity of all studied sites in Burkina Faso, indicating high occupation intensity, which was already concluded from the very rapid build up of sediments (see 2.2.2.2.). However, bone remains were of relatively small size, resulting in a much lower productivity when bone weight is considered. Compared to Saouga 94/120, material was less abundant in the trench at the foot of the same mound (BF 95/7). Identification rates were higher for the latter than for the former, but it is as yet not clear how this should be explained.

5.6.2. Bama Deltaic Complex

A characteristic of many of the Gajiganna sites is the presence of several occupation layers in their stratigraphy, sometimes interrupted by more-or-less sterile bands (see 2.2.3.2.). The presence of the latter resulted in low figures for bone productivity. It was therefore first considered to take only occupation layers into account for the calculations. However, this would be inconsistent with what has been done for the sites from other parts of the study area, where all layers, including the ones with no or very little cultural remains, were taken into account. Bone productivity for Gajiganna phase I contexts is, on average, low. This is probably related mainly to the high mobility postulated for the human groups of that period (see 2.2.3.). NA 93/42 and especially the site Bukarkurari (NA 97/33), with an exceptionally high concentration of bones between -3.1 and -3.3 m, are exceptions, pointing to the existence of more sedentary periods, or groups, as well (see 2.2.3.2.). Nevertheless, exclusively looking at bone productivity can lead to an underestimation of mobility, since sites without fauna, or with faunal samples that were considered too small to be sent to the RMCA for analysis, are not included. Sites of Gajiganna phases IIa/b, are thought to have been inhabited more permanently (see 2.2.3.), but this is not always reflected in higher bone productivities than during the previous phases. Nor is renewed mobility during Gajiganna phase IIc confirmed by the bone productivity. Although there is considerable variation according to the excavated trench, Gajiganna phase III site Zilum (NA 97/37) has low archaeozoological productivity. In this case high mobility cannot be invoked, since it concerns a site characterised by proto-urban features. Instead, a decreased importance of animal food compared to agricultural products has been proposed as an explanation (Breunig *et al.*, 2006). However, different types of waste disposal and deposition compared to the mound sites may also have resulted in lower productivity. Nevertheless, there were other indications for an increased importance of cultivated crops at the site, in the form of numerous granaries and the find of many large storage vessels (see 2.2.3.2.).

Identification rates for mammalian remains from the Gajiganna sites are rather variable. The high rates obtained for NA 91/1A, Labe (NA 97/24) and trench 3 at Zilum (NA 97/37) do not have much significance in view of their rather small sample sizes. It seems that more remains could be identified when sites had proportionally higher amounts of cattle bones, with Kelumeri (NA 96/45) as the most extreme example. Large numbers of cattle bones can either be due to the fact that many were deposited, or to the differential destruction of bones of smaller taxa. It is suspected that the former may have been the case, since destruction of remains of small taxa should also have resulted in lower bone productivity. Gilgila (NA 99/65), the only Gajiganna site with relatively few bones of cattle in comparison to other taxa, had the lowest identification rates. At Zilum, the high identification rate in trench 6 is probably due to better preservation in the pits at the location, where most of its bones come from.

The Iron Age sites in the Bama Deltaic Complex yielded more faunal material by excavated volume than the very youngest investigated Gajiganna site in the area, Zilum (NA 97/37), at least expressed in numbers of bones. However, when looking at bone weight, their archaeozoological productivity is much lower since small taxa abound, including among others, molluscs, intrusives like small frogs or toads and small rodents, and domestic ovicaprines. The identification rate for mammalian remains from Elkido North (NA 99/75) is nearly double that for both of the other Iron Age sites, Labe Kanuri (NA 97/26) and Dorota (NA 97/13). Preservation conditions at the site seem to have been better, although it is not clear why. The subrecent site of Galaga (NA 92/2C) produced the most animal remains of all sites in the Bama Deltaic Complex. Relatively more remains must have stayed well-preserved because of its recent date. Because of good preservation at the site, identification rates for mammals are also relatively high, although the difference from older sites is not very pronounced.

5.6.3. *Firgi* area

Differences in productivity between the Gajiganna sites are only minimal when compared to the strikingly larger productivity of the *firgi* sites. This large productivity is tentatively attributed to the limited habitation surface available in the area, because large parts of it are submerged during several months of the year (see 2.1.2.). At Kursakata (NA 93/46), a very clear shift towards higher bone productivity is visible between the Late Stone Age and Early Iron Age layers, which may be related to the proposed shift to full sedentism around that time. Expressed in numbers of bones by volume, Early Iron Age levels at the site yielded the most fauna of all studied localities, mainly due to the extremely high concentrations of fish remains. Mege (NA 94/7) also has a lot of faunal remains, but from a comparison between bone weights and numbers of bones by volume, it appears that their average size is smaller for the younger layers. This is probably related to a shift in the exploited species spectrum, from a lot of cattle towards more ovicaprines (see 6.7.5.3.). The third *firgi* site, Ngala (NA 93/45), still has high bone productivity, but clearly not as pronounced as at both of the other sites. Trenches at Mege and Kursakata were excavated near the mound top, as opposed to the trench at Ngala, which was dug in the courtyard of the palace at the locality, which may explain the observed differences.

Identification rates for the fish remains from the *firgi* sites are lower than those from the sites in Burkina Faso and also the Bama Deltaic Complex. This is due to the presence of more taxa at the sites, making identifications more difficult. Furthermore, some of the taxa do not preserve as well as clariids, which dominate at sites in Burkina Faso and in the Bama Deltaic Complex, rendering their bones harder to recognise. Identification rates of the mammal bones from the *firgi* sites, varied between about 4 and 60 %. High percentages in phase I and phase II layers at Ngala (NA 93/45) are due to concentrations of intrusive small rodent bones. However, even with a large amount of intrusive rodents, Early Iron Age Kursakata (NA 93/46) has a very low identification rate, as have the older Late Stone Age layers at the site. This seems to be mainly explainable by poor preservation. Identification rates in contemporary layers at Mege are higher, probably because, as mentioned above, bones of cattle abound. As said earlier, high numbers of bones or other large taxa can be due to a predominant exploitation of the animals or to differential preservation. The first possibility is again favoured because bone productivity is not lower than in contexts with fewer cattle, as would be the case when bones of small taxa had been destroyed differentially. Identification percentages of mammalian remains from Ngala are higher than those from other *firgi* sites, but diminish when contexts are older. This is tentatively attributed to a higher proportion of cattle bones and better preservation in the younger layers.

5.6.4. Blé sites

For the Blé sites in Cameroon, information on the volume of the deposits the animal bones stem from is missing and archaeozoological productivity could therefore not be calculated. As for the *firgi* sites identification rates of the fish remains are rather low, due to high species diversity. The lack of fine sieving at the Blé Mound Complex, on the other hand, may have resulted in relatively higher identification rates for mammalian bones, since small (unidentifiable) fragments are underrepresented.

Chapter 6. Palaeo-ecological and palaeo-economical interpretation

In Chapter 4 the faunal remains have been described in taxonomical order, but to allow for a palaeo-ecological and palaeo-economical interpretation they will now be considered in an alternative way. If the intrusive animals are left out, four types of economic activities can be investigated when the animal remains are lumped in the following groups: molluscs (gathering), fish and freshwater turtles (fishing), other reptiles, wild birds and wild mammals (hunting and fowling), and, finally, domestic birds and mammals (stock keeping). For the reconstruction of the former environment the aforementioned animals can be used, but in addition the penecontemporaneous intrusives also yield information on this topic. Information on the biology and ecology of the taxa used for the interpretation has already been described in Chapter 4 and references can mainly be found there. Palaeo-ecological reconstructions from archaeofaunal remains can be hampered by difficulties with discriminating between animals that were caught locally and others that were obtained further away from the settlements. However, through inter- and intra-site comparisons it should be possible to detect the taxa that were brought in from more distant areas.

Before discussing the different economic activities in obtaining animal proteins, including aspects such as dietary contribution, taxa exploited, techniques and gear used, seasonality and cultural aspects, the available arguments on the importance of plant versus animal food in the diet of the sites' inhabitants will first be summarised. After that follows a brief methodological discussion on the reconstruction of former diets and the use of measures of heterogeneity.

6.1. Animal versus plant food

In present sub-Saharan West Africa most diets are dominated by a single crop, e.g., maize (*Zea mays*), sorghum (*Sorghum bicolor*), pearl millet (*Pennisetum americanum*), rice (*Oryza glaberrima*) or yam (*Dioscorea rotunda*), which varies according to the latitude (MacLean and Insoll, 1999; Newman, 2000). Meat only plays a minor role in the total amount of food consumed. Meat consumption in Cameroon in 1981 was, for example, estimated at only nine grams per day, per person (Ntiamoa-Baidu, 1997). Even among pastoralists and specialised fishing groups, crops are a major food resource (e.g., Sundström, 1971, 32-33; Froment, 1988, 43-51). Hunter-gatherers also usually have plants as the key item on the menu, because of their abundance and high nutritional qualities (Guenther, 1997). Nevertheless, hunter-gatherer communities attach more value to meat. For medieval West Africa, Lewicki (1974, 79) has also stressed the minor part played by meat in comparison to vegetable food. However, discussion of animal products should not be limited to meat since the dietary importance of dairy products can be substantial, especially among herders (see 6.7.8.2.). Even if animal products are consumed in smaller quantities than plant food, they contain some essential nutrional elements, e.g., protein and calcium, which are missing in most crops (Draper, 2000). Nonetheless, some plant products can be used to replace animal food. Fruits of the sheabutter tree (*Vitellaria paradoxa*), for example, are a good alternative for animal fats (Pélissier, 1980).

Isotope analysis of human remains can allow the tracing of the relative importance of meat and plant foods archaeologically. Such an analysis was carried out on human bones from Gajiganna phase IIa/b and pointed to a diet mainly consisting of plants (Breunig, 1995). This is in contradiction with an earlier analysis of the teeth of a buried male from Gajiganna A (NA 90/5A). Molar attrition, which was flat rather than oblique, suggested that meat and fish formed the major part of the diet (Breunig *et al.*, 1993a). The only, more unequivocal, indication from the research area of the importance of plant versus animal food products comes from the site of Zilum (NA 97/37). In 5.6, relatively low archaeozoological productivity at the site and finds of granaries and storage vessels was cited as a possible indication for a heavier reliance on plant food during Gajiganna phase III compared to preceding periods (Breunig *et al.*, 2006).

6.2. Introductory remarks on diet reconstruction and measures of heterogeneity

For the palaeo-economical reconstructions, graphs were made by site and phase to estimate the relative dietary contribution of fish and freshwater turtles, other reptiles, wild birds and wild mammals in addition to domestic birds and mammals. In Fig. 35, total Numbers of Identified Specimens (NISP), and in a few special cases Minimal Numbers of Individuals (MNI) (see 3.6.), of the animals considered as food refuse were plotted by group (fishing, hunting and fowling, domestic stock keeping). Molluscs have not been included because, compared to vertebrates, they produce a lot of waste relative to their food value. Taxa with an uncertain status as food waste were not included either. Fauna recovered from archaeobotanical sampling was also omitted, because of the problems with its quantification and because it was not available for all locations. Remains identified as large galliform, African wild cat (*Felis silvestris*) or domestic cat (*F. s.* f. catus), medium-sized carnivore, small bovid, large bovid, and cattle (*Bos primigenius* f. taurus) or buffalo (*Syncerus caffer*), could pertain to either domestic or wild animals. Their numbers were added to the more precisely identified taxa in the same proportions as they occurred in the accurately identified taxa of these groups. For example, when a site yielded 100 bones identified as small bovid, 14 as small or medium-sized antelope and 186 as domestic ovicaprine (*Ovis ammon* f. aries and *Capra aegagrus* f. hircus), then 7 bones of the small bovids were put with the wild mammals and 93 with the domestic mammals. In cases where there were no bones within a certain group, which could be identified either as wild or as domestic, the bones of "wild or domestic" animals were not used to make the graph, e.g., small bovids at the Late Stone Age site of Tin Akof (BF 94/133) and all African wild or domestic cat remains.

It is believed that the graph mainly reflects the frequency with which each group of animals was eaten, but they can only be approximations, because of differences in preservation, recovery and identification chances, depending on the taxon and possible biases due to the quantification method used. However, since the same method was applied on all assemblages and mostly it is the inter- and intra-sites trends that are meaningful, this is a useful approach. The described calculations do not take into account different meat yields of the various taxa, which is necessary to estimate the magnitude of their contribution to the human diet. Some researchers have been working with weights of identified bones by taxon, instead of their numbers, because they assumed a correlation between bone weight and meat yield (see overview in Reitz and Wing, 1999, 225-228). However, the relationship between these two parameters is not straightforward. Alternatively, for the reconstruction of dietary compositions NISP data of the consumed taxa can be multiplied with their average live weight. This was done, for example, for African contexts by Van Neer (2002a) and is also applied here. Following Van Neer, an average weight of 1 kg is accepted for fish, as against 50 kg for hunted mammals and reptiles, 300 kg for domestic cattle and 25 kg for domestic ovicaprines. In addition, an average weight of 10 kg was used for domestic dogs (*Canis lupus* f. familiaris) and of 2 kg for birds. Vigne (1991) also estimated the dietary importance of different taxa of animals based on their live weight. He used a refined version of the method, however, also taking into account the different meat yields according to the animals' slaughtering ages. Because of insufficient data on ages, this could not be done here.

Which of the two factors, frequency of consumption or meat yield, is the most important is debatable. There seems to be a tendency to under-appreciate frequency of consumption, although both should probably be considered equally valuable. Sources yielding small amounts of food relatively constantly during a longer period of time are at least as important as sources that provide a large amount of food, but which must be consumed rapidly to prevent spoilage. However, graphs estimating meat weight are also necessary because otherwise too much emphasis is placed on small animals, such as fish, and trends within other groups of animals may be obscured. It is also important to keep in mind that frequency of consumption and meat yield by animal group do not necessarily equal the importance of the corresponding economic activities, since meat can also be obtained through trade or exchange. In addition, meat brought in without bone cannot be traced archaeologically. Moreover, the cultural importance of food sources may be unrelated to their actual economic importance.

The graphs (Figs 35 and 36) made are compared to similar ones by Van Neer (2002a) for several western and central African archaeological sites, but not including fowling. Jousse (2006) has also investigated the relative importance of fishing, hunting and herding in West African archaeological contexts. Because of problems with the quantification of data compiled from the literature, she used numbers of exploited taxa (species richness) by strategy as a parameter. For hunting, she took only wild ungulates into account, because there are insufficient quality data on other taxa. A problem with her method may be that, numbers of taxa exploited depended more on what was available in the habitats near the sites than on human strategies. Numbers of taxa in an archaeofaunal assemblage also depend on sample size, since with increasing sample size the possibility of adding taxa increases (collector's curve). Some researchers therefore make species richness calculations taking sample sizes into account (e.g., Cruz-Uribe, 1988). Numbers of species are also influenced by the sampling techniques used and the skills of the faunal specialists involved in the study of the animal remains. Since both are consistent for the studied sites – at least by region in the case of sampling – species richness (Dl) was calculated for the fished taxa and the taxa obtained through hunting and fowling, using the following formula: $Dl=(S-1)/\log_n N$, whereby S is the (minimal) number of taxa and N the total numbers of remains (Cruz-Uribe, 1988). Bone remains from archaeobotanical sampling were not included in the calculations.

6.3. Intrusives

As explained in 5.1.1., intrusives can give good indications on the former environment around the sites, provided that they are penecontemporaneous. The presence of a *Lymnaea natalensis* shell at Labe Kanuri (NA 97/26) indicates that the waterbodies around the site may have been stagnant or slow flowing waters, the preferred habitats of the species, rather than seasonal pools. This fits well with the location of Labe Kanuri in the "Jere Bowl", the inland delta of the river Ngadda. The other intrusive freshwater molluscs did not seem to have very specific ecological requirements. *Limicolaria*, a terrestrial snail, aestivates in places with loose soils and prefers damp and shady locations. These habitat requirements seem to fit with most archaeological sites. Bullfrog (*Pyxicephalus edulis*) could be identified mainly from Iron Age sites in Burkina Faso and in Nigerian contexts younger than Gajiganna phases I-II. Its frequency in the studied contexts is influenced by the sampling techniques used during excavation and by differential destruction (see 5.3. and 5.5.), but must also be dependent on former environmental conditions, since bullfrog is mainly confined to arid savannahs. The bullfrog remains thus seem to point to more arid conditions during later phases in both Burkina Faso and Nigeria. Individuals that are perhaps not intrusive, but caught for food (see 5.1.1.), were probably captured in marshy areas during the rainy season. The Agama (*Agama* sp.) bones do not yield any ecological information as lizards of this genus have a wide habitat range. The snake remains, if contemporaneous intrusives, cannot be taken as palaeo-ecological indicators either, because a more precise identification was not possible and, therefore, their possible habitats also remain undetermined. The same also holds true for the songbird bones. White-toothed shrew (*Crocidura* sp.) and many of the small rodent species that have been found occur in a wide range of habitats. Taxa with more specific habitat requirements were all from arid environments: gerbil (*Gerbillus* sp.), attested in Early Iron Age layers at Oursi (BF 94/45), dwarf gerbil (*Desmodilliscus braueri*) in Saouga 94/120 (BF 94/120) and unstriped grass rat (*Arvicanthis niloticus*), recorded at several locations, both in Burkina Faso and Nigeria. The latter rat is water dependent, however. Striped ground squirrel (*Euxerus erythropus*) is a species that is today largely associated with cultivated land, but there seems to be no clear relation at the studied localities between the occurrence of the animal and a high reliance on agricultural products, possibly because the animal should be sought in the actual fields rather than inside the settlement. To conclude, ecological information from intrusives is scanty, but consistently points to arid environments.

6.4. Molluscs

As said, molluscs were not added on the graphs because, compared to vertebrates, they produce a lot of waste relative to their food value. Large shell heaps would have formed had they been important in the human diet. Their dietary contribution at all studied localities must therefore have been small. Sites taking the form of large bone and shell middens are, for example, characteristic for the Hassi-el-Abiod facies in the Malian Sahara (sixth-third millennium BC) (MacDonald, 1997b). A shell heap, mainly composed of Nile oyster (*Etheria elliptica*) and a few *Spathopsis* shells, is also described from the site Tiabel Goudiodé (Fulfulde for shell mound) in the Malian Méma, dating to the mid-third millennium BC (MacDonald, 1994, 89; MacDonald and Van Neer, 1994). Breunig *et al.* (in prep.) also list two shell-middens in their archaeological map of Northeast Nigeria, which, from the associated archaeological material, both seem to date to the Iron Age. The low numbers of molluscan shells at the studied sites may confirm the hypothesis of Connah and McMillan (1995), already cited in 5.1.2., that molluscs were not considered as a very palatable source of food and were mainly consumed during periods of stress. However, there a few factors that may have led to an underestimation of the use of molluscs for food, such as a lower preservation potential than for remains of other animals. There are also recent West African examples of trade in prepared meat of molluscs, which means that even when few molluscan shells were found at a site, their meat may still have been consumed regularly (Lawson, 1963).

Of all the localities studied, *Pila* shells were most common at the sites of the Bama Deltaic Complex, which can probably be explained by the presence of many seasonal pools in the area. The *Lanistes varicus* shell from Saouga 94/120 (BF 94/120) points to the presence of pools with abundant aquatic vegetation near the site. While for most of the freshwater molluscs collected from the studied sites, small ponds, lakes or inundation plains are sufficient to explain their occurrence, *Mutela dubia* can only be found in large waterbodies and Nile oyster needs permanent water. This suggests that the rivers near Saouga (BF 94/120 and BF 95/7) and the Blé mounds were still carrying water year-round at the time of the sites' occupation. Collecting freshwater shells was probably mostly a dry season activity, because they are easier to obtain at low water levels. Nevertheless, Connah and McMillan (1995) write that freshwater molluscs can be collected year-round in present north-eastern Nigeria. Looking at modern parallels, gathering of molluscs may have been an activity mainly carried out by women and children, much like the gathering of wild plants (Deacon, 1984).

6.5. Fish and freshwater turtles

The reason for treating fish and freshwater turtles together is that they must have been caught in a similar environment and using similar techniques. This is also clear from the general observation that high numbers of fish and high numbers of freshwater turtles usually coincide. Ecologically, both groups are markers for the nature of aquatic environments.

			fishing	hunting and fowling	domestic stock keeping
1	BF 94/133	LSA	1	1	0
2	BF 94/96	LSA	0	0	0
3	BF 97/5	LSA	1410	27	0
4	BF 97/5	EIA	892	0	0
5	BF 94/45	LSA	30	10	0
6	BF 94/45	EIA	1474	196	310
7	BF 97/13	EIA	55	4	36
8	BF 97/13	MIA	45	19	274
9	BF 97/13	LIA	55	35	349
10	BF 97/30	LIA	79	27	863
11	BF 96/22	MIA	90	96	312
12	BF 97/31	LIA	129	98	280
13	BF 94/120	LIA	403	313	1285
14	BF 95/7	LIA	178	169	701
15	NA 90/5C	Gaj IIa	0	8	16
16	NA 90/5A	Gaj IIb	415	121	1349
17	NA 90/5BI	Gaj IIa/b	821	41	487
18	NA 90/5BII	Gaj IIc	8	115	241
19	NA 91/1A	Gaj I	0	0	6
20	NA 93/42	Gaj I	71	6	33
21	NA 97/18	Gaj I	32	0	1
22	NA 97/24	Gaj I	0	0	14
23	NA 93/36	Gaj IIa	9	4	8
24	NA 99/65	Gaj IIa	0	0	1
25	NA 99/65	Gaj IIb	61	11	55
26	NA 99/65	Gaj IIc	5	18	56
27	NA 93/10	Gaj IIc	0	0	29
28	NA 97/37	Gaj III	262	55	150
29	NA 97/33	Gaj I	4	11	599
30	NA 97/33	Gaj IIa/b	7	21	244
31	NA 96/45	Gaj IIa/b	40	16	320
32	NA 95/1	Gaj IIa/b	55	51	559
33	NA 97/26	IA	58	32	54
34	NA 99/75	IA	90	28	54
35	NA 97/13	IA	0	12	49
36	NA 92/2C	subrec	124	125	922
37	NA 93/46	LSA	928	0	36
38	NA 93/46	EIA	24942	94	342
39	NA 94/7	I	2261	198	593
40	NA 94/7	II	854	17	55
41	NA 94/7	III	706	28	40
42	NA 94/7	IV	468	7	27
43	NA 93/45	I	2088	36	182
44	NA 93/45	II	4488	67	207
45	NA 93/45	IIIa	1841	23	116
46	NA 93/45	IIIb	1261	23	119
47	NA 93/45	IV	1611	14	132
48	Blé A		129	22	35
49	Blé B		368	53	23
50	Blé C		410	45	20
51	Blé E		849	38	58

Table 7: Relative importance of different economic strategies (NISP) by site and phase. Legend to Figs 35 and 36

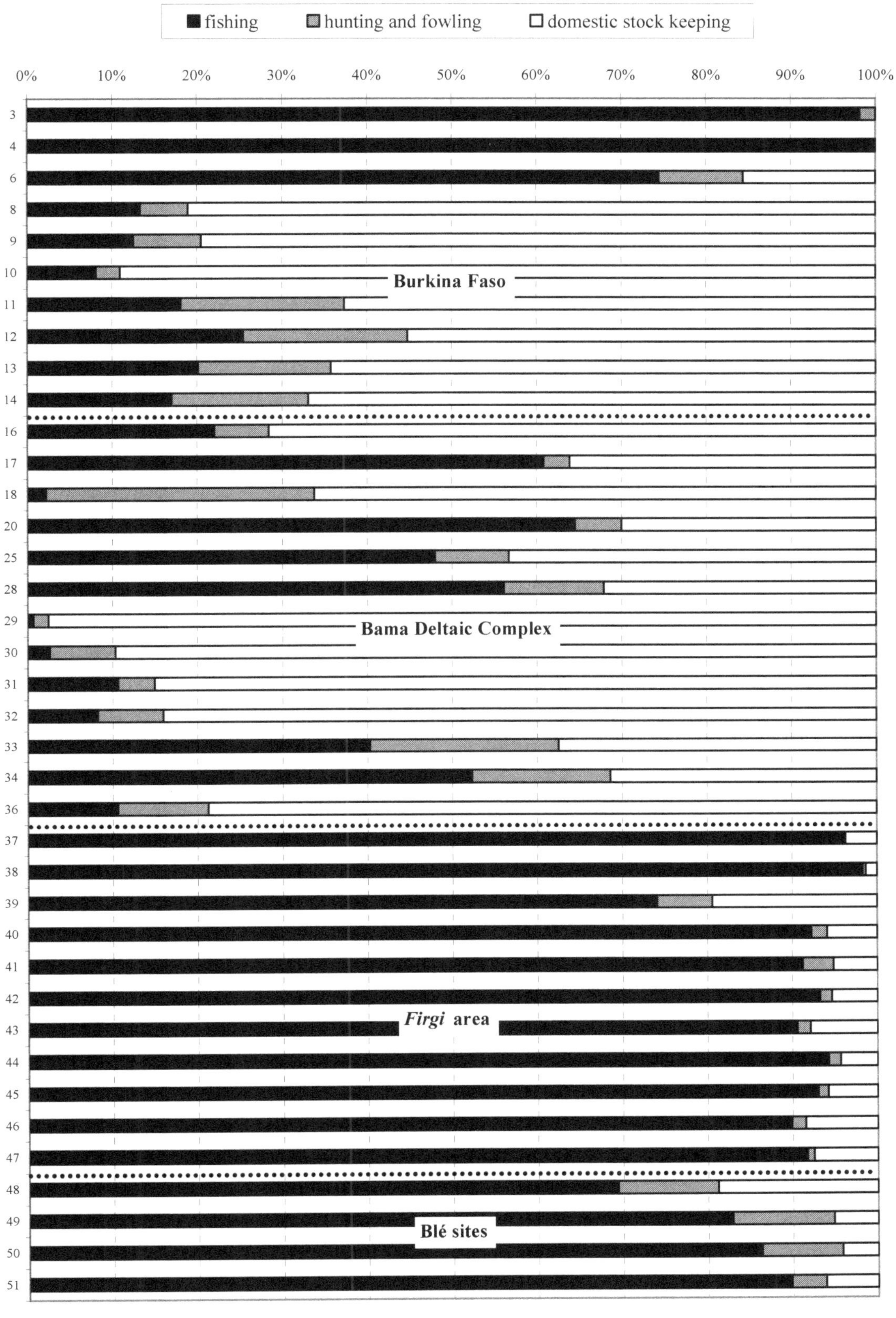

Fig. 35: Relative importance of different economic strategies (NISP) by site and phase (Only samples with more than 100 specimens)

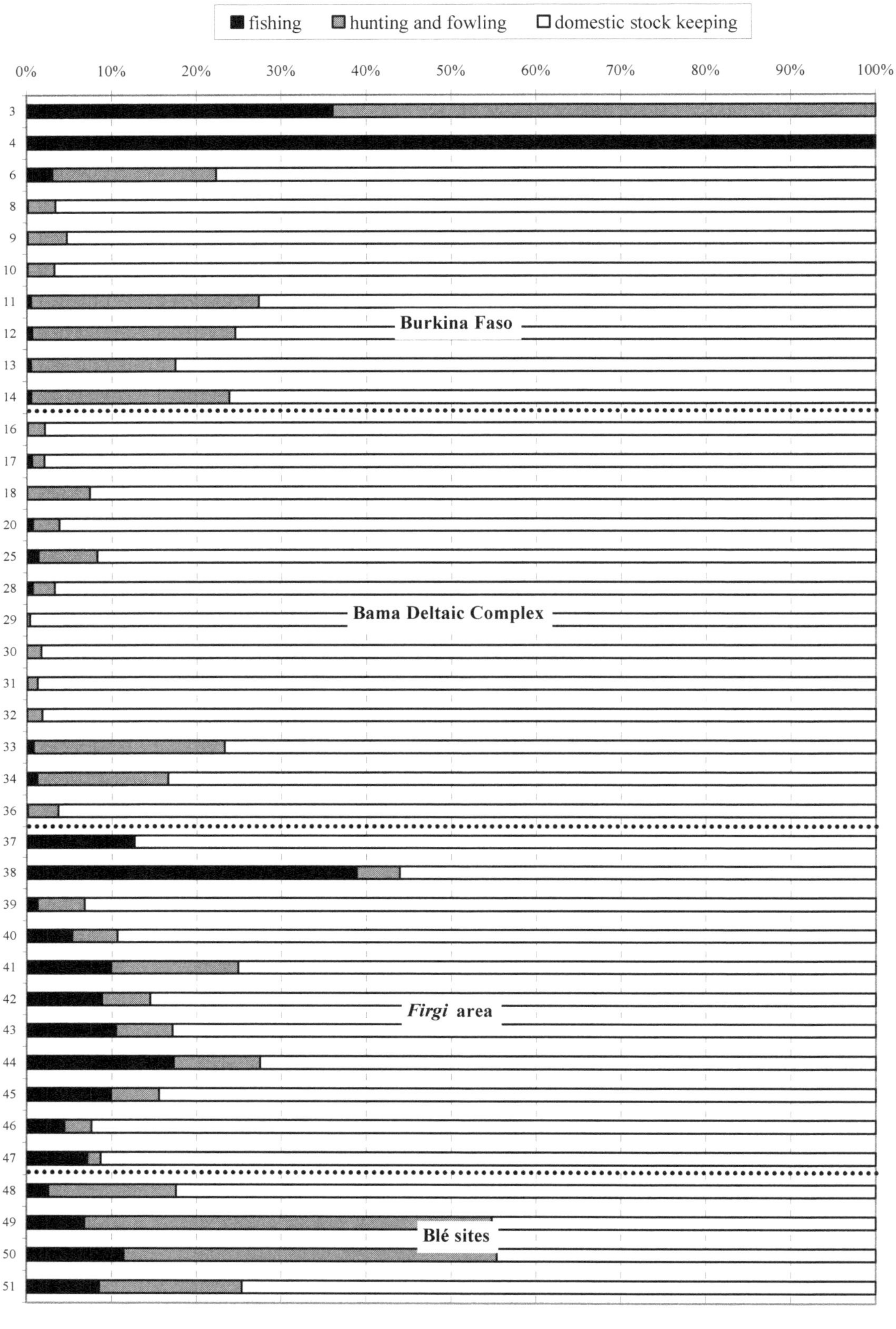

Fig. 36: Relative importance of different economic strategies (NISP x live weight) by site and phase (Only samples with more than 100 specimens)

6.5.1. Frequency of consumption and dietary contribution

Both the identified and the unidentified fish remains were used in the graphs, as opposed to what has been done for the other groups. Combined with the fact that they are easier to recognise from small remains, this may have led to an overestimation of their importance, but on the other hand, fish are more likely to have been missed during sampling and their bones may have been more easily destroyed than those of larger taxa (see 5.3. and 5.5.). In any case, rather than absolute numbers, inter- and intra-sites trends are more important. In what follows, the frequency of fish, as can be derived from the NISP data, is mainly discussed (Fig. 35). When meat value is taken into account, their importance diminishes drastically, because of the low average meat value by fish specimen (Fig. 36).

6.5.1.1. Burkina Faso

The only Late Stone Age site in Burkina Faso with a faunal sample of considerable size is that of Corcoba (BF 97/5), near the shores of the lake of Oursi, where fish and freshwater turtles take up more than 90 percent of the total number of consumed animal remains. The site appears to have been a specialised fishing site. This connects it with fishing sites in the Malian Méma (second millennium BC) (MacDonald and Van Neer, 1994), and further north in the Sahara near Hassi-el-Abiod and Erg Ine Sakane (sixth-third millennium BC) (Van Neer and Gayet, 1988). Fishing seems to have been equally important during the later Early Iron Age occupation at the same location, although there is no cultural continuity between the Late Stone Age and the Early Iron Age in northern Burkina Faso (see 2.2.2.). Like Corcoba, Oursi (BF 94/45), which lies a little closer to the lake, also has a high proportion of fish bones, both in its Late Stone Age and in its Early Iron Age samples. At Oursi village (BF 97/13), further away from the lake, fish again make up more than 50 percent of the total amount of consumed animal remains in the Early Iron Age levels, but less than 20 percent in the subsequent Middle and Late Iron Age contexts. The Early Iron Age sample from the site is rather small, however, and the proportions calculated for it may therefore not be very meaningful. Like the Late Iron Age contexts from Oursi village, the neighbouring Late Iron Age site Oursi hu-beero (BF 97/30) did not yield many fish remains. Despite finer sampling, the sites near Kissi, Middle Iron Age Kissi 22 (BF 96/22) and Late Iron Age Kissi 40 (BF 97/31), yielded only about the same percentage of fish bones as the excavated trenches at Late Iron Age Saouga (BF 94/120 and BF 95/7), around 20 percent. Waterbodies in the vicinity of these two villages are of a different type: at a few hundred metres from Kissi there is a small lake, while 2 km north of Saouga runs the river Gorouol. From the presented data, it seems thus that the importance of fishing has undergone a diachronic shift in Burkina Faso, from a high importance in the Late Stone Age and Early Iron Age to a much-reduced economic contribution in the later Iron Age periods. This shift is probably related to the introduction of domestic stock keeping, which gradually replaced fishing as the main economic activity (see 6.7.4.1.).

6.5.1.2. Bama Deltaic Complex

The Gajiganna A-C sites have yielded variable numbers of bones of fish, presumably captured in pools in the clay depressions in their vicinity. There are none at-all from Gajiganna C (NA 90/5C), but the site's total faunal sample is very small. The larger amount of fish at Gajiganna BI (NA 90/5BI) compared to Gajiganna A (NA 90/5A), around 60 versus around 20 percent, is due to a concentration of fish bones in a pit at the former site. Numbers of fish bones found at Gajiganna BII (NA 90/5BII) are very small, probably related to a (temporary) drying out of the pools near the sites. The fact that the clay depressions could not have contained any water at that time was already indicated by the location of the site on part of them (see 2.2.3.2.). Samples of most other Gajiganna sites of the Gajiganna group are too small to make any reasonable statements on the importance of fish. However, at Gilgila (NA 99/65) a similar trend as seen at Gajiganna B is clear between phase IIb and phase IIc, with a drop in the proportion of fish. At the younger site of Zilum (NA 97/37) numbers of fish have risen again to more than 50 percent. Since the site is located at some distance from the Gajiganna A-C sites, it is not clear if this should be attributed to a geographic, or instead to a diachronic, difference in the environment, if any. The sites of the Gajiganna Culture from the Bama-Konduga group (NA 97/33, NA 96/45 and NA 95/1) all contained few fish remains. The main waterbodies in their vicinity are the rivers Yedseram and Ngadda and their flood plains. Fishing in those environments may have been more difficult than in the seasonal pools of the central Gajiganna area (see 6.5.3.2.), which could explain the smaller economic importance of fish at the sites of the Bama-Konduga group. At the Iron Age sites Labe Kanuri (NA 97/26) and Elkido North (NA 99/75), fish make up about half of the faunal remains, while Dorota (NA 97/13) contains no fish remains at all. Bones at the latter site were retrieved from a refuse pit and may represent a single, or a few, depositional events in which no fish were involved. The subrecent site Galaga (NA 92/2C), in turn, had a faunal sample with a little more than 10 percent of fish and fresh water turtles. In summary, no heavy reliance on fish apparently has to be assumed for any of the sites in the Bama Deltaic Complex.

6.5.1.3. *Firgi* area and Blé sites

In contrast to the studied sites from Burkina Faso and from the Bama Deltaic Complex, all sites in the *firgi* area as well as all Blé sites have yielded comparably high numbers of fish. At Daima, high numbers of fish bones have also been recorded (Connah, 1981, Table 8.3), as well as in the *yaéré* sites studied by Holl (2002). The yearly inundations in the *firgi* area must have been particularly favourable for fishing activities, explaining the high reliance on fish of its inhabitants. However, in dietary terms this reliance must have been smaller than would be suspected from NISP data, in view of the low average meat yield of fish. Perhaps the ancestry of the present Kotoko, an ethnic group of specialised fishers, can be traced back to the occupants of the *firgi* sites (cf. Gronenborn, 2000). As mentioned in 2.1.2., these are traditionally considered to be the oldest inhabitants of the southern Lake Chad region. Of all *firgi* and Blé sites, the most extreme predominance of fish bones can be seen at the site of Kursakata (NA 93/46) where they make up about 96 percent in the Late Stone Age and 98 percent in the Early Iron Age contexts. In Mege (NA 94/7), phase I has, compared to its more recent occupation phases, a smaller component of fish, which is compensated by a larger contribution of herding. This may be connected with a shift from temporary settlement to permanent settlement between phase I and phase II, which rendered stock keeping more difficult (see 6.7.5.3.). Fishing near the Blé sites probably took place in the ancient delta of the Logone and in its inundation plains. Lower proportions of fish than at the *firgi* sites, may be due to coarser sampling. In the sample submitted for analysis from Blé Mound E, the concentration of fish bones in Occupation Horizon III, mentioned in the site's description (see 2.2.2.2.), was not observed.

6.5.1.4. The broader archaeological and ethnographic framework

From the previous paragraphs it appears that the importance of fishing at the studied sites was partly dependent on the type of waterbody in their vicinity. The findings of the present study seem to agree with the topographic preferences recorded by Sundström (1972, 134) for present African fishing groups: periodically flooded, flat-low lying country (*firgi,* Blé) or marginal areas of lakes and rivers (lake of Oursi during the Late Stone Age and Early Iron Age). Other West African archaeological sites, contemporary with the studied sites and in similar environments, also yielded evidence for economies with a large fishing component, e.g., Kobadi (Raimbault *et al.*, 1987; Jousse and Chenal-Vélardé, 2001-2002) and Kolima Sud in the Malian Méma (MacDonald and Van Neer, 1994), Dia (Bedaux *et al.*, 2001) and Jenné-Jeno along the river Niger (Van Neer, 1995), Tulel-Fobo on the Senegal River (Van Neer and Bocoum, 1991) and Koyom in the Chadian part of the Lake Chad area (Rivallain and Van Neer, 1983, 1984). However, other locations, also in the vicinity of large waterbodies, contained only a few fish bones. This was the case, for example, at Gao Saney and Gao-Ancien (MacDonald and MacDonald, 1996), along the river Niger, while more remains were retrieved from Gadei, another area excavated at Gao (Cook, 2000). Cubalel and Siouré, in the Middle Senegal Valley, also had only poor fish yields (MacDonald and MacDonald, unpublished manuscript). Finally at Akumbu in the Méma, the contemporary habitation at the same location of specialist and non-specialist fishers is suggested for the period AD 600-1000 (MacDonald and Van Neer, 1993, 1994). To conclude, one could say that specialised fishers require large waterbodies, but that not all groups in such environments carry out specialised fishing.

6.5.2. Nature of the fishing grounds

Looking at the fish and freshwater turtles found in the faunal assemblages from the studied sites and their species richness (Fig. 37), the nature of the waterbodies in the sites' vicinity will now be discussed in more detail. The fish taxa can be subdivided into three groups according to their main habitat. The first group contains the taxa that are chiefly confined to shallow water and that can survive in adverse conditions with low oxygen contents, high temperatures and high salinity, notably lungfish (*Protopterus annectens*), clariid catfish (Clariidae) and tilapia (tribe Tilapiini). The former two have accessory breathing organs allowing them to take oxygen from the air, while the haemoglobin of the latter has the highest affinity for oxygen of all fish taxa in the region. Lungfish is the only one of the three that can survive complete habitat desiccation, by forming a cocoon. Together with small cyprinids (Cyprinidae), tilapia and clariid catfish are also part of typical relic fish populations in the present Sahara (Van Neer, 1989b; Lévêque, 1990). Fish indicative for marshy, well-vegetated areas form the second group distinguished, including the following taxa: *Polypterus* sp., *Heterotis niloticus*, *Gymnarchus niloticus*, Distichodontidae and Citharinidae, *Auchenoglanis* sp. and *Parachanna obscura*. Some of these taxa are also able to take oxygen from the air, notably *Polypterus* sp., *Heterotis niloticus* and *Parachanna obscura*. Finally, the third group contains the fish that prefer open, deep, water: mormyrids (Mormyridae), tigerfish (*Hydrocynus* sp.), cyprinids with a standard length over 50 cm, *Bagrus* sp. and Nile perch (*Lates niloticus*). They can occasionally be seen on the inundation plain, but only for a very short period of time, when floodwaters reach their maximum level. *Synodontis* remains have also been put in this group, despite the diverse ecological requirements of species of the genus. However, from other archaeozoological studies, it appears that high numbers of *Synodontis* and high frequencies of Nile perch usually coincide, showing that the former were also mainly caught in deep water (Van Neer, 1989c). Of the identified species of freshwater turtles, all can survive periods of low water level by burying themselves in the mud, although the African softshell turtle (*Trionyx triunguis*) mainly occurs in large waterbodies and, especially, in large rivers.

When relative proportions of NISP for each of the three groups of fish are put in graphs, some very distinct geographic and diachronic trends can be seen (Fig. 38). Shallow-water fish generally predominate, which is in the majority of the cases related to a high amount of clariid catfish bones. The preponderance of the latter is partly related to their better preservation chances and to their higher recovery chances in case of coarse sampling techniques, in comparison to bones of other taxa. Clariid cranial roofs moreover usually separate into many, still identifiable pieces, further augmenting the species' relative importance. Large amounts of clariid catfish bones may thus distort the image of the available habitats towards shallow water environments. Other, less obvious taphonomical factors have surely influenced proportions as well. Therefore, as also formulated for the graphs concerning the relative importance of economic strategies, the numbers should only be seen as approximations and they are mostly valuable for intra- and inter-site comparisons.

6.5.2.1. Burkina Faso

The lake near Oursi

At the site of Corcoba (BF 97/5), the species composition clearly differs between the Late Stone Age and the Early Iron Age, with a disappearance of deep-water taxa and a steep drop in the species richness during the latter period. One species found in the Early Iron Age layers, lungfish, is missing from the older contexts. Softshell turtles (Trionychidae) have been found in the Late Stone Age levels, while the Early Iron Age levels contained no turtle remains at all. Between both phases there is a temporal hiatus of about 1000 years, covering roughly the entire first millennium BC. Through diatom analysis, palynological and geomorphological studies, it is known that silting up of the lake of Oursi set in at about the beginning of this intermediate period, caused by aridification or increased erosion related to human activities (Morczinek, 1995; Ballouche and Neumann, 1995; Kahlheber *et al.*, 2001; Albert, 2002, 151-152). The observed changes in the fish spectrum may be related to these phenomena as well and could reflect the disappearance of habitats for deep-water fish. An alternative, but less likely, explanation could be that, in contrast to the Late Stone Age, the site was inhabited in the Iron Age by a group of people lacking the technology for deep-water fishing (see 6.5.3.2.).

The Late Stone Age sample of Oursi (BF 94/45) contains no more than 25 fish bones, all of shallow water taxa; the species richness of the sample is also very poor. The Early Iron Age sample from the site is much larger and also has higher species richness. It has already been mentioned that relative numbers of tilapia at Oursi, both in its Late Stone Age and in its Early Iron Age levels, are underestimated because fine sieving was not systematically carried out at the site (see 5.5.). Nevertheless, besides Corcoba, Oursi is the only one of the investigated sites in northern Burkina Faso with a lot of tilapia remains. The presence of one Nile perch bone in Oursi's Early Iron Age levels is sufficient to suppose that the lake of Oursi still had year-round, deep, well-oxygenated, water at that time. This is furthermore confirmed by the presence of a tigerfish tooth among the fish remains retrieved from archaeobotanical sampling. As will be explained further on, it is believed that both species were obtained locally and were not imported from another waterbody further away. It is not entirely clear how the presence of deep-water fish at Early Iron Age Oursi can be connected to the absence of taxa from such habitats at Early Iron Age Corcoba, but radiocarbon dates for the latter are slightly younger than those for the former. If age is indeed the explanation, then the disappearance of deep-water habitats due to the silting up of the lake must have happened in the early phases of the Iron Age, rather than before the beginning of the period, in contrast to what has been postulated above. When looking at the quantitative progression of a few taxa throughout the Iron Age sequence of Oursi (Fig. 39), it can be observed that the amount of lungfish gradually rises from about -4.0 m upwards, while clariids, tilapia and Adanson's mud turtle (*Pelusios adansonii*) all peak around a depth of 3.4 m and are less common after that. All these taxa are typical of shallow water, but lungfish is the most resistant against droughts. The Nile perch bone came from -3.2 m and the tigerfish bone from -3.4 m. It is most likely that both were brought to the site together with the other fish found in the concentration at this depth, of which it can be assumed that they were caught locally. The increasing proportions of lungfish and the disappearance of deep water species throughout the stratigraphy at Oursi thus seem to support the idea that shrinkage of the lake happened, or at least continued, in the course of the Iron Age, towards the end of the site's occupation.

Throughout its whole sequence, which covers the complete Iron Age, Oursi village (BF 97/13) only yielded shallow water species. However, as deep-water taxa are usually not numerous, their absence at the site may be due to small sample size. A similar spectrum as at Oursi village, but with the addition of one *Gymnarchus niloticus* bone and one Nile perch bone can be seen at the site of Oursi hu-beero (BF 97/30). Both bones are from a context disturbed by erosion, however, and the dating of the finds may thus be unreliable. According to the

			minimal number of species	total number of remains	species richness
1	BF 97/5	LSA	12	925	1,611
2	BF 97/5	EIA	6	850	0,741
3	BF 94/45	LSA	2	25	0,311
4	BF 94/45	EIA	7	1382	0,830
5	BF 97/13	EIA	1	53	0,000
6	BF 97/13	MIA	2	41	0,269
7	BF 97/13	LIA	3	53	0,504
8	BF 97/30	LIA	5	78	0,918
9	BF 96/22	MIA	5	90	0,889
10	BF 97/31	LIA	5	127	0,826
11	BF 94/120	LIA	6	389	0,838
12	BF 95/7	LIA	6	173	0,970
13	NA 90/5A	Gaj IIb	6	415	0,829
14	NA 90/5BI	Gaj IIa/b	4	819	0,447
15	NA 90/5BII	Gaj IIc	2	8	0,481
16	NA 93/42	Gaj I	5	71	0,938
17	NA 97/18	Gaj I	3	32	0,577
18	NA 93/36	Gaj IIa	2	8	0,481
19	NA 99/65	Gaj IIb	2	59	0,245
20	NA 99/65	Gaj IIc	1	5	0,000
21	NA 97/37	Gaj III	6	226	0,922
22	NA 97/33	Gaj I	2	4	0,721
23	NA 97/33	Gaj IIa/b	3	7	1,028
24	NA 96/45	Gaj IIa/b	3	40	0,542
25	NA 95/1	Gaj IIa/b	6	54	1,253
26	NA 97/26	IA	3	58	0,493
27	NA 99/75	IA	5	77	0,921
28	NA 92/2C	subrec	8	113	1,481
29	NA 93/46	LSA	10	647	1,391
30	NA 93/46	EIA	20	17650	1,943
31	NA 94/7	I	9	1824	1,065
32	NA 94/7	II	11	603	1,562
33	NA 94/7	III	11	417	1,658
34	NA 94/7	IV	6	328	0,863
35	NA 93/45	I	16	406	2,497
36	NA 93/45	II	13	920	1,758
37	NA 93/45	IIIa	12	368	1,862
38	NA 93/45	IIIb	12	297	1,932
39	NA 93/45	IV	12	379	1,853
40	Blé A		8	110	1,489
41	Blé B		14	280	2,307
42	Blé C		16	317	2,605
43	Blé E		15	357	2,382

Table 8: Species richness of fish and freshwater turtles by site and phase. Legend to Fig. 37

inhabitants of the present village of Oursi, only lungfish and clariid catfish can be found today in the nearby lake. They also claim that clariids died out at some point, but were re-introduced later on. This is plausible, since the taxon cannot survive complete habitat desiccation and may thus have disappeared during years of severe droughts, unless it was able to retreat into some small, marginal, muddy habitats. From the archaeological sites there are absolutely no indications that clariids disappeared from the lake at any point, but there is a gap in the archaeological sequence for the period between AD 1400 and present. However, pollen analysis has shown a drop in average water levels, combined with larger seasonal differences, for the period after AD 1400 (Ballouche and Neumann, 1995). At present, there are also regulations for fishing in the lake of Oursi, limiting the minimal mesh size of nets used, in order to avoid overfishing, but for the archaeological sites near the lake, there were no clear indications for overfishing, for example in the form of decreasing average fish sizes.

The lake near Kissi

Palaeo-ecological indications on the lake near Kissi are limited because of the small size of fish samples from Kissi 22 (BF 96/22) and Kissi 40 (BF 97/31). Both have limited species richness, with assemblages dominated by shallow water taxa, mainly lungfish and clariid catfish. Tilapia is much less represented, but as fine sieving was carried out at the sites, this cannot be attributed to the sampling techniques used. A possible explanation could be the season in which fishing was predominantly carried out (see 6.5.3.1.). The presence of *Synodontis*, as the only deep-water taxon at the sites, is insufficient to argue for a year-round persistence of deep-water in the lake at time

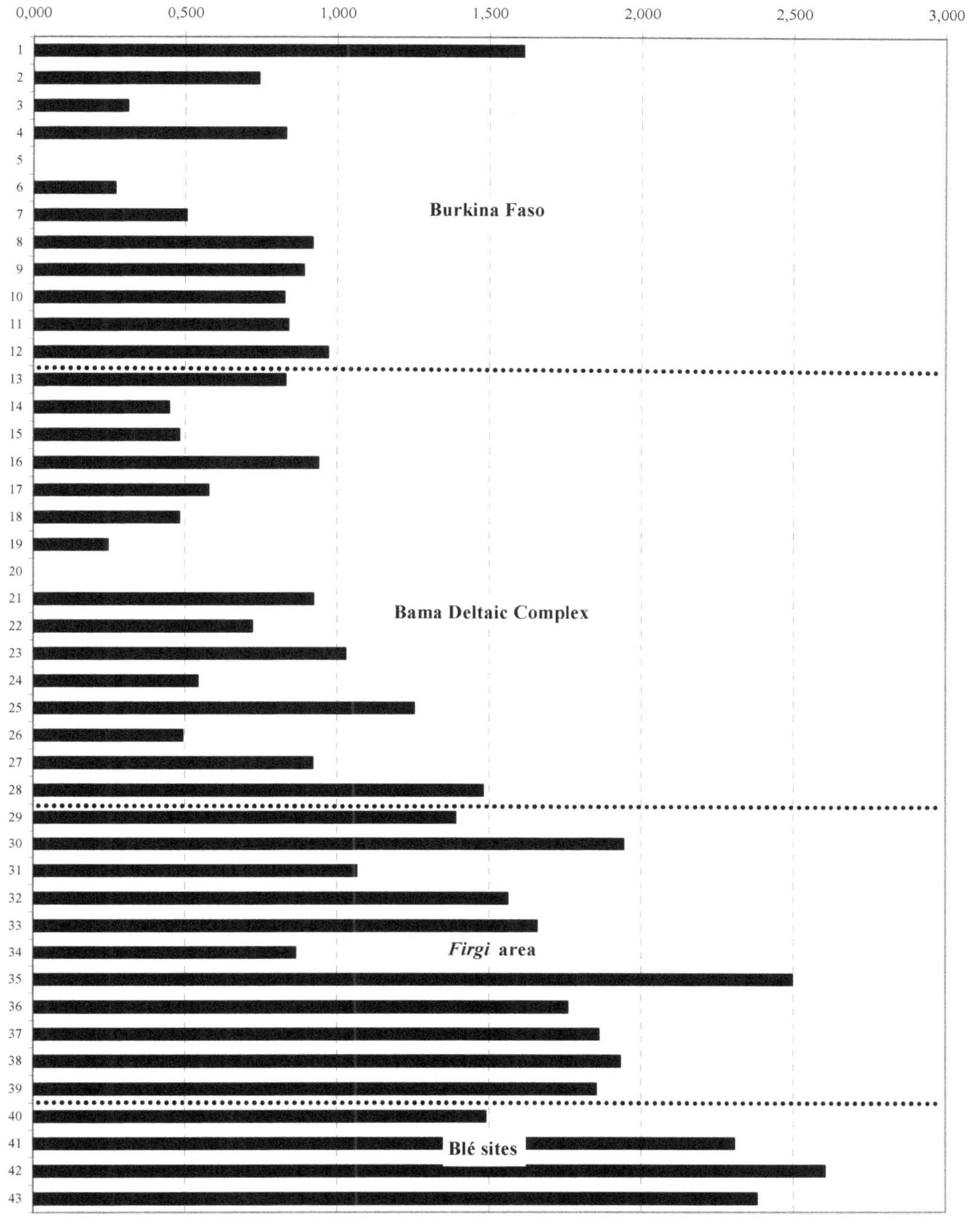

Fig. 37: Species richness of fish and freshwater turtles by site and phase

of their occupation. The lake of Kissi is one of the smaller lakes of northern Burkina Faso and today usually does not carry water during the dry season. During periods of drought the survival of species other than lungfish certainly seems unlikely. From the presence of clariid catfish in the archaeological assemblages, it should therefore be assumed that during the time of the sites' occupation a certain degree of humidity still remained year-round. Data from the present study thus seem to be consistent with the environmental degradation Ballouche (2001) has recorded for the past 500 years in the pollen diagram of the lake of Kissi. For the same period, Albert (2002, 157) also found faster sedimentation rates, indicating a silting up of the lake.

The river Gorouol near Saouga

The river Gorouol apparently provided mainly shallow water habitats, judging from the fish taxa identified from

			shallow water	marshes	deep water
1	BF 97/5	LSA	375	94	46
2	BF 97/5	EIA	815	32	0
3	BF 94/45	LSA	25	0	0
4	BF 94/45	AB LSA	12	0	1
5	BF 94/45	EIA	1285	0	8
6	BF 94/45	AB EIA	486	2	4
7	BF 97/13	EIA	52	0	0
8	BF 97/13	AB EIA	6	0	0
9	BF 97/13	MIA	41	0	0
10	BF 97/13	LIA	51	0	0
11	BF 97/13	AB LIA	11	0	0
12	BF 97/30	LIA	75	1	1
13	BF 97/30	test +AB LIA	6	0	0
14	BF 96/22	MIA	88	0	1
15	BF 97/31	LIA	122	0	1
16	BF 94/120	LIA	362	0	16
17	BF 95/7	LIA	135	0	8
18	NA 90/5A	Gaj IIb	385	4	0
19	NA 90/5BI	Gaj IIa/b	746	64	0
20	NA 90/5BII	Gaj IIc	6	0	0
21	NA 93/42	Gaj I	40	26	0
22	NA 97/18	Gaj I	10	15	0
23	NA 93/36	Gaj IIa	6	0	0
24	NA 99/65	Gaj IIb	58	0	0
25	NA 99/65	Gaj IIc	5	0	0
26	NA 97/37	Gaj III	194	9	0
27	NA 97/33	Gaj I	4	0	0
28	NA 97/33	Gaj IIa/b	5	0	0
29	NA 96/45	Gaj IIa/b	39	0	1
30	NA 95/1	Gaj IIa/b	25	8	2
31	NA 97/26	IA	54	0	0
32	NA 99/75	IA	60	3	0
33	NA 92/2C	subrec	42	11	60
34	NA 93/46	LSA	426	179	19
35	NA 93/46	AB LSA	58	53	10
36	NA 93/46	EIA	11414	4893	508
37	NA 93/46	AB EIA	351	467	2
38	NA 94/7	I	1538	99	14
39	NA 94/7	II	458	62	14
40	NA 94/7	III	401	10	4
41	NA 94/7	IV	314	9	5
42	NA 93/45	I	249	88	58
43	NA 93/45	II	667	132	91
44	NA 93/45	IIIa	279	45	36
45	NA 93/45	IIIb	241	22	13
46	NA 93/45	IV	291	29	25
47	Blé A		66	2	37
48	Blé B		127	22	97
49	Blé C		119	38	135
50	Blé E		239	89	284

Table 9: Relative proportions of fish taxa according to habitat requirements by site and phase. Legend to Fig. 38

both excavations at Saouga (BF 94/120 and BF 95/7), but, again, tilapia is not very well represented. Nonetheless, the river must have carried enough water year-round to explain the presence of Nile perch. This is in agreement with what was postulated earlier based on the find of Nile oyster. Since the river Gorouol only carries water during the rainy season today, higher water levels in the rivers of the Oudalan during the Late Iron Age need to be assumed, consistent with higher rainfall supposed for the area based on archaeobotanical evidence (Neumann *et al.*, 1998; Höhn *et al.*, 2004). No fish species from marshes have been found in the Saouga faunal assemblages. Their absence may be due to small sample sizes rather than to a lack of marshy environments, especially since the presence of *Lanistes* shell indicates that marshes must have been available.

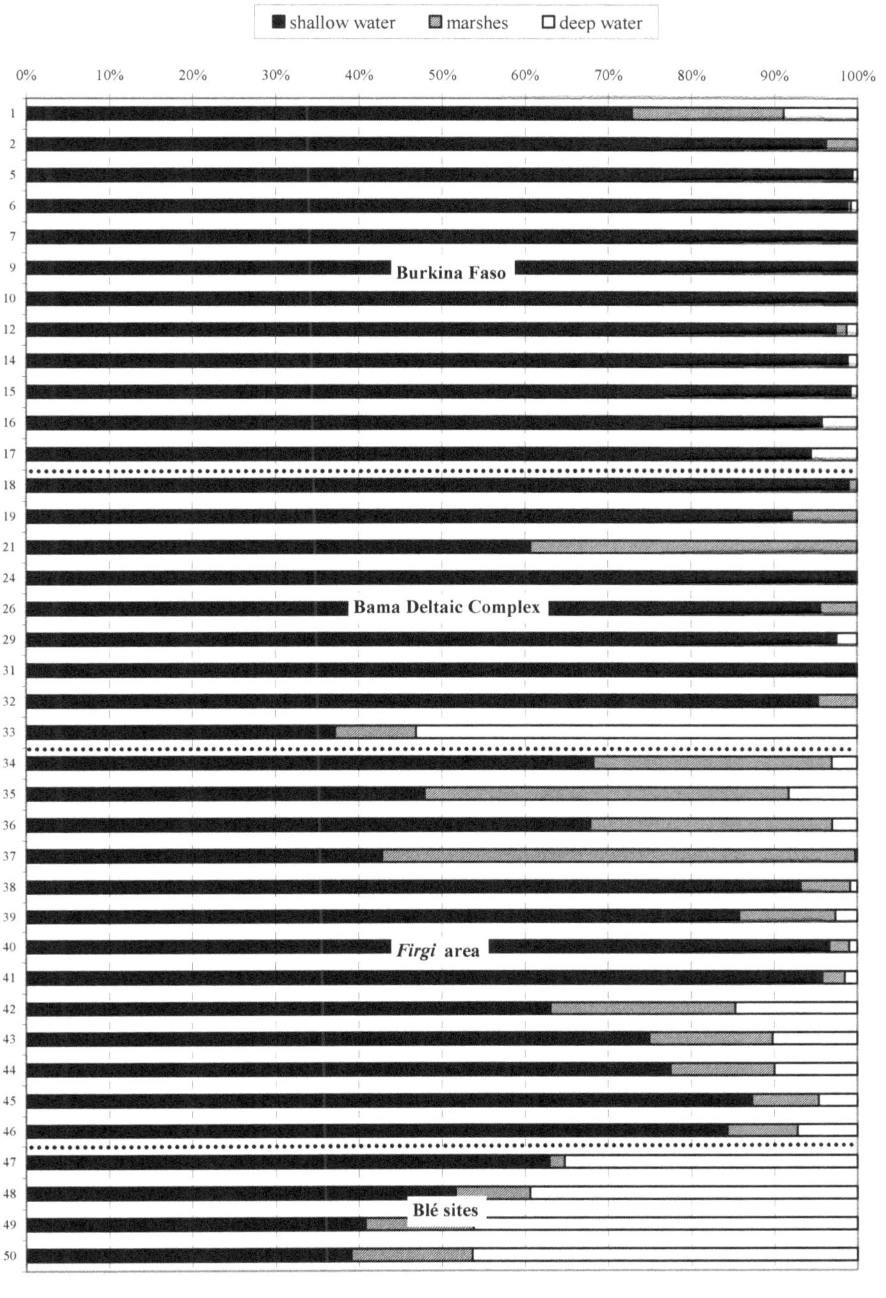

Fig. 38: Relative proportions of fish taxa according to habitat requirements by site and phase (Only samples with more than 40 specimens)

6.5.2.2. Bama Deltaic Complex

The pools in the central Gajiganna area

Today the main waterbodies in the central Gajiganna area are inundated clay depressions, which are believed to be the remnants of a larger lagoon that still existed in the period of the Gajiganna Culture. Judging from the fish assemblages studied from the area, the depressions provided suitable habitats for shallow water taxa, as well as for taxa that need marshy, well-vegetated environments, but deep-water fish could not survive, if the evidence from the subrecent site of Galaga (NA 92/2C) is ignored for now. This seems to indicate that the lagoon was no longer connected to the lake itself by the time the people of the Gajiganna Culture occupied the region. This is especially clear when comparing the fish fauna from sites in the central Gajiganna area to the rich and diverse fauna of the *firgi* (see 6.5.2.3.), but species richness is still greater than that of relic populations in the

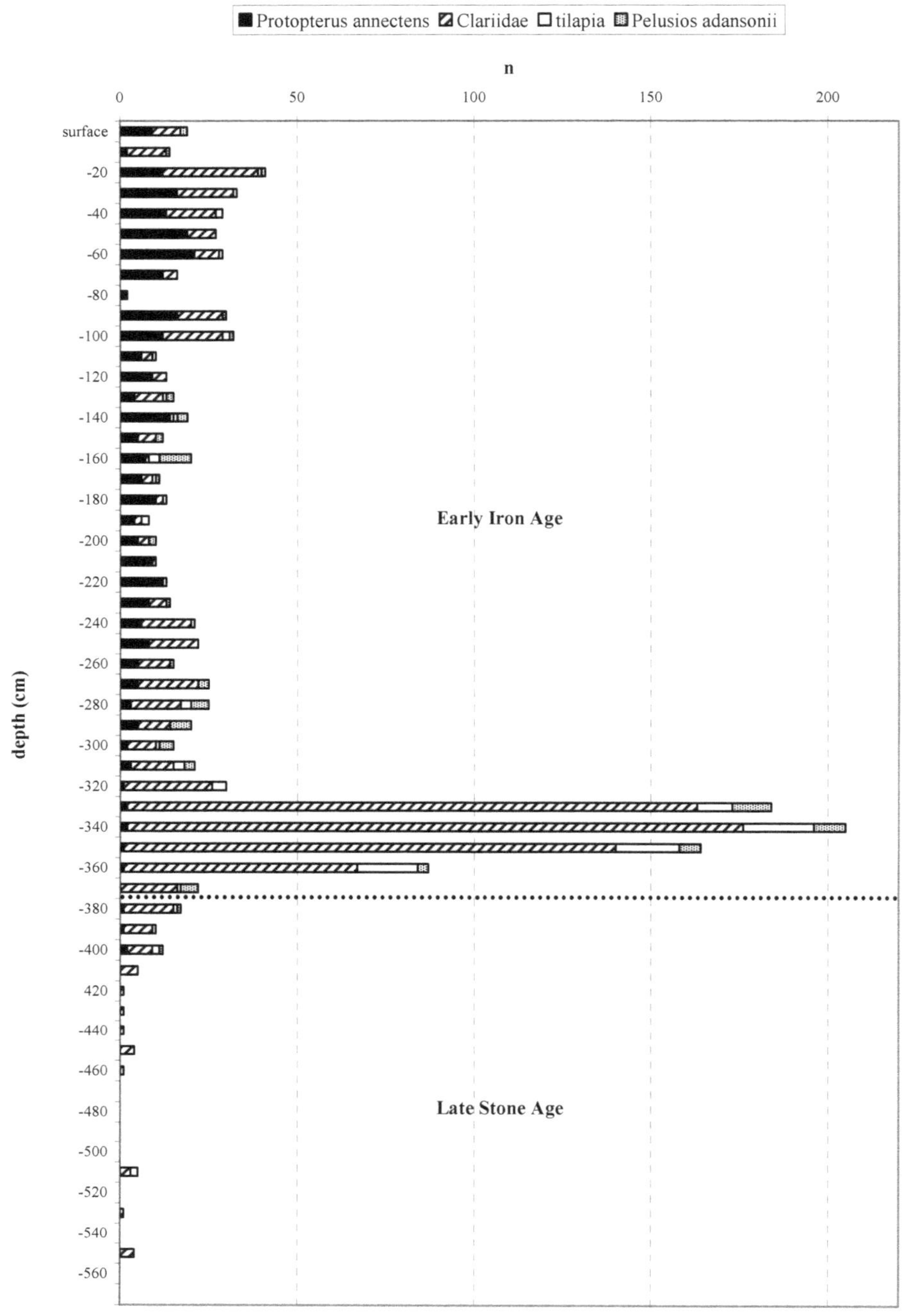

Fig. 39: Numbers of lungfish, clariid catfish, tilapia and Adanson's mud turtle bones from Oursi (BF 94/45) by depth

present Sahara. This shows that conditions from the moment the connection with Lake Chad was broken had never been so extreme as to induce the local extinction of all but the most resistant species. It is unclear if this could mean that the connection with Lake Chad had only been broken off shortly before, but it appears that a relatively large waterbody must have persisted allowing the survival of the different taxa attested. However, there are no arguments to exclude an alternative scenario for the lack of open water species. Such an absence of deep-water taxa could also occur in a situation whereby the pools were still seasonally connected to Lake Chad and people of the Gajiganna Culture were limiting their fishing activities to the dry season. A lack of deep-water species at Karkarichinkat, a site in the Tilemsi Valley in Mali occupied during the late third-early second millennium BC, for example, was used by Smith (1975) as an argument that the location was only visited during the dry season. From finds in the Bama-Konduga group, it can be excluded, however, that the Gajiganna people did not exploit deep-water taxa because they did not possess the necessary techniques to catch them (see 6.5.3.2.).

A few small diachronic trends are visible in the fish composition of sites in the area around Gajiganna, but it is not clear if they are of much significance. While the earliest Gajiganna contexts yielded fair amounts of remains of fish, from shallow water as well as marshes, both contexts from Gajiganna phase IIc contained very little fish bones and only of shallow water taxa. As said earlier, from the physical position of the Gajiganna BII (NA 90/5BII) site it is clear that, at least locally, pools had dried out at that time. The Gajiganna phase III site of Zilum (NA 97/37) has again fish from marshes. Another striking feature of the fish species spectrum from Zilum is the presence of lungfish, which is completely missing from the older Gajiganna sites. Some variation can also be observed between the fish fauna from the Iron Age localities. Labe Kanuri (NA 97/26) only yielded fish from shallow waters, notably lungfish and clariid catfish, the former being about four times as numerous as the latter. They were probably caught in the "Jere Bowl", the inland delta of the River Ngadda, where stagnant, shallow water must have been present all year round, judging from the presence of *Lymnaea* shells (see 6.3.). The fish remains from Elkido North (NA 99/75) also contain, besides lungfish and clariids, three bones of *Polypterus* sp. and *Heterotis niloticus*. The presence of these species suggests that the species richness of the pools in the Bama Deltaic Complex had not diminished by the onset of the Iron Age.

Good information is missing on the fish taxa that can be found in the clay depressions around Gajiganna today. Many of the depressions dry out seasonally and since only few fish taxa can survive such harsh circumstances, they can contain only a limited amount of species. In the pools near the Gajiganna A-C sites, the excavating archaeologists could not locate any fish at all (Breunig, 1995), but this does not necessarily mean that there were none there. They proposed that clariid catfish might have been present, and perhaps lungfish could have been buried in the ground in their cocoons. Taking the presently poor species richness into account and looking at the taxa that have been found at older locations, it seems unlikely that the deep-water fish identified from the subrecent site of Galaga (NA 92/2C) were caught locally. Species richness is also higher than at the other sites in the central Gajiganna area. It seems therefore that the inhabitants of Galaga obtained at least part of their fish from elsewhere, for example through trade (see 6.5.5.).

The Yedseram and Ngadda River near sites of the Bama-Konduga group

The main waterbodies in the vicinity of Gajiganna sites of the Bama-Konduga group are the rivers Yedseram and Ngadda. Shallow water taxa largely prevail in the assemblages of all three sites of this group that were studied. However, some deep-water fish were present at Kelumeri (NA 96/45) and Kariari C (NA 95/1), which are the two locations closest to a river. Deep-water taxa may not have been available in the vicinity of Bukarkurari (NA 97/33), which is situated further away from the two rivers in the area. Gajiganna phase I layers at the latter site are the only contexts from the Bama-Konduga group where lungfish has been found. Judging from the paucity of taxa from well-vegetated areas at sites of the Bama-Konduga group, the rivers may have provided little marshy environments or, when present, these must have been exploited to a limited extent. Archaeobotanical or geomorphological data on the nature of the rivers during the Gajiganna period are missing.

6.5.2.3. *Firgi* area

The flood plains

Compared to the sites from the Bama Deltaic Complex, not only are the much larger fish assemblages from the *firgi* sites striking, but also their different species composition and higher species richness. They have more fish from marshes, and also contain – more importantly – fish from deep-water. The inundations in the *firgi* area thus seem to have provided a variety of aquatic habitats, which the inhabitants of the studied sites apparently also knew how to exploit efficiently. There is, however, considerable variation in the species composition among the studied *firgi* sites, depending on their age and location. Of the three analysed sites from the area, Mege (NA 94/7) has the highest amount of shallow water fish, chiefly clariid catfish. This is possibly related to lower flood levels around the site, presumably due to its larger distance from the rivers bringing in floodwater, and which must have resulted in a predominance of shallow water habitats. Ngala (NA 93/45) and especially Kursakata (NA 93/46), on the other hand, have a lot of fish from marshes and well-vegetated areas. The fish were probably attracted by the plains in the surroundings of the sites, covered by plant growth, when these were submerged during the flood season. For Late Stone Age layers at Kursakata there are archaeobotanical indications that such marshes were available for several months of the year (Klee *et al.*, 2000). From all assemblages studied, the only finds of African softshell turtle were at Ngala and Kursakata, which points to water basins of substantial sizes. Compared to Kursakata, Ngala's fish assemblage contains many remains of deep-water fish. Since the two are neighbouring sites, explanations other than geographical must be sought. Ngala is younger than Kursakata, and perhaps there was a shift at the former site towards more fishing at high water levels, when deep-water fish are more likely to be found on the flood plain (see 6.5.3.1.). Alternatively, Ngala's inhabitants may have practised part of their fishing activities further away from their settlement, for example in the El Beid River. The deep-water fish could also have been obtained through trade or exchange with groups living closer to the river. At the same time, the inhabitants of Ngala seem to have continued exploiting the flood plains and residual pools in the vicinity of their settlement. Increasing relative amounts of shallow water fish at the site through time might be related to progressively lower water levels in its surroundings (Linseele, 2005).

6.5.2.4. Blé sites

The ancient Logone Delta

The Blé sites are also characterised by large fish assemblages, with large species richness and high proportions of deep-water taxa. Fishes are also, on average, larger than at other localities, but this must be mainly due to the lack of sieving during sampling at the sites. Holl (2002, 247) connects the end of occupation at the Blé Mound Complex with a modification of the Logone River drainage system, coinciding with a drastic but short arid climatic phase. However, fish samples of the Blé sites do not allow the tracing of any diachronic changes in the aquatic environment. Their samples are either too small, or they cover only a relatively short time span. Blé Mound A, for example, with the longest chronological sequence, yielded only a little more than 100 fish remains.

6.5.3. Seasonality and fishing techniques

6.5.3.1. Seasonality

As described in 2.1., the climate of the research area is characterised by an alternation of wet and dry seasons, which causes fluctuations in the levels of its waterbodies. Water levels of the lakes in northern Burkina Faso and the clay depressions in the Gajiganna area depend mainly on local rainfall. In contrast, Lake Chad and its tributaries are principally fed by water coming from more southern regions and reach their highest levels during the local dry season. Rising water levels cause the surrounding plains to inundate, where plants and detritus lying on the bottom lead to a food explosion (Blache and Miton, 1962, 15). At this time, fish enter the flood plain for spawning, usually along channels connecting it to the main waterbody. The duration of the fish's stay on the plain depends on the taxon (see summary in Van Neer, 2004). If at all, deep-water species can only be found on the flood plain during the peak of high water. They spawn on the plain or lay their eggs in the main channel. In the latter case, the fry moves into the flood plain where there is enough food and shelter and after a period of rapid growth migrate back to the main channel. During the floods *Heterotis niloticus* and *Gymnarchus niloticus* also make their circular nests for breeding. Shallow water species have a prolonged stay on the alluvial plain. They return to the main water channel with falling water levels and within each species, large individuals go back earlier than smaller ones (Rivallain and Van Neer, 1983). After some time, the water recedes, leaving residuals pools and ponds on the plain. These can either dry out completely or persist until the next floods. They contain mainly young fish that were not able to make it back to the main waterbody.

The success of fishing varies in the course of a year, and depends heavily on the seasonal variations in water levels and the related fish behaviour (Van Neer, 2004). Two periods are particularly rewarding. The first is when the water rises and spawning runs take place towards shallow, marginal, areas. First the clariid catfish and then tilapia move to the floodplain. After breeding, adult fish return to deeper water, but tilapia has a prolonged stay on the flood plain because it is a repetitive breeder. Fish can also be caught relatively easily during their migrations between the main waterbody and the flood plain, or vice versa, by blocking of the channels connecting the two. The inundation season is probably a good period to exploit fish from marshes, as described for the *firgi* sites Kursakata (NA 93/46) and Ngala (NA 93/45) (see 6.5.2.3.). Franke-Scharf *et al.* (2004) mention that present sedentary agriculturalists in the Lake Chad area turn to fishing in the flood season, because work on the fields is then impossible. The Shuwa Arabs of the region also carry out seasonal fishing activities during the flood period (Holl, 2003, 373). However, at the height of the floods, fish are difficult to catch because they are very dispersed. The second peak in fishing is at the end of the floods when residual pools are formed. In the main channel, fishing is also easiest during low water levels, because of reduced turbulence.

Length reconstructions of the fish specimens found may help to determine in what season and which part of the waterbodies people were fishing, because juveniles and adults of most taxa occur in different environments (Van Neer, 1986; Gautier and Van Neer, 1989). However, sampling strategies applied during the excavations at the studied sites may have resulted in slightly distorted size distributions (see 5.5.). Since no fine sieving was carried out at most of the studied localities, the smallest, and therefore the youngest, individuals will be underrepresented. For clariid catfish, small specimens were probably caught in residual pools, while larger ones were fished while spawning. These fish reach maturity at a standard length of between 22 and 65 cm. At the studied sites in Burkina Faso the fishing of clariid catfish seems to have taken place mostly at the end of the flood season, judging from the predominance of relatively small individuals (Table C.11b). The clariids from locations in the Bama Deltaic Complex are also usually rather small, but with larger individuals at the site of Kelumeri (NA 96/45). In the *firgi* area there appears to be a difference between Kursakata (NA 93/46) on the one hand and Mege (NA 94/7) and Ngala (NA 93/45) on the other hand, with larger average and maximal sizes at the latter. The clariids from the Blé sites have sizes comparable to those from Mege and Ngala, although smaller individuals may be missing there because sieving was not carried out. Fishing at Kursakata was possibly more oriented towards the end of the flood season, i.e. on young individuals in residual pools, than at the other *firgi* sites and at the Blé sites.

The differences in yields of tilapia bones described for the sites in Burkina Faso can also probably be explained in terms of seasonality. They were numerous at Corcoba (BF 97/5) and Oursi (BF 94/45), with most remains of relatively large individuals (Table C.16b, maximal size is about 40 cm). This indicates that we are not dealing with young individuals fished from residual pools but with adults that were caught during spawning and during the subsequent period when guarding their nests and caring for young. Since tilapia is a repetitive breeder its abundance at the sites possibly has to be interpreted in terms of fishing activities during consecutive weeks, and possibly months, after the beginning of the floods. Kahlheber (2003, 104) has proposed that Late Stone Age Corcoba (BF 97/5) was occupied at the beginning of the dry season. However, based on the large numbers of tilapia remains, a prolonged occupation from the beginning of the floods seems more likely. This is also in agreement with the paucity of (small) clariids in the layers of the period, showing that no systematic exploitation of residual pools took place. Lower amounts of tilapia at the other sites in Burkina Faso indicate that fishing may have taken place at the end of the floods or during other parts of the year, or that fishing events at the beginning of the floods were only of short duration. As said, relatively small estimated sizes of their clariids speak in favour of fishing in pools left at the end of the floods.

The slack periods in the course of the year make a subsistence based only on fishing virtually impossible, although these periods can (partially) be overcome by preserving and storing fish for later use (see 6.5.4.). It is unlikely that fish form a sufficient food supply year round, especially at places where only the flood plain can be exploited, since between the drying out of residual pools and the next inundation yields are usually too poor (Rivallain and Van Neer, 1984). This may be one of the reasons why specialised fishing communities today usually live in socio-economic symbiosis with groups that can provide them with additional food resources, mainly agriculturalists (Sundström, 1972, 134-136). Nevertheless, it has been argued that specialised fishing requires less mobility than hunter-gathering, herding and even agriculture (Vansina, 1999) and its role in the transition between a mobile and sedentary way of life on the African continent has been emphasised (Phillipson, 1985, 99-109). In this context it was apparently also possible that specialised fishing groups, like the Kotoko, developed urban settlement systems (see 2.1.2.).

Because of the peak periods in fishing activities at the beginning and end of the flood season, it is hard to prove year round fishing from archaeological evidence, because it will always emphasise these peak moments. The only taxon that could still be exploited on floodplains and in pools until deep in the dry season must have been lungfish. It may therefore be no coincidence that the occurrence of lungfish corresponds to locations and periods for which the archaeologists have proposed permanent occupation (see 2.2.2. and 2.2.3.; Table C.2a). The only site with supposedly permanent settlement, but where lungfish is lacking is Mege (NA 94/7). The presence of lungfish thus seems to be indicative for permanent settlement, although its absence cannot be taken as proof for mobile human groups. However, a problem with this line of reasoning is that, rather than the dry season, the flood season is the period when mobile human groups are dispersed. Therefore, to firmly proof permanent settlement, evidence for the presence of the inhabitants during the flood season is needed, which can as yet not be provided by the fish remains. At Kursakata (NA 93/46), a shift from seasonal to permanent settlement was proposed between the Late Stone Age and the Iron Age, based on archaeobotanical and archaeological data (see 2.2.2.2.). However, other than the appearance of lungfish in the Iron Age layers, no changes in the faunal species composition between both periods could be observed. As said in 6.5.2.4., samples at the Blé sites did not allow tracing diachronic changes in the fish fauna. The shift to a seasonal use of the sites towards the end of their sequence therefore cannot be proven.

6.5.3.2. Fishing gear

No finds of fishing gear have been recorded for the studied sites, probably because most was made from perishable material. Judging from the traditional fishing implements that are used today, the techniques were probably well adapted to the diverse aquatic ecosystems in the area and their seasonal variations (e.g., Blache and Miton, 1962; Reed *et al.*, 1967). Terra cotta net weights were recorded at some other West African sites, like Kolima Sud in the Malian Méma (1700-500 BC) (MacDonald and Van Neer, 1994) and subrecent Koyom in southern Chad (Rivallain and Van Neer, 1983). The use of formal net weights seems to be rather exceptional near present Lake Chad and mostly a variety of heavy objects are used (Hopson, 1967). It is therefore suspected that some of the unidentified clay objects from Ngala (NA 93/45) may be net weights, and the same is supposed for a "spindle-whorl like object" from Kursakata (NA 93/46) (Gronenborn, 1998). Looking at the sites cited above, net weights seem to be mainly associated with contexts containing abundant fish remains.

Nowadays all nets are made from nylon, but until recently, fishing nets used to be hand braided with plant fibres (Reed *et al.*, 1967, 145). They are nearly impossible to trace directly in archaeological contexts, but at Kolima Sud, impressions of nets have been found on potsherds (MacDonald and Van Neer, 1994). Nets only select for fish on their size, which usually results in a large number of taxa. It is therefore assumed that nets must have been used at Late Stone Age Corcoba (BF 97/5), and all *firgi* and Blé sites. Nets are probably also the main means with which freshwater turtles were caught. Holl (2002, 211) considers turtles as a by-product of fishing, but looking at their present popularity for food (see 4.4.1.), they might have been deliberately sought after. Clap nets are the most common and widely spread type of fishing gear in present northern Nigeria. Their diameter varies between 50 cm and 2 m, and they are either used alone or together with fish traps and fences (Reed *et al.*, 1967, 145-147). Other types of nets are very variable, but two large groups can be distinguished: beach seines, used in wadeable areas of the floodplain or shallow, large residual pools, and floating nets for the deeper parts of the floodplain and the main channel (Van Neer and Lesur, 2004). For the latter, the use of vessels is necessary. Cast nets are nowadays commonly used in northern Nigeria, throughout all seasons (Reed *et al.*, 1967, 150), but they would be a relatively recent, European introduction (Sundström, 1972, 25).

In many circumstances fish catching in the vicinity of the studied sites could probably have been done without much specialist skills. Spawning clariids can for example be captured with striking or wounding gear, using cover pots or simply by hand. Fish taxa breeding in nests on the floodplain, *Heterotis niloticus*, *Gymnarchus niloticus* and tilapia, can also easily be traced and caught. Techniques applied by the sites' inhabitants in residual pools could have included hand grasping, the use of striking and wounding gear, cover pots, clap nets or stupefaction of the fish by stirring up the mud or using ichthyotoxic plants. Later in the dry season, when the pools had dried out, aestivating lungfish could be obtained by digging them out of their burrows. Possibly, lungfish could also have been found by coincidence by farmers preparing their fields (cf. Daget, 1949). Fishing in the pools of the Gajiganna area and in the small lakes of Burkina Faso may have been comparable to that in residual pools on floodplains, certainly at low water levels during the dry season. The most typical form of fishing that can be found today among agricultural groups is public fishing in communal waters, sometimes in the form of fish festivals (Sundström, 1972, 18-19; Reed *et al.*, 1967, 184). At such occasions, normally during the dry season, a complete village may take part in catching most fish of a swamp or residual pool.

Deep-water fishing is technically more advanced than floodplain fishing (Van Neer, 2004), but was apparently already known during the Early and Middle Holocene wet phase in the Sahara (Van Neer and Gayet, 1988). At locations, which have yielded deep-water species, the use of boats probably has to be supposed, even when these were fished on the flood plain rather than in the main water channel. However, the growth of grasses on inundated plains, like the *firgi* or *yaéré*, probably hampered the use of boats (Bénech *et al.*, 1983). Nevertheless, there are some techniques that allow catching deep-water fish from the shores, like large floating long lines (Reed *et al.*, 1967, 159-160). As mentioned in 4.2.7., the average size of Nile perch individuals caught increases with the distance from the shore where fishing is carried out. Fishermen from the studied sites apparently did not go too far offshore, because very large Nile perch is rare. The dug-out from Dufuna testifies that around 6000 BC sub-Saharan hunter-gatherers had already developed the necessary know-how for the construction of boats. Historical evidence on boats in western Africa is known from the twelfth century AD, when Al-Idrisi mentions their use on the River Niger, both for fishing and transport (Levtzion and Hopkins, 1981, 110). Scarcity of wood near Lake Chad must have induced people to use alternative material and is probably at the base of the traditional reed-boats or *kadei* (Blache and Miton, 1962, 59).

An important group of traditional fishing equipment are the fences used to block of channels, connecting the main waterbody and the inundation plains, which are mostly spread along rivers. Trapped fish are then collected with a variety of fishing nets and traps. Alestidae are part of the typical taxa fished with such barrages (Sundström, 1972, 149). Besides by the differential destruction of their bones (see 5.3.1.), their absence at the studied localities should probably be explained by the lack of suitable channels for Alestidae fishing in the sites' vicinity. Another major traditional fishing technique near Lake Chad, also involving the build of a dam, is the separation of part of the inundated land from deeper parts of the lake, just after the peak of annual flooding (Krings, 2004). After a while, the oxygen content of the water drops and the fish are then attracted by letting fresh water in through a capture basin, where they can easily be caught. Throughout northern Nigeria, the use of different types of (basket) traps is also common (Reed *et al.*, 1967, 161-175). Basket traps are typical of the high water season, which is otherwise the dead season in the fishing cycle.

Spears and harpoons are mainly used in combination with other types of fishing gear, for example to kill large fish that have been trapped (Hopson, 1967). In African archaeological contexts barbed bone points have traditionally been brought into connection with fishing activities (Yellen, 1998). They are considered as a diagnostic element of the "Aqualithic Civilisation of Middle Africa"[12] (Holl, 2005). However, their near exclusive connection with fishing should be reconsidered, since spears can also be used to hunt large mammals or in conflicts between human groups (cf. Connah, 1981, 117). Among the bones objects from the Gajiganna Culture there were many bone points, but few barbed, and a connection with fishing was considered unlikely (Kottusch, 1999). Bone points and barbed hooks were recovered at Kolima Sud (MacDonald and Van Neer, 1994), while bone harpoons were found at Kobadi (Jousse and Chenal-Vélardé, 2000-2001), two Malian sites from the second and first millennium BC with a large fishing component in the economy.

In present northern Nigeria, hooks are usually used as part of multiple long lines, either baited or un-baited (Hopson, 1967). Simple line fishing normally has no commercial importance nowadays but is regarded as a leisure activity for non-professionals or children (Sundström, 1972, 150; Reed *et al.*, 1967, 157). There are a few records of iron fishhooks at archaeological sites in western Africa. At Ogo in Senegal, dated to the tenth-twelfth century AD, a hook was found in the head of a very large catfish (Chavane, 1985). Other examples are known from the Iron Age in the Middle Senegal Valley (McIntosh, 1999). Line fishing may also be documented archaeologically through the find of line sinkers. Peters (1995) for example tentatively interpreted perforated sherds from Mesolithic sites in Sudan that way.

6.5.4. Fish processing and conservation techniques

There are virtually no marks on the fish bones from the studied localities, which can reveal details on the ways fish were processed or conserved. This may be because people used techniques that simply did not leave any traces. At the Blé sites, archaeological features point to large-scale fish smoking activities (Holl, 2002, 73-103; see 2.2.3.2. for a description) and an important set of small jugs was found at Blé Mound E, suggesting the production of fish oil (Holl, 2002, 247). Holl (2002, 245, 247) also describes archaeological features from the Cameroonian part of the southern Lake Chad area, the earliest examples dating to the mid-first millennium AD, which he thinks were used to ferment fish during one or two days before consumption or further treatment. Fermentation apparently gives the fish an extra taste, which is much appreciated in the present Chad Basin (Blache and Miton, 1962, 35).

In present-day lake or riverside villages and fishing camps, part of the fish catches are sold or consumed fresh, shortly after their landing (Blache and Miton, 1962, 35). As described earlier, fishing usually has some peak periods in the course of a year during which there are surpluses, while at other points yields are very poor. Therefore, parts of the fish catches are treated in order to preserve them for times of shortage, and also to allow trading or exchanging them with other groups. The most important techniques for preserving fish in the Chad Basin today are sun-drying, char-burning and smoking, or a combination of these (Neiland, 1992). Figures on how long treated fish can be kept are variable, but it must have been in the range of at least a few months, thus long enough to provide a reserve during the slack fishing season. Most of the loss is caused by insect infestations and, when fish is transported, this may also cause some minor damage (Blache and Miton, 1962, 35).

Before drying, fish can be split open longitudinally, often with the exception of the head (Blache and Miton, 1962, Plate 49; Van Neer, 1995), although generally no particular pre-treatment occurs and fish are just dried whole. Longitudinal splitting seems to be the main part of fish processing that can leave traces, and it is presumably the cause of most of the observed cut-marks: on a lungfish cranial rib, on a clariid Weberian apparatus, on a *Parachanna obscura* ceratohyale and on a Nile perch exoccipitale. The chop-marks on a Nile perch cleithrum and a tilapia anal spine, may have occurred later, for example while subdividing the fish for consumption and removing inedible parts. The arid conditions in the Sahel are very favourable for sun-drying, and it seems to be the most widely spread fish conservation technique in the whole of this zone (Sundström, 1972, 29, 153-154). According to Blache and Miton (1962, 35), mainly small fish are dried in the Chad Basin and salt is never used. However, salting is the only method for fish-preserving reported by medieval Arab travellers visiting West Africa (Lewicki, 1974, 101; Levtzion and Hopkins, 1981, 108). Fish-smoking in kilns, mainly of large fish, is less common nowadays, probably because of wood shortages (Blache and Miton, 1962, 36; Reed *et al.*, 1967, 181). Nevertheless, such fish-smoking structures appeared to be common in the Houlouf area, including at the Blé sites (Holl, 2002, 245-246). Blé Mound E is even supposed to have been a special purpose site for fish-smoking and fish processing. Fish-smoking pits were

[12]This term was first used by Sutton (1974) to refer to Early and Middle Holocene archaeological assemblages in the southern Sahara and Sahel belt with globular pottery carrying wavy-line or dotted wavy-line decoration and with remains of aquatic fauna. He believed those assemblages were the remains of people with an aquatic economy and culture, representing a successful alternative to the adoption of agriculture.

also recorded at the Malian site Toguéré Galia, dating to around the end of the first-beginning of the second millennium AD, archaeologically visible in the form of pits filled with ashes and fish bones (Bedaux *et al.*, 1978). The production of *banda* is a widespread conservation technique in the Chad Basin today, but was only introduced in the 1930s (Krings, 2004). It involves cutting up fish into smaller chunks, which subsequently undergo sun-drying, charring and smoking and are mainly meant for trade with the south of Nigeria (Hopson, 1967). Small fish, which only need to keep for a few days, are put in a single layer on dry grass or on old grass mats (Reed *et al.*, 1967, 151). The grass is then fired in order to roast them slightly. Because of their high fat content, Alestidae are today also boiled to process them into oil (Sundström, 1972, 90, 153-154). The mentioned small jugs from Blé indicate that fish oil production may already have existed in the southern Lake Chad area in prehistoric times. Dried fish of lesser quality is nowadays sometimes pounded in mortars and then fashioned into balls, which can also preserve for a relatively long time (Daget, 1949). This way of processing fish seems to be impossible to trace from archaeozoological evidence. As can be deduced from the information summarised by fish taxon in Chapter 4, in the research area today, fish are usually prepared for consumption by roasting or smoking them, while cooking in pots appears to be done only rarely.

6.5.5. Fish trade

The only one of the studied sites with possible evidence for fish import is subrecent Galaga (NA 92/2C). On palaeo-ecological grounds it was assumed that at least part of the fish taxa, including Distichodontidae or Citharinidae and Nile perch, must have been brought in from elsewhere. It is furthermore striking that the site is one of the only locations of the Bama Deltaic Complex with numerous fish remains, but without turtle bones. This may imply that turtle meat was not a trade product. Presumably the fish imported at Galaga were caught in or near Lake Chad. Around the *firgi* and Blé sites, fishing was apparently very productive and part of the surpluses may have been kept aside for trade or exchange. The fish-smoking features at Blé can be taken as an extra argument for the processing of fish for trade at the sites. Trade with other areas is furthermore confirmed by the presence of a diversified ceramic production (Holl, 2002, 247). However, Holl interpreted the sites as seasonal fishing stations, open to fishing parties from the whole region, and this may also have resulted in a diverse pottery assemblage.

Using historical evidence, Barkindo (1999) describes trade in fish in Borno during the sixteenth century AD, both internal and with groups in the Mandara Mountains, Hausaland and oases in the Sahara. In the eighteenth and early nineteenth century AD the Lake Chad area, and in second place the rivers Benue and Niger, appear to have been the most productive fishing areas of Nigeria from where export was possible (Redmond, 1978). Redmond lists a few limitations to long distance trade in fish: its large volume compared to its value, the risk of rapid decay, and the expense and limited availability of suitable means of transport. He therefore postulates that trade must have been mainly intra-regional. Nonetheless, long-distance trade in fish from the Nile Valley to the Levant has been documented for the Late Chalcolithic, becoming more intense during the Bronze Age and later periods, when fish was also exported towards Anatolia, the Eastern Mediterranean islands and even Italy (Van Neer *et al.*, 2004a). Looking at ethnographic data, local trade or exchange between fishing groups and pastoralists or agriculturalists from the same area was probably more common than trade over long distances by middlemen, (Sundström, 1972, 142-145). However, at Galaga (NA 92/2C) the latter might have been the case, because the site seems to be located far from good fishing grounds. Fish traded or exchanged over larger distances was probably mostly dried, but could also have been in the form of a paste mixed with beans, or in the form of fish balls (Redmond, 1978), although it is clear that the archaeological visibility of such a product must be extremely low. Nowadays, taxa consumed by fishermen themselves differ from those used for commerce. The quality of fish declines with the distance from the fishing grounds and distant consumers therefore use dried fish as a seasoning for their food rather than as a primary ingredient (Sundström, 1972, 32).

6.5.6. Concluding remarks

From the previous paragraphs, it appears that at part of the studied sites we are dealing with specialised fishing groups, both in terms of a large dietary contribution of fish and a high technological level of fishing, e.g., the use of nets and boats. This is presumably the case for Corcoba (BF 97/5), although perhaps only its Late Stone Age occupation, and the *firgi* and Blé sites. Archaeologically, fishing groups have a high visibility because of their relatively large degree of sedentism. Specialised fishers only seem to occur near large waterbodies, since small lakes and pools are not suited for their economic strategies. However, alternative economies are possible in the same type of environment. The inhabitants from the sites in the Bama Deltaic Complex appear to have been opportunistic fishers, catching fish when it was easy, e.g., in the seasonal pools of the central Gajiganna area, but relying much less on fish when they were more difficult to obtain, such as in the rivers of the Bama-Konduga group. The Middle and Late Iron Age people from northern Burkina Faso also seem to have been fishing in an opportunistic way, while the Early Iron Age should probably be seen as a transitional period, still with a relatively high reliance on fish at some locations.

6.6. Hunted reptiles, birds and mammals

6.6.1. Frequency of consumption and dietary contribution

Looking at NISP data (Fig. 35), hunting and fowling usually seems to have been the least important economic strategy. Its relative importance is fairly stable throughout the studied localities, while that of fishing and domestic stock keeping varies more considerably. However, if live weights are taken into account as well (Fig. 36), differences in the economic importance of hunting and fowling between the studied contexts become clearer. Its contribution may generally be underestimated because (parts of) carcasses were left at the killing site and not taken to the settlement (see 5.2.). A greater importance of hunting than is recorded purely from the archaeofaunal remains also seems probable, considering the present popularity of bushmeat. According to Ntiamoa-Baidu (1997), it is eaten in West Africa by all classes of people and is more important than meat of domestics. Lewicki (1974, 91) has also stressed the part played by game in the West African diet in the first half of the second millennium AD, which was according to him even greater than today. Nevertheless, in present northern Burkina Faso the economic role of hunting is very small, which may largely be due to recent game depletion (Barral, 1977, 20-22).

6.6.1.1. Burkina Faso

The image for the Late Stone Age in Burkina Faso is distorted because of low numbers of recovered bones. As explained earlier, the only site with a large faunal sample, Corcoba (BF 97/5), seems to have been a specialised fishing site. Nevertheless, it is assumed that the importance of hunting and fowling was considerable during the Late Stone Age, which must have been one of the main reasons for sustaining a mobile lifestyle. When looking at NISP data from Iron Age contexts in northern Burkina Faso, hunting and fowling was apparently less important at the sites near Oursi than at those near Kissi and Saouga. At the former localities, there was probably less need to supplement the diet in this manner, since the large lake and its environs offered good opportunities for fishing (Oursi BF 94/45) or herding (Oursi village BF 97/13 and Oursi hu-beero BF 97/30). However, the image changes when meat weight is also taken into account because then the inhabitants of Early Iron Age Oursi (BF 94/45) also seem to have obtained much of their animal proteins through hunting.

6.6.1.2. Bama Deltaic Complex

Gajiganna phase I-IIb sites with faunal samples of considerable size all point to a limited importance of hunting and fowling. At Gajiganna phase IIc sites and especially at Gajiganna BII (NA 90/5BII), the strategy seems to have played a more significant role. Besides taphonomical reasons (see 5.2.), this may also be related to a (short) dry event and this will be further argued in 6.6.2.2. The sites of the Bama-Konduga group all have a low proportion of wild mammal, reptile and bird bones in their faunal assemblages. The limited importance of hunting and fowling at the sites is even clearer when also taking bone weight into account. At Gajiganna phase III, Iron Age and subrecent locations in the Bama Deltaic Complex hunting and fowling appears to have been relatively important, with between about 10 and 20 % of their bones being of hunted mammals, reptiles or birds.

6.6.1.3. *Firgi* area and Blé sites

At the *firgi* sites, NISP of hunted animals and wild birds only attain a few percent of all consumed animals. No hunted animals at all have been found in Late Stone Age Kursakata (NA 93/46) and they constituted not even half a percent of NISP in the site's subsequent Early Iron Age layers. Only in phase I at Mege do they make up over six percent of all identified consumed animal specimens. However, as can be expected, when meat weight is taken into account, the importance of hunting and fowling in the *firgi* area clearly increases. It was suggested earlier that the *firgi* sites might perhaps have been inhabited by ancestors of the Kotoko, a fishing group. The Kotoko themselves traditionally put their ancestry with the legendary Sao or So, who are said to have been hunters and fishermen. In any case, unlike the fishing component, the hunting component does not seem to be reflected in the archaeofauna. Compared to the *firgi* sites, the Blé sites have yielded more hunted animals. This cannot solely be due to coarser sampling, since it did not result in more bones of domestic animals, which should be equally affected by it, but must represent a real economic difference.

6.6.1.4. The broader archaeological and ethnographical framework

The savannahs of West Africa are rich in game biomass compared to its desert and forest zones, but its wild game is nevertheless still not as abundant as in the East African grasslands (Bourlière, 1963). The difference in game abundance between West and East Africa may have its origins in the different rainfall regimes of the two areas, unimodal in the former and bimodal in the latter, affecting also the grass cover necessary to sustain large populations of wild herbivores (Le Houérou, 1989, 3). In spite of higher game densities in savannahs than in deserts and rainforests, hunting appears to have been more important at West African archaeological sites in forests than in open environments (Van Neer, 2002a; Gautier and Van Neer, 2005). It therefore seems that, rather than to the potential of hunting, the importance of game is related to difficulties with keeping domestic stock, or with practising fishing. Present African hunter-gatherers are also confined to marginal areas, which are ill-suited for agro-pastoral lifestyles (Guenther, 1997). In early phases of the introduction of domestics, the importance of hunted animals may still have been substantial, as a kind of transitional phase (cf. Zvelebil, 1992). This is possibly how the large hunting component in the economy at Dhar Tichitt (Holl, 1985, 1986, 86-94) should be explained. The site is situated in the Malian Sahara, where the environment was more favourable than today at the time of its occupation in the late third, early second millennium BC. Unfortunately, quantitative data for sites in the Sahara, through which domestics arrived in West Africa (6.7.2.2.), are insufficient to investigate more thoroughly the importance of hunting in the area through time.

At present, West Africa no longer has any hunter-gatherer groups (Hitchcock, 1999), but judging from archaeological sites like Korounkorokalé (MacDonald, 1997a) and Pénténga (Frank *et al.*, 2001), isolated groups probably maintained a hunter-gatherer lifestyle until as recent as the late first millennium AD. Nonetheless, by the time of European contacts hunter-gathering lifestyles were abandoned everywhere in western and northern Africa (Robertshaw, 1999). Hunter-gatherers are today mainly confined to the south and the east of the African continent, with the Kung! and San as the best known examples (Hitchcock, 1999). Some think these are only able to maintain their lifestyle because they live in a kind of symbiosis with agriculturalists. Hunter-gatherer groups are generally flexible (*Ibid.*). They practise livestock keeping and farming to some extent but can turn back to a full hunter-gathering lifestyle when necessary. Among recent pastoral groups, hunting is generally not important (e.g., Evans-Pritchard, 1947, 74). However, Zvelebil (1992) has argued that there are three important reasons why food producing communities continue hunting, either as a risk buffering strategy, for socio-ideological reasons, and in response to a demand of furs or other specific wild animal resources. From the previous two chapters, it is also clear that, protection of crops and domestic herds is another important reason to continue hunting and fowling.

6.6.2. Reconstruction of the terrestrial environment

From the ecological data mentioned in Chapter 4, it appears that a few factors determine what kind of wild reptiles, birds and mammals can be found in a certain habitat. These include the nature of the vegetation, the availability of (drinking) water and the presence of man. Present game populations must be used with caution to estimate their past potential as a food resource because the introduction of firearms, with the coming of the Europeans, has caused a rapid decline. Other factors that caused a recent diminution of wild animals are increased competition with domestics and habitat loss due to expanding agriculture. Increased hunting pressure initially affected large game and has therefore led to a shift of hunting activities towards smaller species. This may perhaps explain why giant pouched rat (*Cricetomys gambianus*), which is a very popular food animal today, was only sporadically found in the studied faunal assemblages. However, the trend towards hunting of smaller animals may have set in much earlier. For South Africa for example, Deacon (1984) situates it as far back as the Pleistocene-Holocene transition (see Stiner *et al.* 1999 for small animal exploitation at Palaeolithic Mediterranean sites). Flannery (1969) has connected the exploitation of small taxa with the beginning of sedentism and labelled it "broad spectrum exploitation". He believed that in the Near East this process preceded and initiated domestication, and later authors have further confirmed this view (e.g., Davis, 2005). Logically, sedentary communities put more pressure on the wild fauna around their habitation sites than mobile groups, because of their permanent presence in a limited habitat. This can lead to a depletion of game in the immediate environment of the site and to changes in the species spectrum. The latter is mainly due to a different sensibility to (over) hunting, depending on the taxon. Among the antelopes found, kob (*Kobus kob*) for example, is particularly vulnerable while bush duiker (*Sylvicapra grimmia*) can withstand heavy exploitation.

Which of the taxa available in a certain environment are caught, depends on the hunting techniques and strategies applied (see 6.6.3.2.), but also on regulations and other culturally determined factors in the human group. Ntiamoa-Baidu (1997) has recorded, for example, that hunter-gatherer communities use a wider range of the available species than pastoral communities. The taxa exploited by the inhabitants of the investigated archaeological sites should therefore also allow us to better understand the nature of hunting and fowling and its economic importance. In the following paragraphs, the hunted taxa are discussed in more detail and it is attempted to reconstruct the former terrestrial environment. To this end, an overview of the taxa found that are tied to specific ecological conditions is given by site and phase in Table 11 (birds) and Table 13 (reptiles and mammals), sorted according to habitat requirement. Table 10 indicates species richness of the hunted animals. Relatively low numbers of bones of wild mammals, reptiles and birds in the studied assemblages, combined with large species diversity, complicate the tracing of geographic and diachronic trends. These trends can, moreover, be related to different catchment areas, rather than to changes in the local environment around the sites.

6.6.2.1. Burkina Faso

Unlike for the fish remains, the complete research area in northern Burkina Faso is discussed together for the wild mammals, reptiles and birds caught, because it is assumed that environmental circumstances must have been similar near contemporary sites. Nevertheless, habitats may have slightly varied locally, for example because of the presence of different types of waterbodies. Even though only three small Late Stone Age assemblages from Burkina Faso are available, it is clear that their species composition deviates markedly from the Iron Age contexts in the area. Two taxa have been found among faunal material from the former period that have not been attested at any of the younger locations, notably savannah cane rat (*Thryonomys swinderianus*) at Corcoba (BF 97/5) and African buffalo (*Syncerus caffer*) at Tin Akof (BF 94/133). Both species occur mainly in areas near water with a dense (grass) cover and do not live in northern Burkina Faso today. Their disappearance by the onset of the Iron Age may point to a reduction of such habitats caused by (gradual) aridification. Alternatively, grass cover could have been reduced because of grazing and trampling by herds of domestic animals. Based on the presence of bohor reedbuck (*Redunca redunca*) and kob at Late Stone Age Corcoba, Jousse (2004a, Table 22) has reconstructed a *"forêt claire"* for the environment near the site. However, palynological studies have shown that the vegetation around Oursi was already Sahelian during the Middle Holocene (Andres *et al.*, 1996). Bohor reedbuck and kob may have been attracted to the area by the seasonally inundated ground around the lake, which also probably provided a locally more lush environment. With the onset of the Iron Age taxa from open, arid savannahs, for example true hare (*Lepus capensis/saxatilis*) and red-fronted gazelle (*Gazella rufifrons*), appear in the faunal assemblages of sites in northern Burkina Faso. This is probably associated with environmental deterioration, although enlargement of the catchment area could equally have resulted in the presence of the species. Summarising, the wild terrestrial fauna from Burkina Faso thus seems to indicate that an important environmental degradation took place in the period between the Late Stone Age and the Iron Age, approximately covering the entire first millennium BC.

The Iron Age sites of Burkina Faso usually contain a relatively large amount of monitor lizard (*Varanus* sp.) remains, which may be read as the exploitation of small taxa by sedentary groups (see 6.6.2.). The only bone identified to species level pertained to *V. exanthematicus*, an inhabitant of open, dry areas. There are no indications for a heavy exploitation of crocodile (*Crocodylus* sp.) during any of the periods under study. Only from one of the sites at Saouga (BF 94/120) were a relatively large number of crocodile bones recovered. From the sites near the lake of Oursi, only Late Stone Age layers at Corcoba (BF 97/5) and Oursi hu-beero (BF 97/30) have yielded crocodile. It is not clear if the observations made are meaningful. When also looking at the Nigerian locations studied, crocodile bones are mainly found at sites near large waterbodies, although small amounts of water are sufficient for the animals' survival.

Most of the sites in Burkina Faso have yielded remains of waterbirds that must have been caught in, or near, the waterbodies in their vicinity. Waterbirds were most common at Oursi (BF 94/45), which is probably related to a high aquatic component in the economy at the site (see 6.5.1.1.). High numbers of ostrich eggshell fragments at Early Iron Age Oursi, when compared to the other studied locations, is remarkable and may point to a desert connection at the site.

Besides Sahelian mammals, most studied Iron Age sites from Burkina Faso also yielded taxa of more lush environments, such as kob and oribi (*Ourebia ourebi*), which no longer seem to occur in the area today. The northern limit of their distribution coincides approximately with the transition between the Sahel and Sudan zone. Where such taxa are lacking, e.g., at Kissi 22 (BF 96/22), this is probably due to small sample size. The terrestrial fauna thus points to the existence of a mosaic landscape during the Iron Age, with a mixture of drier and more humid parts. At some time after the Iron Age, the lush vegetation was apparently reduced. This was also concluded from archaeobotanical research, which provided proof for a higher density of wood taxa during the Iron Age than at present (Höhn, 2002; Kahlheber, 2003, 235; Höhn *et al.*, 2004). Proposed explanations were a higher annual rainfall and less anthropogenic pressure at that time. However, Kahlheber (2003, 235) suggested that changes might have taken place as recently as the past half century.

The terrestrial fauna does not show any clear indications for a degradation of the environment in the course of the Iron Age that may have ultimately led to the abandonment of the settlement mounds. High species richness at two of the Late Iron Age sites, Oursi hu-beero (BF 97/30) and Saouga 94/120 (BF 94/120), may point to the use of alternative resources in times of stress, but Early Iron Age layers at Oursi (BF 94/45) also had high species richness. Nevertheless, there is proof for environmental changes from contemporary sites in the Malian Méma, not far to the north of the research area in Burkina Faso, where the mammal remains pointed to diminishing floodwaters around AD 1000 (MacDonald and Van Neer, 1993).

The terrestrial fauna from Iron Age Burkina Faso generally points to drier circumstances than seem to have

			minimal number of species	total number of remains	species richness
1	BF 94/96	LSA	1	5	0,000
2	BF 97/5	LSA	6	27	1,517
3	BF 94/45	LSA	3	10	0,869
4	BF 94/45	EIA	17	135	3,262
5	BF 97/13	EIA	3	4	1,443
6	BF 97/13	MIA	8	13	2,729
7	BF 97/13	LIA	9	25	2,485
8	BF 97/30	LIA	14	29	3,861
9	BF 96/22	MIA	9	86	1,796
10	BF 97/31	LIA	9	66	1,909
11	BF 94/120	LIA	23	244	4,002
12	BF 95/7	LIA	13	123	2,494
13	NA 90/5C	Gaj IIa	5	7	2,056
14	NA 90/5A	Gaj IIb	14	88	2,904
15	NA 90/5BI	Gaj IIa/b	9	24	2,517
16	NA 90/5BII	Gaj IIc	12	102	2,378
17	NA 93/42	Gaj I	3	6	1,116
18	NA 99/65	Gaj IIb	7	8	2,885
19	NA 99/65	Gaj IIc	7	13	2,339
20	NA 97/37	Gaj III	7	56	1,491
21	NA 97/33	Gaj I	5	10	1,737
22	NA 97/33	Gaj IIa/b	1	12	0,000
23	NA 96/45	Gaj IIa/b	4	8	1,443
24	NA 95/1	Gaj IIa/b	6	36	1,395
25	NA 97/26	IA	4	22	0,971
26	NA 99/75	IA	7	22	1,941
27	NA 97/13	IA	4	6	1,674
28	NA 92/2C	subrec	13	96	2,629
29	NA 93/46	EIA	10	72	2,104
30	NA 94/7	I	5	154	0,794
31	NA 94/7	II	2	11	0,417
32	NA 94/7	III	6	22	1,618
33	NA 94/7	IV	5	6	2,232
34	NA 93/45	I	6	31	1,456
35	NA 93/45	II	11	58	2,463
36	NA 93/45	IIIa	6	19	1,698
37	NA 93/45	IIIb	6	17	1,765
38	NA 93/45	IV	6	12	2,012
39	Blé A		7	19	2,038
40	Blé B		4	46	0,784
41	Blé C		5	34	1,134
42	Blé E		6	32	1,443

Table 10: Species richness of hunted mammals, reptiles and birds by site and phase. Legend to Fig. 40

existed near any of the Nigerian localities throughout the whole of the studied chronological sequence. Some species of arid open areas, notably Sahelian giant tortoise (*Geocheolone sulcata*), sandfox (*Vulpes pallida*) and African hedgehog (*Atelerix albiventris*), have been attested for the Iron Age in Burkina Faso but at none of the Nigerian sites. The sites in Burkina Faso usually also yielded more hare and gazelle bones.

6.6.2.2. Bama Deltaic Complex

The wild mammals from Gajiganna and Iron Age sites in the Bama Deltaic Complex, more specifically baboon (*Papio anubis*) (NA 99/65), vervet or patas monkey (*Cercopithecus aethiops/patas*) (NA 90/5A, NA 99/65) and especially forest duiker (*Cephalopus* cf. *rufilatus*) (NA 90/5BI, NA 99/65, NA 99/75), indicate the presence of at least relic forests at that time, presumably near water edges. Finds of forest duiker were used earlier to propose the existence of denser vegetation during the prehistoric habitation of the area than at present, although archaeobotanical investigations did not yield any forest plants (Breunig, 1995; Breunig *et al.*, 1996). The animal does not occur presently in north-eastern Nigeria, but it probably disappeared only in the course of the last century, since Schultze (1968, 128) still mentions it for Borno during the early twentieth century AD. Other animals identified from the Bama Deltaic Complex, like bustard (Otididae) (NA 90/5BI, NA 92/2C), are indicative for open country. Sahelian species are completely missing at the sites of the Gajiganna Culture, but they appear in the faunal spectrum from the Iron Age onwards. This may be indicative for an environmental degradation

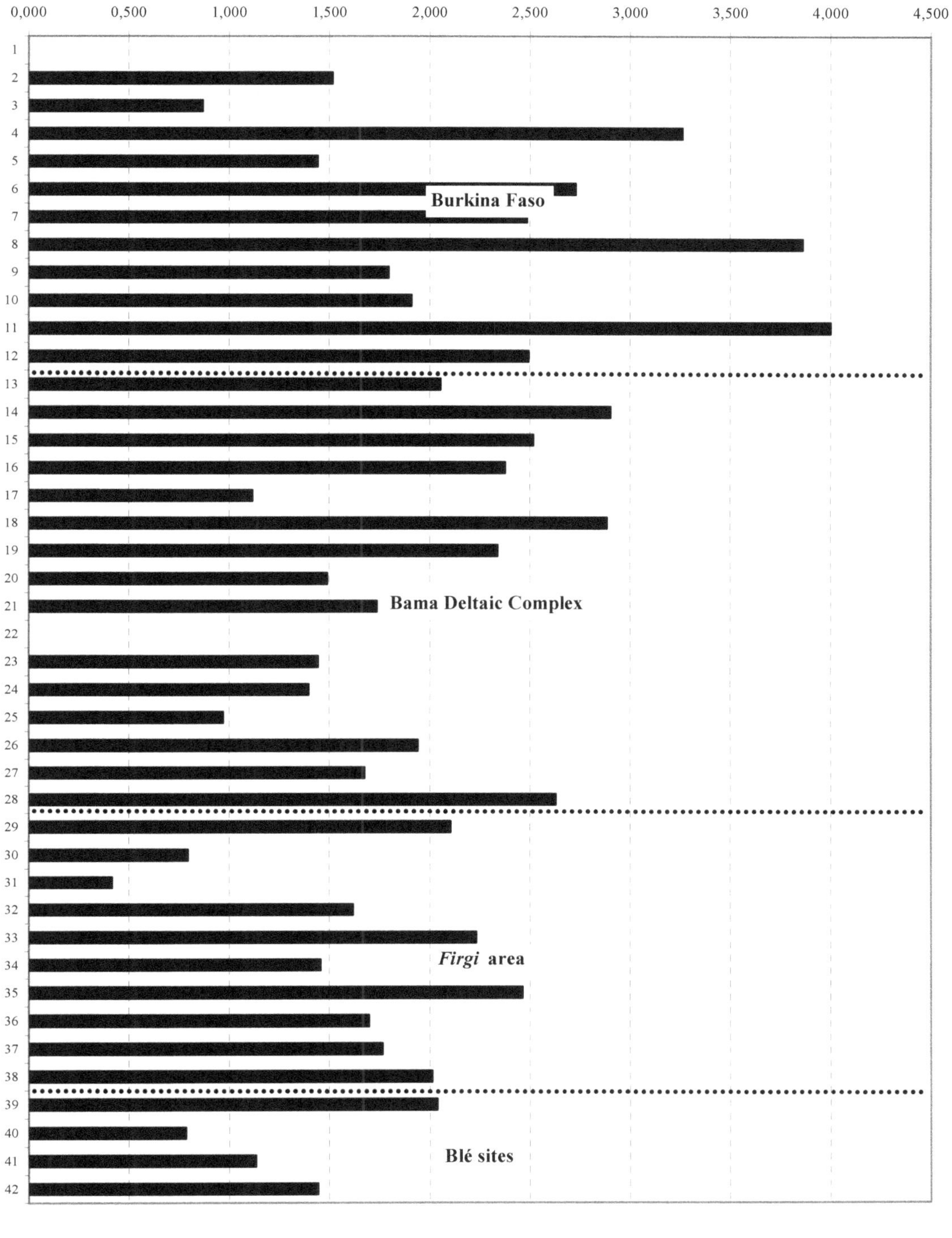

Fig. 40: Species richness of hunted mammals, reptiles and birds by site and phase

that took place in the course of the first millennium BC; contemporary with the one recorded for northern Burkina Faso. In addition, archaeobotanical research has shown an impoverishment of the vegetation in the Bama Deltaic Complex during the last 3000 years, attributed to over-exploitation (Breunig *et al.*, 1996). The consequences of environmental degradation were apparently less extreme than in Burkina Faso: no species seem to have disappeared, possibly because the pools in the clay plains of the Bama Deltaic Complex created better circumstances locally. In summary, it seems that a mixed landscape, with open and forested parts, needs to be proposed for all studied periods in the Bama Deltaic Complex, but with a degradation of the vegetation and an increase of Sahelian faunal elements through time.

The species composition of the site Gajiganna A (NA 90/5A) differs from that of other locations in the Bama Deltaic Complex because of the high amount of warthog (*Phacochoerus africanus*) and bohor reedbuck remains. Perhaps this should be read as habitation during a dry spell, since both species need to drink regularly and may have been attracted by remaining water in arid periods, where they could easily be caught. Breunig *et al.* (1996) also recorded an increase in wild carnivores in the upper layers at Gajiganna A. They saw this as a possible indication for concentration of herbivores around water in a period of increased aridity, which attracted carnivores that then became, in their turn, an easier prey for humans.

		Water →												Terrestrial
		Phalacrocorax africanus	Anatidae	*Ardea* sp.	Ciconiidae	Charadriiformes	*Threskiornis aethiopica*	*Bubulcus ibis*	*Egretta* sp.	*Francolinus* sp.	*Numida meleagris*	*Coturnix* sp.	Cucilidae	Otididae
BF 94/45	LSA	-	8	-	-	-	-	-	-	-	-	-	-	-
BF 94/45	EIA	-	9	1	-	-	-	-	-	-	-	1	-	-
BF 97/13	EIA	-	1	-	-	-	-	-	-	-	-	-	-	-
BF 97/13	MIA	-	-	-	-	-	-	-	-	-	1	-	-	-
BF 96/22	MIA	-	1	-	-	-	-	-	-	-	-	-	-	-
BF 97/31	LIA	-	1	-	-	-	-	-	-	-	-	-	-	-
BF 94/120	LIA	-	3	-	-	-	-	1	-	1	1	-	-	-
BF 95/7	LIA	-	-	1	-	-	-	-	-	-	1	-	-	-
NA 90/5C	Gaj IIa	-	-	-	1	-	-	-	-	-	1	-	-	-
NA 90/5A	Gaj IIb	-	6	2	2	-	-	-	-	1	3	-	-	-
NA 90/5B	Gaj IIa/b	-	1	-	-	-	-	-	-	-	1	-	-	1
NA 90/5BII	Gaj IIc	-	5	-	-	-	-	-	-	-	34	-	-	-
NA 93/42	Gaj I	-	-	-	-	-	-	-	-	-	-	-	-	-
NA 93/36	Gaj IIa	-	-	-	-	-	-	-	-	-	-	-	-	-
NA 99/65	Gaj IIb	-	1	-	-	-	-	-	1	-	3	-	-	-
NA 99/65	Gaj IIc	-	-	-	-	-	-	-	-	-	-	-	-	-
NA 97/37	Gaj III	-	36	-	-	-	-	1	-	-	-	-	-	-
NA 96/45	Gaj IIa/b	-	-	-	-	-	-	-	-	-	1	-	-	-
NA 95/1	Gaj IIa/b	-	-	-	-	-	-	-	-	-	-	-	1	-
NA 99/75	IA	-	6	-	-	-	-	-	-	-	-	-	-	-
NA 97/13	IA	-	-	-	-	-	-	-	-	-	6	-	-	-
NA 92/2C	rec	-	20	-	11	-	1	-	-	2	2	-	-	1
NA 93/46	EIA	1	27	-	-	2	-	-	-	-	-	-	-	-
NA 94/7	I	-	2	-	-	-	-	-	-	-	-	-	-	-
NA 94/7	II	-	-	-	-	-	-	-	-	-	-	-	-	-
NA 94/7	III	-	2	-	-	-	-	-	-	-	5	-	-	-
NA 94/7	IV	-	1	-	-	-	-	-	-	-	-	-	-	-
NA 93/45	I	-	10	-	-	-	-	-	-	-	-	-	-	-
NA 93/45	II	-	13	-	-	1	1	-	-	-	-	-	-	-
NA 93/45	IIIa	-	2	-	-	-	-	-	-	-	-	-	-	-
NA 93/45	IIIb	-	5	-	-	-	-	-	-	-	-	-	-	-
NA 93/45	IV	-	6	1	-	-	-	-	-	-	-	-	-	-
Blé A		-	6	1	-	-	-	-	-	-	-	-	-	-
Blé B		-	1	-	-	-	-	-	-	-	-	-	-	-
Blé C		-	14	-	-	-	-	-	-	-	-	-	-	-
Blé E		-	5	-	-	-	-	-	-	-	-	-	-	-

Table 11: Birds caught by site and phase, sorted according to habitat requirements

A dry event is also proposed as an explanation for the large numbers of bones of hunted animals at Gajiganna BII (NA 90/5BII). Most of the species found need to drink regularly and when water was scarce, they probably gathered near remaining pools, where they were an easy prey. The large amounts of guineafowl (*Numida meleagris*) recorded from the site may also be explained in this way, since the animal's most critical habitat feature appears to be the availability of surface water during the dry season. However, as mentioned in 5.2., taphonomic conditions at Gajiganna BII, as a possible butchery or waste dumping area, may also have influenced its species composition.

At Zilum (NA 97/37), the absence of remains from hunted mammals, apart from two mongoose (Herpestidae) civet or genet (Viverridae) bones, is remarkable. The relatively large proportion of hunting and fowling at the site, in terms of NISP, is almost entirely due to the presence of remains of waterbirds and crocodile. Its species richness for hunted taxa is also low compared to the older sites from the central Gajiganna area. Since some wild mammal bones could be observed in the newly excavated trenches at Zilum, their absence in the assemblage discussed here must be due to small sample size. However, a smaller component of mammal hunting in the economy, compared to the older Gajiganna sites still needs to be supposed, which is perhaps connected to the early-urban nature of the site. It is also striking that Zilum is the only one of the sites in the Bama Deltaic Complex that has yielded crocodile remains. Together with the relatively large amounts of waterbirds, this may indicate increased hunting near aquatic environments.

Gajiganna sites of the Bama-Konduga group not only have smaller proportions of hunted animals in their assemblages compared to those of the central Gajiganna area, but their species richness is also lower. The main animal caught at these sites was apparently bush duiker. Perhaps the difference should be explained by their different environmental setting, without (seasonal) pools and associated locally denser vegetation that attracted various game animals. Unfortunately, there are no archaeobotanical data against which the archaeofauna can be compared.

6.6.2.3. *Firgi* area

The predominantly hunted animal in the *firgi* area seems to have been kob. This species needs a permanent proximity of water and normally does not occur in areas with much seasonal extremes. Together with the large and diverse fish faunal assemblages found in the *firgi* area, its presence stresses the generally wet nature of the environment. This is furthermore emphasised by a few finds of hippo (*Hippopotamus amphibius*), a semi-aquatic mammal. Parallel with the large fishing component in the economy of the *firgi* area (see 6.5.1.3.), many remains of ducks, geese and other waterbirds were found in the sites' assemblages, except in the first two phases at Mege (NA 94/7), which seems to coincide with relatively smaller amounts of fish, at least in Mege phase I. The reed cormorant (*Phalacrocorax africanus*) from Kursakata (NA 93/46), a piscivorous species, may have been attracted by the abundance of fish, which made it more vulnerable to human predation, or it could have got entangled in the fishing nets.

	Eco-climatic zones						
Species	Saharan	Saharo-Sahelian	Sahelian	Sudano-Sahelian	Northern Sudanian	Southern Sudanian	Guinean/Congolian
Gerbillus sp.	+	(+)	-	-	-	-	-
Vulpes pallida	(+)	+	+	(+)	-	-	-
Desmodilliscus braueri	(+)	+	+	+	(+)	-	-
Lepus capensis	+	+	+	+	+	-	-
Hyaena hyaena	+	+	+	+	+	+	-
Felis libyca	(+)	+	+	+	+	+	-
Gazella rufifrons	-	+	+	(+)	-	-	-
Arvicanthis niloticus	-	(+)	+	+	+	-	-
Lepus saxatilis	-	(+)	+	+	+	+	-
Felis caracal	-	+	+	+	+	+	-
Cercopithecus patas	-	-	+	+	-	-	-
Cercopithecus aethiops	-	-	+	+	-	-	-
Crocuta crocuta	-	-	+	+	+	+	-
Euxerus erythropus	-	-	+	+	+	+	-
Felis serval	-	-	+	+	+	+	-
Hippotragus equinus	-	-	+	+	+	+	-
Redunca redunca	-	-	+	+	+	+	-
Damaliscus korrigum	-	-	(+)	+	+	+	-
Papio anubis	-	-	(+)	+	+	+	-
Orycteropus afer	-	-	+	+	+	+	+
Phacochoerus aethiopicus	-	(+)	+	+	+	+	+
Kobus kob	-	-	-	(+)	+	+	+
Alcelaphus buselaphus	-	-	-	(+)	(+)	+	+
Sylvicapra grimmia	-	-	-	(+)	(+)	+	+
Loxodonta africana	-	-	-	+	+	+	+
Hippopotamus amphibius	-	-	-	+	(+)	+	+
Syncerus caffer	-	-	-	+	(+)	+	+
Ourebia ourebi	-	-	-	(+)	+	+	+
Cricetomys gambianus	-	-	-	-	+	+	+
Tragelaphus scriptus	-	-	-	-	(+)	+	+
Cephalophus rufilatus	-	-	-	-	(+)	+	+

Table 12: Mammal species identified and the eco-climatic zones where they occur (after Le Houérou, 1989, Table 2)

Although species richness of hunted animals in the *firgi* is usually high thanks to the variety of bird taxa caught, there is generally less variation in the mammal taxa found than in the Bama Deltaic Complex. Animals may have been more dispersed, hence more difficult to catch, because the area was wetter. Only kob was an easy prey probably because of its territorial behaviour. The territoriality of the animal can even lead to its over-exploitation and ultimate disappearance, as seems to have been the case after phase I at Mege (NA 94/7). A similar heavy exploitation of kob in early habitation phases can be observed at Jenné-Jeno (250 BC-AD 1400) (MacDonald, 1995). Neither at Mege, nor at Jenné-Jeno, kob seems to have been replaced in later phases by other antelope species that were less easy to catch. Nevertheless, species richness is higher in Mege's last two phases, than in its first two, which may indicate that alternative wild animal resources were being exploited after the disappearance of kob. Unlike in the Malian Méma (MacDonald and Van Neer, 1994), a floodplain environment comparable to the *firgi*, larger mammals of wet environments, such as hippo, waterbuck (*Kobus ellipsiprymnus*) and sitatunga (*Tragelaphus spekei*), are missing, or very rare, at the *firgi* sites although they must have been living in the area (Happold, 1987).

The environmental changes during the first millennium BC, postulated for both Burkina Faso and the Bama Deltaic Complex, are not observable in the *firgi*. However, during later phases, phase III at Mege (NA 94/7) and Phase II at Ngala (NA 93/45), Sahelian species are added to the faunal spectrum, which seems to fit with the possible presence of gazelle in Daima II and Daima III layers at Daima (Connah, 1981, Table 8.3). As remarked earlier, such changes may also be due to an enlarged catchment area, rather than to an environmental deterioration.

6.6.2.4. Blé sites

As at the *firgi* sites, bones of waterbirds are relatively numerous at the Blé sites, associated with a heavy exploitation of the aquatic environments (see 6.5.1.3.). The two areas also have a predominance of kob in common, but the low species richness of hunted taxa at the Blé sites, except at Blé Mound E, is different. Nevertheless, compared to the *firgi* area, bohor reedbuck is a new species at the Blé Mound Complex. The animal is typical for unstable grasslands, with alternation of floods and dry periods. Perhaps the large numbers of bohor reedbuck at the Blé sites should be attributed to their different geographical location compared to the *firgi*. They lie in the bend of a fossil river channel and were therefore probably closer to the source of the floods, which may have resulted in larger seasonal extremes. However, as said, kob does not thrive well in such seasonal extremes. Besides bohor reedbuck, buffalo is also a new species when comparing the Blé assemblages with the *firgi*.

6.6.3. Seasonality and hunting techniques

6.6.3.1. Seasonality

As already partially indicated in the previous paragraphs, there is seasonal variation in the presence and abundance

		Congolian ⟶																				Sahelian
		Cephalophus cf. *rufilatus*	*Tragelaphus scriptus*	*Ourebia ourebi*	*Syncerus caffer*	*Thryonomys swinderianus*	*Hippopotamus amphibius*	*Loxodonta africana*	*Sylvicapra grimmia*	*Sylvicapra grimmia/Ourebia ourebi*	*Kobus kob*	*Phacochoerus africanus*	*Papio anubis*	*Alcelaphus buselaphus/Damiliscus lunatus*	*Redunca redunca*	*Hippotragus equinus*	*Cercopithecus aethiops/patas*	*Lepus capensis/saxatilis*	*Gazella rufifrons*	*Vulpes pallida*	*Atelerix albiventris*	*Geocheolone sulcata*
BF 94/133	LSA	-	-	-	1	-	-	-	-	-	-	-	-	-	-	-	-	-	-	-	-	-
BF 97/5	LSA	-	-	-	-	7	-	-	-	-	3	-	-	-	2	-	-	-	-	-	-	-
BF 94/45	LSA	-	-	-	-	-	-	-	-	-	-	-	-	-	-	-	-	-	-	-	-	-
BF 94/45	EIA	-	-	1	-	-	-	-	2	13	1	2	-	-	5	-	-	15	30	-	1	-
BF 97/13	EIA	-	-	-	-	-	-	-	-	-	-	-	-	-	-	-	-	-	-	-	-	-
BF 97/13	MIA	-	-	-	-	-	-	-	-	1	-	1	-	-	1	-	-	4	1	1	-	-
BF 97/13	LIA	-	-	-	-	-	-	-	-	1	-	1	-	-	1	-	-	7	2	-	-	-
BF 97/30	LIA	-	-	-	-	-	-	1	-	1	-	1	-	-	-	-	-	6	-	-	-	1
BF 96/22	MIA	-	-	-	-	-	-	-	-	2	-	1	-	-	-	-	-	13	6	-	3	-
BF 97/31	LIA	-	-	-	-	-	-	-	2	-	-	-	-	-	1	-	-	17	6	-	-	-
BF 94/120	LIA	-	-	2	-	-	-	-	5	5	1	4	-	-	-	3	-	50	25	1	1	-
BF 95/7	LIA	-	-	1	-	-	-	-	1	8	-	2	-	-	-	-	-	19	4	2	4	-
NA 90/5C	Gaj IIa	-	-	-	-	-	-	-	2	-	-	-	-	-	1	-	-	-	-	-	-	-
NA 90/5A	Gaj IIb	-	-	-	-	-	-	-	2	-	-	19	-	-	32	-	4	-	-	-	-	-
NA 90/5B	Gaj IIa/b	3	-	7	-	-	-	-	-	-	-	3	-	-	6	-	-	-	-	-	-	-
NA 90/5BII	Gaj IIc	-	1	-	1	-	-	-	9	-	-	8	-	-	1	1	-	-	-	-	-	-
NA 93/42	Gaj I	-	-	-	-	-	-	-	-	-	-	1	-	-	-	-	-	-	-	-	-	-
NA 99/65	Gaj IIb	1	-	-	-	-	-	-	1	-	-	-	-	-	-	-	2	-	-	-	-	-
NA 99/65	Gaj IIc	-	-	-	-	-	-	-	1	-	-	-	2	-	3	-	3	-	-	-	-	-
NA 97/33	Gaj I	-	-	-	-	-	-	-	-	1	-	2	-	-	-	-	-	-	-	-	-	-
NA 97/33	Gaj IIa/b	-	-	-	-	-	-	-	2	10	-	-	-	-	-	-	-	-	-	-	-	-
NA 96/45	Gaj IIa/b	-	-	-	-	-	-	-	-	2	-	1	-	-	-	-	-	-	-	-	-	-
NA 95/1	Gaj IIa/b	-	-	-	-	-	-	-	12	17	-	3	-	-	1	-	-	-	-	-	-	-
NA 97/26	IA	-	-	-	-	-	-	-	-	-	-	-	-	-	-	-	-	1	1	-	-	-
NA 99/75	IA	3	-	-	-	-	-	-	-	-	-	-	-	-	1	-	-	-	1	-	-	-
NA 97/13	IA	-	-	-	-	-	-	-	-	-	-	-	-	-	1	-	-	1	-	-	-	-
NA 92/2C	subrec	-	-	-	-	-	-	-	1	-	-	1	-	-	-	-	-	7	9	-	-	-
NA 93/46	EIA	-	-	1	-	-	1	-	-	4	2	1	-	-	-	-	-	-	-	-	-	-
NA 94/7	I	-	-	-	-	-	-	-	-	1	55	1	-	-	-	-	-	-	-	-	-	-
NA 94/7	II	-	-	-	-	-	-	-	-	4	3	-	-	-	-	-	-	-	-	-	-	-
NA 94/7	III	-	-	-	-	-	-	-	4	-	1	-	-	-	-	-	-	-	8	-	-	-
NA 94/7	IV	-	-	-	-	-	-	-	-	1	-	-	-	-	-	-	-	-	-	-	-	-
NA 93/45	I	-	-	-	-	-	-	-	-	-	7	-	-	-	1	-	-	-	-	-	-	-
NA 93/45	II	-	-	-	-	-	-	-	-	-	6	-	-	-	-	-	-	1	1	-	-	-
NA 93/45	IIIa	-	-	-	-	-	1	1	-	-	4	2	-	-	-	-	-	-	-	-	-	-
NA 93/45	IIIb	-	-	-	-	-	-	-	-	-	5	-	-	-	-	-	-	-	-	-	-	-
NA 93/45	IV	-	-	-	-	-	-	-	-	-	1	-	-	-	-	-	-	-	-	-	-	-
Blé A		-	-	-	-	-	-	-	-	-	4	1	-	-	1	-	-	-	-	-	-	-
Blé B		-	-	-	-	-	-	-	-	1	29	-	-	-	6	-	-	-	-	-	-	-
Blé C		-	-	-	-	-	-	-	-	-	7	-	-	-	4	-	-	-	-	-	-	-
Blé E		-	-	-	3	-	-	1	-	-	11	1	-	-	2	-	-	-	-	-	-	-

Table 13: Reptiles and mammals caught by site and phase, sorted according to habitat requirements (see Table 12)

of wild animals, which affects their chances of being caught. Some of the bird species identified from the studied sites are seasonal visitors to the area. Spur-winged goose (*Plectropterus gambensis*), identified at Gajiganna A (NA 90/5A), Gajiganna BII (NA 90/5BII), Elkido North (NA 99/75), phase I and III at Mege (NA 94/7), all phases at Ngala (NA 93/45) and all Blé sites, can for example be found between February and May, while quail, found at Early Iron Age Oursi (BF 94/45), is a winter visitor to western Africa. A cuckoo (Cuculidae) bone found at Kariari C (NA 95/1), may point to rainy season habitation, because this is the part of the year when most species of the family visit the arid parts of Africa. Many herbivores are water dependent and therefore most easily hunted near remaining water in the dry season, while in the wet season the animals are dispersed and less visible because vegetation hampers pursuit (e.g., Evans-Pritchards, 1947, 73; Schnelle, 1971, 47). Concentrations of herbivores in turn attract carnivores, which are thus also most vulnerable for human predation during the dry season. Judging from the predominance of water dependent species at the sites, hunting was probably a mainly dry season activity in Burkina Faso and the Bama Deltaic Complex. Presently hunting does mainly take place during the dry season (Aiyedun, 1991, 1996). An additional reason to the ones already mentioned for dry season hunting, is that this part of the year is a dead period for agricultural communities, except in areas where floodplain cropping is carried out. In the *firgi* area and at the Blé sites hunting may have occurred spread throughout the year, since it concentrated on kob, which was presumably living near the settlements year-round. Nevertheless, there are also examples in the literature for hunting during high floods, when animals are concentrated in dry areas (Brelsford, 1946, 129). As mentioned before, the floods in the *firgi* area and near Blé occur during the local dry season.

Because game and other wild resources are not evenly distributed through time and space, hunter-gatherer groups are usually highly mobile and have therefore a very limited archaeological visibility in the landscape. This was formulated earlier as an explanation for poor archaeological finds from the Late Stone Age in Burkina Faso. Another point related to seasonality made by Deacon (1984), based on ethnographical evidence from Botswana, is that small animals yield less meat than larger ones, but that they are killed more often and catches are more evenly spread throughout the year. This brings us back to the discussion at the beginning of this chapter, that frequency of consumption may be more important than meat yield.

6.6.3.2. Hunting gear

Only a few direct finds of hunting equipment can be reported because most gear was presumably made from organic, and thus perishable, material. At least part of the microliths from the Dori facies and part of the retouched arrowheads from the Tin Akof facies in Late Stone Age Burkina Faso were probably used as arrow or spearheads. From the Iron Age in the area, there are finds of pieces of iron arrows and spears that have been interpreted as hunting gear (Magnavita *et al.*, 2002; von Czerniewicz, 2004, 137). At the Gajiganna sites in the Bama Deltaic Complex possible finds of hunting equipment include bone harpoons and points, besides stone arrowheads (Breunig, 1995).

Looking at historically and ethnographically documented African hunting techniques, it seems that we should distinguish between two main groups: the active and the passive. The main gear used during active hunting includes bows and arrows (sometimes poisoned), spears, nets, knives, cutlasses and machetes (Evans-Pritchard, 1947, 73; Lewicki, 1974, 91-92; Levtzion and Hopkins, 1981, 83, 185, 283; Aiyedun, 1996). Active hunting can be done either in a group or individually, and may be accompanied by hunting dogs. Agriculturalists in West Africa today practise collective hunts near their villages, while individual hunting happens further away (Schnelle, 1971, 47). Passive hunting involves the use of traps, pitfalls or snares (Geis-Tronich, 1991, 227-232). Present West African farmers, put their traps near to the fields, because they need to be checked regularly (Schnelle, 1971, 47). Catching animals with traps is not time-consuming and can even be done during the rainy season in combination with the work on the fields (Schnelle, 1971, 85). An important difference between active and passive hunting, which can allow distinction between them archaeologically, is that the latter cannot make a selection for age. When selection is possible, people will prefer to take adult, large wild ungulates in order to get the best output for the invested energy, while passive hunting usually results in a higher proportion of young animals (see summary in Marshall and Stewart, 1994). There are also examples of deliberate catching of local small wild ungulates with nets or spears, especially of territorial animals or animals that are found in small groups. Only a few young wild mammals have been found in the studied assemblages, although this may be because of differential destruction, or to the fact that the bones of very young individuals are more difficult to identify.

During the hunt, vegetation is sometimes set on fire to smoke out animals (e.g., Ntiamoa-Baidu, 1997). This technique was possibly also used by Late Stone Age hunter-gatherer groups in West Africa and may have had a serious impact on the natural landscape. Archaeobotanical indications have been found for vegetation changes during the period caused by fires, but it appeared to be impossible to distinguish between natural and anthropogenic fires (Salzmann, 2000; Neumann *et al.*, 2004).

There are some sources on hunting strategies and techniques specifically suited for the catch of certain animals. In medieval West Africa, crocodiles seem to have been mainly hunted with spears and hippopotami with javelins, presumably with a harpoon-like edge (Lewicki, 1974, 95-96). According to the local people from Oursi, fowling can most easily be done with snares. The Gulmanceba from southern Burkina Faso predominantly use pitfalls for catching gazelles (Geis-Tronich, 1991, 227). Hares are supposed to be mainly actively chased (Kingdon, 1974, 351). This brings up the point that some species are perhaps missing from the archaeofaunal samples because the hunting techniques of the human groups were not suited to catching them, or because they perhaps preferred to get the animals that could be caught with simple means and minimal effort. The latter may be another reason, besides taphonomy (see 5.2.), for the rarity of large mammals in the studied assemblages, although some species, like hippopotamus, giraffe (*Giraffa camelopardalis*) and elephant (*Loxodonta africana*), must have been relatively common in the area.

6.6.4. Game processing and conservation techniques

Processing and conservation of game meat is assumed to have been largely similar to that of domestic animals, and the subject will therefore be discussed more elaborately there. According to Ntiamoa-Baidu (1997), bushmeat today is mainly eaten fresh, salted, dried or smoked. Smoking appears to be the most widespread form of preservation and smoked bushmeat is available in urban markets in most countries in sub-Saharan Africa. Geis-Tronich (1991, 226-227) mentions that smoked meat can be kept at least two months, even up to ten months when repeated smoking is carried out.

6.6.5. Trade

At the studied sites, there is little indication of trade in game products, since all species found could well have been available locally. Traded products may have consisted of meat or rather secondary products, like hides or skins and ivory. Both elephant and hippo ivory were part of the commodities of trans-Saharan trade. However, as mentioned in 4.6., the former was mainly exported from forested zones, south of the research area. Other items that may have been traded over longer distances are the glands of civets, used for perfume making, but such a practice would not be archaeologically visible.

6.6.6. Cultural aspects

At present, hunting appears to be an activity that does not enjoy the same esteem among different human groups. It usually has a high social value among farmers, while pastoralists view it as a minor activity, focusing on particular species, and many African pastoral groups do not hunt at all (FAO, 2001). The latter moreover sometimes have food taboos concerning wild animals (Bower, 1997). This cultural difference may have an ecological basis, since the dry season, which was described earlier as the best part of the year for the hunt, is a dead period among agricultural groups, while it is the most strenuous period for pastoralists (see 6.7.6.). A large pastoral component may explain the relatively low amount of hunted animals in sites of Gajiganna phases I-II and especially those of the Bama-Konduga group (see 6.7.4.2.). The prestige attached to hunting is also variable according to the techniques used, with active hunting valued higher than passive hunting, and the species caught (Geis-Tronich, 1991, 226-227). Hunting, particularly active hunting, is traditionally an activity attached to the males within a society (Deacon, 1984). Among the Mossi, a present group of farmers in Burkina Faso, hunting is exclusively done by men, for example, who spend more time on hunting than they do on stock keeping (Froment, 1988, Table 12).

Possible food prohibitions concerning wild game among the groups living at the studied sites are very hard to trace archaeologically. The faunal assemblages did not yield any indications for food restrictions concerning the animals dealt with in this chapter, not even for periods with presumed Islamic influences. On the contrary, at the Iron Age sites near Oursi, crocodile, an animal regarded as unclean by Islam, appears at the end of the sequence, when there must have been contacts with northern Africa (see 2.2.2.1.). At Ngala (NA 93/45), the presence of crocodile in phases I-IIb was seen as a contra-indication for Islamic influences (Linseele, 2005). Wild animals must also have played a role in the human symbolic world, since they were depicted in figurines and their bones were sometimes bored through to be used as pendants.

6.6.7. Concluding remarks

For most of the studied localities it is probably appropriate to speak of opportunistic hunting, for which the absence of large game may be symptomatic. Hunting and fowling was not the main focus of economic activities and people seem to have been catching mostly those animals that they could get without much effort. Variations in the importance of hunting between the different sites studied should probably be connected to local difficulties in practising other economic strategies, like fishing or stock keeping. Sites where hunting was the prime economic activity are missing among the assemblages, with the probable exception of Late Stone Age sites in Burkina Faso. Their absence may be due to poor archaeological visibility because of high mobility. The impact of hunting on game populations in the area around the sites seems to have been local and mainly confined to one species, kob. No evidence was found for the drastic reduction of wild fauna that Jousse (2006) has postulated for what she called the "Neolithic/historic transition", around the beginning of our era. However, for all parts of the studied area, shifts in the species spectrum towards more taxa of dry, Sahelian environments can be observed during that period.

6.7. Domestic animals

"Pastoralists" is an important term that is often used in relation to domestic livestock. However, there is no generally accepted definition for it and its use can cause confusion. This part on domestic animals therefore starts with a definition of the term. Before discussing the economic importance of domestic species and the information on the palaeo-environment they occupy, a summary will first be given of all available data on early food production in Africa in general, and West Africa in particular. As stressed in the first chapter, this was formulated as one of the major research themes at the beginning of this study.

6.7.1. A definition of pastoralists

In the strictest sense, the term "pastoralists" refers to highly mobile human groups for whom stock keeping is the principle economic activity and where domestic animals are also very important culturally. In the broadest sense it is synonymous with domestic stock keepers. There are several definitions in between these two extremes. Recently, Gifford-Gonzalez (2005) has for example defined pastoralists as "groups who depend primarily on the products of their hoofed animals, and who organise their settlement and mobility strategies to suit the dietary needs of their livestock". According to Chang and Koster (1986), anthropologists have generally focussed on two characteristic variables while defining pastoralists: the degree of dependence on domestic animals and the degree of mobility. These two parameters have been used, for example, in the classification made by Gallais (1975, 188-189) (Fig. 41). Bouquet (1990, 228, 245) seems to be correct in his criticism that there are zones of overlap between the categories of this classification. Nevertheless, much confusion can be avoided when, while speaking about pastoralism, the degree of mobility and the degree of dependence on livestock of the human community concerned is specified. Unfortunately, from descriptions of human groups in the literature, ethnographic or archaeological, it is often not clear how these should be classified. "Transhumance" fits in the category of "partly sedentary". In FAO (2001) it is defined as "the regular movement of herds among fixed points in order to exploit the seasonal availability of pastures". Transhumant pastoralists often have a permanent homestead and base, where the older members of the community stay year-round. Generally, mobility increases the greater the economic importance of livestock keeping, but like Blench (1999, 54) has illustrated, this is not necessarily the case. An important factor in the archaeological context is that the degree of mobility determines visibility.

A criticism of the proposed definition might be that it neglects cultural aspects. Robertshaw (1990, 299) writes that, "the identification of a particular society as 'pastoral' rests not on any strict definition but rather on a recognition that it considers its proper business to be the tending of livestock". Herders from present-day northern Kenya have, for example, been recorded as turning to other economic activities out of necessity, but still considering themselves as pastoralists, and returning to stock keeping as soon as circumstances allowed (Sobania, 1988). However, contrary to the strictly economical side and mobility, the cultural aspects of herding are almost impossible to trace archaeologically. It therefore seems justified not to include the aspect in an archaeological definition.

Degree of mobility →

Degree of dependence on livestock ↑

	« highly mobile »	« mobile »	« partly sedentary »	« sedentary »
« pastoralists »				
« livestock keepers »				
« livestock keepers-farmers »				
« farmers-livestock keepers »				
« farmers »				

Fig. 41: Classification system for food producing communities (Gallais, 1975, 188-189)

6.7.2. Earliest food production

6.7.2.1. Herding before farming

Because domestic stock needs continuous attention and can never be abandoned, pastoralism is thought to be a more difficult transition to make for egalitarian hunter-gatherers than agriculture (Marshall, 2000). Nevertheless, available evidence indicates that domestic stock keeping is the oldest form of food production on the African continent, probably first appearing in the north-east during the ninth millennium BC or, when only secure evidence is accepted, by the seventh millennium BC (Gautier, 1984b, 2002; Marshall, 1998; Marshall and Hildebrand, 2002). A possible explanation can be found in di Lernia (2002), who stresses that in present northern Africa domestic animals are a more stable source of food than cultivated crops and are less sensitive to droughts. Stock keeping was for the earliest African herders probably only one of many ways of getting food, besides gathering plants, fishing and hunting (Marshall, 1998). The main advantage of keeping domestic animals over pure hunting and gathering must have been their predictable availability (Marshall and Hildebrand, 2002). This also fits with what Smith (1992, 9) has written, that the main benefit of herders over hunter-gatherers is their disposition over sustainable yields from their domestic herds, in the form of meat and milk. Pastoralism without agriculture has no ethno-historical parallels. Pastoralists nowadays either obtain grain from farming communities or grow crops themselves seasonally (McIntosh, 1997). In the Near East, non-cultivating pastoralists only emerged as a secondary development, after the establishment of settled village life, mixed farming and animal husbandry (Harris, 1996).

In comparison to other parts of the world, agriculture emerged only late in sub-Saharan Africa. Neumann (2003, 2005) has argued that the African savannah is so rich in wild plants that for a long time there was no need for domestication. She thinks that the particular role of nomadic pastoralism, fitting well with the exploitation of wild plant foods, was moreover another important factor in postponing agriculture for several millennia. Unlike in the Near East, there was no single centre for crop domestication in Africa, because of the lack of an area of overlap for the ancestors of the cultivated plants. Multiple local experiments in agriculture must have occurred over a vast area. Most available data on the origins of agriculture in sub-Saharan Africa come from the arid zones because the grain crops stemming from those areas are archaeobotanically easier to trace than the tuber crops, originating from more southerly, forested areas (Neumann, 2005). There is up to now no evidence for agriculture in central, eastern, and southern Africa before the middle of the first millennium BC, while Near Eastern crops were already being grown in Egypt at least three millennia earlier. The earliest indigenous African cultivated crop is pearl millet (*Pennisetum americanum*), which appeared in West Africa at the beginning of the second millennium BC and remained the only domestic crop during the entire millennium (Kahlheber and Neumann, in press). There are so far no archaeological sites with evidence of pearl millet in an intermediary stage between the wild and domestic form, but the domestication area should probably be sought in the south-western Sahara. Kahlheber and Neumann (in press) proposed that, during the second millennium BC, agriculture was integrated into the existing pastoral and foraging systems, to further minimise subsistence risks and to increase the predictability of resources. They furthermore argued that cultivation of pearl millet must have been possible with simple tools and without much agricultural experience and could even have been integrated into a mobile way of life. The importance of agriculture in the early phases appears to have been variable from site to site (Kahlheber and Neumann, in press); with relatively intensive pearl millet cultivation at the Kintampo site Birimi in Ghana (second millennium BC), for example (D'Andrea *et al.*, 2001).

6.7.2.2. First domestic animals in sub-Saharan West Africa

With the possible exception of the guineafowl and recent domestication experiments, none of the domestic animals of present sub-Saharan West Africa has a local wild ancestor. Domestic species were thus all introduced into the area from elsewhere at some point in the past. It is generally accepted that the first domestics arrived from north-eastern Africa by way of the Sahara. Available archaeological evidence suggests that this happened with the onset of the dry phase after the mid-Holocene that pushed Saharan pastoral groups southwards (see 2.2.1.). Agriculture may also have been introduced in the Sahel by these herders coming from the Sahara (Smith, A.B., 1980, 1984b). About three thousand years prior to beginning of food production, Central Saharan influences had already reached the Chad Basin in the form of pottery, collected from the site Konduga on the Bama Ridge (Breunig and Neumann, 2002a). From this single archaeological find, two important conclusions can be drawn, first of all that the introduction of ceramics in the area was not associated with sedentary, food-producing, economies. Similarly, finds of pottery were also made at Late Stone Age, hunter-gatherer sites in Burkina Faso. The second important point that can be made from the finds at Konduga is that some Saharan influences had apparently already reached sub-Saharan areas long before aridification set in.

The earliest evidence for domestic livestock in northern Burkina Faso is as recent as the beginning of the Iron Age, when domestic cattle (*Bos primigenius* f. taurus), sheep (*Ovis ammon* f. aries) and goat (*Capra aegagrus* f.

hircus) appear simultaneously. No single identification of a domestic animal could be made for the Late Stone Age locations, not even at Corcoba (BF 97/5), the only site of that period with a relatively large archaeofaunal sample. However, it should be remarked, that the Early Iron Age deposits from this site did not contain any bones of domestic species either. At Windé Koroji Ouest (MacDonald, 1996), a settlement mound in the Malian Gourma that neighbours the research area in northern Burkina Faso and is broadly contemporary with its Late Stone Age sites, domestic animals were present, but only in small numbers. Their absence at the studied sites in Burkina Faso may therefore just be a consequence of low visibility. However, there are reasons to doubt the identification of domestic animals at Windé Koroji Ouest (see 6.7.3.1.), although the cultural material of the area has affinities with that of the Tilemsi Valley, including Karkarichinkat, where domestic animals are known from at least the second millennium BC (Smith, 1974, 1975). The available evidence thus more convincingly points towards a true absence of domestic livestock in Late Stone Age northern Burkina Faso. For that period, however, cultivated crop plants have been identified, although their economic importance must have been very limited (Kahlheber, 2003; Kahlheber and Neumann, in press). Between the Late Stone Age and the Iron Age in Burkina Faso, there is an archaeological gap of about a millennium. Only one site, Kissi 49, was dated to this intermediate period. The site yielded very little faunal remains that were sent to the RMCA, and only after the practical work for this study was closed. However, no evidence for domestic animals is present in its assemblage either. In the Malian Gourma the Zampia tradition, which is the linear successor of the Windé Koroji tradition and has a date around 800 BC, falls chronologically inside the "dark millennium" from Burkina Faso (MacDonald, 1994, 227-251, 1996). Evidence for the tradition mainly comes from tumulus fields, where remains of some sacrificed cattle and ovicaprines have been found.

In the southern Lake Chad area, domestic animals are present from its oldest occupation, by people of the Gajiganna Culture, onwards. For Gajiganna phase I, only Bukarkurari (NA 97/33), on the Bama Ridge, yielded a large archaeofaunal sample, which is dominated by domestic animals, mainly cattle. With a date of around 1600/1500 BC these domestics belong to the oldest that have been attested in sub-Saharan West Africa (cf. Jousse, 2004b). An earlier date, minimal 2400 BC, has been obtained for the presence of ovicaprines at the site of Blabli, in northern Cameroon but, as said before, neither the date nor the identifications are very reliable (David and Sterner, 1987, 1989). From the same site a cattle figurine and possibly even cattle bones were also retrieved. An important difference between the southern Lake Chad area and the research area in northern Burkina Faso is that, in the former cultivated plants appear only around 1200 BC, thus later than domestic stock keeping.

The study area well illustrates that, within western Africa, the introduction of the earliest domestics took place at highly variable dates. In certain areas it happened only very late, for example in the Chaîne de Gobnangou in southern Burkina Faso, where domestic plants and animals, with the possible exception of domestic dogs, were apparently not known before AD 1000 (Frank *et al.*, 2001). Similarly, there is evidence from Korounkorakolé in Mali for human groups maintaining a hunter-gathering lifestyle until as recently as the mid to late first millennium AD (MacDonald, 1997a). This urged MacDonald to propose a model for the peopling of West Africa with a long-term autochthonous human presence south of the Sahara. In the southern Lake Chad area domestic stock keeping could perhaps have spread at an early date and on a relatively large scale, because of the absence of autochthonous populations, as the area was uninhabitable until shortly before Saharan pastoralists arrived. Inversely, the presence of indigenous hunter-gatherer groups may have retarded the introduction of domestic livestock in areas such as northern Burkina Faso. However, this remains speculative since former hunter-gatherer populations of West Africa are poorly documented and archaeological evidence in the area for the period before 2000 BC is scarce. Moreover, domestic animals appear at an early date, during the second millennium BC, at sites of the Kintampo Complex in the forest margin of Ghana, where available evidence indicates that they were incorporated into a local tradition (Anquandah, 1993; Stahl, 1993; Gifford-Gonzalez, 2005). Another reason for the early presence of domestics in the southern Lake Chad area might be its attractive geographical circumstances, comparable to the Inland Niger Delta of Mali and the Senegal Valley, the other large floodplains in arid West Africa (McIntosh, 1999). In the Inland Niger Delta, domestic animals date to the second-first millennium BC, with early finds of cattle but no ovicaprines at Kobadi (ca. 1600 BC) (Jousse and Chenal-Vélardé, 2001-2002). The presence of this favourable area to its north may also have slowed down the introduction of domestic animals into northern Burkina Faso, because there was probably no ecological pressure on the herders to move further south. A remarkable site is Hosséré-Djaba, dated to the nineteenth century AD and situated in the Sudan zone of northern Cameroon (Lesur and Langlois, 2002). The site was inhabited by sedentary agriculturalists but its faunal assemblage did not contain any domestic animals at all. Its habitants were therefore characterised as hunter-agriculturalists. The excavating archaeologist, O. Langlois (pers. comm.), was not surprised by the archaeozoological results since there are ethnographic examples in the area of groups with similar economic strategies.

A question that arises in regard to the first archaeological appearance of domestic animals at West African sites is if this corresponds with the actual date of their first introduction. Linguists especially defend the point of view that they were introduced earlier. It is true that archaeology may suffer from a problem of visibility, but on the other hand dating methods in linguistics are not accurate enough to take linguistic dates as firm evidence. Blench (1993) sees no reason why domestics would not have been introduced south of the Sahara before the onset of drier conditions after the mid-Holocene. This may be a pertinent remark, especially since Saharan influences have been attested in sub-Saharan West Africa long before the first archaeological finds of domestic animals (see above). However, a presence of domestic animals in sub-Saharan western Africa before 4000 BC seems unlikely, because that is around the date when they first appear in its Saharan parts (Jousse, 2004b). Their introduction must also have been influenced by the position of the tsetse belt. Domestic animals could probably only enter sub-Saharan West Africa thanks to the southwards shift of this belt after increased aridification, or once trypano-tolerant breeds had developed (see 6.7.3.3.).

6.7.3. Appearance and development of domestic animal types

6.7.3.1. Discussion by species

For the reconstruction of the first appearance of the different species and types of domestic animals in sub-Saharan West Africa, archaeozoological data were mainly used and they were confronted with linguistic and ethno-historical information. In Chapter 4, clay figurines of domestic animals found at sites all over West Africa, and beyond, were mentioned. These may also help in describing the history of the different types of domestics, although their use as evidence in such contexts must be looked at with the necessary scepticism. The figurines are often very stylised and their attribution to a certain (domestic) species or type sometimes seems to have a poor basis. It is not clear why, for example, a statuette from Daima III layers at Daima was identified as a suid (Connah, 1981, 184-185), in spite of its spiky protuberances, unless these are an exaggeration of the bristly hair of wild suids.

Domestic fowl (*Gallus gallus* f. domestica)

Two possible routes have been proposed for the introduction of domestic fowl into West Africa. The first one is southward through the Sahara from North Africa, where the bird was known since at least the Phoenician colonization around AD 800, although it is not clear if it would have been possible at all to carry domestic fowl over the arduous trans-Saharan route (MacDonald, 1992, 1995). The second possible route is east-west, along the Sahelo-Sudanic environmental belt, with early East African finds also dating to the mid-first millennium AD (*Ibid.*; Blench and MacDonald, 2000). Linguistic evidence strongly suggests multiple introductions, through both routes described (Blench and MacDonald, 2000).

By the early second millennium AD domestic fowl seems to have been spread over settlements all throughout modern Mali (MacDonald and MacDonald, 2000). Data from the studied sites moreover show that the species was also well known in northern Burkina Faso by that time. The oldest bone remains of domestic fowl from the region date to the Late Iron Age, although looking at bird eggshell fragments found, it is not excluded that the bird had already been introduced during the Middle Iron Age (see 5.5.). The oldest domestic fowl remains from the Chad Basin date to around the end of the first millennium AD. The animal was for example identified in phase II layers at Ngala (NA 93/45), chronologically situated in the eighth-tenth/eleventh century AD. The two remains of "domesticated fowls" mentioned by Connah (1981, Table 8.3) from Daima III layers at Daima, could therefore be domestic fowl as well.

Williamson (2000) has brought into attention the discrepancy between the deep embedding of domestic fowl in West African culture, where its most important traditional uses are ritual, and its relatively recent attestation in archaeological contexts. She therefore believes that domestic fowl is older in West Africa than suggested by archaeological evidence Other linguists have also defended this view, e.g., Blench (1995), who points to the old roots of the words for domestic fowl. Equally based on linguistic data, but also on early Egyptian finds, Donkin (1991, 57) argues that domestic fowl was already being spread over the African continent from at least the first millennium BC. Besides the query over the date of the bird's introduction into western Africa, there is also no satisfactory answer yet on the question of why earlier peoples preferred domestic fowl to the indigenous helmeted guineafowl. However, MacDonald (1992, 1995) proposed that it might be because they are easier to control.

The small size of West African domestic fowl, both archaeological and recent, compared to its European counterpart has been stressed in the literature and has been attributed to unfavourable environmental living conditions, poor nutrition and inattention to breeding for size (MacDonald, 1992, 1995; MacDonald and MacDonald, 2000). Unfortunately, no individual measurements on West African archaeological or recent domestic fowl bones are available to compare against the measurements from the studied assemblages. However, MacDonald (1992) gives some size intervals for well-preserved coracoids, GL: 45-52 mm (n=3), scapulae, Dic: 9.6-10.6 mm (n=5) and tarsometatarsi, GL: 59-70 mm (n=4), from archaeological sites in the Méma and the Inland Niger Delta (undated). These seem to be roughly in the same range as recorded in this study (Table C.36). As mentioned in 4.5.5.2., MacDonald also gives size ranges for modern (West African?) domestic fowl. Compared to these, most of the measurements obtained from the studied sites seem to fall in the lower size limits, although some carpometacarpals appear to be rather large.

Helmeted guineafowl (*Numida meleagris*)

The helmeted guineafowl has a separate position compared to the other West African domestics, because it is the only species that may have been domesticated locally. Direct evidence for the time and place of its domestication is thus far lacking, however, but it must have happened at some time before the fifth century BC, when domestic guineafowl appears in Europe, where it is described by classical Greek authors (Mongin and Plouzeau, 1984). It is not very clear which subspecies may have been domesticated. Donkin (1991, 16) argues that the Greeks must have been acquainted with both the East African (*Numida meleagris meleagris*) and the West African (*N. m. sabyi*) variety of guineafowl, and perhaps also with *N. m. galeata* from further south. On the other hand, Urban *et al.* (1986, 8-10) claim that only the West African subspecies of helmeted guineafowl was domesticated, although it is unclear on what kind of evidence their assertion is based. Gautier (1990, 4-5) proposes the North African Mediterranean as the domestication centre for guineafowl, but this cannot be proven without further research into the taxonomy and status of past North African guineafowl populations (wild versus feral, see 4.5.5.2.). This option seems more plausible than a West African origin for early European guineafowl, because, although sporadic contacts between West Africa and Europe may already have occurred during the first millennium BC, no important cultural influences resulted from these early contacts (Masonen, 1997).

Studying the origins of domesticated guineafowl is complicated by the fact that the wild and domestic form may be indistinguishable from their bones. Moreover, in present western Africa one can hardly speak of true domestic guineafowl, because the living conditions of animals kept in captivity are very close to those in the wild. Before the present study, archaeological evidence for helmeted guineafowl in West Africa was relatively poor. The animal was best represented at sites in the Middle Senegal Valley, dating to the first millennium AD, but, strangely enough, no domestic fowl remains were associated with it (MacDonald and MacDonald, 2000; Van Neer and Bocoum, 1991). The studied sites from the southern Lake Chad area have yielded guineafowl bones from as far back as the second millennium BC, the oldest dating to Gajiganna phase IIa. However, there is usually no evidence at the sites for a heavy exploitation of guineafowl, which could be an indication for taming or domestication. Only one site, Gajiganna BII (NA 90/5BII), is an exception. At that site, dating to Gajiganna phase IIc, 34 out of 36 identified bird bones are of guineafowl. As explained in 2.2.3.1., Gajiganna phase IIc is considered as a period of increased aridification and several of the animal taxa found at Gajiganna BII, including guineafowl, are known to congregate near water in periods of drought. The tendency of guineafowl to form large flocks during dry periods may perhaps have stimulated the taming or domestication of the bird, certainly during periods when the existing modes of subsistence of human groups came under pressure and they must have been looking for alternative food resources. Ongoing research in the Nigerian part of the southern Lake Chad area is increasingly showing that the Gajiganna phase III was a period of innovation (Magnavita *et al.*, 2004) and perhaps the final step towards domestication or taming of guineafowl should also be sought there. However, from the site Zilum (NA 97/37), one of the key-sites of this innovative era, no large galliform remains at all have been identified up to now, although this could change when more faunal material from the site is studied. At this point it is thus not clear if Gajiganna BII is an isolated case of increased exploitation of guineafowl, or if it is perhaps at the beginning of the bird's domestication process. The scenario proposed by Donkin (1991, 66) to explain the origins of guineafowl keeping in West Africa is completely different to the hypothetical one just described. He thinks the practice of guineafowl rearing may have gradually been propagated to the area from North East Africa, together with domestic fowl (after ca. 1000 BC), or alternatively and perhaps more likely, that the introduction of domestic fowl to sub-Saharan Africa prompted local interest in the guineafowl.

Dog (*Canis lupus* f. familiaris)

Problematic for the reconstruction of the history of domestic dogs in western Africa is the difficulty in distinguishing their remains from those of jackals (*Canis aureus/adustus*) (see 4.6.5.2.). MacDonald and MacDonald (2000) cited remains from Jenné-Jeno, dated to about 200 BC, as the oldest secure evidence for domestic dogs south of the Sahara. In Burkina Faso, there were no canid remains at all at sites dating before the beginning of the Iron Age, but in the southern Lake Chad area, domestic dogs may have been present from Gajiganna phase IIb onwards. An earlier presence of the animals in sub-Saharan West Africa than demonstrated up to now from archaeological remains seems to make sense, since they are thought to have spread as shepherd dogs together with domestic cattle and ovicaprines (Gallant, 2002, 51; Gautier, 2002), which are present in the area from the second millennium BC onwards. Domestic dogs may be missing or rare from second and first millennium BC archaeological contexts, which mainly present food refuse, because they were perhaps not eaten in these early phases. This seems to fit with the find of three buried individuals in the Sahara in Niger, dated to 2600-1300 BC (Paris, 1984, 66, Figs 52 and 53; Paris, 2000).

Cat (*Felis silvestris* f. catus)

As explained in 4.6.5.2., the reconstruction of the history of the domestic cat in (West) Africa from archaeological remains is hampered by the difficulties in distinguishing it from its wild form. Blench (2000b) proposes that the animal may have spread as a commensal into sub-Saharan Africa from about 1000 BC onwards, both across the Sahara and down the Nile Valley. He thinks Muslim traders probably also carried it through the Sahara about two millennia later. However, Lagercrantz (1950, 65) argues that the distribution of domestic cats over Africa outside Egypt is purely post-Arabian. This could fit with data from southern Africa, where the oldest evidence for the animals is as recent as the colonial period (Plug and Badenhorst, 2001, 84).

Horse (*Equus ferus* f. caballus)

In the course of the second millennium BC horses were introduced into Egypt by the Hyksos (Chaix, 2000). By 700 BC, horses and chariots are depicted in Saharan rock art and the study of these depictions indicates an introduction from the north (Muzzolini, 2000). Nonetheless, the oldest reliable bone finds of domestic horse in sub-Saharan western Africa, including the buried animals from Aissa Dugjé (MacEachern *et al.*, 2001), do not date before the second half of the first millennium AD. In the course of the second millennium AD, they would then gradually have become more common, appearing also at the studied sites in northern Burkina Faso, and spreading further south to the tropical forest zones (Law, 1980, 8-23). Some have argued that in the first millennia BC and AD, a small type of horse, ancestral to the modern pony breeds, was already present in western Africa (Law, 1980, 3, 1995; Blench, 2000c). The larger horses from the research area, the Barb and Dongolawi, are then considered to have arrived with Arab contacts. A first line of evidence for early West African horses, are medieval Arabic historical sources reporting the presence of very small horses in several West African regions (Levtzion and Hopkins, 1981, 81, 185, 263), suggesting that the introduction of these animals was pre-Arab. This is in agreement with the existence in some areas of more archaic equestrian equipment than the Islamic bitted bridle, saddle and stirrups (Law, 1980, 93-96, 1995). Linguistic evidence also supports an early presence of horses (Law, 1980, 5-7; Blench, 1993). Finally, the adaptation of present pony types to the West African environmental conditions is sometimes used as an indication for their ancient origins. The only possible bone finds that could support the early presence of horses in sub-Saharan Africa are the equid teeth from Rop Rockshelter in Nigeria, identified as wild ass or donkey and dated to the first millennium BC (Sutton, 1985). Blench (2000c) considers the identifications unlikely and thinks that, either the site's stratigraphy is misleading, or the teeth are in fact pony teeth.

Donkey (*Equus africanus* f. asinus)

This animal is usually somewhat neglected in the literature in favour of its relative, the domestic horse. The reconstruction of the history of the donkey in West Africa also suffers from a lack of bone remains. None at all could be identified with certainty from either part of the research area, for example. Based on their absence in Saharan rock art, Muzzolini (2000) argues for a relatively recent westwards diffusion of domestic donkeys after their domestication. This fits with the view of Blench (2000c), who considers it likely that donkeys were only spread over West Africa with the rise of long-distance caravan trade with northern Africa. On the other hand, he also mentions that there are indications for an east-west link suggesting that donkeys could have reached Lake Chad across the Sahel, coming from East Africa, where domestic donkeys were known by the end of the second or the first millennium BC (Marshall, 2000).

Pig (*Sus scrofa* f. domestica)

In 4.6.8., it was mentioned that most evidence points towards a European introduction of domestic pig in West Africa, although bone finds at one Early Iron Age site in South Africa suggests that the animal may already have been propagated over the African continent around AD 600 (Badenhorst and Plug, 2001, 113). Pigs are generally not kept by nomadic pastoralists and, therefore, they probably did not spread across the Sahara from North Africa together with the domestic cattle and ovicaprines. Linguists have argued for a pre-European introduction and proposed that domestic pigs may have been taken from north-eastern Africa down the Nile, spreading further overland to west-central Africa along a corridor from Darfur to Lake Chad and reaching central Nigeria in medieval times (Blench, 2000d). However, the fauna from the studied sites failed to provide new arguments for an early introduction of the domestic pig into sub-Saharan Africa.

Dromedary (*Camelus thomasi* f. dromedarius)

Because remains of the animal are missing from the studied sites, dromedary has not been discussed in the faunal descriptions in Chapter 4. The animal was probably domesticated around 3000 BC on the Arabian Peninsula (Gautier, 1990, 4-5) and the oldest African finds are from first millennium BC Qasr Ibrim in Egypt (Rowley-Conwy, 1988). In sub-Saharan West Africa, more precisely at Siouré in the Middle Senegal Valley, dromedary bones have been found in contexts dating as far back as AD 250-400 (MacDonald and MacDonald, 2000). These remains are used as proof for the existence of trans-Saharan trade from that period onwards (e.g., Magnavita *et al.*, 2002). Words for dromedary in present Nigeria and Cameroon come from Berber, not from Nilo-Saharan languages (Blench, 1995), which seems to be consistent with a northern route of introduction for the animal to the area. Nonetheless, some authors favour an east-west introduction through the savannah corridor over a trans-Saharan route (Bulliet, 1990).

The absence of dromedary at the studied localities, although some of them are younger than the animal's date of introduction into West Africa, may be due to taphonomical reasons. Most of the investigated bone assemblages probably represent consumption refuse (see 5.1.). Even if dromedaries were kept, they may not have been eaten but their carcasses left far away from the habitation areas. In addition, dromedary may be missing because of unsuitable habitats as it thrives better in more arid areas further north. The southern distribution limit for dromedaries is situated at the southern end of the Sahel zone, because of the diseases further south (Wilson, 1984, 17). Ecological difficulties in keeping dromedaries in the research area can also be illustrated by the Shuwa Arabs, who were initially camel nomads, but had to turn to cattle nomadism when they arrived in the southern Lake Chad region (Braukämper, 2004). Nevertheless, traders from northern areas must have passed through on their dromedaries.

Dromedaries are nowadays mainly kept for their milk and their slaughter has the same value load as that of cattle (Dahl and Hjort, 1976, 200). The dromedary is also an important "cultural symbol" associated with Muslims (Insoll, 1999, 102-103). Bones of the animal were found amongst consumption refuse at Tegdaoust, in layers dating to the ninth-thirteenth century AD (Bouchud, 1993).

Sheep (*Ovis ammon* f. aries) or goat (*Capra aegagrus* f. hircus)

As apparent from the descriptions in 4.6.10.1., the oldest evidence for domestic ovicaprines on the African continent, from the early sixth millennium BC, postdates the first (controversial) appearance of domestic cattle by two millennia. Nonetheless, cattle, sheep and goat seem to have arrived simultaneously in sub-Saharan West Africa, coming from north-eastern Africa via the Sahara. The numerous different ovicaprine breeds of present-day West Africa must have developed, or have been introduced, separately. Analysis of bone sizes was proposed as a possible means of contributing to the reconstruction of their very poorly documented history, since both sheep and goat types can be divided into two groups, one of small animals and another of larger animals. As described in 4.6.10.1., the former are at present mainly associated with humid environments and sedentism and the latter with arid areas and nomadic pastoralism. Size can, therefore, also give information on the palaeo-economy and palaeo-ecology of the studied sites.

For the size analysis, measurements on bones identified to species level have been used, both from material analysed in the framework of this study and from archaeozoological reports on other West African sites. The latter are mainly included for comparative reasons. Recent ovicaprine skeletons from the research area were not available, but some measurements on skeletal elements of Cameroon dwarf goats have been published by Brink and Holt (1992). Owing to the lack of sufficient metrical data on recent animals, all attributions of archaeological remains to types based on bone sizes should remain tentative because size variation within one type, e.g., according to sex, is not known. LSI's (see 3.8.) have been calculated separately for sheep and goat, by site and phase, sometimes representing only very small samples of measured elements (Figs 44 and 45). Reference measurements are taken from Uerpmann and Uerpmann (1994). The graphs are used in combination with withers height estimations that could be calculated from complete long bones. To study sizes of sheep relative to goats, length and breadth measurements on tali and first phalanges of the two species have additionally been plotted (Figs 42 and 43).

Sheep were usually less common at the studied sites than goats (see 6.7.5.) and only a limited amount of measurements are available for them. From measurements on tali, sheep appear to have been larger on average than goats. A similar, but less clear, picture can be seen on the graph with first phalanx measurements. Judging from the great variation in LSI values for sheep from the Iron Age in northern Burkina Faso, in comparison to other areas, the presence of a smaller and a larger type probably needs to be assumed. Withers height estimations were not possible. Maximal LSI values for the Iron Age sheep from Burkina Faso are as high as, or even slightly higher, than for the tall animals from Cubalel and Siouré in the Middle Senegal Valley and from Akumbu (AD 1000-1200) and Gao in Mali, although measurements on tali and first phalanges

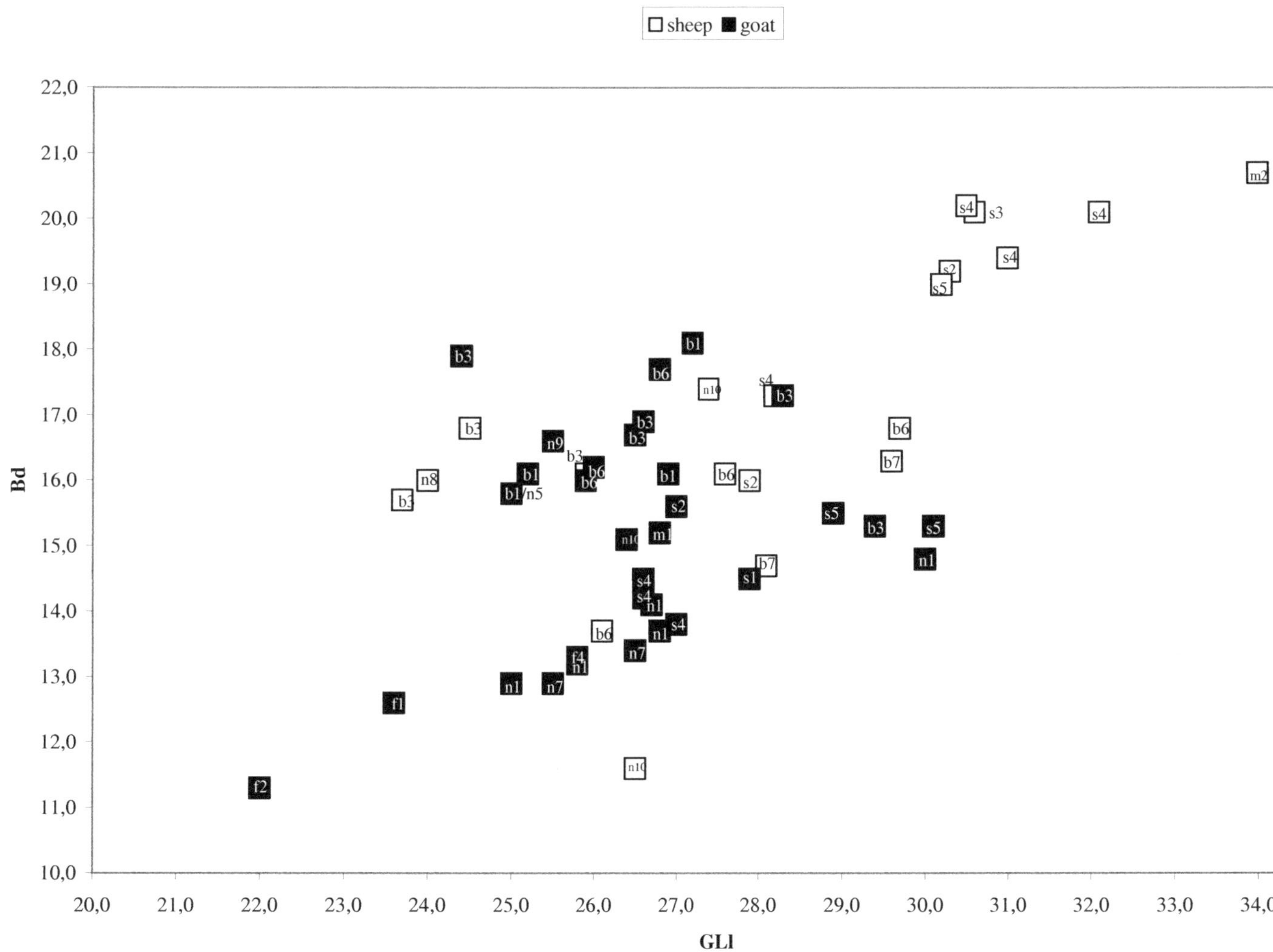

b1	BF 94/45	EIA
b2	BF 97/13	MIA
b3	BF 97/30	LIA
b4	BF 96/22	MIA
b5	BF 97/31	LIA
b6	BF 94/120	LIA
b7	BF 95/7	LIA
n1	NA 90/5A	Gaj IIb
n2	NA 99/65	Gaj IIb
n3	NA 97/37	Gaj III
n4	NA 97/33	Gaj I
n5	NA 97/33	Gaj II
n6	NA 96/45	Gaj II
n7	NA 95/1	Gaj II
n8	NA 99/75	IA
n9	NA 97/13	IA
n10	NA 92/2C	subrec
f1	NA 93/46	EIA
f2	NA 93/45	I
f3	NA 93/45	II
f4	NA 93/45	IIIa
f5	NA 93/45	IIIb
s1	Cubalel and Siouré	AD 0-250
s2	Cubalel and Siouré	AD 250-400
s3	Cubalel and Siouré	AD 400-600
s4	Cubalel and Siouré	AD 600-950
s5	Cubalel and Siouré	AD 250-400
m1	Akumbu	AD 600-1000
m2	Gao Saney	AD 900-1200
c1	Cameroon dwarf (front)	
c2	Cameroon dwarf (hind)	

Fig. 42: Length (GLl) and breadth (Bd) measurements (mm) on ovicaprine tali
Cubalel and Siouré (MacDonald and MacDonald, 2000), Akumbu (MacDonald and MacDonald, 2000), Gao Saney (MacDonald and MacDonald, 2000), Cameroon dwarf (Brink and Holt, 1992)

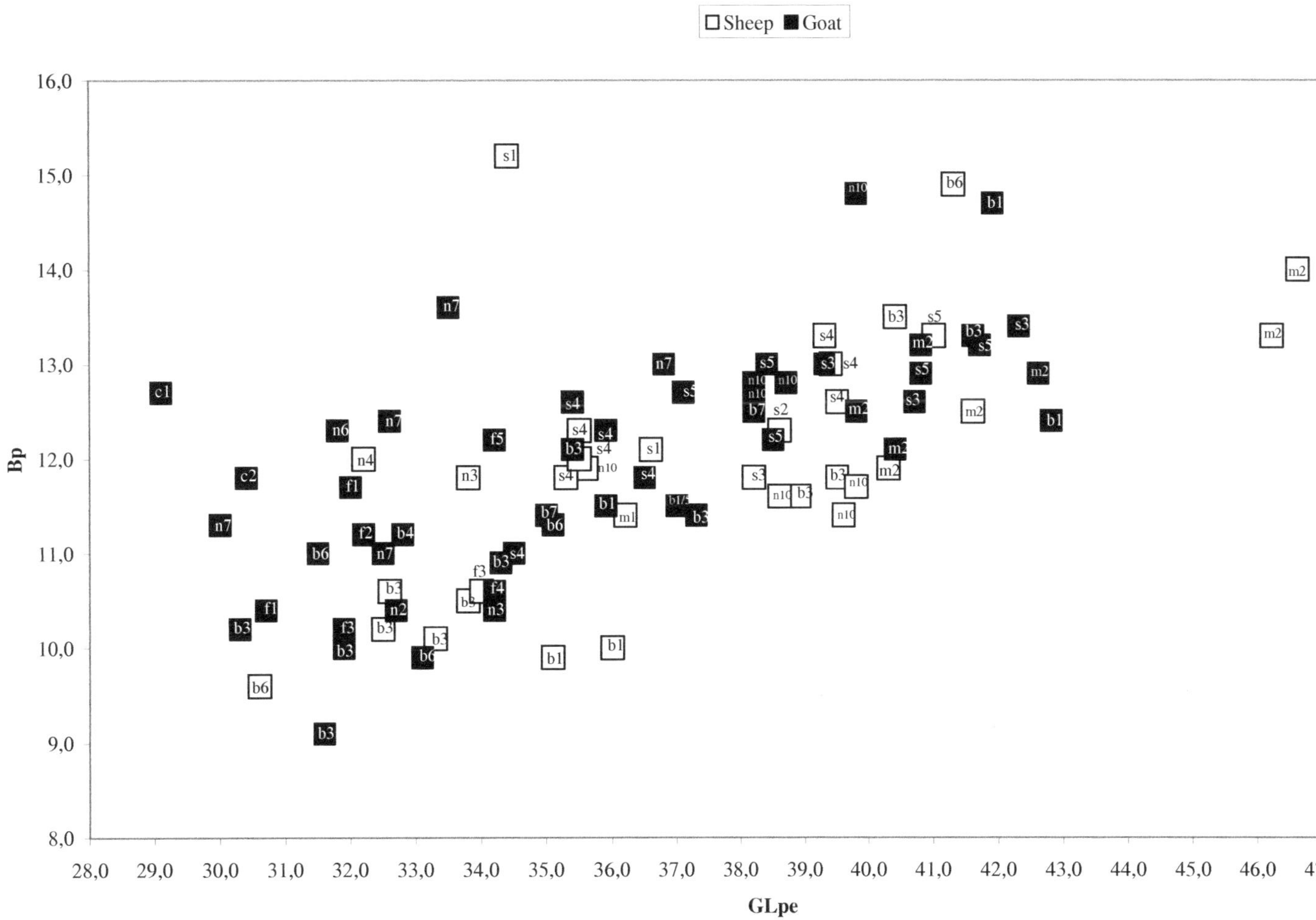

b1	BF 94/45	EIA
b2	BF 97/13	MIA
b3	BF 97/30	LIA
b4	BF 96/22	MIA
b5	BF 97/31	LIA
b6	BF 94/120	LIA
b7	BF 95/7	LIA
n1	NA 90/5A	Gaj IIb
n2	NA 99/65	Gaj IIb
n3	NA 97/37	Gaj III
n4	NA 97/33	Gaj I
n5	NA 97/33	Gaj II
n6	NA 96/45	Gaj II
n7	NA 95/1	Gaj II
n8	NA 99/75	IA
n9	NA 97/13	IA
n10	NA 92/2C	subrec
f1	NA 93/46	EIA
f2	NA 93/45	I
f3	NA 93/45	II
f4	NA 93/45	IIIa
f5	NA 93/45	IIIb
s1	Cubalel and Siouré	AD 0-250
s2	Cubalel and Siouré	AD 250-400
s3	Cubalel and Siouré	AD 400-600
s4	Cubalel and Siouré	AD 600-950
s5	Cubalel and Siouré	AD 250-400
m1	Akumbu	AD 600-1000
m2	Gao Saney	AD 900-1200
c1	Cameroon dwarf (front)	
c2	Cameroon dwarf (hind)	

Fig. 43: Length (GLpe) and breadth (Bp) measurements (mm) on ovicaprine first phalanges
Cubalel and Siouré (MacDonald and MacDonald, 2000), Akumbu (MacDonald and MacDonald, 2000), Gao Saney (MacDonald and MacDonald, 2000), Cameroon dwarf (Brink and Holt, 1992)

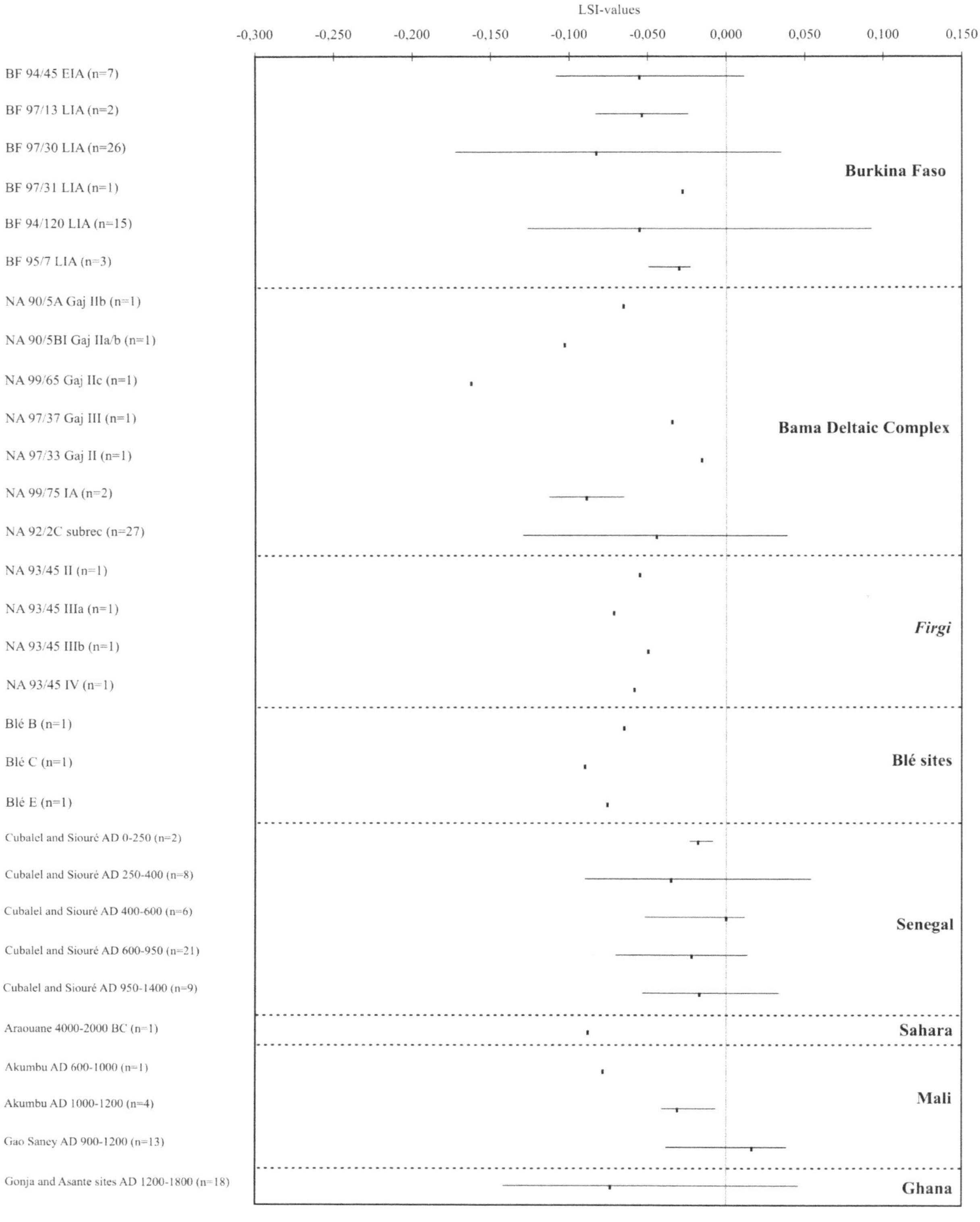

Fig. 44: LSI values of sheep by site and phase
Cubalel and Siouré (MacDonald and MacDonald, 2000), Araouane (Jousse, 2004a, Table 46), Akumbu (MacDonald and MacDonald, 2000), Gao Saney (MacDonald and MacDonald, 2000), Gonja and Asante sites (Gautier and Van Neer, 2005)

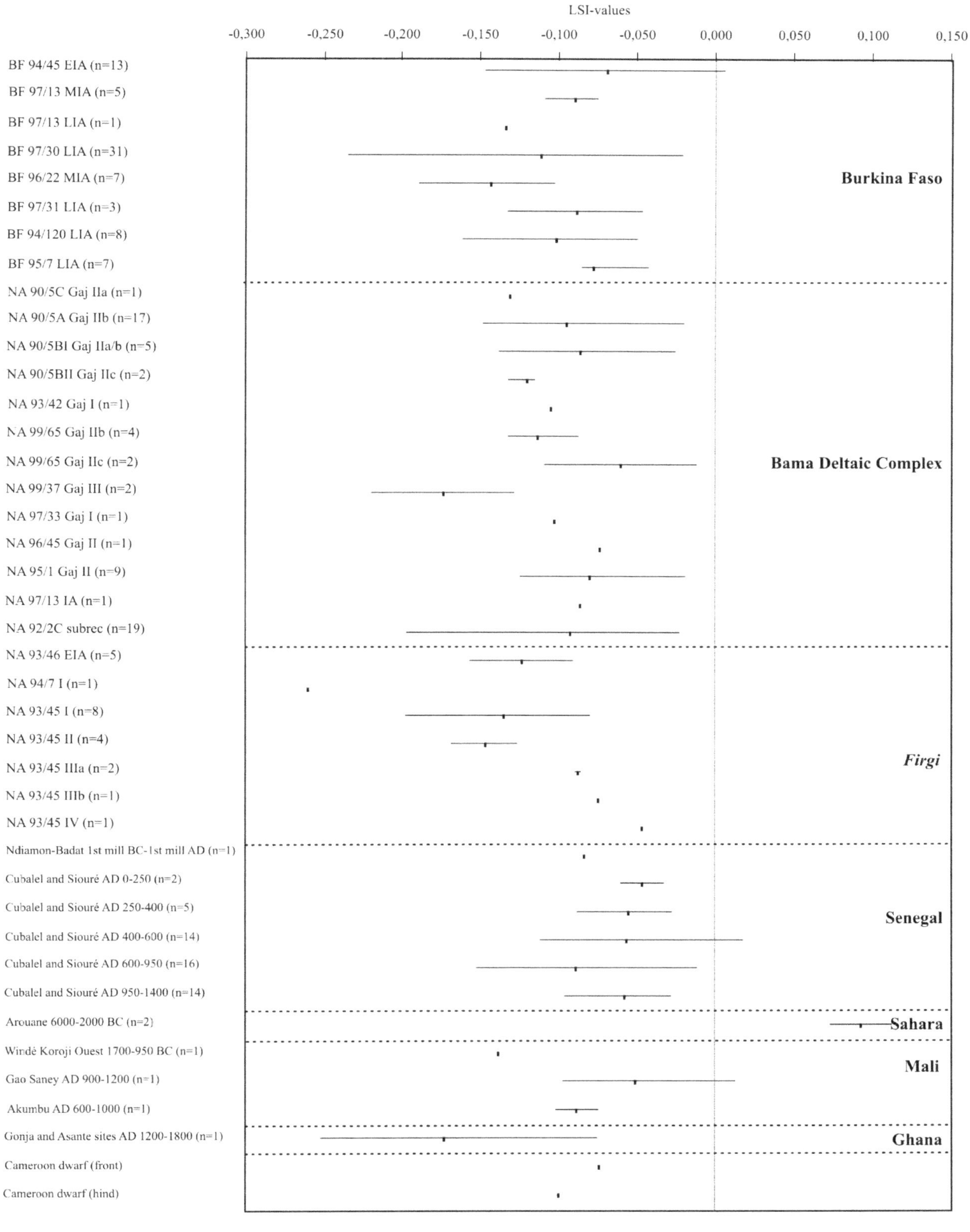

Fig. 45: LSI values of goats by site and phase

Ndiamon-Badat (Van Neer, 1997), Cubalel and Siouré (MacDonald and MacDonald, 2000), Araouane (Jousse, 2004a, Table 46), Windé Koroji Ouest (MacDonald, 1996), Gao Saney (MacDonald and MacDonald, 2000), Akumbu (MacDonald and MacDonald, 2000), Gonja and Asante sites (Gautier and Van Neer, 2005), Cameroon dwarf (Brink and Holt, 1992) point towards smaller animals. The large size of the mentioned animals from Mali and the Middle Senegal Valley may be a consequence of the northern location, near the desert edge, of the sites where they were found. Some of the sheep of Iron Age northern Burkina Faso are as small as the smallest individuals from the Ghanaian Gonja and Asante sites, which are situated in southern savannah and forested areas. They were possibly part of flocks kept in the vicinity of the settlements, while the larger sheep may then be due to a nomadic component in the economy.

For locations of the Gajiganna Culture, only a few LSI values, not more than one by site and phase, could be calculated for sheep. The animals from Gajiganna A (NA 90/5A), Gajiganna BI (NA 90/5BI) and especially Gilgila (NA 99/65) are small, while those from Zilum (NA 97/37) and Bukarkurari (BF 97/33) are relatively large. For the more recent periods in the Bama Deltaic Complex, only two sites allowed the calculation of LSI values. The Iron Age site Elkido North (NA 99/75) has rather small sheep, while average sizes at subrecent Galaga (NA 92/2C) are considerably larger. At the latter site an animal with an estimated withers height of 71 cm was found, which fits with the sheep of "extraordinary" size mentioned by Schultze (1968, 166) for early twentieth century Borno. Nevertheless, smaller sheep also seem to have been present at Galaga. Judging from measurements on ovicaprine first phalanges, its sheep appear to have generally been more slender than its goats. Data from Galaga also indicate that the (sub) recent sheep types of the Bama Deltaic Complex may not have been the same as during the Gajiganna Culture and the Iron Age.

For the *firgi* area and the Blé sites, LSI values that could be calculated for sheep are as rare as for the Gajiganna sites, but they show more homogeneity, on average being rather low. The estimated height at the withers of 54 cm from Blé Mound B indicates small animals, but which are not dwarf. The small sizes of the sheep may be related to permanent settlement in a humid environment.

Even though a detailed discussion of the sizes of ovicaprines from West African sites other than the ones that are part of this study was not intended, the very small goat talus from Windé Koroji Ouest (MacDonald and MacDonald, 2000) requires some special attention. It is the only measurable element of this species found at the site (GLl: 23.3 mm, GLm: 21.4 mm). With the exception of a specimen from phase I at Ngala (NA 93/45), the talus is smaller than any such elements from the studied sites identified as goat (Table C.84). Nevertheless the LSI value obtained for it is not exceptionally low. The talus falls in the size range for bush duiker (Van Neer, 1989a, Table 44), but no depiction is available that could allow the exclusion of an attribution to this species. A correct identification of this specimen is important to prove or refute the early presence of domestics in the Malian Gourma (see 6.7.2.2.).

The goats from the sites were, as said, on average smaller than the sheep. When looking at LSI values and first phalanx measurements of goats from Burkina Faso, those from Early Iron Age Oursi (BF 94/45), the only context from that period with measurable elements, seem to be on average larger than those from more recent contexts and also even than the sheep from the same location and date. Perhaps the large goats at Oursi are related to the site's desert connection, which was proposed earlier based on the large quantity of ostrich eggshell fragments retrieved from it (see 6.6.2.1.). Goats from Kissi 22 (BF 96/22) have the smallest average sizes of all sites in Burkina Faso, but minimal LSI values are lower at Oursi hu-beero (BF 97/30). Large size variation at the latter site may be due to a larger sample of measurable elements, but could also be indicative for the presence of a small as well as a large breed. From the LSI values, as well as the measurements on first phalanges and tali, most Late Iron Age goats from northern Burkina Faso are relatively large, of comparable sizes to the sheep from the period, although they are not as large as the goats from Gao Saney, for example. Very large goats were recorded at Araouane in the Malian Sahara, which may be an extreme example of the large size of desert animals. Large goats in Late Iron Age Burkina Faso may be related to mobile herding systems in an arid environment, but not as arid as further north in Gao. The smaller goats, some approaching the size of the presumed dwarf animals from the Gonja and Asante sites in Ghana (cf. Gautier and Van Neer, 2005), would then indicate more sedentary animals. From the graph of first phalanx measurements, specimens from Cameroonian dwarfs appear to be relatively sturdy. This probably explains why the small size of the dwarfs cannot be deduced from the LSI values, because only breadth measurements on phalanges could be used. Badenhorst and Plug (2003) also mention problems with recognising dwarf types from phalanges.

Goats from sites of the Gajiganna Culture, including those of the Bama-Konduga group, are all relatively large. Only the animals from Zilum (NA 97/37) are clearly smaller, probably in the size range of dwarf animals. Not more than two LSI values could be calculated for the graphs, but among newly excavated material from the site some very small elements were noticed, which removes possible doubts on the presence of dwarfs. An introduction from outside seems unlikely, since dwarf goats are not very well suited for travelling over long distances (MacDonald, 1995). Their sudden appearance then indicates that dwarfing was not a gradual process but may have happened quickly. Perhaps it should be connected with the shift to full sedentism, in a relatively humid environment, recorded for the period of habitation of the site (see 2.2.3.). The dwarf goats from Zilum are slightly younger than those from Kolima Sud, dated to the early first millennium BC, for which no measurements could be used on the graphs made (MacDonald, 1994, 104; MacDonald and Van Neer, 1994). The oldest West African examples of dwarf goats are mentioned for Kintampo, a Ghanaian site of the second millennium BC (Carter and Flight, 1972), but the only available measurement is on a mandible. Moreover, the differentiation of goats from sheep is questioned by Gautier and Van Neer (2005). For the region of Galaga (NA 92/2C) and the period of its occupation, there are historical descriptions of goats with very thin bodies and short legs (Schultze, 1968, 166). LSI values for goats from the site show considerable variation, which possibly points to the presence of a small and a larger type.

Goats from the *firgi* sites seem to be rather small, although average LSI values are usually higher than for Zilum (NA 97/37). The only LSI value for Late Stone Age goats in the *firgi* indicates an extremely small animal. However, this evidence is too meagre for establishing size differences between goats of the Late Stone Age and the Iron Age, at the transition of which most changes in the area seem to have taken place (see 2.2.3.1.). At Jenné-Jeno (250 BC-AD 1400) in the Inland Niger Delta, which has comparable geographical circumstances to the *firgi*, the goats of the initial occupation and later are dwarf (MacDonald, 1995). Measurements on ovicaprines at the site were not included on the graphs, since they were not identified to species level. The scanty LSI values for goats from phases IIIa-IV at Ngala (NA 93/45) seem to indicate that goats in the *firgi* were gradually becoming larger in later periods, which may be due to reduced flood levels as proposed earlier from the species composition of the fish fauna (see 6.5.2.3.). A withers height estimation (53 cm) for phase II at the site, points towards small but non-dwarf animals. At the subrecent site Koyom in southern Chad (Rivallain and Van Neer, 1983) a complete goat metatarsal allowed calculation of a withers height estimate of 46 cm, which is in the range for dwarf types.

Cattle (*Bos primigenius* f. taurus)

It was mentioned in 4.6.10.2. that archaeozoological evidence from sub-Saharan West Africa places the arrival of domestic cattle in the area in the early second millennium BC, which is five to seven millennia after their earliest appearance on the African continent. Nevertheless, the linguist Blench (1993) argues that cattle were part of the cultural repertoire of the Niger-Congo speakers at the beginning of their expansion in West Africa, which he believes cannot be later than 5000 BC. As for the ovicaprines, there are several types of cattle, which must have arrived or developed in western Africa at different times. Taurine/Sanga cattle probably arrived from northern Africa through the Sahara or via the Atlantic coast, although most evidence points to the former route (Jousse, 2004b). Zebu cattle, on the other hand, are assumed to have been propagated to western Africa from the east of the continent, along the savannah corridor, and their spread is probably related to Arabian migrations over the African continent.

In 4.6.10.2. it was described that, with the exception of a putative zebu thoracic vertebra from Galaga (NA 92/2C), determination of the cattle types at the archaeological sites studied was not possible on osteomorphological grounds, but metrical data were proposed as possible indications. Inspired by MacDonald and MacDonald (2000), three size groups were distinguished among the present cattle types of the research area: small (West African Dwarf Shorthorn or WAD), medium-sized (N'Dama) and large (Kuri, zebu) cattle. Unfortunately, for none of these cattle types were any recent skeletons available, or osteometrical data on them from the literature. Genetics could also be a means of determining archaeological cattle types. However, mainly recent cattle have been genetically analysed up to now and there are often problems with the preservation of DNA in archaeological contexts (Edwards *et al.*, 2004). Moreover, types like Kuri and N'Dama cattle are genetically very similar, although they have very different appearances (see 4.6.10.2.). Besides attribution to a certain type, metrical data can also be used to follow the size evolution of cattle through time determined by, amongst other reasons, local geographic circumstances or deliberate breeding strategies. They can also allow the further testing of the hypothesis that taurine/Sanga cattle gradually became smaller during their spread from north-eastern to western Africa by way of the Sahara (Grigson, 2000; Jousse, 2004b).

MacDonald and MacDonald (2000) have already tried to link cattle remains from West African archaeological sites to present breeds and compared lengths of first phalanges from several sites and periods to this end. However, as already indicated in 3.8., small skeletal elements have only a limited influence on the total size of an animal (von den Driesch and Boessneck, 1974). Unfortunately, these elements predominate in poorly preserved contexts, like the ones that are part of this study. Despite the formulated objections, it was therefore decided to use measurements on some small elements after all, more precisely length and breadth measurements on tali and first phalanges (Figs 46 and 47), albeit in combination with LSI's (see 3.8.) (Fig. 48) and withers height estimations. Grigson (2000) has also worked with LSI's while studying supposed native African cattle. To facilitate comparisons with her results, the same reference animal was used, a female aurochs from Holocene Denmark (Degerbøl and Fredskild, 1970, Tables 11-22), as well as the same suite of measurements (Grigson, 1989). As for the ovicaprines, measurements from the studied faunal material were compared against measurements gathered from the literature. In addition,

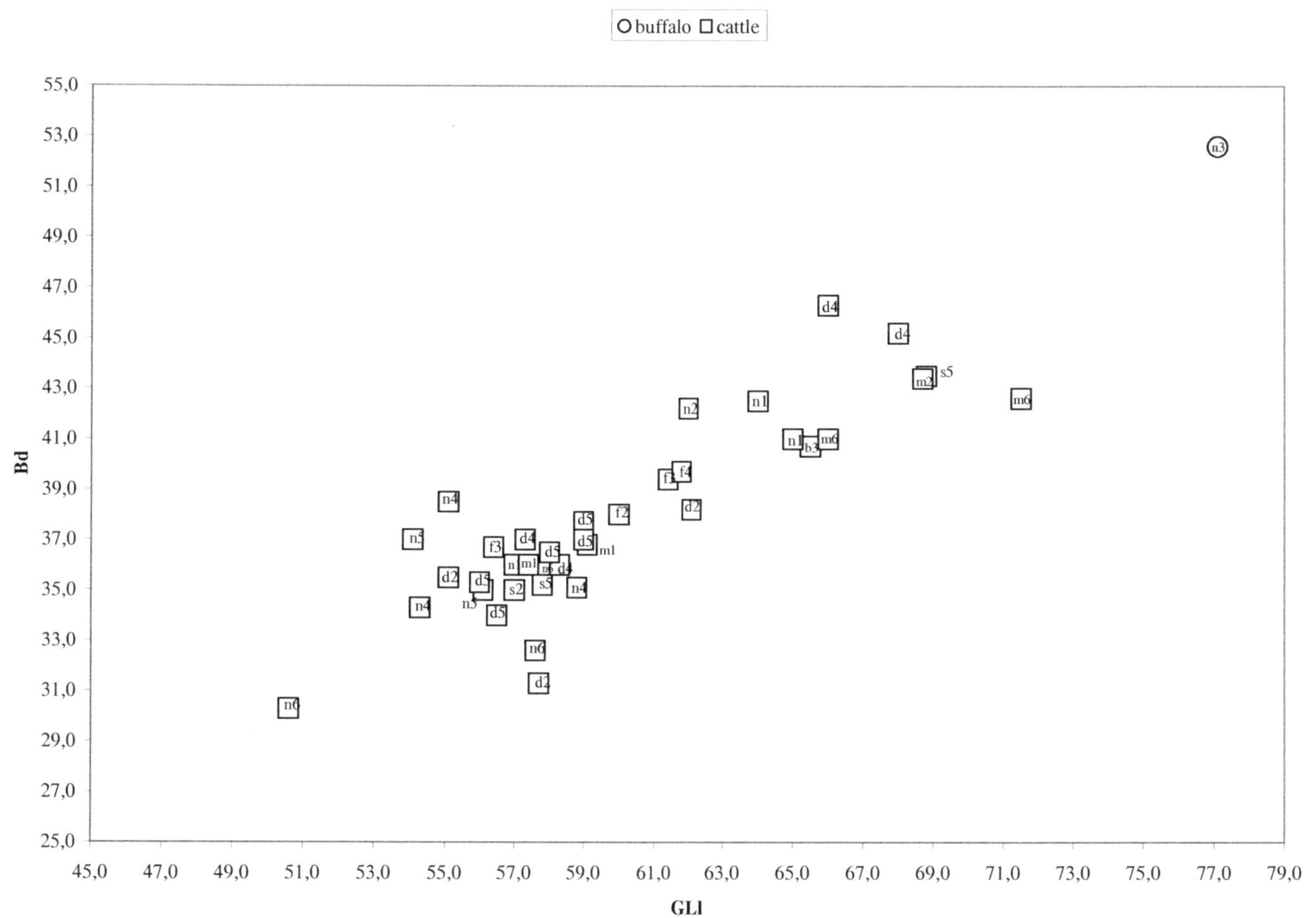

b1	BF 94/45	EIA
b2	BF 97/30	LIA
b3	BF 94/120	LIA
n1	NA 90/5A	Gaj IIb
n2	NA 90/5BI	Gaj II
n3	NA 90/5BII	Gaj IIc
n4	NA 97/33	Gaj I
n5	NA 97/33	Gaj II
n6	NA 96/45	Gaj II
n7	NA 95/1	Gaj II
n8	NA 92/2C	subrec
f1	NA 93/46	EIA
f2	NA 94/7	I
f3	NA 93/45	II
f4	NA 93/45	IIIa
f5	NA 93/45	IIIb
E	Blé E	
s1	Ndiamon Badat (m)	1st mill BC-1st mill AD
s2	Faboura	1st mill AD
s3	Cubalel and Siouré	AD 0-250
s4	Cubalel and Siouré	AD 250-400
s5	Cubalel and Siouré	AD 400-600
s6	Cubalel and Siouré	AD 600-950
d1	Adrar Bous	5500-3500 BC
d2	Araouane	4000-2000 BC
d3	Erg Ine Sakane	4000-1500 BC
d4	Boû Khzâmâ	2300-2000 BC
d5	West Tessalit	3300-2000 BC
m1	Kolima Sud	1700-500 BC
m2	Akumbu	AD 600-1000
m3	Jenné-Jeno	200 BC-AD 400
m4	Jenné-Jeno	AD 400-800
m5	Jenné-Jeno	AD 800-1400
m6	Gao Ancien and Saney	6th-16th century AD
m7	Windé Koroji Ouest	2100-900 BC
g	Gonja and Asante sites	AD 1200-1800

Fig. 46: Length (GLl) and breadth (Bd) measurements (mm) on cattle and buffalo tali
Ndiamon-Badat (Van Neer, 1997), Faboura (Van Neer, unpublished data), Cubalel and Siouré (MacDonald and MacDonald, 2000), Adrar Bous (Carter and Clark, 1976), Araouane, Erg Ine Sakane, West Tessalit, Boû Kzâmâ (Jousse, 2004a, Table 48), Kolima Sud (MacDonald and MacDonald, 2000), Akumbu (MacDonald and MacDonald, 2000), Jenné-Jeno (MacDonald, 1995), Gao (MacDonald and MacDonald, 2000), Windé Koroji Ouest (MacDonald, 1996), Gonja and Asante sites (Gautier and Van Neer, 2005)

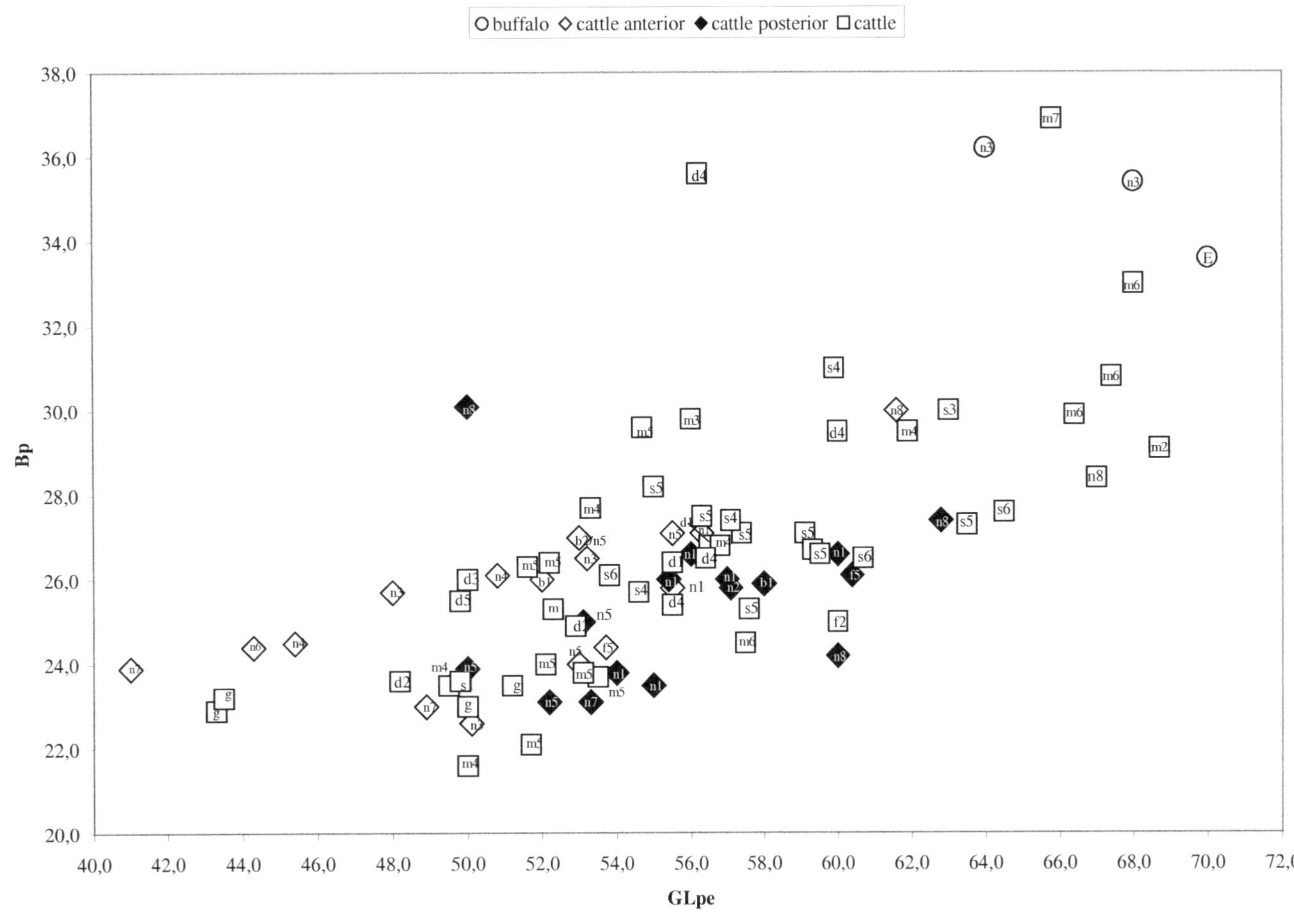

Fig. 47: Length (GLpe) and breadth (Bp) measurements (mm) on cattle and buffalo first phalanges
Ndiamon-Badat (Van Neer, 1997), Faboura (Van Neer, unpublished data), Cubalel and Siouré (MacDonald and MacDonald, 2000), Adrar Bous (Carter and Clark, 1976), Araouane, Erg Ine Sakane, West Tessalit, Boû Kzâmâ (Jousse, 2004a, Table 48), Kolima Sud (MacDonald and MacDonald, 2000), Akumbu (MacDonald and MacDonald, 2000), Jenné-Jeno (MacDonald, 1995), Gao (MacDonald and MacDonald, 2000), Windé Koroji Ouest (MacDonald, 1996), Gonja and Asante sites (Gautier and Van Neer, 2005)

b1	BF 94/45	EIA
b2	BF 97/30	LIA
b3	BF 94/120	LIA
n1	NA 90/5A	Gaj IIb
n2	NA 90/5BI	Gaj II
n3	NA 90/5BII	Gaj IIc
n4	NA 97/33	Gaj I
n5	NA 97/33	Gaj II
n6	NA 96/45	Gaj II
n7	NA 95/1	Gaj II
n8	NA 92/2C	subrec
f1	NA 93/46	EIA
f2	NA 94/7	I
f3	NA 93/45	II
f4	NA 93/45	IIIa
f5	NA 93/45	IIIb
E	Blé E	
s1	Ndiamon Badat (m)	1st mill BC-1st mill AD
s2	Faboura	1st mill AD
s3	Cubalel and Siouré	AD 0-250
s4	Cubalel and Siouré	AD 250-400
s5	Cubalel and Siouré	AD 400-600
s6	Cubalel and Siouré	AD 600-950
d1	Adrar Bous	5500-3500 BC
d2	Araouane	4000-2000 BC
d3	Erg Ine Sakane	4000-1500 BC
d4	Boû Khzâmâ	2300-2000 BC
d5	West Tessalit	3300-2000 BC
m1	Kolima Sud	1700-500 BC
m2	Akumbu	AD 600-1000
m3	Jenné-Jeno	200 BC-AD 400
m4	Jenné-Jeno	AD 400-800
m5	Jenné-Jeno	AD 800-1400
m6	Gao Ancien and Saney	6th-16th century AD
m7	Windé Koroji Ouest	2100-900 BC
g	Gonja and Asante sites	AD 1200-1800

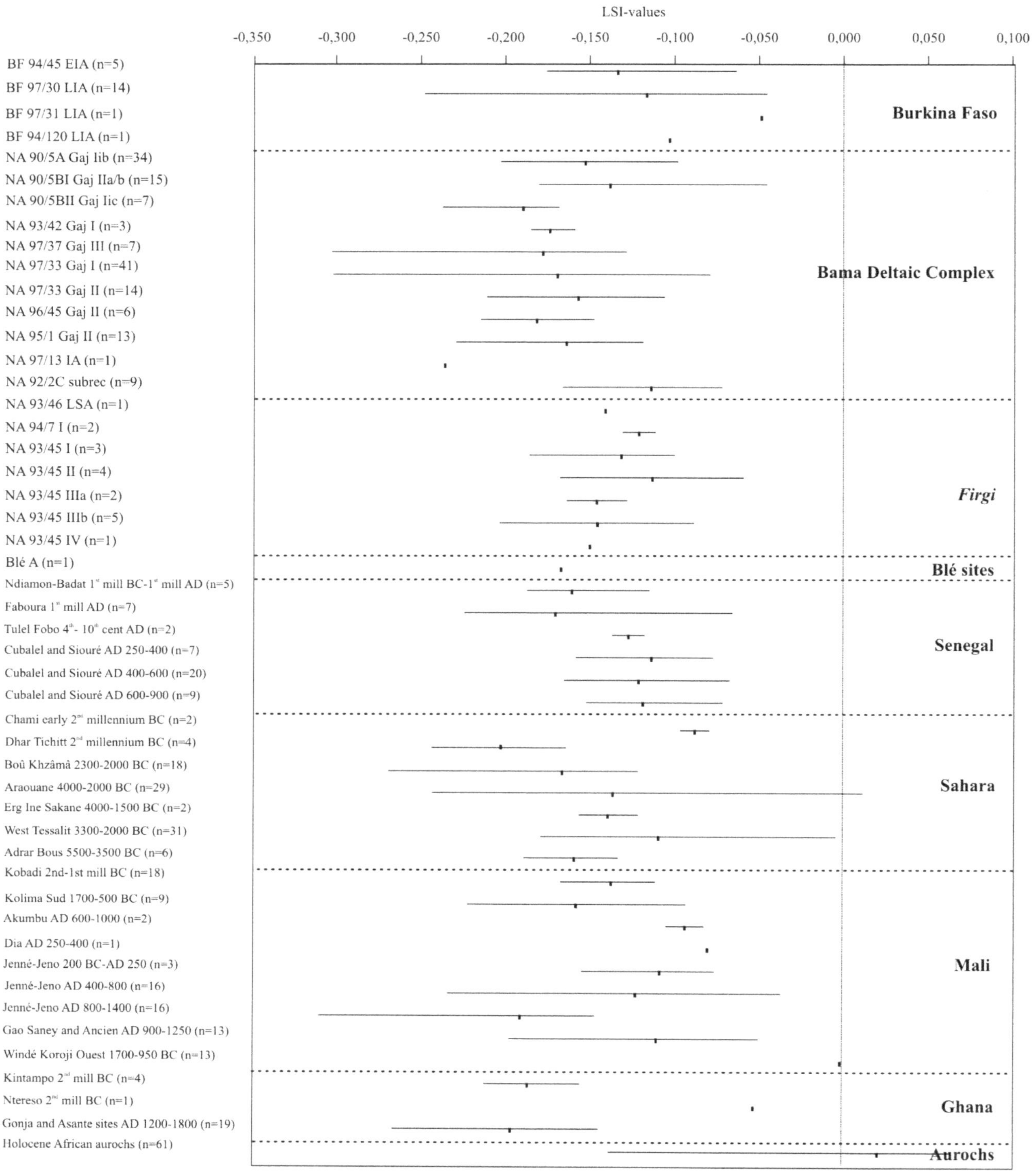

Fig. 48: LSI values of cattle by site and phase

Ndiamon-Badat (Van Neer, 1997), Faboura (Van Neer, unpublished data), Tulel Fobo (Van Neer and Bocoum, 1991), Cubalel and Siouré (MacDonald and MacDonald, 2000), Dhar Tichitt (Holl, 1986, 93), Chami (Bouchud *et al.*, 1991), Adrar Bous (Carter and Clark, 1976), Araouane, Erg Ine Sakane, West Tessalit, Boû Kzâmâ (Jousse, 2004a, Table 48), Kobadi (Raimbault *et al.*, 1987; Jousse and Chenal-Vélardé, 2001-2002), Kolima Sud (MacDonald and MacDonald, 2000), Akumbu (MacDonald and MacDonald, 2000), Dia-Shoma (MacDonald and MacDonald, 2000), Jenné-Jeno (MacDonald, 1995), Gao (MacDonald and MacDonald, 2000), Windé Koroji Ouest (MacDonald, 1996), Kintampo (Carter and Flight, 1976), Ntereso (Carter and Flight, 1976), Gonja and Asante sites (Gautier and Van Neer, 2005), Holocene African aurochs (Linseele, 2004) comparisons were made with metrical data assembled in Linseele (2004) on remains of African aurochs, which must be the ancestor of any indigenous African cattle types.

The LSI value obtained for a cattle first phalanx from Windé Koroji Ouest indicates an exceptionally large animal, which could perhaps be buffalo instead. However, some cattle specimens from the Sahara gave similarly high LSI values. Nevertheless, the raised suspicion is further confirmed on the graph with first phalanx measurements, where dimensions taken on the specimen from Windé Koroji Ouest fall inside the range recorded for buffalo. A first phalanx from Gao also approaches sizes recorded for buffalo, but in view of the date of the site we could be dealing with zebu cattle, which belong to the largest types of cattle. Also striking from the same graph of first phalanx measurements is the robustness of a specimen from Araouane, in the Malian Sahara. As was also apparent from a distal tibia from Oursi hu-beero (BF 97/30), elements from West African archaeological sites with cattle morphological traits can apparently overlap in size with buffalo (see 4.6.10.2.), and large size alone is therefore not sufficient to exclude an identification as cattle. However, other than those mentioned, no measurements on cattle bones are available from Windé Koroji Ouest that could offer more certainty about the presence of cattle at the site.

When comparing data from this study to LSI analyses of cattle remains from East and North East Africa (Grigson, 2000; Lesur, 2004, 332-344), they generally point towards smaller animals. This may be due to the different rainfall regimes in the two areas, which have already been cited earlier. Circumstances must have been more favourable in East Africa, with its bimodal rainfall system that also provided better pasture.

Until recently, the cattle skeleton from Adrar Bous represented the oldest find in Saharan West Africa for which measurements were available. Judging from the LSI graph and comparing it to data from Grigson (2000), the animal is already distinctively smaller than its North African (wild and domestic) predecessors. With a withers height estimated at 105 cm, it falls well within the range of the modern N'Dama and at the upper limit of the WAD. Carter and Clark (1976) wrote that the cranial morphology of the Adrar Bous animal is close to the latter, but according to Gautier (1987a) nothing can be said about its horn characteristics. The latter author argues that the animal might represent a cow. New measurements on cattle from the West African Sahara are now available through the work of Jousse (2004a, Table 48). The material comes from four areas, near Araouane, Erg Ine Sakane, West Tessalit and Boû Khzâmâ and dates approximately between 4000 and 2000 BC. The size variation of the cattle from these sites is striking and indicates very diverse cattle populations in the early phases of cattle introduction into Saharan West Africa. A more refined diachronic and geographical analysis is needed to investigate the possible causes for this extreme size variation. In any case it seems that the hypothesis of a gradual size decrease of cattle during its spread to western African should perhaps be reconsidered, or at least more thoroughly investigated.

Very few cattle remains were available from Burkina Faso for which LSI values could be calculated, but animals seem to have been rather large throughout the entire Iron Age in the area. They fall in about the same size range as the putative zebu from the Nigerian site Galaga (NA 92/2C). According to the present knowledge on the introduction of zebu cattle into West Africa, the Early Iron Age seems too early for the presence of the animal in Burkina Faso. However, excluding it *a priori* would be circular reasoning and West African finds of dromedary from the first half of the first millennium AD (see 6.7.3.1.), indicate that far-reaching inter-regional contacts were already possible at the time. Nevertheless, since zebu cattle were probably introduced along the savannah corridor in an east-west direction, early finds are rather to be expected more to the east. Perhaps we are, instead, dealing with N'Dama cattle from the upper end of the animal's size range. Gautier and Van Neer (2005) are of the opinion that the Kintampo cattle might be early N'Dama, but these are smaller than the Iron Age animals from Burkina Faso. However, the limited metrical data available indicate that the cattle from Kintampo are considerably smaller than those from Ntereso, a site from the same period and region, which Gautier and Van Neer think may be explained in terms of sexual dimorphism. It seems that size differences between material from both Kintampo and Ntereso may perhaps be too great to attribute them to sexual dimorphism, unless we are dealing with animals from the two extreme ends of the size range. In the Late Iron Age of northern Burkina Faso the presence of zebu seems more plausible. For Oursi hu-beero (BF 97/30) we probably need to suppose the presence of a smaller cattle type, besides possible zebu, judging from the large variation in LSI values. Although data from other Iron Age contexts in Burkina Faso are very poor, it seems that the smaller animals are missing there. During phase III at Jenné-Jeno (AD 400-800) there also appears to have been a small and a larger cattle type. The site was a trade centre, and the animals of the larger breed may have been obtained through contacts with nomadic people, as already suggested by MacDonald (1995). In the subsequent phase IV, apparently only small animals were being killed at the site.

In the Bama Deltaic Complex cattle sizes seem to have been relatively uniform during the Gajiganna Culture and perhaps continuing up to the Iron Age, but just one

LSI value could be calculated for this period. Only at Gajiganna A (NA 90/5A) and BI (NA 90/5BI) do cattle seem to have been larger. The reconstructed withers height of an individual from Gajiganna A (107/113 cm) falls in the size range of present N'Dama cattle. It is not clear if we should then suppose the presence of dwarf cattle for the other sites in the Bama Deltaic Complex. The size difference between Gajiganna A and BI and the other Gajiganna sites is much unexpected and it is unclear how it should be explained. Perhaps a different slaughtering pattern, with predominantly males in the former and mainly females in the latter, could have resulted in the observed size pattern, although the question then remains why such a different slaughtering pattern would have existed. Such a scenario would also put the proposed explanations for size differences between cattle of other localities at stake, because these could then perhaps also be due to sexual dimorphism. All LSI values for the subrecent site Galaga (NA 92/2C) indicate large animals and the presence of only one breed, presumably zebu or a zebu crossbreed, needs to be assumed. Nevertheless, measurements on cattle first phalanges from the site show considerable variation in robusticity. As for the ovicaprines, the cattle metrical data from Galaga deviate from those of older sites, in the Bama Deltaic Complex and beyond, and warn against projecting present breeds back into time.

LSI values for cattle from the *firgi* sites are relatively high, although not much higher than those from Gajiganna A (NA 90/5A) and BI (NA 90/5BI). We may perhaps be dealing with a second wave of introduction of domestic cattle, which could represent ancestors of the present Kuri cattle. If these early cattle from the *firgi* are indeed linked to the present Kuri, then an introduction from outside seems more likely than their local development, since such a development probably did not happen quickly in view of the very different appearance of Kuri compared to other cattle types. An introduction from outside is opposite to the view of Blench (1993), but would fit better with the possible occurrence of Kuri cattle in areas other than the Chad Basin (Alberro and Haile-Mariam, 1982). The idea that the earliest cattle from the *firgi* are the ancestors of the present Kuri is reinforced by the diachronic size continuity of cattle, which is especially clear from the site Ngala (NA 93/45). However, the present distribution of the Kuri cattle in the Lake Chad area suggests that we may, instead, have to search for them, and their ancestors, closer to the lake shore or on islands in the actual lake. For later periods, the larger cattle in the *firgi* could perhaps also represent zebu type animals, especially since a zebu clay figurine was found in Daima III layers at Daima (Connah, 1981, Fig. 8.9, 183). The possible introduction of zebu cattle at this time could be connected to the penetration of Arabs into the area by the ninth century AD (Braukämper, 2004).

Other domestic species

There are no indications from the archaeofauna that the inhabitants of the studied sites kept domestic species other than the ones mentioned above. It is not excluded that muscovy duck (*Cairina moschata*), brought from South America to West Africa by the Portuguese, was kept at the subrecent site Galaga (NA 92/2C). In medieval times geese, the local spur-winged goose (*Plectropterus gambensis*) or, less likely, North African greylag goose (*Anser anser*), may have been raised in captivity in the research area. Pigeon (*Columbia livia* f. domestica) keeping is also possible for that period. Indications from the studied archaeofauna for the date of arrival of domestic rabbit (*Oryctolagus cuniculus* f. domestica) in northern Burkina Faso and the southern Lake Chad area, probably in the late nineteenth-early twentieth century AD, are missing. In recent years experiments have been conducted in sub-Saharan Africa with the domestication of savannah cane rat (*Thryonomys swinderianus*), giant pouched rat (*Cricetomys gambianus*) and various species of antelopes. Perhaps such experiments have also taken place in much older times, but there is no archaeological evidence pointing in that direction, neither from the studied sites, nor from any other in West Africa.

6.7.3.2. Introduction waves and routes

As can be deduced from the discussion by species, multiple, later introductions must have taken place after the very first appearance of domestic animals in sub-Saharan West Africa. Mobile pastoralists may have had a significant role in the spread of new species and types. It is proposed here that the introductions happened in four main waves, labelled "Late Stone Age", "Iron Age", "Arab" and "European wave". Besides introductions from outside, there are also indications for a possible local domestication of guineafowl, and the development of dwarf types probably also happened locally. In pre-European times, exotic domestic animals seem to have arrived in West Africa via two routes, one north-south through the Sahara, and one east-west, along the savannah corridor. Archaeological evidence has put the main cultural influences for the whole region and period of investigation to the north, in the Sahara (see 2.2.2. and 2.2.3.).

The first introduction wave, the "Late Stone Age wave", second millennium BC-mid-first millennium BC, contains all the earliest introductions, notably the first cattle, ovicaprines and probably also domestic dogs. In addition, the beginning of pearl millet cultivation is placed here. Animals were propagated from north-eastern Africa by way of the Sahara, and arrived in the different parts of sub-Saharan Africa at very variable dates. It was proposed earlier that the introduction of domestics in northern Burkina Faso may have been retarded by the location of the Niger bend not far to its north, and perhaps also by the presence of an autochthonous hunter-gatherer population that was reluctant to adopt new subsistence strategies. Nevertheless, the cultivation of pearl millet did find its way to northern Burkina Faso during the Late Stone Age. Once south of the Sahara, domestics seem to have spread very rapidly, judging from finds at the Ghanaian sites Ntereso and Kintampo, dated to the early second millennium BC (Carter and Flight, 1972; Stahl, 1985). As explained earlier, this first wave is believed to have been climatically induced and, from the archaeological material, it is clear that both people and animals were migrating (see 2.2.2. and 2.2.3.). The nomadic pastoralists from the Sahara may have been moving like present-day Peul, adjusting their transhumance patterns to subtle ecological changes and undertaking more drastic movements in case of famine (Stenning, 1959, 22; FAO, 2001). The migration of people and their livestock after the mid-Holocene must have been made possible by a contemporary southward shift of the tsetse-zone (Hassan, 2000, 2002), or by a development of trypano-tolerant breeds by that time (see 6.7.3.3.). The southwards migration from the Sahara possibly happened through corridors, like the Tilemsi Valley, linking the Sahara and Sahel zone geographically (Smith, A.B., 1980).

The "Iron Age wave", mid-first millennium BC-early first millennium AD, appears to have left the least archaeozoological evidence. Judging from the metrical data, new types of cattle and ovicaprines may have arrived during this period. No bone remains of new species have been attested, but if the early date of introduction for domestic horse is correct, the animal had already arrived during the "Iron Age wave". These early horses are the possible ancestors of the present pony breeds and size reduction presumably happened only at some point after their introduction. More clear than the introduction of new domestic animal species during this period is the appearance of new cultivated crops: *Hibiscus sabdariffa*, Bambara groundnut (*Vigna subterranea*), cowpea (*Vigna unguiculata*), okra (*Abelmoschus esulentus*) and sorghum (see overview in Kahlheber and Neumann, in press). The region of origin of these crops is not well known, but okra may have been domesticated locally in the Chad Basin. Like the "Late Stone Age wave", The "Iron Age wave" also seems to be associated with the arrival of new cultural groups from the Sahara (see 2.2.2. and 2.2.3.).

The "Arab wave", first millennium AD-early second millennium AD, is then assumed to have comprised domestic fowl, the ancestors of the large horse breeds, donkey, dromedary and finally, zebu cattle. This does not necessarily imply that the animals were brought in by Arab traders, but their introduction is associated with increased inter-regional contacts during this period. Nevertheless, the new species seem to have arrived earlier in trade centres, e.g., Jenné-Jeno (MacDonald, 1995), than in the hinterland. New plant species that appear during this period, either as traded items or grown locally, are: wheat (*Triticum aestivum/T. turgidum* conv. *durum*), date (*Phoenix dactylifera*), cotton (*Gossypium* sp.) and watermelon (*Citrullus lanatus*) (see overview in Kahlheber and Neumann, in press). Unlike for the two previous waves, there may have been less question of true population movements. The animals introduced during the "Arab wave" arrived both by way of the Sahara (domestic fowl, horse, donkey and dromedary) and from East Africa along the savannah corridor (domestic fowl, donkey, zebu cattle). For East African influences, the Lake Chad area has a more favourable location than northern Burkina Faso and this may explain the slightly earlier presence of domestic fowl. We could also expect an earlier introduction of donkey and zebu cattle in the area, although there were no bone remains to prove this. Importantly, the archaeozoological data from the southern Lake Chad area, through the presence of new domestic species, point to inter-regional contacts already by about AD 800 (Connah, 1981, 255), rather than later, from around the fourteenth century AD onwards (Gronenborn, 2000, 263).

The fourth and last introduction wave, the "European wave", contains all the late introductions by Europeans, such as pigs, muscovy duck and rabbit, and also including the commensal rodents black rat (*Rattus rattus*) and brown rat (*R. norvegicus*). The latter two arrived in southern Africa in pre-colonial times, probably from the Arabian Peninsula via the east of the African continent (Ervynck, 1989, 119), but in western Africa indications are missing for an early east-west spread of the animals. Since introductions of domestics by Europeans occurred from the Atlantic coast, inwards, it is not surprising that such species have not been attested at any of the studied sites, not even at subrecent Galaga (NA 92/2C). They must have arrived in the research area only at a very late date.

In the model proposed, the east-west dispersal of domestic animals, along the savannah corridor, is chiefly associated with the "Arab wave". Domestic fowl and zebu cattle, which were probably brought in that way, may have arrived by boat on the East African coast, coming from the Asian continent. Finds of banana (*Musa* sp.) in contexts dated to the first millennium BC in the Cameroonian rainforest, indicate that contacts over the Indian Ocean and subsequent east-west dispersal over the African continent, may already have existed at that time (Mbida *et al.*, 2000; Mbida *et al.*, 2001). Moreover, at least four African crops were known in South Asia by the mid-second millennium BC (Blench, 2003; Fuller, 2003). There is no archaeological or textual evidence for these early contacts between

Asia and Africa and this suggests that an earlier presence of zebu cattle in western Africa than indicated by the available data is also possible. However, these must then have been present in very small numbers, because they remained unremarked in the archaeozoological and iconographic record of West Africa before the second millennium AD.

6.7.3.3. Adaptation to new environments

The major factor limiting and slowing down the spread of domestic animals over Africa must have been diseases, although the impact of these was probably very variable across the continent (Gifford-Gonzalez, 1998, 2000). In sub-Saharan West Africa the main disease was presumably trypanosomiasis, spread by the tsetse fly. The tsetse belt must have formed a true barrier until trypano-tolerant breeds had developed (Hassan, 2000, 2002). To some extent people could have developed measures to protect their herds against tsetse flies and other harmful insects, for example by keeping the animals inside houses during the day and letting them graze during the night (de Leeuw *et al.*, 1972, 61), or by visiting infested areas only during the dry season. The presence of diseases must also have been problematic for the humans themselves. Talbot (1980) mentions, for example, that surface waters of the modern savannah carry with them some highly undesirable effects, including river blindness (onchocerciasis). Besides diseases, there must have been other ecological constraints in the new environments, including higher humidity (Smith, 1984a). Floodplains, like the *firgi*, probably also posed particular problems, such as diseases of the foot when animals had to spend much time on wet or submerged ground (Evans-Pritchard, 1947, 57).

Some time after initial introduction, animals developed that were relatively well adapted to the newly inhabited areas, but it is not clear if this happened more rapidly for certain species than for others. According to some, dwarfism is a consequence of adaptation to (unfavourable) local circumstances and would be especially associated with trypano-tolerance (Epstein, 1971, 276; Clutton-Brock, 2000; MacDonald and MacDonald, 2000). As proposed for the dwarf goats from Zilum (NA 97/37), dwarfing may have happened relatively quickly. Looking at the Kuri cattle from the Lake Chad area, the adaptation of exotic domestics to their new habitat does not necessarily lead to dwarfism. When certain types of domestic animals are well adapted to their environment, this is also used as an argument that they must have been living there for a long time. It was mentioned, for example, that the pony breeds of horses in western Africa are much better adapted to local conditions than the larger types, which may indicate that they arrived there much earlier and had more time to adjust to the new circumstances.

6.7.4. Frequency of consumption and dietary contribution

All remains of consumed domestic animals have been included in the graphs made to estimate the relative importance of the different economic strategies (Figs 35 and 36). This means that the domestic dogs from the Iron Age sites in northern Burkina Faso were also included, although these may not have been an ordinary source of food (Linseele, 2003). When exclusively looking at bone remains, the relative importance of domestics in the diet of the sites' inhabitants is underestimated, because these only reflect the consumption of the animals' meat. Domestic animals provide additional food products, however, such as milk, blood and eggs. In present Burkina Faso, for example, the amount of milk consumed annually is nearly double that of the meat of all domestic animals put together (FAO, 2004).

6.7.4.1. Burkina Faso

As said earlier, domestic animals appear in northern Burkina Faso with the onset of the Iron Age. However, no remains of domestic species have been found at all in the Early Iron Age contexts at Corcoba (BF 97/5). The studied faunal sample is perhaps not representative for the total economy of the site's inhabitants. It comes from a special pit context and may be the result of a single depositional event. At Early Iron Age Oursi (BF 94/45), domestic stock keeping comes in second place after fishing, looking only at NISP data, but is the prime food resource when meat yields are also taken into account. The presence of domestic animals at Oursi during the Early Iron Age is also proven indirectly through archaeobotanical finds of several fodder plant species, some of which must have been collected inside the lake (Kahlheber, 2003, 118-119). At all younger sites from Burkina Faso, domestic animals are dominant, both expressed in NISP and when also considering live weight. However, domestic animals are less numerous at the sites near Kissi (BF 96/22 and BF 97/31) and Saouga (BF 94/120 and BF 95/7) than at those near Oursi (BF 97/13 and BF 97/30). This is possibly due to less favourable environments near the former two sites, which did not have waterbodies as large as the lake of Oursi in their vicinity.

From the described faunal evidence, it seems that an increased reliance on domestic animals between the earlier and later Iron Age periods needs to be supposed, parallel with a greater importance of agriculture found by the archaeobotanists (Kahlheber, 2003, 223; Kahlheber and Neumann, in press). At Saouga (BF 94/120 and BF 95/7), pearl millet production intensified in the course of its short occupation period, which was no longer than 100-250 years (Neumann *et al.*, 1998; Kahlheber, 1999). For this time span, archaeobotanical data also point to the expansion of pasture areas to new zones (Kahlheber, 2003, 223). This may be due to increased agriculture or to growing herd sizes, but the archaeofaunal assemblages from Saouga do not allow determination if the latter was the case. After the end of the Iron Age in northern Burkina Faso a shift to full nomadic pastoralism probably took place (see 2.2.2.). People may have kept cattle of the zebu type, which was presumably known in the area by at least the Late Iron Age (see 6.7.3.1.).

6.7.4.2. Bama Deltaic Complex

At the Gajiganna sites the proportion of domestic animals is rather variable. The most remarkable trend is the difference between sites of the Gajiganna group and those of the Bama-Konduga group, with a higher importance of herding at the latter. Maybe there were fewer problems with humidity and humidity related diseases in this subarea. Bukarkurari (NA 97/33), in the Bama-Konduga group, is the only site with faunal samples from both Gajiganna phase I and phase II. Herding seems to have been slightly more important during the former period than during the latter. The importance of domestics at the Iron Age and subrecent sites in the Bama Deltaic Complex is relatively high, especially at Dorota (NA 97/13), since the site had no fish remains (see 6.5.1.2.).

6.7.4.3. *Firgi* area and Blé sites

Expressed in NISP, domestic stock does not seem to have been important in the *firgi*, which is possibly related to the difficulties with keeping herds around the sites during the annual, large-scale, inundations. However, the problems for herding must have been compensated by high productivity fishing (see 6.5.1.3.). Numbers of bones of domestic animals at the *firgi* sites appear to be inversely related to numbers of fish bones, while the importance of hunting remains consistently low. Domestic livestock is least important at Kursakata (NA 93/46), where, in addition, a decrease between the Late Stone Age and the Early Iron Age can be observed. Phase I at Mege (NA 94/7) has the largest percentage of domestic animals of all *firgi* contexts, but during the site's following phases the percentage decreases considerably. Both at Kursakata and at Mege the observed shift coincides with cultural and environmental changes, i.e. the assumed beginning of full sedentism and large-scale agriculture and the onset of a drier phase (see 2.2.3.1.). Compared to Daima I and II, there seems to have been a decreased interest in stock keeping during Daima III at Daima (Connah, 1981, 191-194, Table 8.2, Table 8.3), but no such changes could be observed in contemporary contexts at the studied sites.

Determining the actual dietary importance of domestic stock compared to fish in the *firgi*, purely from archaeofaunal remains, is not possible. Although the numbers of their bones are limited, domestic animals must have represented a large amount of food, when taking their meat weight into account. It is therefore probably justified to speak of a large herding component, as well as a large fishing component, in the economy of the *firgi*, especially in comparison with a site such as Koyom in southern Chad, dating to the eighteenth-nineteenth century AD, where also a lot of fish but almost no domestic species were found (Rivallain and Van Neer, 1983, 1984). If the connection between the *firgi* sites and the ancestors of the present Kotoko, an ethnic group of specialised fishers, is correct (see 6.5.1.3.), then these also obtained much animal protein from domestic animals, although fishing is emphasised in the Kotoko oral traditions.

Looking at NISP data, the importance of domestic livestock at the Blé sites is comparable to that at the *firgi* sites. However, in reality it may have been smaller, since sampling techniques were not as fine as in the *firgi* and small taxa, such as fish, must thus be underrepresented (see 4.5). When meat weight is also considered, the importance of domestics at the Blé mounds is smaller than in the *firgi*, but its relatively low contribution of herding is compensated by a larger proportion of hunting and fowling (see 6.6.1.3.).

6.7.4.4. The broader archaeological and ethnographical framework

In West Africa nowadays, the economic importance of livestock within a household generally rises with declining rainfall, culminating in an almost total dependence in (sub-) desert zones (Amanor, 1995; FAO, 2001). However, it should be added that pastoralists have in recent periods been increasingly pushed into marginal areas, further north, because of expanding agricultural populations (Blench, 1985). Existing pastoral nomadic groups of arid West Africa would hardly be traceable in archaeological contexts, because they are very mobile and because they do not have their own ceramic industries (MacDonald, 1999). Some livestock breeds are well adapted to high-humidity conditions and are kept by forest-dwelling communities, but never in significant numbers and as prestige animals rather than as important food providers (FAO, 2001). A similar pattern has been attested archaeologically, with domestics being only a secondary source of food in wooded areas (Van Neer, 2002a).

There are hardly any data on the relative importance of domestic animals in the economy of Saharan people before they retreated to more southern zones after the mid-Holocene. The reasons for this lacuna are twofold. First of all Saharan sites are usually shallow, yielding only small bone assemblages, and secondly faunal reports for the area often do not contain quantitative data. In archaeological contexts in the Mauritanian and Malian Sahara, remains of domestic and wild animals are found together, indicating that domestic stock keeping did not totally replace the previously existing economic strategies (Jousse, 2004a). In some areas, such as at Dhar Tichitt in Mauritania, (late third-early second millennium BC) (Holl, 1985, 1986, 86-94), domestic livestock seems to have been only a minor source of food during the early phases of its introduction. Jousse (2006) situates an intensification of pastoral activities at the "Neolithic/Historic transition", around the beginning of our era. This view broadly corresponds with that of DeCorse and Spiers (2001) who speak of the beginning of sedentism, demographic growth and the intensification of food production for that period. However, as will be argued further on, available archaeological evidence seems to point to an intensification chiefly of farming. The Gajiganna sites studied also indicate that in some areas there was a heavy reliance on domestic stock from early phases onwards.

An ethnographical observation that may have important archaeological implications is that survival solely on livestock products is not possible. African pastoralists often supplement their diet with agricultural produce obtained either from farmers, or by seasonally growing crops themselves (Dahl and Hjort, 1976, 155). As will be explained in 6.7.8.2., milk is the staple animal food of African pastoral groups. A heavier reliance on livestock is possible in East compared to West Africa, because in the former milk production is more constant throughout the year, thanks to its bimodal rainfall regime. In East African archaeological contexts an almost exclusive dependence on domestics for obtaining animal proteins has been attested from around the end of the second millennium BC onwards (Gifford-Gonzalez, 1998), although the role of wild taxa varies with the geographical location of the sites (Marshall and Stewart, 1994). The development of these pastoral communities probably happened relatively late, a millennium after the first introduction of domestics, because of the presence of animal diseases (Gifford-Gonzalez, 1998).

6.7.5. Herd composition. Environmental and other implications

A few herd characteristics are assembled under the term "herd composition": relative proportion of the different species, including also non-pastoral ones like domestic dog and domestic fowl, ages and sexes of the animals and their total number. When studying herd composition it is also important to look at the types of domestic species present. As explained earlier, small ovicaprines are, for example, mostly associated with sedentary farmers, while larger types can usually be found in the flocks of nomadic pastoralists. The aspect of herd composition that can most easily be investigated for the studied sites is species composition. Preservation conditions were unfortunately not good enough to make age and sex profiles (see 3.7.) and, except in special circumstances, the total size of a herd is generally difficult to determine from archaeological evidence.

Factors that determine the composition of a herd can be divided into two groups. The first group contains the ecological factors, most importantly the presence of sufficient drinking water, the availability of suitable pasture and mineral licks, besides the absence of diseases. The second group concerns the anthropogenic factors, such as the mobility of a community, determining its ability to look for areas with the best possible circumstances for its herds, and the purposes for which the animals are kept. In addition, cultural reasons may be involved. Cattle usually enjoy a higher esteem than ovicaprines, for example, and for Islamic feasts sheep are preferred to goats. Most attention of human groups usually goes to their cattle and ovicaprines, while domestic fowl and domestic dogs are left to find their own food. Nevertheless, in present north-eastern Nigeria, large numbers of sheep and goats are maintained as domestic semi-scavengers in urban areas. Moreover, management systems for taurine/Sanga cattle may vary from hand-fed in permanent confinement to free range with almost no input, and even scavenging (Blench, 1999, 40, 56). The low inputs needed for taurines/Sanga's is mainly due to their extreme adaptation to the environment they live in, allowing them to survive on a wide variety of food. This is in contrast to zebu breeds, for which the owners need to find the appropriate pasture composition. Donkeys need very low management inputs, while much more (labour) investment is needed to keep horses under West African ecological conditions.

In present sub-Saharan Africa, herds of domestic animals usually consist of both cattle and ovicaprines, mostly to spread risks because the species have different susceptibilities to diseases (Dahl and Hjort, 1976, 237; Robertshaw, 1990, 24; Smith, 1992, 125). Small stock, for example, is said to be more vulnerable than large stock (Dahl and Hjort, 1976, 232-233). Another strategy against epizootics, or other disasters such as droughts, is maximisation of herd size (Dahl and Hjort, 1976, 115; FAO, 2001). The more animals there are to start with, the more will be left after a disaster. However, having large herds can also be a way to express social status (Reid, 1996; Clutton-Brock, 2000). A second reason for keeping mixed herds is the maintenance of good pasture and maximal use of available resources (Robertshaw, 1990, 24; Smith, 1992, 125). Cattle and goats are complementary in their pasture requirements. Cattle need good quality grass, while the pasture components rejected by them are used by goats (Dahl and Hjort, 1976, 253). Sheep have similar requirements to cattle, but on tall grass pastures cattle are likely to thrive better than sheep, and on short grass pastures the opposite is the case. In general, ovicaprines, and especially sheep, need to drink more often than cattle, but their intake is smaller and they can therefore be run from sources that cattle cannot (Dyson-Hudson, 1966, 45; Dahl and Hjort, 1976, 249). During periods of green pasture they can go up to two weeks or longer without drinking. Keeping mixed herds also allows better spread of their yields over the year. Seasonal shortages of cattle milk can, for example, be overcome by using goat milk. Goats breed more indiscriminately throughout the year and are a reliable source of milk during the dry season, when cattle milk is scarce (Dahl and Hjort, 1976, 92-93, 213). During periods when herds of cattle are taken to distant pasture and water, goats are often kept close to home compounds to provide the remaining family with milk (Dahl and Hjort, 1976, 200). A final reason for keeping both small and large livestock is their different growth rates, rapid for the former and slow for the latter (Dahl and Hjort, 1976, 237). Sheep and goats are often kept together and regarded as one unit (Dahl and Hjort, 1976, 200), and they are generally associated with sedentary farmers and villages. Nevertheless, the two have different characteristics. Sheep are, for example, better suited for travel over long distances (MacDonald, 1995; Blench, 1999, 49). However, the main pastoral nomadic species is cattle, and especially zebu type cattle (FAO, 2001), although the species kept much depends on ecological circumstances. In the Holocene of the Acacus and surrounding areas, for example, ovicaprines appeared to be more common in pastoral economies in mountainous parts, while cattle were dominant in lowland areas (di Lernia, 2002).

As for the determination of the relative proportion of the different economic strategies, graphs have been made to investigate the relative importance of domestic fowl, domestic dog, ovicaprines and cattle by site and phase. On a first graph the NISP data were plotted, and on a second one the animals' live weight was also taken into account (Figs 49 and 50). A difficulty with the interpretation of the graphs is that they reflect animal consumption and not the composition of the living herds. It is expected, for example, that they will be distorted towards ovicaprines compared to cattle, because these have a faster turnover. Moreover, domestic equids are not included, because they were presumably not eaten, but remains of them were very rare. As mentioned for the other graphs, we should thus mainly concentrate on inter- and intra-site differences. In general, the economic importance of domestic fowl appears to have been very limited, especially considering its poor meat yield. On another graph, numbers of sheep bones were compared to numbers of goat bones, identified by site and phase (Fig. 51). This was preferred to plotting relative numbers, because ovicaprine bones identified to species level are rare. Usually a predominance of goat was found, in contrast to MacDonald and MacDonald (2000), who mainly studied archaeological sites from the Middle Senegal Valley and near the River Niger. These areas lie closer to the desert edge and may have had more mobile herding systems, in which sheep are also today normally predominant over goats (Bourn *et al.*, 1994). The findings obtained are also opposed to the archaeological evidence from southern Africa, where goats were apparently much less important as livestock animals than sheep (Clutton-Brock, 1993). There are very few bone remains of goats from archaeological sites in this part of the African continent and depictions of the animal are lacking in rock art. Redding (1984) has made models to explain sheep-goat ratios at Near Eastern sites, but stressed that these should not be used in other geographic areas. No similar models exist for Africa, and mainly the information described above and in 4.6.10. will be used to explain observed ratios.

6.7.5.1. Burkina Faso

As mentioned in 2.1.1., the Oudalan province of Burkina Faso, and especially its northern half, is nowadays a very favourable area for stock keeping. In the case of cattle, this is mostly true for the zebu type and we cannot, therefore, project the present situation back to periods before its introduction. The date of this event remains unknown but it may have happened during the Late Iron Age. All locations in Burkina Faso with domestic animals have yielded more ovicaprines than cattle, which corresponds to the proportion between the animals in the area today. Their inhabitants were probably sedentary agriculturalists that were not able to look for good pasture and enough water to sustain large cattle herds. However, the presence of both larger and smaller domestic ovicaprines indicates that the Iron Age landscape of northern Burkina Faso may have known a nomadic as well as a sedentary component (see 6.7.3.1.). This agrees with Pelzer *et al.* (2004) who think that, before the abandonment of the settlement mounds, there must have already been an increasing nomadic element. With the exception of Oursi hu-beero (BF 97/30), all studied Iron Age sites in northern Burkina Faso yielded more goat than sheep remains, which may be due to poor pasture areas in their vicinity. It was proposed earlier that the predominance of sheep at Oursi hu-beero could perhaps be explained by a higher social status of its inhabitants (Linseele, in press), since today the animals also have a greater value than goats on average (e.g., Geis-Tronich, 1991, 464).

At the Middle and Late Iron Age sites in northern Burkina Faso, the relative proportion of domestic dog remains is substantial. It is especially high for the trench that was dug at the bottom of the mound at Saouga (BF 95/7). However, the food value of dogs by individual is small, as reflected in Fig. 50. Increasing amounts of dog throughout the Iron Age occupation in northern Burkina

			Domestic fowl	Domestic dog	Sheep/goat	Cattle
1	BF 94/45	EIA	0	1	201	108
2	BF 97/13	EIA	0	1	27	8
3	BF 97/13	MIA	0	39	159	76
4	BF 97/13	LIA	0	34	212	103
5	BF 97/30	LIA	28	65	656	114
6	BF 96/22	MIA	0	18	275	19
7	BF 97/31	LIA	0	31	217	32
8	BF 94/120	LIA	81	176	879	149
9	BF 95/7	LIA	46	253	356	46
10	NA 90/5C	Gaj IIa	0	0	6	10
11	NA 90/5A	Gaj IIb	0	0	500	849
12	NA 90/5BI	Gaj IIa/b	0	0	61	426
13	NA 90/5BII	Gaj IIc	0	0	84	155
14	NA 91/1A	Gaj I	0	0	0	6
15	NA 93/42	Gaj I	0	0	2	31
16	NA 97/18	Gaj I	0	0	0	1
17	NA 97/24	Gaj I	0	0	0	14
18	NA 93/36	Gaj IIa	0	0	0	8
19	NA 99/65	Gaj IIa	0	0	0	1
20	NA 99/65	Gaj IIb	0	0	45	10
21	NA 99/65	Gaj IIc	0	0	27	23
22	NA 93/10	Gaj IIc	0	0	4	25
23	NA 97/37	Gaj III	0	0	31	119
24	NA 97/33	Gaj I	0	0	78	521
25	NA 97/33	Gaj IIa/b	0	0	43	202
26	NA 96/45	Gaj IIa/b	0	0	127	193
27	NA 95/1	Gaj IIa/b	0	0	113	446
28	NA 97/26	IA	0	0	39	15
29	NA 99/75	IA	0	0	37	17
30	NA 97/13	IA	0	0	43	6
31	NA 92/2C	subrec	18	0	622	282
32	NA 93/46	LSA	0	0	16	20
33	NA 93/46	EIA	0	0	242	100
34	NA 94/7	I	0	0	46	547
35	NA 94/7	II	0	0	8	47
36	NA 94/7	III	2	0	22	16
37	NA 94/7	IV	0	0	13	14
38	NA 93/45	I	20	0	117	45
39	NA 93/45	II	40	0	114	53
40	NA 93/45	IIIa	5	0	64	47
41	NA 93/45	IIIb	3	0	31	85
42	NA 93/45	IV	9	0	59	64
43	Blé A		0	0	23	12
44	Blé B		2	0	14	7
45	Blé C		0	0	16	4
46	Blé E		4	0	32	22
	All sites		**258**	**618**	**5660**	**5098**

Table 14: Relative importance of domestic fowl, domestic dogs, ovicaprines and cattle by site and phase. Legend to Figs 49 and 50

Faso have been connected earlier to cultural influences from North African Berbers, who are notorious dog-eaters (Linseele, 2003). Perhaps it is also connected to food stress, caused by the circumstances that resulted in the abandonment of the mounds, including political instability, environmental changes and maybe also conflicts between agriculturalists and pastoralists (see 2.2.2.).

6.7.5.2. Bama Deltaic Complex

All Gajiganna sites have yielded more cattle than ovicaprines; only Gajiganna phases IIa and IIb at Gilgila (NA 99/65) had inverse proportions. The Bama Deltaic Complex apparently had enough pasture and water to sustain large cattle herds. Groups were probably sufficiently mobile to follow the best possible circumstances for the herds and to avoid diseases. Nevertheless, except during Gajiganna phase I, mobility,

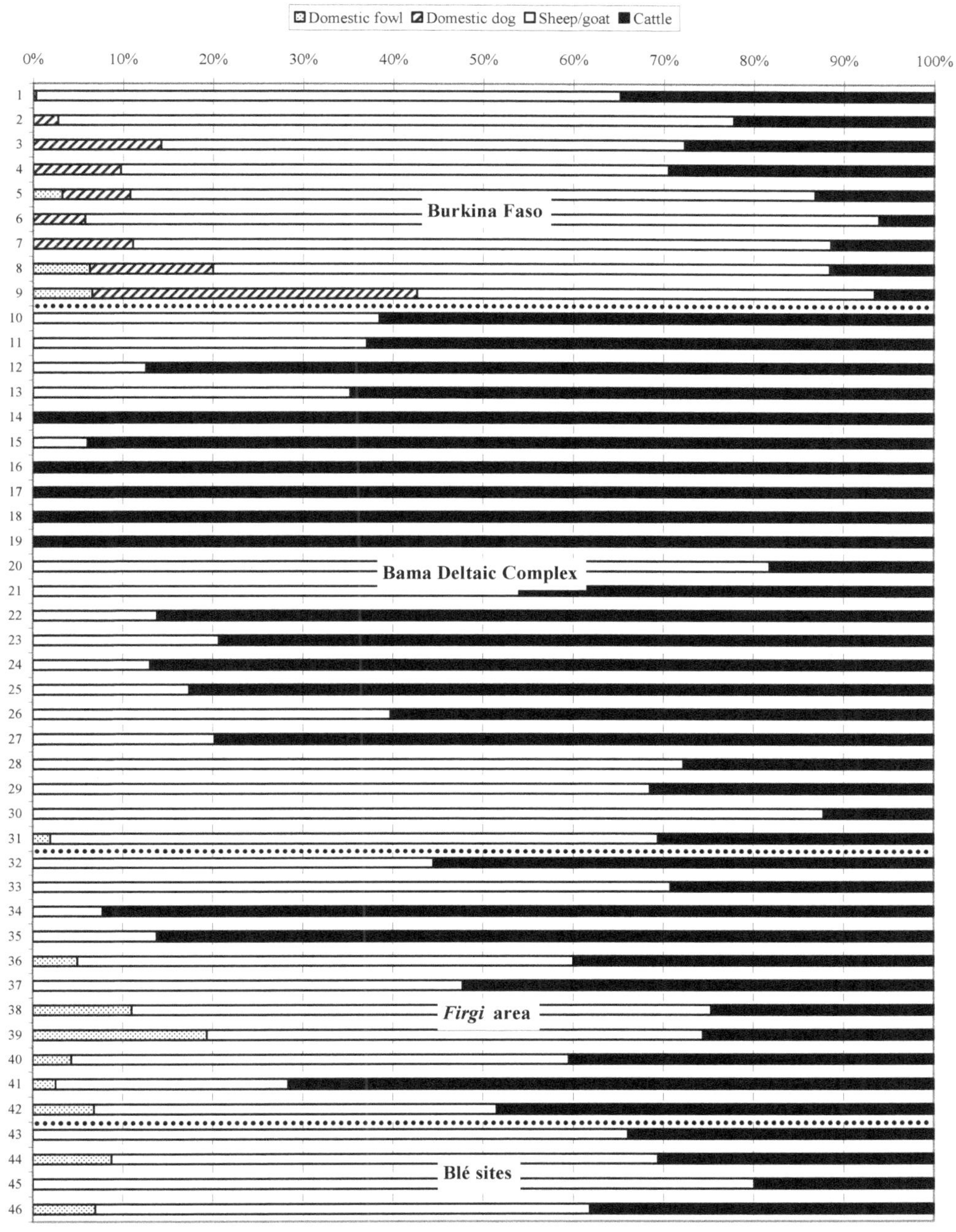

Fig. 49: Relative importance of domestic fowl, domestic dogs, ovicaprines and cattle (NISP) by site and phase

must have been relatively limited and settlements were at least semi-permanent (see 2.2.3.). The people from Zilum (NA 97/37) also appear to have been capable of keeping herds with a lot of cattle, even though they were probably fully sedentised. Perhaps mobility was not needed because of favourable circumstances near the site. Alternatively, only a small part of the human group could have practised mobile herding, or cattle could have been entrusted to pastoral nomadic groups for herding (see 6.7.6.). Dwarf goats at the site indicate that at least part of the livestock herds was kept near the settlement. An additional explanation for the large percentages of cattle in early phases at the Bama Deltaic Complex may be the presence of bourgou grasses on the clay patches in the area. Imprints of such grasses, including *Echinocloa* cf. *pyramidalis*, *E.* cf. *stagnina*, and *Oryza* cf. *longistaminata*, have been found in potsherds of the Gajiganna Culture (Klee *et al.*, 2004). Bourgou grasses are said to allow a large proportion of cattle in the herds while small stock does not thrive well in the humid

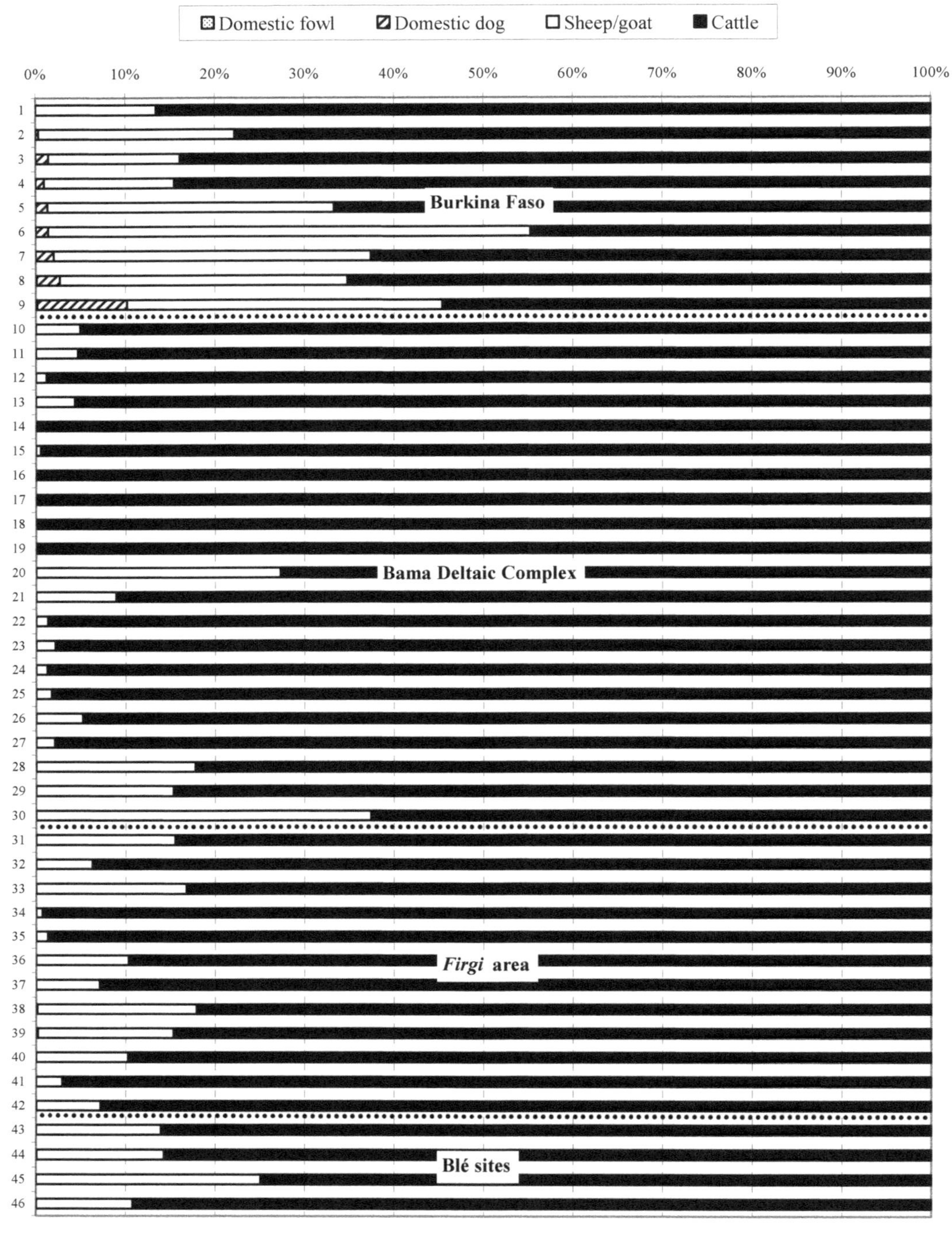

Fig. 50: Relative importance of domestic fowl, domestic dogs, ovicaprines and cattle (NISP x live weight) by site and phase

circumstances of this kind of pasture (Gallais, 1975, 190). However, this does not provide a satisfying explanation for the extreme predominance of cattle at the Gajiganna sites in the Bama-Konduga group. Clay patches with bourgou grasses were probably missing there, but no archaeobotanical data are available that could allow the verification of this hypothesis. In the living herds of the Bama-Konduga group, the preponderance of cattle was possibly even more pronounced, since, as said, the animals have a slower turnover than ovicaprines.

The period in the Bama Deltaic Complex between the Gajiganna Culture and the Iron Age still remains archaeologically undocumented (see 2.2.3.). A shift to nomadic herding may have taken place, which resulted in poor archaeological visibility. In contrast to the

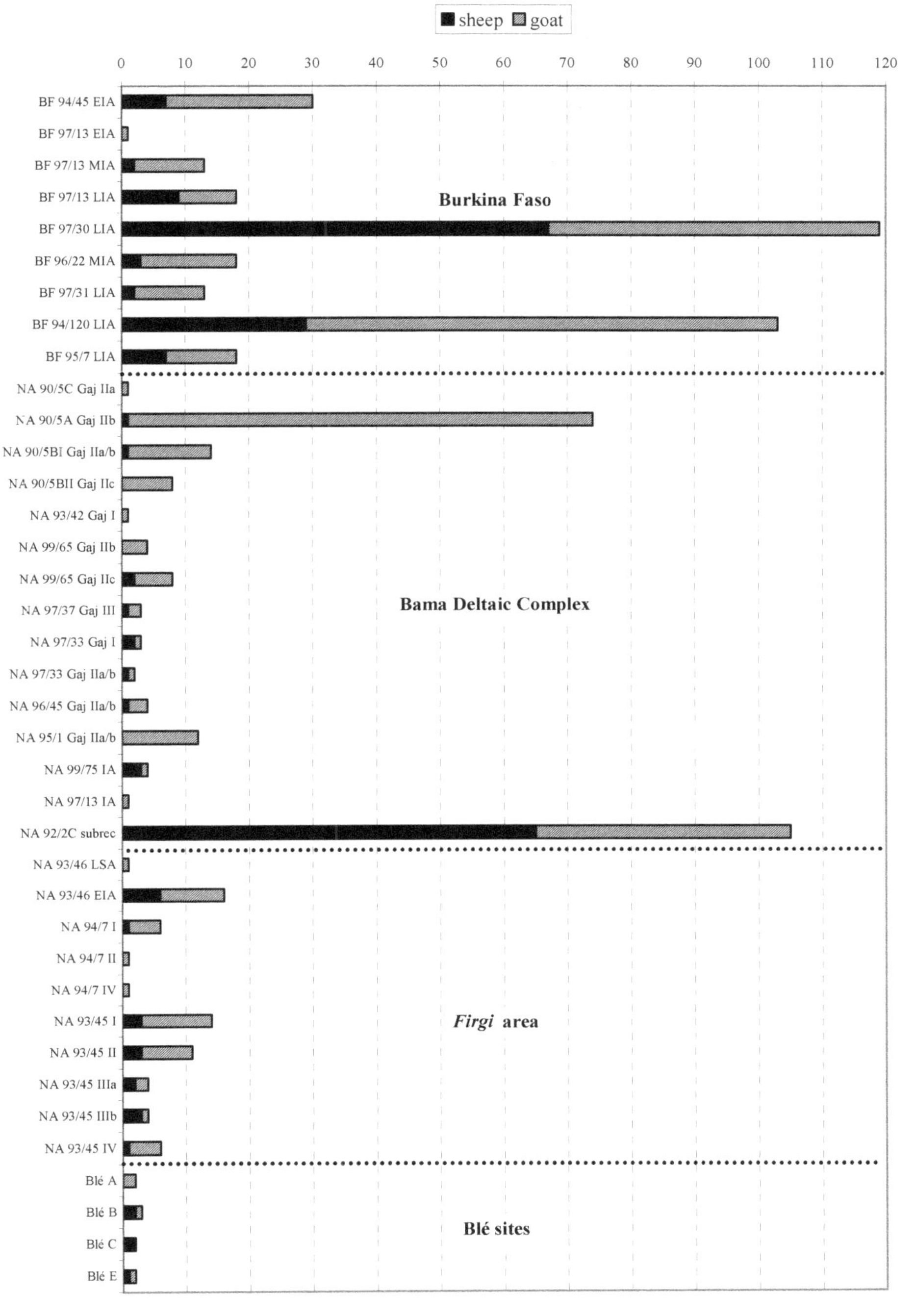

Fig. 51: Numbers of sheep and goat bones by site and phase

Gajiganna sites of the Bama Deltaic Complex, ovicaprines dominate among the livestock at the Iron Age sites and the subrecent site Galaga (NA 92/2C). From the Iron Age onwards, the area may no longer have been favourable enough for keeping large cattle herds, certainly not by sedentary communities. The large ovicaprines at Galaga possibly point towards a pastoral nomadic component in its economy. Galaga deviates from the older sites in the Bama Deltaic Complex by its predominance of sheep over goat, which can probably also be connected to pastoral nomadic economic elements, since sheep are better suited for mobile herding than goats. In 4.6.10.1., an extremely long and slender sheep metatarsal from the site was mentioned, which may point to harsh circumstances and animals facing malnutrition. Because of the large amounts of archaeological material recovered from Galaga, we probably need to assume that only part of its inhabitants practised pastoral nomadic activities or that livestock was obtained through cooperation or exchange with pastoral nomads.

6.7.5.3. *Firgi* area and Blé sites

At the *firgi* and Blé sites the importance of cattle compared to ovicaprines is variable. At Kursakata (NA 93/46), cattle remains are proportionally more important in the site's Late Stone Age than in its Early Iron Age layers. It is assumed that the change is related to the shift to sedentism between the two periods. The mobile communities of the Late Stone Age were able to keep large cattle herds, because they could take their animals away from the area during the inundation season, when living surfaces and pasture areas were insufficient. Cattle in the area today are also mainly kept in transhumant systems, driven westwards to more elevated areas when water levels begin to rise (Sturm *et al.*, 1996; Braukämper, 1995; de Leeuw *et al.*, 1972, 15). At Mege (NA 94/7), proportions evolve from an extreme predominance of cattle in phases I and II, to a majority of ovicaprines in phase III, and in its final occupation phase there is again a slight preponderance of cattle. The first shift may be related to decreasing mobility as at Kursakata. The second change is possibly due to the arrival of cattle keeping Shuwa Arab populations, who were living at Mege at the time of its abandonment in 1983. At Daima, a predominance of cattle was recorded for all three of its occupation phases, but gradually becoming less extreme (Connah, 1981, Table 8.2). Changing proportions at the site are also reflected in the species depicted in clay figurines (Connah, 1981, 192). At Ngala (NA 93/45) an inverse trend can be observed, where an initial preponderance of ovicaprines changes in the last two phases to a majority of cattle remains. This may be due to decreasing flood levels, leaving more surface free for cattle herding during the inundation season. Decreasing flood levels around the site were already hypothesised earlier from the composition of its fish fauna and increasing goat sizes (see 6.5.2.3. and 6.7.3.1.). Ovicaprines from the *firgi* sites are usually small, pointing to sedentary stock. This also fits with the fact that all sites with relatively large numbers of ovicaprines that were identified to species level had a predominance of goats. For the Blé sites, samples of domestic animals are too small to trace meaningful trends in the species composition of their livestock herds.

6.7.5.4. Ethno-historical data on culling and castration

As said, very little data could be gathered on the age and sex composition of the herds kept at the studied sites. For both ovicaprines and cattle, individuals in all age classes between foetuses and animals of mature size were found, but there is no information on the age distribution of the older animals. Some historically documented and recent culling practices can possibly be projected back into the past. Al-Bakri, writing in the eleventh century AD, mentions that male goats are slaughtered young and that only the females are allowed to live (Levtzion and Hopkins, 1981, 82). In addition, present flocks of sheep and goats usually consist predominantly of females, which are kept for reproduction (Dahl and Hjort, 1976, 88). Smith (1992, 109) mentions that of those males in present African cattle herds that are not slaughtered young, only half are kept for breeding, while the rest are castrated. Castration reduces the animals' aggression and increases the quality of their meat, but was probably rarely practiced in sub-Saharan Africa in pre-modern times because of the health risks involved for them (FAO, 2001). Nevertheless, indications for castration of cattle have been found at sites of the Handessi horizon in the Wadi Hariq in northern Sudan, dated to 2200-1600 BC (Jesse *et al.*, 2004) and is also practiced with very traditional means by the present Nuer (Evans-Pritchard, 1947, 33-34). The Middle Senegal sites Cubalel and Siouré (AD 0-1400) are up to now the only West African sites that yielded sufficient well-preserved material to calculate age curves (MacDonald and MacDonald, 2000, unpublished manuscript). At these sites, a shift could be observed for ovicaprines around AD 600, when animals were no longer kept alive as long as possible, but culling of younger animals was apparently practiced to maximise meat and milk production. On the other hand, herding practices remained unchanged for cattle, aiming at maintaining maximal herd size by avoiding the culling of younger animals.

6.7.6. Seasonality and stock keeping strategies

Seasonal movements, in search of water, mineral licks and pasture and to avoid diseases, are usually necessary to keep large herds of cattle, and the animals are therefore associated with nomadic pastoralists rather than sedentary agriculturalists. However, in some present African communities only part of the group is sent away with the flocks, or else the cattle are entrusted to nomadic groups to herd them for at least part of the year (Blench, 1999, 15). It is not clear how far the latter strategy can be pushed back into time, but present pastoral groups in both parts of the research area arrived there only during the past few centuries (see 2.1.) Finds of foetuses may perhaps be used as proof that the human groups were keeping stock themselves, since pregnant animals need especially good care. Ovicaprine foetuses were relatively common, found at a few Iron Age sites in Burkina Faso, several Gajiganna sites, one Iron Age site from the Bama Deltaic Complex and at the three *firgi* sites (Table C.82). Foetuses of cattle were less numerous and were confined to a few Gajiganna sites, one Iron Age site from the Bama Deltaic Complex and phase I at Mege (NA 94/7) (Table C.97). A large pastoral component in these contexts with cattle foetuses was already supposed earlier, among other things from high amounts of cattle bones. Only for the Iron Age site Elkido North (NA 99/75) was the find of a cattle foetus bone more unexpected. The foetuses mainly indicate that ovicaprines were more often kept close to the compounds than cattle, which were probably taken elsewhere during large parts of the year, very much like the situation today. The presence at one locality of a smaller and a larger type of the same domestic species found, possibly sheep in Iron Age Burkina Faso and goats at Oursi hu-beero (BF 97/30) and Galaga (NA 92/2C), may point to one type kept locally and another one kept in mobile herds.

Almost all domestic livestock species give birth during the rainy season (Smith, 1992, 109). Only goats breed more indiscriminately throughout the year and also often have their young during the dry season (Dahl and Hjort, 1976, 92-93). Ovicaprine foetuses therefore do not give information on the season of habitation. The cattle foetuses found cannot be used to determine seasonality either. No shaft lengths of long bones could be measured (cf. Habermehl, 1975, Table 5), hence it could not be determined how long after conception abortion occurred.

Since water is probably the main factor limiting the spread of domestic herds, it is important to know from which date artificial wells were known, because these must have allowed the keeping of livestock in areas that were more arid than its natural range. At Nabta Playa, in the South Western Desert of Egypt, a well has been found that was not directly dated but probably dates as far back as the seventh millennium BC (Wasylikowa *et al.*, 1993). For the southern Lake Chad area, Connah (1981, 161-162) has argued that (well) digging in the heavy clay soils must have been extremely difficult before iron tools were known. Nevertheless, evidence from Zilum (NA 97/37) has now shown that even with only stone tools large-scale digging was possible (see 2.2.3.2.). In other words, wells deliberately dug by humans seem possible from about the earliest spread of livestock on the African continent. A drawback of artificially created wells is that the concentration of stock around them can lead to local overgrazing (Le Houérou, 1989, 148). In both parts of the research area natural surface waters near the sites must have provided sufficient drinking water during large parts of the year. Only at the height of the dry season these may no longer have been sufficient. Seasonal extremes were probably greatest in the *firgi* area, with an alternation of inundations and water shortages. However, judging from the presence of kob, mainly in the early phases of occupation, there must have been a year-round presence of at least minimal quantities of surface water.

Livestock grazing near the sites probably happened in extensive systems like today, i.e. without closed pastures, both in the natural savannah and on harvested fields. As described in 2.1.1., good pasture areas in northern Burkina Faso can be found on top of the dunes. Archaeobotanical research in northern Burkina Faso has shown that the inhabitants of the sites near Oursi also let their herds graze on the aquatic and semi-aquatic vegetation of the lake, while near Saouga the inundation area of the river Gorouol was not used for pasture (Kahlheber, 2003, 177). In the *firgi* area, and possibly also in early phases in the Bama Deltaic Complex, pasture must have been limited to sandy areas during high water. After the retreat of the floodwaters, bourgou grasses on the clay plains must have provided attractive pasture areas. Millet and sorghum stems are important additional food for livestock in present arid West Africa (Connah, 1981, 42; FAO, 2004). The main importance of crop residues is for carrying the livestock through the dry season (de Leeuw *et al.*, 1972, 51). Dahl and Hjort (1976, 151) have written that such alternative or complementary strategies to free-ranging can allow the smoothing out of the adverse effects of a single rainy season. In Burkina Faso possible archaeobotanical evidence for grazing and browsing plants, intentionally brought to the sites, has been found from the Early Iron Age onwards (Höhn, 2002; Kahlheber, 2003, 118-119; Höhn *et al.*, 2004). *Faidherbia albida* must have been an especially important plant for herders, because it carries leaves during the dry season, thus providing forage when other plants are bare (Höhn *et al.*, 2004).

For stock keepers of the studied sites, the dry season must have been the most strenuous period of the year, although the flood season in the *firgi* probably also brought specific problems and this may equally have been the case in early phases in the Bama Deltaic Complex. During these periods (cattle) herds could probably not be kept near permanent settlements in the research area. Herders and their flocks nowadays concentrate near remaining water and pasture during dry parts of the year, while they are more dispersed in the rainy season. In arid zones of present West Africa, the main cattle sales usually take place at the beginning of the dry season, or in periods of drought, because the herders are not able to keep their total herds alive (Amanor, 1995; FAO, 2001).

In the study area nowadays, people keep their livestock animals inside houses or other special structures during parts of the day, or year, for several reasons. The Shuwa Arabs (Braukämper, 1993, 1995) and also the Nuer (Evans-Pritchard, 1947, 66) do it mainly to protect them against diseases, but it can also be against heat, predators, or perhaps hostile human groups. The Gulmanceba of Burkina Faso even have separate rooms for each animal species (Geis-Tronich, 1991, 63). From the sites studied, the best evidence for animals kept inside house structures are the equid dung and, especially, the concentration of ovicaprine droppings recovered from floor levels at Late Iron Age Oursi hu-beero (BF 97/30) in Burkina Faso (Linseele, in press). The latter were found in parts of the house that were still in use when it burned down. As indicated in 2.2.3.2., a dung layer in phase I layers at Mege (NA 94/7) was interpreted as the remains of a kraal. Holl (2002, 150) furthermore mentioned archaeological features at Houlouf (AD 0-1800), resembling present-day chicken coops.

6.7.7. Stock keeping versus agriculture

In the first chapter it was mentioned that, for all parts of the research area, a shift to full farming took place with the onset of the Iron Age, around the beginning of our era. Nevertheless, at the *firgi* sites Kursakata (NA 93/46) and Mege (NA 94/7), the importance of agriculture seems to have been limited throughout their chronological sequence (Klee and Zach, 1999). In the Bama Deltaic Complex, on the other hand, the beginning of full farming should probably already be placed during Gajiganna phase III (Magnavita, 2003, 169). Whereas pearl millet was the only cultivated crop during the last two millennia BC, a more diversified agricultural system developed in arid West Africa by the beginning of the first millennium AD, with the appearance of new crops such as sorghum and legumes (Neumann, 2003; Kahlheber and Neumann, in press). In the *firgi* crop diversification apparently only happened by AD 800, when the oldest finds of sorghum were attested (Connah, 1981, 188-189; Zach *et al.*, 1996; Klee and Zach, 1999). While a combination of small-scale pearl millet cultivation and semi-nomadic herding is possible, full farming requires full sedentism, and agricultural communities therefore have a high archaeological visibility. Agricultural communities are also very expansive and agriculture knows the most rapid spread of all types of economy (Zohary and Hopf, 2000), which may be another reason why they are so prominent in the landscape. The beginning of sedentism and full farming may have been the basis of the creation of the social structures that led to the development of the medieval West African empires (Kahlheber and Neumann, in press).

The beginning of full farming inevitably must have had repercussions on domestic stock keeping. The sedentary communities were usually not able to sustain large herds of cattle, reflected archaeozoologically in the shift towards a predominance of ovicaprine bones (see 6.7.5.). However, in the model by Robertshaw and Collett (1983) for changes in pastoral subsistence, the arrival of farmers in neighbouring areas allows pastoralists to evolve from mixed herding, farming and hunting to specialised herd management with acquirement of agricultural products through trade. It may well be possible that for the studied area, during the Iron Age and later, we are only seeing the agricultural aspect of the archaeological landscape and that we are missing the pastoral nomadic aspect. This is probably simply a consequence of the high archaeological visibility of agricultural communities compared to the very low visibility of mobile pastoralists. However, what may argue against this reasoning is that, as mentioned earlier, present pastoral groups in both parts of the research area arrived there only during the past few centuries. Blench (1999, 40) has argued that, with expanding agriculture, pressure on natural grasses rose, favouring the re-growth of woody vegetation in the spaces remaining between the fields. This must in turn have been favourable for browsing animals, like domestic goats, which seems to fit with the predominance of goats over sheep observed at most of the sites dating to the Iron Age or later. In older phases, where cattle were usually predominant in the herds, goats were also more common than sheep, but this may be due to the fact that cattle and sheep are competitors for good grazing pasture.

Agricultural intensification must also have involved changes in the seasonal round, variable according to the crops cultivated and the farming strategies used. In rain-fed farming systems, for example, the dry season is the dead period of the year. In *masakwa* (see 2.1.2.) and recession farming, i.e. farming on floodplains once floodwaters start to retreat, the main activities take place around the flood season, which in the *firgi* occurs during the local dry season. At the studied sites we are probably mainly dealing with rain-fed agricultural systems, except in the *firgi* from the time that *masakwa* farming was developed onwards, at the latest in the sixteenth century AD (Gronenborn, 2001). A shift to full farming presumably also led to changes in the choices for the implantation of the settlements. Breunig and Neumann (2002a) were even able to trace the change from pastoralism to agro-pastoralism between Gajiganna phase I and phase II (see 2.2.3.) They observed that sites of the former phase had more water in their surroundings than those of the latter, correlated with a shift from stress on watering areas for the herds, to sandy areas needed to grow pearl millet. The geomorphological information available on the studied sites is insufficient to further investigate the correlation between their implantation and a predominance of either agriculture or stock keeping.

As described in 2.1., both northern Burkina Faso and the southern Lake Chad area are at the present border between pastoral nomadic communities in the north and sedentary farmers in the south, corresponding approximately with the 400 mm rainfall isohyet. At lower rainfall levels, a farming based subsistence is not possible, although other authors put the limit at 200-300 mm (Breman and de Wit, 1983; Breman, 1992). There is a minimum amount of rainfall required for crop growing, unless compensated by the presence of flood waters, and enough drinking water is needed year-round near the permanent settlements. With fluctuating rainfall through time, the optimal zones for both strategies must also have shifted. The agricultural zone probably moved southwards parallel with increasing aridification. Rather than speaking of the northern part as an optimal zone for stock keeping, it is probably more correct to say that stock keeping was gradually pushed there by expanding agriculture (see 6.7.4.4.). The present situation can probably not entirely be projected back in time, since herding in arid areas must only have become especially favourable after the introduction of zebu cattle, which thrive better in dry conditions than other cattle types.

Agricultural and pastoral economies can be complementary in several ways (e.g., Krings, 1980, 70), some of which have already been indicated. Pastoral herds can graze on the stubble fields after the harvest and stems of cultivated plants can be used as dry season fodder reserves. Present farmers much welcome grazing herds onto their fields because the animals' manure increases fertility (e.g., Bouquet, 1990, 246). Dung can also be carried from stables or other livestock structures to the cultivated land. As said earlier, present farmers moreover give part of their livestock to nomadic groups for herding, and both groups exchange food products. However, conflicts may arise between farmers and herders as well, mainly for land (Krings, 1980, 40). As said earlier, this could be one of the factors at the root of the abandonment of the settlement mounds in northern Burkina Faso. Areas suited for agriculture are often equally good pasture areas. In northern Burkina Faso, the dunes are favourable for (rain-fed) pearl millet cultivation, but attract herds for wet season pasture as well. Flood plains in the southern Lake Chad area are taken by farmers for the growing of sorghum (Franke-Scharf *et al.*, 2004), but also offer good pasture for domestic flocks at receding water levels.

6.7.8. Meat versus secondary products

Meat is only a small part of the useful products that can be obtained from domestic stock. For present East African nomadic pastoralists, for example, exploitation of the yields of their stock, like milk and blood, is more important than the stock itself (Dyson-Hudson, 1966, 42). In traditional societies, almost all post-abattoir products have some economic value: blood is dried and used as a fertilizer, while horns and bones are cleaned and ground up as animal feed (FAO, 2001). In what follows, the main products that can be obtained from domestic animals are discussed: meat, milk, blood, bone and marrow, dung, skin, wool and power.

6.7.8.1. Meat

Ethnographic evidence indicates that small livestock is nowadays primarily kept for meat, while the main purpose of cattle is producing milk for consumption (Dahl and Hjort, 1976, 200). There are differences according to the breeds, however. Among ovicaprines, mainly the long-legged ones are being used for meat (de Leeuw *et al.*, 1972, 54), although in present north-eastern Nigeria, both savannah and dwarf goats are valued for their meat rather than for their milk (Connah, 1981, 43). Taurine/Sanga cattle, with the exception of Kuri, are also more suited for meat than for milking, in contrast to the zebu breeds (Porter, 1991, 199-212). Europeans usually like sheep meat better than goat meat and this seems to have pushed up prices for the animal in present sub-Saharan Africa (Dahl and Hjort, 1976, 206; Geis-Tronich, 1991, 464). However, more importantly for the studied archaeological contexts, sheep are more prestigious than goats in Islamic ceremonies, which forces prices further up around the time of religious festivals (Bourn *et al.*, 1994; Blench, 1999, 48). Cattle usually have more social value than small livestock and are therefore mainly killed in cases of emergency: during droughts, to meet needs for cash, to prevent the spread of diseases, to anticipate natural death, or for sick people or pregnant women (Stenning, 1959, 5; Dahl and Hjort, 1976, 111, 162-163). The second most important set of reasons for their slaughter are ceremonial purposes. When an animal is killed, its meat is often shared communally, which seems to be more practical and less labour consuming than treating it for preservation. Besides their meat, the fat of slaughtered domestic animals is also used, for example as cooking oil or for curing rheumatism or colds (Aiyedun, 1991).

High amounts of cattle remains at the Gajiganna sites, especially at those of the Bama-Konduga group, may indicate that the animals were kept for their meat rather than for secondary products like milk. This would correspond to the qualities associated with taurine/Sanga cattle, the presumed type present at those sites. Milking may also have been only a later development, although there are indications that it was practised in Africa at a very early date (see 6.7.8.2.). Typically, animals kept for their meat are killed once they have reached their maximal weight, because this gives most output for the least energy input. It is mainly males that are slaughtered at this young age, since females are needed for reproduction. Of the ovicaprines and cattle from the studied sites, many seem to have been killed before they reached maximal weight (see 4.6.10.), which could perhaps be a sign of harsh conditions, in which animals could not be kept alive until they reached optimal slaughter age. As said, age distributions for older animals, after they had reached maximal weight, remain unknown.

6.7.8.2. Milk

As mentioned earlier, among present-day African pastoralists the milk of domestic animals is a more important source of food than their meat. In contrast the use of milk is uncommon among sedentary populations, as indicated by the high rates of lactose intolerance that exist among them (Cerný, 2002). It is only fresh milk that gives them intestinal problems, however, because fermented milk products, such as butter and cheese, do not contain any lactose. Species that are milked nowadays primarily include cattle, besides ovicaprines and dromedary, but yields depend much on the breed. Of the present West African cattle breeds only zebus and Kuri cattle are milked, while other types are only rarely used for milk (Blench, 1999, 56).

Cerný (2002) has advanced the hypothesis that cattle raising in Africa was, from the very beginning, oriented towards milk production and that milking in Africa predated that in Europe and the Near East. Sherratt (1981, 1983) has argued from textual and pictorial evidence that, in the latter areas, the first use of milk began around the fourth millennium BC, about four millennia after the oldest evidence for domestic cattle. A project is now running to test this idea through analysis of milk fat residues in prehistoric pottery (Evershed *et al.*, 2004). The hypothesis of early milking in Africa is reinforced by milking scenes in Saharan rock art, probably dating from at least 4000-3000 BC, or perhaps even from about 5500 BC (Simoons, 1971). As an indication for cattle milking in West Africa at an early date, Holl (1986, 144) mentions some depictions of cows with exaggerated udders in rock art at Dhar Tichitt in Mauritania (late third-early second millennium BC).

The milk production of a cattle herd rises with the number of adult females, their fecundity and the quality of the available forage (Dahl and Hjort, 1976, 142). The latter two are largely related to the quantity and periodicity of rainfall. Because West Africa has only one rainy season, in contrast to the east of the continent, milk yields are low. In addition, only part of the yield can be used for human consumption, because enough needs to be reserved for the calves (Le Houérou, 1989, 139). Domestic ovicaprines are also kept for milking besides for their meat, but goats seem to be better milk producers than sheep (Dahl and Hjort, 1976, 210). This is related to the more indiscriminate breeding of goats throughout the year than sheep, which makes them a reliable source of milk during the dry season (Dahl and Hjort, 1976, 213, 235). However, milk can be kept between a few months and half a year, even up to one year, when curdled or dried (Dahl and Hjort, 1976, 159), and a variety of techniques exist among African pastoral groups for processing milk into longer lasting products like butter or cheese (Casimir and Bollig, 1994). There is historical mention of milk being made into butter in medieval West Africa and milk was apparently also drunk sour (Lewicki, 1974, 111, 124-127). From archaeofaunal remains milking can be demonstrated through high neonatal cull and a predominance of mature females, but such detailed data on ages and sex could not be obtained from the studied sites.

6.7.8.3. Blood

The practice of taking blood from living cattle is mostly associated with present East African pastoral groups, especially near Lake Victoria (Lagercrantz, 1950, 51-54). There are no West African records for the practice, except from the Kel Air in Niger. Blood extraction is mainly from bulls (Dyson-Hudson, 1966, 19). The activity usually takes place during the dry season, but blood does not represent more than two percent of the total energy obtained from the herds (Marshall, 1990). The blood is eaten alone, in liquid or clotted form, or mixed with milk (Gifford-Gonzalez, 2005). There does not seem to be a way to determine from archaeological bone remains if the inhabitants of the studied site practiced blood extraction or not.

6.7.8.4. Bone and marrow

As explained earlier, the inhabitants of the studied sites in the southern Lake Chad area made many of their tools out of bone, at least before the introduction of iron. Bones of domestics were mostly used, probably because these were most easily available. Apart from other taphonomic factors, the fragmentation of the majority of the long bones found may perhaps be explained by the use of marrow. This could have been extracted in a similar manner as is done by present East African cattle keeping peoples, who prepare soup by fragmenting long bones mid-shaft and adding boiling water in order to facilitate marrow extraction (see summary in Marshall, 1990). As mentioned earlier, bones can also be ground up to be used as animal feed (FAO, 2001). As a final use for animal bones, Aiyedun (1991) mentions recent Nigerian examples of powdering them into chalk.

6.7.8.5. Dung

The most common use for the dung of domestic livestock in the research area today seems to be as manure on the fields (e.g., Holl *et al.*, 1991 for ovicaprine droppings). The animals can either be allowed to graze on the stubble fields to fertilise them, or dung, for example from stables, can be carried to the cultivated land. According to Le Houérou (1989, 138) dung is not used for fuel in Sahelian Africa. Nevertheless, 65 % of the archaeobotanical samples from Early Iron Age Oursi (BF 94/45) and 88 % of those from Oursi village (BF 97/13) yielded charred fragments of dung resembling ovicaprine droppings, which could perhaps point to their use as fuel (Kahlheber, 2003, 202). Animal dung, mixed with grass and mud, is also occasionally used today as a building material (Le Houérou, 1989, 138; Aiyedun, 1991).

6.7.8.6. Skin

Domestic species of which the skin is removed to be used as leather include sheep, goat, cattle and, less commonly, also donkey (e.g., Geis-Tronich, 1991, 355-366). Pastoralists are usually more concerned about the survival of their animals than the quality of their hides (FAO, 2001). Good quality hides and skins, according to present, Western, standards are therefore mainly sedentary products. Hides and skins are an important export product from the Sahel and enjoy a good reputation; Sokoto Red goats of the Sahel are especially renowned for their high-quality leather (Blench, 1999, 49). There is a strong cultural association between Islam and the Sokoto Red (*Ibid.*). Their skins were an important trans-Saharan trade commodity, used for binding Korans, and Muslim leatherworkers could be found in all major markets of Sahelian West Africa. There are historical references to the famous leather from western Africa and there are also records that it was exported from sixteenth century Borno (Levtzion and Hopkins, 1981, 133; Barkindo, 1999). At Oursi hu-beero (BF 97/30) finds of *Acacia nilotica* have been thought to be associated with leather fabrication, because the seedpods of the tree can be used as a tanning agent (Kahlheber, 2003, 160).

Animal leather has multiple purposes in present western Africa. Cattle hides, for example, are used for shoes, for drums, for making bag containers and for storing farm crops (Aiyedun, 1991). Goatskins are popular among nomads today as bags for carrying water (Krings, 1980, 56) and are also mentioned for medieval times (Levtzion and Hopkins, 1981, 169). In addition, there are Arabic historical sources on the use of leather garments (Levtzion and Hopkins, 1981, 113, 114, 120, 185). On top of their use for leather, there is presently also a substantial market in West Africa in hides for human consumption (Aiyedun, 1991; FAO, 2001).

6.7.8.7. Wool

At one of the Kissi 3 burials (fifth-seventh century AD) textile made from animal fibres has been found (S. Magnavita, pers. comm.). At Gadei, one of the excavations in the Gao region (7th- 16th century AD), a piece of textile made from unspecified wool has also been recovered (Fuller, 2000). There is, moreover, historical mention of medieval West African clothes made from wool (Levtzion and Hopkins, 1981, 112, 123, 124, 127, 193). It is not clear from what animal species the wool may have been taken, but, as mentioned in the faunal description, woolly sheep have a very limited spread in present western Africa and they are presumably late, marginal, introductions (Blench, submitted).

6.7.8.8. Power

The species among the identified domestic animals, which have to be considered for use as riding animals, for traction or as beasts of burden, probably include cattle, donkey and horse. Starkey (2000) mentions that these animals, and also dromedary, have been used for centuries, or even millennia, in sub-Saharan Africa for riding and transport, but the origins and spread of this use are not well known. While very few remains of domestic equids were found, cattle are well represented in the studied animal bone assemblages. No remains showed pathologies that can be brought into connection with their use for power (see 4.6.10.2.), although cattle may have been used in the southern Lake Chad region to bring in stone from remote areas (cf. Holl, 1988, 285). At present, cattle are the major work animals world-wide (Starkey, 2000). In West Africa today they can be mainly found among agro-pastoralists or farmers (FAO, 2001). Farmers in semi-arid areas are known to use cattle for ploughing or carting produce during the rainy season, after which they give the animals to specialised herders for the rest of the year. However, Starkey (2000) argues that animal traction, for pulling ploughs but also for wheeled transport, was only introduced in sub-Saharan Africa during the colonial period. This seems to be contradictory with the depiction of wheeled vehicles, pulled by cattle, in Mauritanian rock dating to the first millennium BC (Vernet, 1993, 322-325). Most present draught cattle belong to the zebu type, or are zebu crossbreeds (FAO, 2001). Zebus are preferred to other types because they are easier to train and because they are larger and stronger.

6.7.9. Meat processing and conservation techniques

Butchery and processing of animals for consumption can leave traces on their bones in the form of cut, chop and shaving marks (see 3.9.). These marks differ microscopically, depending on whether stone or metal tools were used (Greenfield, 1999). Therefore, a change should be visible between the Late Stone Age and the Iron Age and later phases, and a study to investigate this will possibly be conducted at the J.W. Goethe-University, Frankfurt. A problem for the older phases is that traces may no longer be visible because preservation conditions have altered bone surfaces. One may also wonder if metal was already used for ordinary tasks from its early introduction onwards. Its precious nature for the inhabitants of Early Iron Age and even younger sites in West Africa can be guessed, for example, from relatively rare iron finds (e.g., McIntosh and Bocoum, 2000). Nonetheless, in the oldest Iron Age phases in Burkina Faso, knifes and awls already appear that may have been used for butchery and food preparation (see 2.2.2.). At Jenné-Jeno, MacDonald (1995) has found bones cleaved in two with metal implements from the site's oldest occupation phase onwards (250 BC-AD 400).

		domestic fowl	large gallif.	dog	ovicapr.	small bovid	cattle
BF 97/5	EIA	-	-	-	-	-	-
BF 94/45	EIA	-	0,0	0,0	2,4	2,2	3,7
BF 97/13	EIA	-	-	0,0	0,0	0,0	0,0
BF 97/13	MIA	-	0,0	0,0	3,8	0,0	1,3
BF 97/13	LIA	-	0,0	7,1	3,2	1,3	0,0
BF 97/30	LIA	0,0	4,2	0,0	2,4	0,7	3,5
BF 96/22	MIA	-	0,0	0,0	0,0	0,0	0,0
BF 97/31	LIA	-	0,0	0,0	1,4	0,0	3,1
BF 94/120	LIA	0,0	2,9	0,0	2,2	0,8	0,0
BF 95/7	LIA	25,0	0,0	1,7	0,9	0,0	0,0
NA 90/5C	Gaj IIa	-	-	-	0,0	0,0	0,0
NA 90/5A	Gaj IIb	-	-	0,0	0,0	0,0	0,1
NA 90/5BI	Gaj IIa/b	-	-	-	0,0	0,0	0,7
NA 90/5BII	Gaj IIc	-	-	-	0,0	0,0	1,3
NA 91/1A	Gaj I	-	-	-	-	-	0,0
NA 93/42	Gaj I	-	-	-	0,0	0,0	0,0
NA 97/18	Gaj I	-	-	-	-	-	0,0
NA 97/24	Gaj I	-	-	-	-	-	0,0
NA 93/36	Gaj IIa	-	-	-	-	0,0	0,0
NA 99/65	Gaj IIa	-	-	-	-	-	0,0
NA 99/65	Gaj IIb	-	0,0	-	0,0	0,0	0,0
NA 99/65	Gaj IIc	-	-	0,0	0,0	0,0	0,0
NA 93/10	Gaj IIc	-	-	-	0,0	-	0,0
NA 97/37	Gaj III	-	-	33,3	5,9	-	5,0
NA 97/33	Gaj I	-	-	-	0,0	0,0	0,6
NA 97/33	Gaj IIa/b	-	-	-	0,0	0,0	0,0
NA 96/45	Gaj IIa/b	-	-	-	0,0	0,0	0,0
NA 95/1	Gaj IIa/b	-	-	-	1,3	1,9	0,4
NA 97/26	IA	-	-	-	0,0	0,0	0,0
NA 99/75	IA	-	-	-	3,8	0,0	0,0
NA 97/13	IA	-	-	-	0,0	0,0	0,0
NA 92/2C	subrec	0,0	0,0	0,0	3,1	11,1	4,6
NA 93/46	LSA	-	-	-	0,0	0,0	0,0
NA 93/46	EIA	-	-	-	0,7	0,0	1,0
NA 94/7	I	-	-	0,0	0,0	0,0	0,5
NA 94/7	II	-	-	-	0,0	0,0	0,0
NA 94/7	III	0,0	0,0	-	0,0	0,0	6,3
NA 94/7	IV	-	50,0	-	0,0	0,0	7,1
NA 93/45	I	-	5,3	0,0	5,3	0,0	8,9
NA 93/45	II	-	2,9	-	4,2	3,9	7,5
NA 93/45	IIIa	-	-	-	3,1	0,0	4,3
NA 93/45	IIIb	-	-	-	6,7	4,3	12,9
NA 93/45	IV	-	12,5	-	4,2	7,7	4,7
Blé A		-	0,0	-	10,5	16,7	16,7
Blé B		0,0	0,0	-	16,7	0,0	28,6
Blé C		-	0,0	-	30,0	5,9	0,0
Blé E		0,0	33,3	-	12,5	14,3	33,3

Table 15: Percentages of bones of domestic species with butchery marks by site and phase

In Table 15, the percentage of bones with butchery marks is depicted by site and phase for the following taxa: domestic fowl, large galliforms, domestic dog, ovicaprines, small bovids and cattle. With the exception of one small bovid bone from Tin Akof (BF 94/133), none of these taxa was found in Late Stone Age contexts from Burkina Faso, but for that period and region no traces of butchery were found at all. For the research area as a whole, traces prevail in Iron Age and younger contexts, with a particularly high level of butchery marks at Galaga (NA 92/2C), all phases at Ngala (NA 93/45) and the Blé sites. The relatively frequent occurrence of traces at Zilum (NA 97/37), compared to the older Gajiganna sites, was also notable.

Traces left on carcasses can also reveal details of the different stages in animal processing. Most of the marks observed on the bones are probably the result of skinning (e.g., cut-marks on phalanges and skull parts), dismemberment of the carcass (e.g., cut and chop-marks on joints) and its subdivision into smaller units (e.g., chopped long bones, ribs or vertebrae). Amounts of traces found are generally limited. Those that are probably related to the subdivision of the carcass are especially rare and, instead, this may have been done simply by cracking and cutting between the joints. The observed distribution of butchery marks is probably also distorted by poor preservation with, for example, only a few long bone shafts present. Subrecent site Galaga (NA 92/2C) yielded many cut and chopped ribs of small bovids compared to older sites. This may be due to changing butchery techniques in recent periods. Zimmermann (1994) noted that present-day butchery practices have undergone influences of the colonial and post-colonial policy, but that traditional methods still persist in ritual contexts. Cut-marks observed on distal tibiotarsals of large galliforms, probably result from cutting off the feet, which do not have much food value.

Pastoralists usually match the species they slaughter to the occasion and consume all meat before it goes bad (FAO, 2001). They occasionally smoke meat, especially when it is intended for the market, or keep it by salting or sun-drying (FAO, 2001; Dahl and Hjort, 1976, 169). S.E. Smith (1980) also described how the Kel Tamashek dry meat by cutting it from the bones, dividing it into thin strips and then hanging it in their tents. Sun-drying, after cutting the meat in slices or strips, is the only preservation method recorded for medieval West Africa by Arab authors (Levtzion and Hopkins, 1981, 76, 109, 119, 122). When suitably treated, meat can be preserved up to twelve months (Dahl and Hjort, 1976, 169).

Fig. 52: Cattle meat drying near Katsina (Nigeria)

Meat preparation by roasting should be possible to recognise archaeozoologically from bones with burned ends. These kinds of traces have been found in limited amounts at the site of Akumbu in Mali (AD 400-1400), especially on bones of large galliforms and ovicaprines (MacDonald and Van Neer, 1993). At the studied locations, traces of roasting could not be clearly recognised, but as said, there were problems of distinguishing (low temperature) burning from other bone colourations (see 3.9.). The fragmentary nature of the material suggests that meat may have been mainly stewed in pots, as was also concluded for the aforementioned site of Akumbu and equally suggested for Gao Ancien (6th-16th century AD) (Barrett-Jolley, 2000). However, it is not clear to what extent modes of food preparation were similar for all investigated regions and periods. McIntosh (1995, 157-165) interpreted a change from a preponderance of shallow, small-volume, bowls and a relatively low number of small cooking pots in the earlier phases of Jenné-Jeno (250 BC-AD 400) to a much greater frequency of wide-mouthed cooking pots in the later phases (AD 400-1400), as a change from baking, grilling and roasting to boiling and stewing. The latter two modes of food preparation are in agreement with modern cooking techniques of Malian sedentists such as the Bambara (MacDonald and Van Neer, 1993). The Kel Tamashek, nomadic herders, additionally also prepare their meat through roasting, by covering it with sand and building a fire on top (Smith, S.E., 1980). Such a technique would probably be impossible to recognise from archaeological bone remains. Al-Bakri, writing in the eleventh century AD, mentions how desert nomads ground strips of dried meat into powder, over which they poor melted fat or clarified butter (Levtzion and Hopkins, 1981, 109).

6.7.10. Trade

As indicated earlier, mobile pastoral nomads must have had a considerable role in the trade and exchange of goods and ideas, perhaps already from very early periods onwards. Besides live animals, traded items derived from domestic stock were probably mainly products that do not easily putrify such as hides and preserved meat. As mentioned earlier, animal skins from the West African Sahel were exported to northern Africa in medieval and later times because of their quality. A slave-chain found at Oursi hu-beero (BF 97/30) suggests that the site's inhabitants took part in slave trade across the Sahara (Petit and Hallier, pers. comm.). The area near Oursi may have been particularly attractive for passing trade caravans, thanks to its large lake that provided drinking water for the traders and their pack animals, probably mainly dromedaries, the main species associated with trans-Saharan trade. Typical for present-day West Africa is the internal north-south trade in cattle, mainly males under three years old, from more arid to the humid zones, probably related to the difficulties in livestock keeping in southern areas (Amanor, 1995). It is not clear how far this internal trade dates back in time, but the situation must have been different before the introduction of zebu type cattle, thriving best in arid zones. Before the colonial era, horses were also a major item of north-south trade (Law, 1980, 54-61, 1995). Trade and exchange probably also took place on an even smaller geographical scale, between pastoral and agricultural groups living in the same area. Cooperating specialist groups may have had a single ethnic identity, or may have belonged to ethnically different societies living in symbiosis, although ethnohistorically the latter is most common (MacDonald, 1999). Contacts between different economic groups may also have been less peaceful. The present Buduma in the Lake Chad area are feared for their cattle raids, for example (see 2.1.).

6.7.11. Cultural aspects

As usual, the cultural aspects are the most difficult ones to trace from the archaeological remains. Late Stone Age cattle burials in the Sahara in Niger and southern Algeria testify that early domestics may have had an important symbolic value in human societies (Paris, 1997, 2000). Clay figurines from the studied sites, and other archaeological excavations in West Africa and beyond, also testify to the role domestic animals may have played in the symbolic world of the humans.

The West African domestic fowl is often used today in ritual contexts, which was one of the reasons for linguists to assume an older date for its introduction than the one proven archaeozoologically (see 6.7.3.1.). At the studied locations, there are absolutely no indications for the use of the animal in special contexts. However, since the faunal material has usually not been found where it was initially dumped (see 5.2.), the sites are not well suited to trace such aspects.

The domestic dog remains from localities in Burkina Faso have been the subject of a separate publication (Linseele, 2003). As explained, they were very common in Iron Age, and especially Late Iron Age, contexts and were interpreted as evidence for dog eating. Although the practice of cynophagy in West Africa is probably at least two millennia old, influences from North African Berbers have been proposed for the observed rise during the Late Iron Age. The habit does not seem to have been equally popular over the entire western part of the African continent. At the studied sites from the southern Lake Chad area, for example, domestic dog remains were not common and the animals were probably not consumed. Some domestic dog remains, and at least one complete individual, have moreover been found associated with the buried horses at Aissa Dugjé, in northern Cameroon, dating to the late first-early second millennium AD (MacEachern *et al.*, 2001). Perhaps domestic dogs had a different function in the Lake Chad area to that in northern Burkina Faso. Ethnographic evidence shows for example that hunting dogs usually enjoy a high esteem and are well treated and not consumed (Frank, 1965, 18).

The association in West Africa of horses with elite social status, warfare and political power has already been stressed in the faunal descriptions in 4.6.8. Cattle can also be symbols of richness and wealth, and having large herds is usually a sign of high status, although the Kotoko despise domestic stock keeping (Lebeuf, 1969, 20). High amounts of cattle at Ngala (NA 93/45), in the periods when a palace of a local ruler is known to have existed at the locality, have tentatively been connected to a higher social status of the inhabitants at that time (Linseele, 2005). However, such an explanation is perhaps too romantic and should probably be replaced or at least supplemented by the ecological explanation proposed earlier, i.e. decreasing flood levels, which also fits with other evidence from the site (see 6.5.2.3., 6.7.3.1. and 6.7.5.3.). MacDonald (1994, 57) thinks that by 100 BC cattle were, as today, used as a form of "currency", e.g., for bride prices or to trade for agricultural goods. As said, slaughter of cattle nowadays often takes place in ceremonial contexts, while ovicaprines, with the exception of sheep in Islamised societies, have much less symbolic value.

6.7.12. Concluding remarks

At none of the studied sites do we seem to be dealing with highly mobile groups with a very large dependence of domestic livestock, although the north of Burkina Faso may have been inhabited by such groups after the abandonment of the settlement mounds in the fourteenth century AD and this was perhaps also the case in the Bama Deltaic Complex between the Gajiganna Culture and the beginning of the Iron Age. Only for Gajiganna phase I, we must assume the presence of nomadic herders, keeping mainly cattle. Human groups of Gajiganna phase II were characterised as agro-pastoralists (see 2.2.3.). It has been argued in the previous paragraphs that in order to keep large flocks of cattle, they probably maintained a certain degree of mobility. The semi-permanent nature of the settlements indicates that this may have been in the form of transhumance. With the beginning of our era, and in the Bama Deltaic Complex even earlier, full-farming communities appeared. For the farming communities stock keeping was not a priority and this is probably reflected in the observed changes in the herd composition compared to previous periods. For the period, the specialised herders are probably missing from the archaeological record because of their poor visibility.

Chapter 7. Summary and conclusions

Below the palaeo-ecological and palaeo-economic trends observed through time for the different parts of the research area are first summarised and then discussed within the archaeological and archaeobotanical framework given in the first chapter. Following this summary, the most important economic changes that have occurred in the later Holocene of sub-Saharan West Africa are then described. Possible causes for these changes, but equally for periods of economic continuity, will be defined. This part is followed by a supra-regional discussion of some important themes that have emerged in the course of this study, notably risk management, specialised versus generalised economies, the relation between food strategies, mobility and archaeological visibility and, finally, seasonal rounds. To conclude, an evaluation is made of the extent to which the research questions formulated at the beginning have been answered satisfactorily and suggestions are given for further research.

7.1. Change and continuity. Possible causes

7.1.1. Overview by region

7.1.1.1. Burkina Faso

The available evidence for the Late Stone Age in northern Burkina Faso, dated between 2200 BC and 1000 BC, points towards mobile hunter-gatherers, practising small-scale cultivation of pearl millet (*Pennisetum americanum*) at the latest near the end of the period. The possibility that domestic animals may have been present in small numbers, judging from a few records of cattle (*Bos primigenius* f. taurus) and ovicaprines at contemporary Windé Koroji Ouest in the neighbouring Malian Gourma area, should probably be revised since identifications from the site are not convincing. Continuing a Saharan tradition, the hunter-gatherers of Late Stone Age Burkina Faso also (seasonally) carried out specialised fishing. At that time, the lake of Oursi must still have carried enough water year-round to sustain populations of Nile perch (*Lates niloticus*), a species confined to deep water. Wild mammals found indicate that the terrestrial environment was more lush than today. More specifically, grass cover near the lake of Oursi was probably denser than in later periods.

After a millennium for which there is almost no archaeological information, the Iron Age in the north of Burkina Faso sets in around the beginning of our era. Compared to the Late Stone Age, drastic changes have taken place. The Iron Age marks the beginning of full farming and sedentism, but pearl millet remains the main crop grown on the sandy soils of the area. The first evidence for domestic livestock in northern Burkina Faso, cattle, but predominantly ovicaprines, also appears at the beginning of the Iron Age. Nevertheless, hunting and fishing continued, although their importance varied locally. A general degradation of the environment could be observed, with shrinkage of the lake near Oursi, the disappearance of mammalian taxa of dense grass cover and the appearance of Sahelian mammals. It is not entirely clear if the environmental changes are related to climatic changes or to anthropogenic activities, and if they happened either before or in the early part of the Iron Age. Climate and anthropogenic factors probably both played a role and the two may have reinforced each other. With the exception of a gradual intensification of stock keeping and agriculture between the Early and Late Iron Age, the Iron Age of Burkina Faso is mainly a stable period with a lot of continuity, both economically and ecologically (cf. Kahlheber, 2003, 223). A similar continuity in the course of the Iron Age has been noted in the Inland Niger Delta (McIntosh, 1999), with which the research area in northern Burkina Faso shows a lot of cultural parallels. However, compared to present northern Burkina Faso, the reconstructed aquatic and terrestrial environments point to more favourable ecological circumstances during the Iron Age. There were no faunal remains available for the period between the end of the Iron Age and present, but archaeobotanical research has shown that significant environmental changes may have taken place only in the last 50 years. After the abandonment of the Iron Age sites, around the middle of the second millennium AD, people presumably turned to nomadic pastoralism.

7.1.1.2. Bama Deltaic Complex

While domestic animals, sheep (*Ovis ammon* f. aries), goat (*Capra aegagrus* f. hircus) and especially cattle, dominate faunal assemblages from Gajiganna phase I, no cultivated crops have been identified from sites of that period. The data agree with the label "pastoral" that was earlier attributed to the phase. Archaeological sites of Gajiganna phase I (1800-1400 BC) are shallow, indicating high mobility. The only site with a thicker stratigraphy is Bukarkurari (NA 97/33), a site of the Bama-Konduga group, outside the core area of the Gajiganna Culture. Habitation with a more permanent character has been assumed for the latter site. Ecological conditions on and near the Bama Ridge may have been more favourable for habitation. More precisely, humidity was probably lower than in the central Gajiganna area, where many pools existed in its clay patches, which were the remnants of a former lagoon. However, faunal assemblages of Gajiganna phase I are insufficient to confirm this palaeo-ecological model. Moreover, no

	habitation	aquatic environment	terrestr. environment	new crops	herd composition
Burkina Faso					
AD 1400 to subrecent	temporary				cattle, ovicaprines?
Iron Age	permanent	diminished water levels	Sahelian species	sorghum	goat, sheep, cattle
"Dark Millennium"	temporary?				?
Late Stone Age	temporary	deep water	dense grass	pearl millet	?
Bama Deltaic Complex					
Subrecent	permanent		Sahelian species		sheep, goat, cattle
Iron Age	permanent		Sahelian species	sorghum	goat, sheep, cattle
Gajiganna III	permanent	pools richer fauna than today - not connected to Lake Chad	short wet phase?	cowpea	cattle, goat, sheep
Gajiganna IIc	semi-permanent (dry season)		short dry phase?		cattle, goat, sheep
Gajiganna IIa/b	semi-permanent (dry season)		aridification?	pearl millet	cattle, goat, sheep
Gajiganna I	temporary (dry season)				cattle, goat, sheep
Firgi					
Late Iron Age to recent	permanent	Gradually diminishing floodlevels	Sahelian species	sorghum	cattle, goat, sheep
Early Iron Age	permanent				goat, sheep, cattle
Late Stone Age	semi-permanent (low water)	high floods *firgi* inaccesible?		pearl millet	cattle, goat, sheep
Blé	temporary?			?	ovicaprines, cattle

Fig. 53: Summary of palaeo-economical and palaeo-ecological data by area and period

archaeobotanical data are available from the Bama-Konduga group to compare against those of the central Gajiganna group.

The oldest remains of pearl millet identified from the Bama Deltaic Complex date to Gajiganna phase IIa/b (1500-1000 BC), although the quantitative importance of the crop may still have been limited at that time. As in the previous phase, domestic herds consisted mainly of cattle. The characterisation of the period as "agropastoral" seems justified, although domestic flocks were economically probably still more important than cultivated crops. The terrestrial wild fauna indicates that the Gajiganna area may have witnessed a period of increasing aridification during phase IIa/b. The duration of this dry period remains unknown. During Gajiganna phase IIc (1000-800 BC) the importance of pearl millet cultivation had possibly increased in comparison to the previous phases. This does not seem to fit with the return to more mobility that was proposed by the archaeologists; the analysis of archaeozoological productivity does not lend support to increased mobility either. The wild terrestrial fauna of phase IIc seems to point to a short dry event. Gilgila (NA 99/65) is the only Gajiganna phase II site that does not fit in the general image just sketched, notably by its predominance of ovicaprines over cattle.

During Gajiganna phase III (600-400 BC), new crops, e.g., cowpea (*Vigna unguiculata*), and new cropping systems were introduced into the Bama Deltaic Complex. The shift to full farming in the area can be placed here, resulting probably also in a decreased importance of stock keeping. No changes could be observed in the species composition of the herds compared to the older phases, but dwarf goats had appeared by this time. Hunting moreover seems to have lost economic importance during Gajiganna phase III. Archaeobotanical and archaeozoological data indicate that people in the Bama Deltaic Complex may at that time have profited from a moister phase during a generally arid period.

During all phases of the Gajiganna Culture, fishing was apparently carried out in the pools of the area, where fish must have been easy to catch. It was concluded that these pools were probably no longer in connection with Lake Chad. However, the species present, including taxa that do not occur in the pools of the area today, indicate that the disconnection might have happened only shortly before, or that no arid spell occurred that was severe enough to cause local extinction of species typical of permanent waters.

In the period between the Gajiganna Culture and the Iron Age, a return to nomadism, with presumably a higher pastoral component in the economy, may have occurred. At the beginning of the Iron Age, in the early first millennium AD, changes in comparison to the Gajiganna Culture became apparent. Sorghum (*Sorghum bicolor*) is for example added to the spectrum of cultivated crops. Faunal assemblages from this period in the Bama Deltaic Complex show similarities with those of Iron Age Burkina Faso: the main domestic species are ovicaprines and the terrestrial wild mammals are indicative for a landscape with a Sahelian component. Fishing in the pools of the area continued and these still contained more species than today. With the exception of subrecent Galaga (NA 92/2C), no sites from the period between the Iron Age and present were available for archaeozoological study. Faunal remains from the site are in some respects clearly different from those of older locations in the area: some types of domestic animals appear to be different and part of its fish must have been obtained through trade. Evidence from the site thus warns against uncritical use of ethnographic data for archaeozoological interpretation, since the (sub) recent situation can deviate from that in the past in several ways.

7.1.1.3. *Firgi* area

The fauna as well as botanical remains from Late Stone Age (1300/800 – 800/500 BC) contexts in the *firgi* area point towards seasonal habitation during the dry parts of the year. Habitation during the inundation season was probably still too difficult at this time. Small-scale cultivation of pearl millet was practised, besides domestic stock keeping, fishing and some hunting. Mobility probably allowed the human groups to keep livestock herds mainly composed of cattle. Between the Late Stone Age and the Early Iron Age in the *firgi* discontinuity has been noticed in several ways. The beginning of the Iron Age marked a shift to sedentism and full farming, induced by increased aridification. Nevertheless, the importance of domestic crops seems to have been relatively limited at sites for which archaeobotanical data are available, notably Kursakata (NA 93/46) and Mege (NA 94/7). Reduced mobility resulted in decreasing numbers of cattle in the herds, since insufficient space must have been available for them during the annual floods. Very few data are available for the Late Stone Age, but types of domestic animals in the *firgi* seem to have been generally different from those in the Bama Deltaic Complex. During its entire occupational sequence, the area was characterised by a large aquatic component in its economy, which distinguishes its sites from those in northern Burkina Faso and in the Bama Deltaic Complex. Between the Iron Age and present the archaeological record of the *firgi* shows remarkable continuity (see Breunig, 2004) and only a few minor changes have been observed. Even the important political changes the *firgi* underwent in the sixteenth century AD, with their integration in the Kanem-Borno Empire, did not bring any significant changes in subsistence strategies, neither did the introduction of sorghum, attested by AD 800, or that of the *masakwa* technique for its cultivation, at the latest in the sixteenth century AD. However, it was argued that a very gradual aridification took place, leading to less extreme inundations. This is reflected in increasing amounts of shallow-water species in the fish catches and in increasing proportions of cattle in the domestic flocks.

7.1.1.4. Blé sites

Faunal assemblages of the Blé sites show a lot of similarities with those of the *firgi* area. Their most important common feature is the large and diverse fish fauna. However, the inhabitants of Blé probably practised more deep-water fishing and also the importance of hunting was greater at the sites, while domestic stock keeping seems to have been less important. If the sites were indeed visited seasonally by fishing parties from surrounding areas, then these may have preferred exploiting local wild resources to bringing many of their livestock animals with them. Because of a lack of archaeobotanical remains from the Blé sites, data on cultivated crops and their importance are missing.

7.1.2. Supra-regional trends. The role of climate, environment, politics and ethnicity

There appear to have been two turning points of economic change in the course of the prehistory of sub-Saharan West Africa (cf. Breunig and Neumann, 2002b). The first one took place in the course of the second millennium BC with the introduction of the first domestic livestock and the earliest pearl millet cultivation. The second one is marked by the beginning of full farming, at the onset of the Iron Age, between the mid-first millennium BC and the beginning of our era. In sub-Saharan West Africa, both the domestic livestock kept and the crops grown seem to have been introduced from outside. However, the date of introduction of these novelties shows considerable regional variation. There is no evidence for domestic livestock during the first stage in northern Burkina Faso and the beginning of full farming happens relatively late in the region. The Nigerian Lake Chad area, on the contrary, seems to have been at the cradle of the innovations, since both stock keeping, in a first phase without pearl millet cultivation, and full farming appear there at an early date. The first millennium BC can be seen as a transitional phase between the Late Stone Age and the Iron Age, in which the innovations of the Iron Age are rooted (Breunig and Neumann, 2002b; Kahlheber and Neumann, in press). Both in Burkina Faso and in the Bama Deltaic Complex, there is an archaeological gap for this period, although in the latter region it is gradually being filled by new finds. In the *firgi* area all sites showed a discontinuity around the Late Stone Age-Iron Age transition, at approximately 500 BC. The Iron Age itself, on the other hand, was a stable period over large parts of West Africa in which the trade centres and cities of medieval West Africa could develop.

As already indicated in 2.2.1., short dry spells are often considered as triggers for cultural changes. Data from this study suggest that changes happened, rather, as a response to long periods of adverse conditions. Both periods of economic change described above coincide chronologically with times of drastic climatic aridification, although these do not seem to have had the same impact everywhere. The investigated area in Burkina Faso and the one in the southern Lake Chad area are situated approximately at the same latitude, but the presence of Lake Chad in the latter seems to have had buffering effects on environmental deterioration. This is especially clear in the *firgi*, where the arid spell of the first millennium BC apparently had less influence than elsewhere, because of their annual floods. The connection between the introduction of the first domestic species in sub-Saharan West Africa and increasingly dry circumstances seems clear, since these appear to have driven Saharan pastoralists southwards into the Sahel. On the other hand, the beginning of full farming in the research area occurred immediately after rather than during an arid period. However, as said, the roots of this innovation should probably be sought in the centuries or millennia before its actual appearance. Apparently, economic changes may also have accelerated the environmental changes caused by climatic aridification, but no evidence has been found for environmental degradation solely caused by humans (cf. Neumann *et al.*, 2004). The stability of the Iron Age and later periods is in turn connected to a relatively stable climate, although a gradual trend towards further aridification could be traced and adverse circumstances characterised the thirteenth-fourteenth centuries AD. The former may have been part of the factors causing the abandonment of the settlement mounds in northern Burkina Faso. Environmental deterioration seems to have accelerated in the course of the last century, with wild fauna suffering increasingly from overhunting due the introduction of firearms and from expanding livestock herds and agriculture. In the course of the occupation of the studied sites, there were probably also shorter periods of drought, which are mainly known for periods with historical sources. However, these were hard to trace from the archaeozoological remains because they do not seem to have had many repercussions on the species composition of the (wild) fauna around the settlements. In addition, the dating of the archaeological layers at the studied sites is not fine enough to document small-scale diachronic changes and mixing of material with different dates may also have happened. Nonetheless, for the sites Gajiganna A (NA 90/5A) and Gajiganna BII (NA 90/5BII), a short dry event has been proposed, while a short wet phase is assumed during the occupation of Zilum (NA 97/37), although the exact duration of these events could not be determined.

The differences between Burkina Faso and the southern Lake Chad area and, in extension, also between other West African areas, in dates of appearance of economic and other novelties, must have been largely related to local environmental circumstances. The Lake Chad area is one of the large West African flood plains, with favourable conditions thanks to the availability of surface water during large parts of the year and to the presence of good pasture in the form of bourgou grasses. These conditions probably attracted prehistoric human groups and their flocks. The north of Burkina Faso, on the other hand, lies in the hinterland of another such flood plain, the Niger bend. This may have retarded innovations since there was probably no need or reason for human communities from that area to move further south.

Climate and environment alone are insufficient to explain all economic changes, since in the same conditions different survival strategies are possible. This can, for example, be derived from the long continuation of the autochthonous hunter-gathering component in sub-Saharan West Africa after the appearance of food-producing groups. It seems, therefore, that human groups to some extent had a free choice between different possible economic strategies for survival in the area that they were living in, and different ethnic groups made other choices. In later phases politics and related factors must also have been involved, although these may ultimately be rooted in the reigning environmental conditions. The abandonment of the settlement mounds of northern Burkina Faso, for example, may have been caused by an unstable political climate, amongst other reasons. Thanks to flourishing inter-regional contacts from the later first millennium AD onwards, new domestic animals found their way to western Africa. Both northern Burkina Faso and the southern Lake Chad area were situated in the hinterland of the large medieval West African trade centres and knew these domestics relatively late. New domestic species appear to have been included into the existing economic systems without drastically changing them. Especially for zebu cattle, probably introduced somewhere in the beginning of the second millennium AD, this seems remarkable, since these animals have very different characteristics to the other cattle types that were already known in the area at that time. It is therefore assumed that zebu type cattle only found their way to western Africa very gradually and through interbreeding with the existing types (cf. Loftus and Cunningham, 2000). Together with North African trade commodities, Islam also spread to western Africa, but the rural areas remained only slightly touched by it until the nineteenth and twentieth centuries AD when large-scale conversions took place. Therefore, Islamic influences probably could not be traced in the archaeofaunal assemblages of the research area.

7.2. Risk strategies

According to Wilkinson and Stevens (2003, 140), "risk strategies refer to those schemes that people employ to minimise the impact of unpredictable events". Stock keepers in the research area mainly had to cope with very irregular rainfall in Sahelian climatic regimes on the one hand, and with diseases and epizootics on the other. As stressed before, a climatic regime with very irregular rainfall, divided over just one rainy season, is typical for the research area in West Africa and differentiates it from regions in the east of the continent, situated at the same latitude. Halstead and O'Shea (2004) identified four types of risk strategies within traditional societies: mobility, diversity, storage and exchange. When normal risk strategies are insufficient, in situations of more extreme or longer crises, human groups are forced to make changes to their economic strategies in order to survive. This probably also lies at the heart of the two major economic innovations that occurred in sub-Saharan West Africa (see 7.1.2.).

Mobility allows easy adaptation to fluctuating (climatic) circumstances, because it permits visiting those areas, which at a given time offer the best conditions. However, mobility is not possible for groups practising full farming because they cannot abandon their fields. They presently occupy, for them, the best-suited zones of western Africa, while they pushed out mobile groups with other economic strategies, such as mobile herding, into more marginal ecological zones. The degree of mobility of the sites' inhabitants is discussed in more detail further on in this chapter, in 7.4.

The inhabitants of nearly all of the studied sites consumed animal proteins obtained through gathering and fishing, as well as hunting and fowling; this even after the introduction of domestic species, although the importance of the different types of animal proteins in the diet varied diachronically and with local geographic circumstances. The West African climatic regime probably did not allow sole reliance on the yields of domestic livestock herds. Diversification of economic resources is also a way of spreading risks in a difficult environment, since it is unlikely that they will all be affected in the same way when small or large setbacks occur. This is also the reason why people kept mixed herds of both ovicaprines and cattle. Evidence from the studied sites and other West African locations indicates that hunting and fishing are more important, where circumstances for herding are difficult, e.g., in rainforests, or in areas with limited drinking water. No indications could be found for an increased reliance on wild resources during arid spells, but as said, the faunal material was not well suited for tracing small diachronic trends. In addition, for risk minimisation, the gathering of wild plants continued after the introduction of domestic crops, even after the Iron Age shift to full farming and up to the present-day.

Keeping herds of domestics as large as possible can be seen as a form of "storage on the hoof". Looking at modern parallels, this may predominantly have been done for cattle herds, while herds of ovicaprines were being culled to maximise production. In the previous chapter, several techniques were also discussed for preserving fish, meat and milk, which can allow storage of them for the less prosperous times of the year. However, these preserved foodstuffs do not seem suitable for overcoming longer than seasonal periods of shortage.

Exchange and cooperation is a fourth way to anticipate risk situations. The present Pokot of Kenya, for example, distribute their livestock over the herds of several families, because the chances that all are hit by disasters at the same time are small (Bollig, 1994). Fish import at Galaga (NA 92/2C) and its possible export from the *firgi* and Blé sites, was probably part of the economic strategies of these human groups to exchange their excess products with other sources of food that they did not, or could not, exploit themselves. Before the introduction of a money economy in the research area, internal trade might have been mainly a survival strategy. Risk management may also be part of the factors that are the basis of the cooperation between different economically specialised ethnic groups, characterising the West African Sahel today (McIntosh, 1993; see 7.3.).

7.3. Specialists versus generalists

It is not clear if the inhabitants of the studied sites obtained all their animal products through their own economic activities or also through trade or exchange. In any case, it seems that all wild resources used, with the exception of part of the fish from Galaga (NA 92/2C), were available locally. As explained, cooperation between economically specialised ethnic groups is one of the main features of the research area today. However, it should be stressed that the present ethnic situation was formed only in the past few centuries. There are no archaeological indications from the study area for economically specialised groups that were living in close proximity to each other, but with usually only one trench dug at each site, this is almost impossible to trace. Nonetheless, it is argued that different economic groups were present in the area, but that only part of them left archaeologically visible traces (see 7.4.). The possible occurrence of two different types of goats at Oursi hu-beero (BF 97/30) and Galaga (NA 92/2C) may indicate contact or cooperation between sedentary farmers and nomadic pastoralists. Both kinds of economically specialised groups are mentioned historically for the Senegal Valley during the twelfth century AD (Levtzion and Hopkins, 1981, 184, 400). Cooperation between different types of specialists must have been stimulated by the fact that they are often attracted to similar environments. Floodplains, for example, offer good conditions for fishers and herders, as well as farmers. For the research area today, and probably also in the past, we cannot speak of specialisation in the sense that human groups carry out only one type of economic activity. Nevertheless, stress is normally on one particular economic strategy, both in terms of labour investment and in social value. Economic specialisation also does not necessarily concern the total human group, but parts of it, or certain individuals may practise specialised activities.

Throughout the discussion in the previous chapter, it has been stressed that none of the present specialist groups of hunters, fishers or herders, could survive without agricultural products obtained from farmers. It was therefore probably the shift to full farming at the beginning of the Iron Age that allowed the development of economic specialisation. Increasing specialisation lead

	« highly mobile »	« mobile »	« partly sedentary »	« sedentary »
« pastoralists »	BF after AD 1400, Bama between Gajiganna and IA	Gajiganna I		
« livestock keepers »				
« livestock keepers-farmers »			Gajiganna IIa/b, *firgi* LSA	
« farmers-livestock keepers »				BF Iron Age, Gajiganna III, Bama Iron Age, Bama subrecent, *firgi* Iron Age
« farmers »				

BF=Burkina Faso, Bama=Bama Deltaic Complex

Fig. 54: Subdivision of the economies of the research area according to the classification system for food producing communities

to higher interdependence between different groups and inversely, more cooperation between groups allowed further specialisation. It is believed that the gradual introgression of zebu blood into the cattle of sub-Saharan western Africa was an additional factor for stimulation, notably in the development of specialised nomadic pastoralism. McIntosh (1997) has an almost opposite view on the origins of specialisation and has postulated that specialisation preceded agriculture and was even the basis of grain domestication. His argument is based on first millennium AD sites in the Middle Niger Valley, that often take the form of clustered settlements (McIntosh, 1993). He sees this as an indication for several groups living together, which developed economic specialisation as a consequence or already had different economies by that time. In his "Pulse Model", McIntosh projects this situation back to the second and first millennium BC. Palaeoclimatic stress and increased resource unpredictability in that period would have set conditions that encouraged occupational specialisation, when several groups were attracted to the same, remaining, favourable environments. However, in the research area, full farming usually does not seem to have been a local development and its introduction probably meant the start of specialisation. This may have happened through the mechanisms that McIntosh has described, i.e. by different groups being attracted to the same favourable spots in periods of environmental stress.

7.4. Food strategy, mobility and archaeological visibility

The faunal assemblages studied usually showed considerable uniformity by period and by region. This seems to indicate that there was in each case only one economic component in the landscape, but probably there is, rather, only one component visible, or at least prominent, archaeologically. Sedentary groups must have left more traces than mobile ones. Nowadays the former mainly include farmers, although fishing may also allow a high degree of sedentism. However, year-round settlement at a single spot seems to be excluded for fishing groups, because of the slack seasons for them in the course of a year. Permanent settlement is usually hard to distinguish archaeologically from semi-permanent settlement or from frequent, seasonal, re-use of the same location. In 6.5.3.1. the presence of lungfish (*Protopterus annectens*) was proposed as a possible argument for permanent occupation.

Fig. 54 is an attempt to fit the human groups documented for each period and region into the classification system for pastoralists of Gallais (1975, 188-189), according to their degree of mobility and their degree of dependence on livestock. For obvious reasons, the Late Stone Age hunter-gatherers from northern Burkina Faso have not been included on the figure. We should probably think of them as "mobile" groups, since they only left shallow sites but did use ceramics, whose weight seems to exclude very high mobility. All studied sites in Burkina Faso, after the beginning of the Iron Age, are probably from sedentary farmers, keeping some livestock as well. Large ovicaprine types suggest that the landscape also had a nomadic component, although we are probably missing much of the archaeological evidence for nomadic pastoralists that were living in the area. Surface sites recorded for the Iron Age of northern Burkina Faso might be remnants of their temporary settlements. After AD 1400, the area was probably only visited by highly mobile pastoralists, who did not leave any archaeological traces at all.

During Gajiganna phase I, which was named pastoral, people were probably mobile, but the use of ceramics seems to imply that mobility was limited. At one site, on the Bama Ridge, there was also evidence for habitation of a more permanent nature. Archaeological material is more frequent for Gajiganna phase II, which is presumed to have been agro-pastoral. Habitation in one place now seems to have been possible during longer periods. Both for Gajiganna phase I and phase II, the sites recorded are probably the dry season camps, when people concentrated around remaining water bodies. Visiting the area with domestic flocks was possibly avoided during the rainy season because of the disease risk. Only in the Bama-Konduga area might settlement also have been possible during the rainy season. Human groups probably returned to the same spot year after year, where they also buried their dead. The intermittent nature of the habitation is also visible in the irregular spread of the fauna and other finds over the sites' stratigraphies. During the wet season people must have been more dispersed, leaving fewer traces. Alternatively they could have carried out transhumant movements between the dry season camps in the Bama Deltaic Complex and hitherto unknown wet season camps elsewhere. Studied sites from the Bama Deltaic Complex, dating to Gajiganna phase III or younger, are probably all of sedentary farmers also practising some stock keeping. Nevertheless, evidence from subrecent Galaga (NA 92/2C) indicates that the landscape may also have known a pastoral nomadic component.

For the Late Stone Age in the *firgi*, an agro-pastoral economy with semi-sedentary settlement is proposed and, archaeologically, probable dry season occupation has been documented. From the Iron Age onwards, habitation in the area was probably permanent, with an increased importance of agriculture. Nevertheless, agriculture may have played a smaller role during the Iron Age in the *firgi* than in northern Burkina Faso and the Bama Deltaic Complex. Occupation at the Blé sites is presumed to have evolved from permanent to seasonal settlement, by fishing parties from surrounding areas. The importance of agriculture at the sites remains unknown, but if they were indeed temporary fishing camps, then its inhabitants probably had little time and interest to practice crop cultivation. Although the importance of herding at Blé was limited, while that of fishing was high, the hypothesis of seasonal fishing camps could not be supported with the archaeofauna. Nevertheless, the numerous fish-smoking features on the sites' surfaces suggest that excess fish was being processed on the spot for future consumption and probably also to be traded.

7.5. Seasonal rounds

Throughout the previous chapter, it appeared that the main economic activities of human groups shift with seasonal physical changes in the exploited environment, and consequent changes in animal behaviour and resource availability. However, thanks to preservation techniques, animal food products can be consumed outside the periods of their actual acquisition. Figs 55 and 56 summarise during which months of the year the different sources were available in the research area and when economic activities must have mainly taken place. In northern Burkina Faso and the Bama Deltaic Complex, local rainfall was probably the main determinant, while in the *firgi* and near Blé the floods, occurring during the local dry season, must have been more influential.

In rain fed cultivation, e.g., of pearl millet, crops are sown from May to June, with the beginning of the rainy season, while the harvest takes place between October and December (Klee *et al.*, 2004). In *masakwa* type farming, activities shift to August-September, when seeding and subsequent transplantation of seedlings to the clay fields take place (Sturm *et al.*, 1996). Harvest then follows around January. Detailed discussions of the seasonality in economic activities, focussed on obtaining animal proteins, can be found in the previous chapter. An important feature is that periods of shortage in one food resource are compensated by better yields in another one. The peak fishing seasons, for example, fall in the months when cattle are not lactating. In northern Burkina Faso and Bama Deltaic Complex, remaining water probably attracted game at the end of the dry season, when fishing was no longer rewarding. Before the appearance of permanent settlements with the beginning of the Iron Age, the Bama Deltaic Complex and *firgi* where probably visited only during the dry season. Especially in the *firgi* the end of the floods must have been a rewarding period for fishing and good pasture was then available in the form of bourgou grasses, while disease risk was minimal.

7.6. Prospects for further research

As a last and final part of this study, it is evaluated if the research questions formulated at the beginning have been answered satisfactorily and where further research would be desirable. Moreover, some new lines of research, which would also be worth investigating, will be suggested.

One of the major research themes was the beginning of food production in sub-Saharan West Africa. The assembled data on domestic animals have mainly underscored the regional variability of their first appearance, the earliest species including cattle, sheep and goat. The Lake Chad area was, together with the other large flood plains in West Africa, such as the Niger bend, a favourable area where domestic stock appeared at an early date. Its oldest traces have been found at sites dating to the beginning of the Gajiganna Culture, around 1800 BC. The people of the Gajiganna Culture represent the first colonists of the ground that became free in the southern Lake Chad area with the retreat of Lake Mega

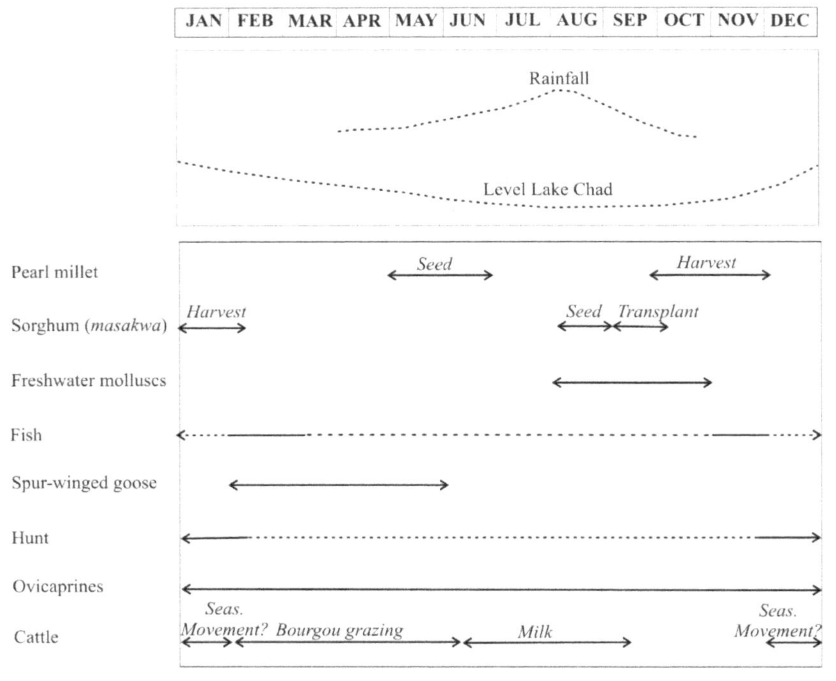

Fig. 55: Reconstruction of seasonal rounds in the *firgi* area

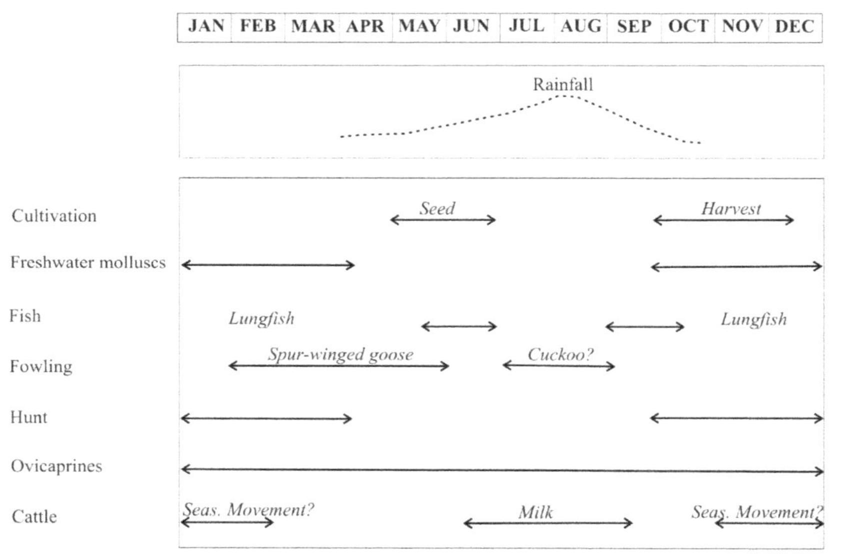

Fig. 56: Reconstruction of seasonal rounds in Burkina Faso and the Bama Deltaic Complex

Chad, after the mid-Holocene. They were Saharan pastoralists, who were driven southwards because of the desiccation of their environment and who brought their herds with them. In the north of Burkina Faso, domestic animals have only been identified at the Iron Age sites (AD 0-1400), although cultivated pearl millet was found in contexts from at least 1000 BC. The putative early appearance of domestic cattle and ovicaprines at Windé Koroji Ouest (2200-950 BC) in the neighbouring Malian Gourma was furthermore questioned. The oldest reliable identifications of domestics from that area date to the first millennium BC. The early presence of domestic animals in northern Burkina Faso and adjacent parts of Mali remains an issue that has to be more thoroughly investigated and for which new excavations are probably required. In any case, it is clear that during the last two millennia BC domestic animals in the area cannot possibly have played the same role as in the southern Lake Chad region, unless they were kept by highly mobile groups. The probably retarded adoption of herding in northern Burkina Faso was mainly attributed to the presence of the more favourable area of the Niger bend to its north.

After the earliest introductions, others followed of new species; domestic fowl (*Gallus gallus* f. domestica), horse (*Equus ferus* f. caballus) and donkey (*Equus africanus* f. asinus), as well as new types of ovicaprines and cattle, most importantly zebu cattle. Perhaps also domestic cat (*Felis silvestris* f. catus) and in later phases domestic pig (*Sus scrofa* f. domestica), besides some minor domestic species, were present in the research area. The assembled data largely conform with existing archaeozoological data on the first appearance of the different domestic species in sub-Saharan West Africa, except that also domestic dog (*Canis lupus* f. familiaris) was added to the earliest introductions. Linguistic and ethno-historical data often place the arrival of domestic animals earlier, and this may be indicative of the problem of visibility from which archaeozoology may suffer. It was postulated that the arrival of domestics in the area happened in four main waves; a "Late Stone Age", an "Iron Age", an "Arab" and a "European" one. As in previous studies, two routes were discerned for pre-European introductions, one from the north, through the Sahara, and the other one from the east of the continent, along the savannah corridor. The relatively early finds of domestic fowl in the southern Lake Chad area, eighth-tenth/eleventh century AD, support the existence of an east-west route of introduction for the animal.

Research into the appearance of domestic species and types, and into their size evolution, was hampered by a lack of sufficient osteomorphological and osteometric data on recent specimens. The gathering and publication of more such data is therefore pleaded for. However, from an investigated subrecent site (Galaga, NA 92/2C) it appears that present types and breeds might differ from their older counterparts. Despite the described difficulties, a local and rapid development of dwarf types of domestics was proposed, based on finds of dwarf goats at Zilum (NA 97/37). The gradual size decline of taurine/Sanga cattle during their spread from north-eastern to western Africa through the Sahara, that was proposed earlier, should probably be reconsidered in view of the great size variation demonstrated for early Saharan cattle, with some specimens reaching sizes as large as the African aurochs. More detailed studies are required to correlate these variations with diachronic, geographic, or perhaps other factors.

As hoped, this study also yielded new elements in the discussion on possible domestication centres for guineafowl, the only animal that may have been domesticated locally in sub-Saharan West Africa. It was suggested that domestication might have happened during the Late Stone Age Gajiganna Culture of the southern Lake Chad area, but the data were not conclusive. The tendency of guineafowl to form large flocks near remaining water during dry periods was possibly a stimulating factor in its domestication process, wherever this may have taken place.

Stress in this work was on tracing diachronic and geographic trends, because available material covered an extensive time period and came from a large geographic area. Such trends could indeed be recorded for the studied sites, in their archaeozoological productivity as well as in their faunal composition. However, explaining these trends appeared to be more problematic. Several hypotheses, founded with ethnographical and other types of data, were formulated, although alternative explanations are also possible. The usually small size of the surface excavated at each locality, made the assemblages ill-suited for detailed intra-site studies. Most sites are, moreover, settlement mounds without clear occupation horizons and therefore without much potential for spatial analysis. Only the site Zilum (NA 97/37), a flat site with early urban features, showed indications for spatial specialisation. Excavations at the site are currently still going on and a spatial analysis is planned when more of its faunal material has been studied. Zilum belongs to Gajiganna phase III, which seems to have brought many innovations to the Bama Deltaic Complex. In the fauna, especially the appearance of dwarf goats was noted, besides the reduced economic importance of mammal hunting. The new material of Zilum will also serve to further test these observations and perhaps to gather more new information on this innovative era.

The reconstruction of the former diet suffered, as is usual, from problems inherent to archaeozoological data, and therefore mainly inter- and intra-site differences in consumed animals have been studied. A more precise reconstruction of the actual diet would require isotope analysis of human remains. The present study aimed to explain why and how the economic specialisation by ethnic group, characteristic for the research area today, developed. It was proposed that it must have begun with the onset of full farming, around the beginning of our era. However, within the same area no traces could be found of contemporary groups with different economic specialisations. It was argued that this is probably a consequence of archaeological visibility, since some types of economies, e.g., farming, leave more traces, mainly related to the degree of sedentism they require. Perhaps excavations at more of the shallow sites could allow the gathering of further information on economic specialisation, although highly mobile groups probably left no traces at all.

For the planned detailed studies into the strategies and equipment used for the collection of animal food, fishing, hunting, fowling and herding, mainly ethnographical evidence had to be relied on. Almost no fishing or hunting gear was preserved in the archaeological contexts. Poor bone preservation excluded the use of age and sex curves for domestic animals, which are possible means of documenting herd management strategies. Food processing and preservation techniques were partly reconstructed from traces of butchery, but again mainly using ethnographic data. In some cases it was possible to make some guesses on cultural aspects, such as food taboos and social status. Even though trade in animals and animal products probably took place in the research area, there were hardly any indications for it. It is argued here that the research area was mainly a region of export for animal products, e.g., skins of domestics, live cattle and horses, and fish, which is much harder to prove from archaeological evidence than the import of exotic commodities.

As expected, the fauna provided some indications of diachronic changes in the former environment. Differentiation between climatologically and anthropogenically induced changes was not possible, but it is suspected that both occurred and may have reinforced each other. The general trend observable for all parts of the research area was increased aridification between the oldest occupation and present. Hypothetical reconstructions were made of the seasonal rounds of the sites' inhabitants, correlated with seasonal environmental changes. Our image of the research area is distorted in favour of those groups that had an economy, which allowed them to sustain a sedentary lifestyle, more precisely farmers. However, this is the faith of all archaeological research, having to work with what has been preserved.

References

Aiyedun, K. (1991). Utilization of meat animal by-products in Niger State: an ethnoarchaeological approach. *Nyame Akuma* **35**: 30-32.

Aiyedun, K. (1996). The social and cultural habits of hunters in various parts of Niger State: an ethnoarchaeological study. *Nyame Akuma* **45**: 65-71.

Alberro, M. and Haile-Mariam, S. (1982). The indigenous cattle of Ethiopia. *World Animal Review* **41**: 2-10 and **42**: 27-34.

Albert, K.D. (2002). *Die Altdünenlandschaft im Sahel NE-Burkina Fasos – Geomorphogenese und Geomorphodynamik einer semiariden Kulturlandschaft*, Ph.D. thesis, Johann Wolfgang Goethe-University, Frankfurt am Main. http://publikationen.ub.uni-frankfurt.de/volltexte/2003/310/

Albert, K.-D., Andres, W. and Lang, A. (1997). Palaeodunes in NE Burkina Faso; pedo- and morphogenesis in a chronological framework provided by luminescence dating. *Zeitschrift für Geomorphologie. N. F.* **41**: 167-182.

Albert, K.-D., Hallier, M., Kahlheber, S. and Pelzer, C. (2000). Montée et abandon des collines d'occupation de l'Age de Fer au Nord du Burkina Faso. *Berichte des Sonderforschungsbereichs 268* **14**: 335-351.

Amanor, K.S. (1995). Dynamics of herd structures and herding strategies in West Africa: a study of market integration and ecological adaptation. *Africa* **65**: 351-394.

Ambrose, S.H. (1997). African Neolithic. In Vogel, J.O. (ed.), *Encyclopedia of precolonial Africa: archaeology, history, languages, cultures and environments*, Altamira Press, Walnut Creek/London/New Delhi, pp. 381-385.

Andres, W., Ballouche, A. and Müller-Haude, P. (1996). Contribution des sédiments de la Mare d'Oursi à la connaissance de l'évolution paléoécologique du Sahel du Burkina Faso. *Berichte des Sonderforschungsbereichs 268* **7**: 5-15.

Angel, F. (1950). *Vie et mœurs des serpents*, Payot, Paris.

Angelici, F.M., Luiselli, L., Politano, E. and Akani, G.C. (1999). Bushmen and mammal-fauna: a survey of the mammals traded in bush-meat markets of local people in the rainforests of southeastern Nigeria. *Anthropozoologica* **30**: 51-58.

Anquandah, J. (1993). The Kintampo complex: a case study of early sedentism and food production in sub-Sahelian West Africa. In Shaw, T., Sinclair, P., Andah, B. and Okpoko, A. (eds), *The archaeology of Africa. Food, metals and towns*, Routledge, London and New York, pp. 255-260.

Arkell, A.J. (1949). *Early Khartoum. An account of the excavation of an early occupation site carried out by the Sudan government antiquities service in 1944-1945*, Oxford University Press, London.

Armitage, P. and Chapman, H. (1979). Roman mules. *London Archaeologist* **13**(3): 339-346.

Ba, C., Descamps, C. and Thilmans, G. (1997). Fouille d'un tumulus à Ndiamon-Badat (Iles du Saloum, Sénégal). *Fouille et dégradations dans les îles du Saloum* **3**: 1-14.

Badenhorst, S. and Plug, I. (2003). The archaeozoology of goats, *Capra hircus* (Linnaeus, 1758): their size variation during the last two millennia in southern Africa (Mammalia: Artiodactyla: Caprini). *Annals of the Transvaal Museum* **40**: 91-121.

Baker, J.R. and Brothwell, D.R. (1980). *Animal diseases in archaeology*, Academic Press, London.

Baker, P. and Gautier, A. (1997). Site E71K12 and the Late Paleolithic annual round along the Nile in Upper Egypt. In Wendorf, F., Schild, R., Baker, P., Gautier, A., Longo, L. and Mohamed, A. (eds), *A Late Paleolithic kill-butchery-camp in Upper Egypt*, Department of Anthropology, Institute for the Study of Earth and Man/Southern Methodist University, Dallas and Institute of Archaeology and Ethnology/Polish Academy of Sciences, Warsaw, pp. 43-56.

Balasse, M. and Ambrose, S.H. (2005). Distinguishing sheep and goats using dental morphology and stable carbon isotopes in C4 grassland environments. *Journal of Archaeological Science* **32**: 691-702.

Ballouche, A. (2001). Un diagramme pollinique de la Mare de Kissi (Oudalan, Burkina Faso). Nouveaux éléments pour l'histoire anthropique de la végétation Sahélienne. In Kahlheber, S. and Neumann, K. (eds), Man and environment in the West African Sahel – an interdisciplinary approach. *Berichte des Sonderforschungsbereichs 268* **17**: 129-135.

Ballouche, A. and Neumann, K. (1995). A new contribution to the Holocene vegetation history of the West African Sahel: pollen from Oursi, Burkina Faso and charcoal from three sites in northeast Nigeria. *Vegetation History and Archaeobotany* **4**: 31-39.

Barkindo, B. (1999). Kanem-Borno: its relations with the Mediterranean Sea, Bagirmi and other states in the Chad Basin. In Ogot, B.A. (ed.), *Africa from the sixteenth to the eighteenth century*. General history of Africa, Abridged Edition, Unesco Publishing, Paris, James Currey, London, California University Press, Berkeley, pp. 178-186.

Barral, H. (1967). Les populations d'éleveurs et les problèmes pastoraux dans le Nord-Est de la Haute-Volta (cercle de Dori – subdivision de l'Oudalan) 1963-1964. *Cahiers ORSTOM. Série Sciences Humaines* **4**(1): 3-30.

Barral, H. (1977). *Les populations nomades de l'Oudalan et leur espace pastoral*, Travaux et Documents de l'ORSTOM 77, ORSTOM, Paris.

Barrett-Jolley, S. (2000). The faunal remains from Gao Ancien (excluding fish). In Insoll, T. (ed.), *Urbanism, archaeology and trade: further observations on the Gao region (Mali): the 1996 fieldseason results*, BAR International Series 829, Oxford, pp. 45-55.

Bartosiewicz, L., Van Neer, W. and Lentacker, A. (1997). *Draught cattle: their osteological identification and history*, Annales Sciences Zoologiques 281, Musée royal de l'Afrique centrale, Tervuren.

Barth, H. (1857-1858). *Travels and discoveries in north and central Africa: being a journal of an expedition undertaken under the auspices of H. B. M.'s government, in the years 1849-1855* (5 vols.), Longman, London.

Bawden, M.G. (1972). Geomorphology. In Aitchison, P.J., Bawden, M.G., Caroll, D.M., Glover, P.E., Klinkenberg, K., de Leeuw, P.N. and Tuley, P. (eds), *The land resources of North East Nigeria, Vol. 1, The environment*, Land Resource Study 9, Surbiton, pp. 60-70.

Beadle, L.C. (1981). *The inland waters of tropical Africa. An introduction to tropical limnology*, Longman, London and New York.

Bedaux, R.M.A., Constandse-Westermann, T.S., Hacquebord, L., Lange, A.G. and van der Waals, J. (1978). Recherches archéologiques dans le Delta intérieur du Niger (Mali). *Palaeohistoria* **20**: 91-220.

Bedaux, R.M.A., MacDonald, K.C., Person, A., Polet, J., Sanogo, K., Schmidt, A. and Sidibé, S. (2001). The Dia archaeological project: rescuing cultural heritage in the Inland Niger Delta (Mali). *Antiquity* **75**: 837-848.

Behrensmeyer, A.K. (1978). Taphonomic and ecologic information from bone weathering. *Paleobiology* **4**(2): 150-162.

Beja-Pereira, A., England, P.R., Ferrand, N., Jordan, S., Bakhiet, A.O., Abdalla, M.A., Mashkour, M., Jordana, J., Taberlet, P. and Luikart, G. (2004). African origins of the domestic donkey. *Science* **304**: 1781.

Bénech, V., Durand, J.-R. and Quensière, J. (1983). Fish communities of Lake Chad and associated rivers and floodplains. In Carmouze, J.-P., Durand, J.-R. and Lévêque, C. (eds), *Lake Chad. Ecology and productivity of a shallow tropical ecosystem*, Dr. W. Junk Publishers, The Hague/Boston/Lancaster, pp. 293-356.

Benkheira, M.H. (2000). *Islâm et interdits alimentaires. Juguler l'animalité*, Presses Universitaires de France, Paris.

Blache, J. and Miton, F. (1962). *Première contribution à la connaissance de la pêche dans le bassin hydrographique Logone-Chari-Lac Tchad*, Mémoires ORSTOM 4(1), Paris.

Blache, J., Miton, F., Stauch, A., Iltis, A. and Loubens, G. (1964). *Les poissons du Bassin du Tchad et du Bassin adjacent du Mayo Kebbi. Etude systématique et biologique*, Mémoires ORSTOM 4(2), Paris.

Blench, R.M. (1985). *Pastoral labour and stock alienation in the subhumid and arid zones of West Africa*, ODI Pastoral Development Network Paper 19, Overseas Development Institute, London.

Blench, R.M. (1993). Ethnographic and linguistic evidence for the prehistory of African ruminant livestock, horses and ponies. In Shaw, T., Sinclair, P., Andah, B. and Okpoko, A. (eds), *The archaeology of Africa. Food, metals and towns*, Routledge, London and New York, pp. 71-103.

Blench, R.M. (1995). A history of domestic animals in northeastern Nigeria. *Cahiers des Sciences Humaines* **31**(1): 181-238.

Blench, R.M. (1999). *Traditional livestock breeds: geographical distribution and dynamics in relation to the ecology of West Africa*, ODI Working Paper 122, Overseas Development Institute, London.

Blench, R.M. (2000a). A survey of ethnographic and linguistic evidence for the history of livestock in Africa. In Blench, R.M. and MacDonald, K.C. (eds), *The origins and development of African livestock: archaeology, genetics, linguistics and ethnography*, UCL Press, London, pp. 18-27.

Blench, R.M. (2000b). African minor livestock species. In Blench, R.M. and MacDonald, K.C. (eds), *The origins and development of African livestock: archaeology, genetics, linguistics and ethnography*, UCL Press, London, pp. 314-338.

Blench, R.M. (2000c). A history of donkeys, wild asses and mules in Africa. In Blench, R.M. and MacDonald, K.C. (eds), *The origins and development of African livestock: archaeology, genetics, linguistics and ethnography*, UCL Press, London, pp. 339-354.

Blench, R.M. (2000d). A history of pigs in Africa. In Blench, R.M. and MacDonald, K.C. (eds), *The origins and development of African livestock: archaeology, genetics, linguistics and ethnography*, UCL Press, London, pp. 355-367.

Blench, R.M. (2003). The movement of cultivated plants between Africa and India in prehistory. In Neumann, K., Butler, A. and Kahlheber, S. (eds), *Food, fuel and fields*, Progress in African Archaeobotany, Africa Praehistoria 15, Heinrich-Barth-Institut, Köln, pp. 273-292.

Blench, R.M. (submitted). *Archaeology, language and the African past.*

Blench, R.M., Jode, A. de, Hall, S.J.G. and Wint, W. (1992). *National Livestock Resource Survey* (6 vols.), Report to Federal Livestock Department, Abuja, Nigeria.

Blench, R.M. and MacDonald, K.C. (2000). Chickens. In Kiple, K.F. and Ornelas, K.C. (eds), *The Cambridge world history of food* (2 vols.), Cambridge University Press, Cambridge, pp. 496-499.

Boahen, A.A. (1990). *Africa under colonial domination 1880-1935*, General history of Africa, Abridged Edition, Unesco Publishing, Paris, James Currey, London, California University Press, Berkeley.

Boessneck, J. and Driesch, A. von den (1982). *Studien an subfossilen Tierknochen aus Ägypten,* Münchner Ägyptologische Studien 40, Deutscher Kunstverlag, München.

Boessneck, J., Driesch, A. von den and Ziegler, R. (1989). Die Tierreste von Maadi und Wadi Digla. In Rizkana, I. and Seeher, J. (eds), *Maadi III. The non-lithic small finds and the structural remains of the Predynastic settlement*, Verlag Philipp von Zabern, Mainz am Rhein.

Boessneck, J., Müller, H. and Teichert, M. (1964). Osteologische Unterscheidungsmerkmale zwischen Schafe (*Ovis aries* Linne) und Ziege (*Capra hircus* Linne). *Kühn-Archiv* **78**(1-2): 1-129.

Boettger, C.R. (1958). *Die Haustiere Afrikas*, G. Fischer Verlag, Jena.

Bohlken, H. (1961). Haustiere und Zoologische Systematik. *Zeitschrift für Tierzüchtung und Züchtungsbiologie* **76**: 107-113.

Böhme, W., Meinig, H. and Rödel, M.-O. (1996). New records of amphibians and reptiles from Burkina Faso and Mali. *British Herpetological Society Bulletin* **56**: 7-26.

Bollig, M. (1994). Krisenmanagement und Risikominimierung bei den pastoralnomadischen Pokot Nordwestkenias. In Bollig, M. and Klees, F. (eds), *Überlebungsstrategien in Afrika*, Heinrich-Barth-Institut, Köln, pp. 125-157.

Bouchud, J. (1993). Paléofaune de Tegdaoust. In Devisse, J. and Robert, S. (eds), *Tegdaoust III. Recherches sur Aoudaghost: Campagnes 1960-1965*, Editions Recherche sur les Civilisations, Paris, pp. 355-363.

Bouchud, J., Brébion, P. and Saban, R. (1981). Etude des faunes holocènes provenants de la zone aride du Sahara atlantique. In Roubet, C., Hugot, H.J. and Souville, G. (ed.), *Préhistoire Africaine. Mélanges en Hommage au Doyen Lionel Balout*, A.D.P.F., Paris, pp. 237-244.

Bouquet, C. (1990). *Insulaires et riverains du lac Tchad. Etude géographique, Vol. 2*, l'Harmattan, Paris.

Bourges, C., MacEachern, S. and Reeves, M. (1999). Excavations around Aissa Hardè, 1995 and 1996. *Nyame Akuma* **51**: 6-13.

Bourlière, F. (1963). Observations on the ecology of some large African mammals. In Howel, F.C. and Bourlière, F. (eds), *African ecology and human evolution*, Viking Fund Publications in Anthropology 36, Wenner-Gren Foundation, New York, pp. 43-54.

Bourn, D.M., Wint, W., Blench, R.M. and Woolley, E. (1994). Nigerian livestock resources survey. *World Animal Review* **78**: 49-58.

Bower, J.R.F. (1997). Eastern African pastoralists. In Vogel, J.O. (ed.), *Encyclopedia of precolonial Africa: archaeology, history, languages, cultures and environments*, Altamira Press, Walnut Creek/London/ New Delhi, pp. 205-210.

Bradley, D.G., MacHugh, D.E., Cunningham, E.P. and Loftus, R.T. (1996). Mitochondrial diversity and the origins of African and European cattle. *Proceedings of the National Academy of Sciences USA* **93**: 5131-5135.

Bradley, D.G. and Loftus, R. (2000). Two eves for taurus? Bovine mitochondrial DNA and African cattle domestication. In Blench, R.M. and MacDonald, K.C. (eds), *The origins and development of African livestock:*

archaeology, genetics, linguistics and ethnography, UCL Press, London, pp. 224-250.

Brain, C.K. (1974). Some suggested procedures in the analysis of bone accumulations from southern African Quaternary sites. *Annals of the Transvaal Museum* **29**: 1-8.

Braukämper, U. (1993). Umweltanpassung Arabisch-sprachiger Rindernomaden (Shuwa) im Nigerianischen Tschadsee-Gebiet. *Berichte des Sonderforschungs-bereichs 268* **3**: 53-60.

Braukämper, U. (1995). Probleme der Nutzung von Weideressourcen bei den Shuwa-Arabern im Nigerianischen Tschadbecken. *Berichte des Sonderforschungsbereichs 268* **5**: 147-162.

Braukämper, U. (2004). Towards a chronology of Arabic settlement in the Chad Basin. In Krings, M. and Platte, E. (eds), *Living with the lake: perspectives on cultures, economies and histories of Lake Chad*, Studien zur Kulturkunde 21, Rüdiger Köppe Verlag, Köln, pp. 148-170.

Brelsford, W.V. (1946). *Fishermen of the Bangweulu swamps. A study of the fishing activities of the Unga tribe*, Rhodes Livingstone Paper 12, Livingstone.

Breman, H. (1992). Agro-ecological zones in the Sahel: potentials and constraints. In Blokland, A. and van der Staaij, F. (ed.), *Sustainable development in semi-arid Sub-Saharan Africa*, Development Cooperation Information Department DVL/OS of the Ministry of Foreign Affairs, The Hague, pp. 19-35.

Breman, H. and De Wit, C.T. (1983). Rangeland productivity and exploitation in the Sahel. *Science* **221**(4618): 1341-1347.

Breunig, P. (1994). Early prehistoric art in Borno (NE Nigeria). *Sahara* **6**: 98-102.

Breunig, P. (1995). Gajiganna und Konduga. Zur Frühen Besiedlung des Tschadbeckens in Nigeria. *Beiträge zur allgemeinen und vergleichenden Archäologie* **15**: 3-48.

Breunig, P. (1996). The 8000-year-old dugout canoe from Dufuna (NE Nigeria). In Pwiti, G. and Soper, R. (eds), *Aspects of African Archaeology. Papers from the 10th congress of the PanAfrican Association for Prehistory and Related Studies*, University of Zimbabwe publications, Harare, pp. 461-468.

Breunig, P. (2001). Die Gajiganna-Kultur des Tschadbeckens (Nordost-Nigeria). Chronik einer archäologischen Feldarbeit in Westafrika. In Gehlen, B., Heinen, M. and Tillman, A. (eds), *Zeit-Räume. Gedenkschrift für Wolfgang Taute*, Archäologische Berichte 14, Selbstverlag der Deutschen Gesellschaft für Ur- und Frühgeschichte, Bonn, pp. 283-302.

Breunig, P. (2004). Environmental instability and cultural change in the later prehistory of the Chad Basin. In Krings, M. and Platte, E. (eds), *Living with the lake: perspectives on cultures, economies and histories of Lake Chad*, Studien zur Kulturkunde 121, Rüdiger Köppe Verlag, Köln, pp. 52-72.

Breunig, P., Ballouche, A., Neumann, K, Rösing, F.W., Thiemeyer, H., Wendt, P. and Van Neer, W. (1993a). Gajiganna. New data on early settlement and environment in the Chad Basin. *Berichte des Sonderforschungsbereichs 268* **2**: 51-74.

Breunig, P., Eichhorn, B., Kahlheber, S., Linseele, V., Magnavita, C., Posselt, M., Neumann, K. and Rupp, N. (2006). G(l)anz ohne Eisen: große Siedlungen aus der Mitte des 1. Jahrtausend v. Chr. im Tschadbecken von Nordost-Nigeria. In: Wotzka, H.-P. (ed.), *Grundlegungen. Beiträge zur europäischen und afrikanischen Archäologie für Manfred K. H. Eggert*, Francke, Tübingen, pp. 255–270.

Breunig, P., Garba, A., Gronenborn, D., Van Neer, W., Wendt, P. (1993b). Report on excavations at Gajiganna, Borno State, Northeast Nigeria. *Nyame Akuma* **40**: 30-41.

Breunig, P., Garba, A. and Waziri, I. (1992). Recent archaeological surveys in Borno, Northeast Nigeria. *Nyame Akuma* **37**: 10-17.

Breunig, P. and Neumann, K. (2002a). From hunters and gatherers to food producers: new archaeological and archaeobotanical evidence from the West African Sahel. In Hassan, F.A. (ed.), *Droughts, food and culture. Ecological change and food security in Africa's later Prehistory*, Kluwer Academic/Plenum Publishers, New York/Boston/Dordrecht/London/Moscow, pp. 123-155.

Breunig, P. and Neumann, K. (2002b). Continuity or discontinuity? The 1st millennium BC-crisis in West African prehistory. In Jennerstrasse 8 (eds), *Tides of the Desert-Gezeiten der Wüste*, Heinrich-Barth-Institut, Köln, pp. 491-505.

Breunig, P. and Neumann, K. (2004). Zwischen Wüste und Regenwald. Besiedlungsgeschichte der westafrikanischen Savanne im Holozän. In Albert, K.-D., Löhr, D. and Neumann, K. (eds), *Mensch und Natur in Westafrika*, Abschlussband des Sonderforschungs-bereichs 268, Wiley-VCH Verlag, Weinheim, pp. 93-138.

Breunig, P., Neumann, K. and Van Neer, W. (1996). New research on the Holocene settlement and environment of the Chad Basin in Nigeria. *African Archaeological Review* **13**(2): 111-145.

Brink, J. and Holt, S. (1992). A small goat, *Capra hircus*, from a Late Iron Age site in the Eastern Orange Free State. *South African Field Archaeology* **1**: 88-91.

Brinkhuizen, D.C. (1989). *Ichtyo-archeologisch onderzoek: methoden en toepassingen aan de hand van Romeins vismateriaal uit Velsen (Nederland)*, Ph.D. thesis, University of Groningen, Groningen.

Brown, D.S. (1980). *Freshwater snails of Africa and their medical importance*, Taylor & Francis Ltd., London.

Brown, L.H., Urban, E.K. and Newman, K. (1982). *The birds of Africa, Vol. 1*, Academic Press, New York and London.

Brooks, G.E. (1998). Climate and history in West Africa. In Connah, G. (ed.), *Transformations in Africa. Essays on Africa's later past*, Leicester University Press, London, pp. 139-159.

Brunk, K. and Gronenborn, D. (2004). Floods, droughts, and migrations. The effects of Late Holocene lake level oscillations and climate fluctuations on the settlement and political history in the Chad Basin. In Krings, M. and Platte, E. (eds), *Living with the lake: perspectives on cultures, economies and histories of Lake Chad*, Studien zur Kulturkunde 21, Rüdiger Köppe Verlag, Köln, pp. 101-132.

Bruton, M.N. (1979a). The survival of habitat desiccation by air breathing clariid catfishes. *Environmental Biology of Fishes* **4**(3): 273-280.

Bruton, M.N. (1979b). The breeding biology and early development of *Clarias gariepinus* (Pisces: Clariidae) in Lake Sibaya, South Africa, with a review of breeding in species of the subgenus *Clarias* (*Clarias*). *Transactions of the Zoological Society of London* **35**: 1-45.

Bruton, M.N. and Boltt, R.E. (1975). Aspects of the biology of *Tilapia mossambica* Peters (Pisces: Cichlidae) in a natural freshwater lake (Lake Sibaya, South Africa). *Journal of Fish Biology* **7**: 423-445.

Budgett, J.S. (1900). Observations on *Polypterus* and *Protopterus*. *Proceedings of the Cambridge Philosophical Society* **10**: 236-240.

Bulliet, R.W. (1990). *The camel and the wheel* (2nd edition), Columbia University Press, New York.

Carmouze, J.-P. and Lemoalle, J. (1983). The lacustrine environment. In Carmouze, J.-P., Durand, J.-R. and Lévêque, C. (eds), *Lake Chad. Ecology and productivity of a shallow tropical ecosystem*, Dr. W. Junk Publishers, The Hague/Boston/Lancaster, pp. 27-63.

Carter, P.L. and Clark, J.D. (1976). Adrar Bous and African cattle. In Abebe, B., Chavaillon, J. and Sutton, J.E.G. (eds), *Actes du Congrès Panafricain de Préhistoire et des Etudes Quaternaires*, 7ième session, Addis Abéba, 1971, pp. 487-493.

Carter, P.L. and Flight, C. (1972). A report on the fauna from the sites of Ntereso and Kintampo rock shelter six in Ghana: with evidence for the practice of animal husbandry during the second millenium B.C. *Man* **7**(2): 277-282.

Casey, J., Sawatzky, R., Godfrey-Smith, D.I., Quickert, N., D'Andrea, A.C., Wollstonecroft, M. and Hawkins, A. (1997). Report of investigations at the Birimi site in northern Ghana. *Nyame Akuma* **48**: 32-38.

Casimir, M.J. and Bollig, M. (1994). Milchverarbeitung und Nahrungsmittelknappheit bei afrikanischen Pastoralisten. In Bollig, M. and Klees, F. (eds), *Überlebungsstrategien in Afrika*, Heinrich-Barth-Institut, Köln, pp. 227-253.

Cerný, V. (2002). L'économie pastorale, la production laitière et l'apparition de la tolérance au lactose chez les populations africaines et européennes. *Actes du colloque Méga-Tchad, « Ressources vivrières et choix alimentaires dans le bassin du lac Tchad », Nanterre, 20-22 novembre 2002.* http://www.uni-bayreuth.de/afrikanistik/mega-tchad/Table/Colloque2002/Cerny.pdf

Chaix, L. (2000). A Hyksos horse from Tell Heboua (Sinai, Egypt). In Mashkour, M., Choyke, A.M., Buitenhuis, H. and Poplin, F. (eds), Archaeozoology of the Near East. *Proceedings of the fourth international symposium on the archaeozoology of southwestern Asia and adjacent areas*, ARC Publicatie 32, Groningen, pp. 177-186.

Chaix, L. and Grant, A. (1987). A study of a prehistoric population of sheep (*Ovis aries* L.) from Kerma (Sudan). *Archaeozoologia* **1**: 93-107.

Chang, C. and Koster, H.A. (1986). Beyond bones: toward an archaeology of pastoralism. *Advances in archaeological method and theory* **9**: 97-148.

Chang, T.K. and Landauer, W. (1950). Observations on the skeleton of African dwarf goats. *Journal of Morphology* **86**(2): 367-376.

Channing, A., Du Preez, L. and Passmore, N. (1994). Status, vocalisation and breeding biology of two species of African bullfrogs (Ranidae: *Pyxicephalus*).

Journal of the Zoological Society of London **234**: 141-148.

Chavane, B. (1985). *Villages de l'Ancien Tekrour. Recherches archéologiques dans la moyenne vallée du fleuve Sénégal*, Editions Karthala, Paris.

Chenal-Vélardé, I. (1996). Etude taphonomique, observations ethnologiques et interprétations archéologiques: essai sur les techniques de boucherie à Hamdallahi (Mali, XIXe siècle). *Anthropozoologica* **23**: 85-95.

Chenal-Vélardé, I. (1997). Les premières traces de boeuf domestique en Afrique du Nord: état de la recherche centré sur les données archéozoologiques. *Archaeozoologia* **9**: 11-40.

Chippaux, J.-P. (2001). *Les serpents d'Afrique occidentale et centrale*, Collection Faune et Flore tropicales 35, Editions de l'IRD, Paris.

Close, A. (2002). *Sinai, Sahara, Sahel: the introduction of domestic caprines to Africa.* In Jennerstrasse 8 (eds), Tides of the desert-Gezeiten der Wüste, Heinrich-Barth-Institut, Köln, pp. 459-469.

Clutton-Brock, J. (1993). The spread of domestic animals in Africa. In Shaw, T., Sinclair, P., Andah, B. and Okpoko, A. (eds), *The archaeology of Africa. Food, metals and towns*, Routledge, London and New York, pp. 61-70.

Clutton-Brock, J. (2000). Cattle, sheep, and goats south of the Sahara: an archaeozoological perspective. In Blench, R.M. and MacDonald, K.C. (eds), *The origins and development of African livestock: archaeology, genetics, linguistics and ethnography*, UCL Press, London, pp. 30-37.

Connah, G. (1981). *Three thousand years in Africa. Man and his environment in the Lake Chad region of Nigeria*, Cambridge University Press, Cambridge.

Connah, G. (1987). An optimal zone: the West African savanna. In Connah, G. (ed.), *African civilizations*, Cambridge University Press, Cambridge, pp. 97-120.

Connah, G. and Freeth, S.J. (1989). A commodity problem in prehistoric Borno. *Sahara* **2**: 7-20.

Connah, G. and McMillan, N.F. (1995). Mollusca utilization in prehistoric Borno: a case of human preference? *Sahara* **7**: 29-38.

Cook, H. (2000). The fish bones from Gadei. In Insoll, T. (ed.), *Urbanism, archaeology and trade: further observations on the Gao region (Mali): the 1996 fieldseason results*, BAR International Series 829, Oxford, pp. 38-44.

Copley, H. (1952). *The game fishes of Africa*, H.F. and G. Witherby Ltd., London.

Crowder, M. (1977). *West Africa. An introduction to its history*, Longman, London.

Crowley, T.E. and Pain, T. (1970). *A monographic revision of the African landsnails of the genus* Limicolaria, *Schumacher. (Mollusca-Achatinidae)*, Annales du Musée royal de l'Afrique centrale, Sciences Zoologiques 177, Tervuren.

Crowley, T.E. and Pain, T. (1989). Report on the molluscan remains found in the excavations at Matupi cave, Zaire. Appendix to Van Neer, W., *Contribution to the archaeozoology of Central Africa*, Annales Sciences Zoologiques 259, Musée royal de l'Afrique centrale, Tervuren, pp. 131-136.

Cruz-Uribe, K. (1988). The use and meaning of species diversity and richness in archaeological faunas. *Journal of Archaeological Science* **15**: 179-196.

Cyffer, N., Löhr, D., Platte, E. and Tijani, A.I. (1996). Adaptation and delimitation. Some thoughts about the Kanurization of the Gamergu. *Berichte des Sonderforschungsbereichs 268* **8**: 49-66.

Czerniewicz, M. von (2004). *Studien zur Chronologie der Eisenzeit in der Sahel-zone von Burkina Faso/Westafrika*, unpublished Ph.D. thesis, Johann Wolfgang Goethe-University, Frankfurt am Main.

Czerniewicz, M. von, Pelzer, C. and Petit, L. (2004). *The last moments in the Iron Age house of Oursi hu-beero: new evidence.* Paper presented at the 17th Biennial Conference of the Society of Africanist Archaeologists, Bergen, Norway, June 26-29.

Daget, J. (1949). La pêche dans le Delta Central du Niger. *Journal de la Société des Africanistes* **19**: 1-79.

Daget, J. (1954). *Les poissons du Niger supérieur*, Mémoires de l'Institut Français d'Afrique Noire 36, IFAN, Dakar.

Daget, J. and Iltis, A. (1965). *Poissons de Côte d'Ivoire (eaux douces et saumâtres)*, Mémoires de l'Institut Français d'Afrique Noire 74, IFAN, Dakar.

Dahl, G. and Hjort, A. (1976). *Having herds. Pastoral herd growth and household economy*, Liber Tryck, Stockholm.

D'Andrea, A.C., Klee, M. and Casey, J. (2001). Archaeobotanical evidence for pearl millet (*Pennisetum glaucum*) in sub-Saharan West Africa. *Antiquity* **75**: 341-348.

David, N. and Sterner, J. (1987). The Mandara archaeological project, 1984-1987. *Nyame Akuma* **29**: 2-8.

David, N. and Sterner, J. (1989). The Mandara archaeological project, 1988-89. *Nyame Akuma* **32**: 5-9.

Davis, S.J.M. (1987). *The Archaeology of animal bones*, Batsford, London, Yale University Press, New Haven.

Davis, S.J.M. (2005). Why domesticate food animals? Some zoo-archaeological evidence from the Levant. *Journal of Archaeological Science* **32**: 1408-1416.

Davies, O. (1980). The Ntereso cultures in Ghana. In Schwartz, B.K. and Dumett, R. (eds), *West African culture dynamics: archaeological and historical perspectives*, Mouton, The Hague, pp. 205-225.

Deacon, J. (1984). Later Stone Age people and their descendants in southern Africa. In Klein, R.G. (ed.), *Southern African prehistory and paleoenvironments*, A.A. Balkema, Rotterdam and Boston, pp. 221-328.

De Beer, G.R. (1937). *The development of the vertebrate skull*, Clarendon Press, London.

de Benoist, J.R. (1983). Tendances actuelles de l'Islam en Afrique de l'Ouest. *Etudes* **359**(6): 667-682.

de Buffrénil, V. (1993). *Les varans africains (*Varanus niloticus *et* Varanus exanthematicus*). Données de synthèse sur leur biologie et leur exploitation*, Secrétariat de la Convention sur le commerce international des espèces de faune et de flore menacés d'extinction (CITES), Genève.

De Cock, G. (2002). *Archeozoölogische studie van de historische site NA 92-2C in Galaga, Noordoost Nigeria (19de eeuw),* Unpublished M.Sc. thesis, Catholic University Leuven, Leuven.

DeCorse, C.R. (1997). Western African historical archaeology. In Vogel, J.O. (ed.), *Encyclopedia of precolonial Africa: archaeology, history, languages, cultures and environments*, Altamira Press, Walnut Creek/London/New Delhi, pp. 545-549.

DeCorse, C.R. and Spiers, S. (2001). West African Iron Age. In Peregrine, P. and Ember, M. (eds), *Encyclopedia of Prehistory, Vol. 1, Africa*, Kluwer Academic/Plenum, New York/Boston/Dordrecht/London/Moscow, pp. 313-318.

Degerbøl, M. and Fredskild, B. (1970). *The Urus (*Bos primigenius *Bojanus) and Neolithic domestic cattle (*Bos taurus domesticus *Linné) in Denmark,* Kongelige Danske Videnskabernes Selskab Biologiske Skrifter 17(1), Kobenhavn.

Descamps, C., Thilmans, G., Thommeret, J., Thommeret, Y. and Hauptmann, E.F. (1977). Données sur l'âge et la vitesse d'édification de l'amas coquiller de Faboura (Sénégal). *Bulletin ASEQUA* **51**: 27-32.

Diallo, H. (1990). Introduction à l'étude de l'histoire de l'Islam dans l'Ouest du Burkina Faso: des débuts à la fin du XIXe siècle. *Islam et Sociétés au Sud du Sahara* **4**: 33-45.

Donkin, R.A. (1991). *Meleagrides, an historical and ethnographical study of the guinea fowl*, Ethnographica, London.

Dorst, J. and Dandelot, P. (1970). *A field guide to the larger mammals of Africa*, Collins, London.

Doutresolle, G. (1947). *L'élevage en Afrique Occidentale Française*, Larose, Paris.

Dowsett, R.J. and Dowsett-Lemaire, F. (1993). *A contribution to the distribution and taxonomy of Afrotropical and Malagassy birds*, Tauraco Research Report 5, Tauraco Press, Liège.

Draper, H.H. (2000). Human nutritional adaptation: biological and cultural aspects. In Kiple, K.F. and Ornelas, K.C. (eds), *The Cambridge world history of food* (2 vols.), Cambridge University Press, Cambridge, pp. 1466-1476.

Driesch, A. von den (1976). *A guide to the measurement of animal bones from archaeological sites*, Peabody Museum Bulletin 1, Harvard University, Harvard.

Driesch, A. von den (1983). Some archaeozoological remarks on fishes in ancient Egypt. In Grigson, C. and Clutton-Brock, J. (eds), *Animals and archaeology. 2. Shell middens, fishes and birds*, BAR International Series 183, Oxford, pp. 87-110.

Driesch, A. von den and Boessneck, J. (1974). Kritische Anmerkungen zur Widerristhöhenberechnung aus Längenmassen vor- und frühgeschichtlicher Tierknochen. *Säugetierkundliche Mitteilungen* **22**(4): 325-348.

Driesch, A. von den and Boessneck, J. (1985). *Die Knochenfunde aus der neolitischen Siedlung von Merimde-Benisalâme an westlichen Nil Delta*, Institut für Paläoanotomie, Domestikationsforschung und Geschichte der Tiermedizin, München and Deutsches Archäologisches Institut, Abteilung Kairo.

Ducos, P. (1968). *L'origine des animaux domestiques en Palestine*, Publications de l'Institut de Préhistoire de l'Université de Bordeaux, Mémoire 6, Delmas, Bordeaux.

Dusart, J. (1963). Contribution à l'étude de l'adaptation des Tilapia (Pisces Cichlidae) à la vie en milieu mal oxygéné. *Hydrobiologia* **21**(3-4): 328-341.

Dyson-Hudson, N. (1966). *Karimojong politics*, Oxford University Press, Oxford.

Efremov, J.A. (1940). Taphonomy: new branch of paleontology. *Pan-American Geologist* **74**(2): 81-93.

El Fasi, M. and Hrbek, I. (1988). Stages in the development of Islam and its dissemination in Africa. In El Fasi, M. and Hrbek, I. (eds), *Africa from the seventh to the eleventh century*, General history of Africa, Abridged Edition, Unesco Publishing, Paris, James Currey, London, California University Press, Berkeley, pp. 56-91.

Elgood, J.H. (1982). *The birds of Nigeria. An annotated check-list*, British Ornithologists' Union Checklist 4, Zoological Society, London.

Epstein, H. (1971). *The origin of the domestic animals of Africa* (2 vols.), Africana Publishing, New York.

Ervynck, A. (1989). *Archeozoölogisch onderzoek van de zwarte rat (*Rattus rattus*) en de bruine rat (*Rattus norvegicus*)*, Unpublished Ph.D. thesis, University of Amsterdam, Amsterdam.

Estes, R.D. (1991). *The behaviour guide to African mammals*, University of California Press, Berkeley.

Evans-Pritchard, E.E. (1947). *The Nuer: a discription of the modes of livelihood and political institutions of a Nilotic people,* Oxford University Press, Oxford.

Evershed, R.P., Sherratt, G., Payne, S., Copley, M.S. and Coolidge, J. (2004). The emergence of dairying in early farming practices of the Fertile Crescent and the Balkans. http://catal.arch.cam.ac.uk/catal/Archive_rep04a/r04_35.html

Evin, J., Marechal, J. and Marien, G. (1983). Lyon natural radiocarbon measurements IX. *Radiocarbon* **25**(1): 59-128.

Ewards, C.J., MacHugh, D.E., Dobney, K.M., Martin, L., Russell, N., Horwitz, L.K., McIntosh, S.K., MacDonald, K.C., Helmer, D., Tresset, A., Vigne, J.D. and Bradley, D.G. (2004). Ancient DNA analysis of 101 cattle remains: limits and prospects. *Journal of Archaeological Science* **31**: 695-710.

Fagg, B. (1965). Carbon dates from Nigeria. *Man* **65**: 22-23.

Fagg, B. (1972). Rop rock shelter excavations 1944. *West African Journal of Archaeology* **2**: 1-12.

FAO (2001). *Pastoralism in the new millennium*, FAO Animal Production and Health Paper 150. http://www.fao.org/documents/show_cdr.asp?url_file=/DOCREP/005/Y2647E/y2647e00.htm

FAO (2004). *Livestock Sector Brief. Burkina Faso*, Livestock Information Sector Analysis and Policy Branch. http://www.fao.org/ag/againfo/resources/en/publications/sector_briefs/lsb_BFA.pdf

Fish, G.R. (1956). Some aspects of the respiration of six species of fish from Uganda. *Journal of Experimental Biology* **33**: 168-195.

Flannery, K.V. (1969). Origins and ecological effects of early domestication in Iran and the Near East. In Ucko, P.J. and Dimbleby, G.W. (eds), *The domestication and exploitation of plants and animals*, Aldine, Chicago, pp. 73-100.

Flight, C. (1968). Kintampo 1967. *West African Archaeological Newsletter* **8**: 15-20.

Frank, B. (1965). *Die rolle des Hundes in afrikanischen Kulturen*, Franz Steiner Verlag, Wiesbaden.

Frank, T., Breunig, P., Müller-Haude, P., Van Neer, W., Neumann, K., Vogelsang, R. and Wotzka, H.-P. (2001). The Chaîne de Gobnangou, SE Burkina Faso: archaeological, archaeobotanical, archaeozoological and geomorphological studies. *Beiträge zur allgemeinen und vergleichenden Archäologie* **21**: 127-190.

Franke-Sharf, I., Kirscht, H., Krings, M., Platte, E. and Thiemeyer, H. (2004). Seespiegelschwankungen und wechselnde Nutzungspotentiale. In Albert, K.-D., Löhr, D. and Neumann, K. (eds), *Mensch und Natur in Westafrika*, Abschlussbuch des Sonderforschungs-bereichs 268, Wiley-VCH Verlag, Weinheim, pp. 196-220.

Froment, A. (1988). *Le peuplement humain de la boucle du Niger*, Collection Travaux et Documents 215, Editions de l'ORSTOM, Paris.

Fry, C.H., Keith, S. and Urban, E.K. (1988). *The birds of Africa, Vol. 3*, Academic Press, New York and London.

Fuller, D.Q. (2000). The botanical remains. In Insoll, T. (ed.), *Urbanism, archaeology and trade: further observations on the Gao region (Mali): the 1996 fieldseason results*, BAR International Series 829, Oxford, pp. 28-35.

Fuller, D.Q. (2003). African crops in prehistoric South Asia: a critical review. In Neumann, K., Butler, A. and Kahlheber, S. (eds), *Food, fuel and fields*, Progress in African Archaeobotany, Africa Praehistoria 15, Heinrich-Barth-Institut, Köln, pp. 239-271.

Gabriel, B. (1987). Palaeoecological evidence from neolithic fireplaces in the Sahara. *African Archaeological Review* **5**: 93-103.

Gallais, J. (1975). *Pasteurs et paysans du Gourma: la condition sahélienne*, Centre National de Recherche Scientifique, Paris.

Gallant, J. (2002). *The story of the African dog*, University of Natal Press, Pietermaritzburg.

Garson, A.G. (1980). Comment upon the economic potential of fish utilization in riverine environments and potential archaeological biases. *American Antiquity* **45**(3): 562-567.

Gasse, F. (2000). Hydrological changes in the African tropics since the Last Glacial Maximum. *Quaternary Science Reviews* **19**: 189-211.

Gautier, A. (1983). Animal life along the prehistoric Nile: the evidence from Saggai 1 and Geili (Sudan). *Origini* **12**: 50-115.

Gautier, A. (1984a). How do I count you, let me count the ways. Problems of archaeozoological quantification. In Grigson, C. and Clutton-Brock, J. (eds), *Animals and archaeology. 4. Husbandry in Europe*, BAR International Series 227, Oxford, pp. 237-251.

Gautier, A. (1984b). Archaeozoology of the Bir Kiseiba region, Eastern Sahara. In Close, A. (ed.), *Cattle-keepers of the Eastern Sahara: The Neolithic of Bir Kiseiba*, Department of Anthropology, Institute for the Study of Earth and Man, Southern Methodist University, Dallas, pp. 49-72.

Gautier, A. (1987a). Prehistoric men and cattle in North Africa: A dearth of data and a surfeit of models. In Wendorf, F. and Schild, R. (ass.), Close, A.E. (ed.), *Prehistory of arid North Africa. Essays in honor of Fred Wendorf*, Southern Methodist University Press, Dallas, pp. 163-187.

Gautier, A. (1987b). Taphonomic groups: how and why? *Archaeozoologia* **1**(2): 47-52.

Gautier, A. (1990). *La domestication et l'homme créa ses animaux*, Editons Errance, Paris.

Gautier, A. (1993). "What's in name?" A short history of the Latin and other labels proposed for domestic animals. In Clason, A., Payne, S. and Uerpmann, H.-P. (eds), *Skeletons in her cupboard, Festschrift for Juliet Clutton-Brock*, Oxbow Monograph 34, Oxbow books, Oxford, pp. 91-98.

Gautier, A. (2002). The evidence for the earliest livestock in North Africa: or adventures with large bovids, ovicaprids, dogs and pigs. In Hassan, F.A. (ed.), *Droughts, food and culture. Ecological change and food security in Africa's later Prehistory*, Kluwer Academic/Plenum Publishers, New York/Boston/ Dordrecht/London/Moscow, pp. 195-207.

Gautier, A. and Van Neer, W. (1989). Animal remains from the Late Paleolithic sequence at Wadi Kubbaniya. In Wendorf, F. and Schild, R. (eds), *The prehistory of Wadi Kubbaniya, Vol. 2.*, Southern Methodist University Press, Dallas, pp. 119-161.

Gautier, A. and Van Neer, W. (2005). The continuous exploitation of wild animal resources in the archaeozoological record of Ghana. *Journal of African Archaeology* 3(2): 195-212.

Gayet, M. and Van Neer, W. (1990). Caractères diagnostiques des épines de quelques silures africains. *Journal of African Zoology* **104**: 241-252.

Geis-Tronich, G. (1991). *Materielle Kultur der Gulmanceba in Burkina Faso*, Studien zur Kulturkunde 98, Rüdiger Köppe Verlag, Köln.

Gentry, A.W. (1978). Bovidae. In Maglio, V.J. and Cook, H.B.S. (eds), *Evolution of African mammals*, Harvard University Press, Cambridge, pp. 540-572.

Gentry, A.W., Clutton-Brock, J. and Groves, C.P. (2004). The naming of wild animal species and their domestic derivatives. *Journal of Archaeological Science* **31**: 644-651.

Gentry, A.W. and Gentry, A. (1978). Fossil Bovidae (Mammalia) of Olduvai Gorge, Tanzania, Part I. *Bulletin of the British Museum (Natural History). Geology* **29**(4): 289-446.

Gifford-Gonzalez, D. (1998). Early pastoralists in East Africa: ecological and social dimensions. *Journal of Anthropological Archaeology* **17**: 166-200.

Gifford-Gonzalez, D. (2000). Animal disease challenges to the emergence of pastoralism in sub-Saharan Africa. *African Archaeological Review* **17**(3): 95-139.

Gifford-Gonzalez, D. (2005). Pastoralism and its consequences. In Stahl, A.B. (ed.), *African archaeology. A critical introduction*, Blackwell, Malden, pp. 187-224.

Grant, A. (1982). The use of tooth wear as a guide to the age of domestic ungulates. In Wilson, B., Grigson, C. and Payne, S. (eds), *Ageing and sexing animal bones from archaeological sites*, BAR International Series 109, Oxford, pp. 91-108.

Greenfield, H.J. (1999). The origins of metallurgy: distinguishing stone from metal. Cut-marks on bones from archaeological sites. *Journal of Archaeological Science* **26**: 797-808.

Grigson, C. (1980). The craniology and relationships of four species of *Bos*. V. *Bos indicus* L. *Journal of Archaeological Science* **7**: 3-32.

Grigson, C. (1989). Size and sex – morphometric evidence for the domestication of cattle in the Near East. In Milles, A., Williams, D. and Gardner, N. (eds), *The beginnings of agriculture,* BAR International Series 496, Oxford, pp. 77-109.

Grigson, C. (1991). An African origin for African cattle? – some archaeological evidence. *African Archaeological Review* **9**: 119-144.

Grigson, C. (2000). *Bos africanus* (Brehm)? Notes on the archaeozoology of the native cattle of Africa. In Blench, R.M. and MacDonald, K.C. (eds), *The origins and development of African livestock: archaeology, genetics, linguistics and ethnography*, UCL Press, London, pp. 38-60.

Gronenborn, D. (1996). Kundiye: archaeology and ethnoarchaeology in the Kala-Balge area of Borno State, Nigeria. In Pwiti, G. and Soper, R. (eds), *Aspects of African archaeology, Papers from the 10th congress of the PanAfrican Association for Prehistory and Related Studies*, University of Zimbabwe publications, Harare, pp. 449-459.

Gronenborn, D. (1997). An ancient storage pit in the SW Chad Basin, Nigeria. *Journal of Field Archaeology* **24**: 431-439.

Gronenborn, D. (1998). Archaeological and ethnohistorical investigations along the southern fringes of Lake Chad, 1993-1996. *African Archaeological Review* **15**: 225-257.

Gronenborn, D. (2000). *Mai Mbauji. Eine Studie über Entstehung und Wandel eisenzeitlich-historischer Fürstentümer im südlichen Tschadbecken (7./8. Jahrhundert n. Chr. bis ca. 1925)*, Unpublished Habilitation thesis, Johann Wolfgang Goethe-University, Frankfurt am Main.

Gronenborn, D. (2001). Masakwa in the Chad Basin – an examination of the archaeological and historical sources. In Kahlheber, S. and Neumann, K. (eds), Man and environment in the West African Sahel – an interdisciplinary approach. *Berichte des Sonderforschungsbereichs 268* **17**: 73-84.

Gronenborn, D., Van Neer, W. and Skorupinski, T. (1995). Kleiner Vorbericht zur archäologischen Feldarbeit südlich des Tschad-Sees. *Berichte des Sonderforschungsbereichs 268* **5**: 27-39.

Grouzis, M. (1988). *Structure, productivité et dynamique des systèmes écologiques sahéliens (Mare d'Oursi, Burkina Faso)*, Collection Etudes et Thèses, Editions de l'ORSTOM, Paris.

Gruber, A. (1990). *Vergleichend morphologische Untersuchungen an Einzelknochen in Ägypten vorkommender Ciconiidae*, Unpublished Ph.D. thesis, Ludwig-Maximilians-University, Munich.

Guenther, M.G. (1997). African foragers. In Vogel, J.O. (ed.), *Encyclopedia of precolonial Africa: archaeology, history, languages, cultures and environments*, Altamira Press, Walnut Creek/London/New Delhi, pp. 179-184.

Guérin, C. and Faure, M. (1991). III. La Faune. Les mammifères holocènes du site KNT 2 (région de Tombouctou, Mali). In Raimbault, M. and Sanogo, K. (eds), *Recherches archéologiques au Mali*, Editions Karthala, Paris, pp. 372-377.

Guérin, C. and Faure, M. (1996). Chasse au chacal et domestication du boeuf dans le site néolithique d'Asa Koma (République de Djibouti). *Journal des Africanistes* **66**: 299-311.

Habermehl, K.-H. (1975). *Die Altersbestimmung bei Haus- und Labortieren*, Verlag Paul Parey, Hamburg and Berlin.

Habermehl, K.-H. (1985). *Die Altersbestimmung bei Wild- und Pelztieren*, Verlag Paul Parey, Hamburg and Berlin.

Hallier, M. and Petit, L. (2001). Fouille d'une maison de l'Age du Fer dans le nord du Burkina Faso. *Nyame Akuma* **56**: 2-3.

Halstead, M. and O'Shea, J.M. (2004). *Bad year economics: cultural responses to risk and uncertainty*, Cambridge University Press, Cambridge.

Halstead, P., Collins, P. and Isaakidou, V. (2002). Sorting the sheep from the goats: morphological distinctions between the mandibles and mandibular teeth of adult

Ovis and *Capra*. *Journal of Archaeological Science* **29**: 545-553.

Haltenorth, T.S., Diller, H. and Smeenk, C. (1979). *Elseviers gids van de Afrikaanse zoogdieren,* Elsevier, Amsterdam and Brussel.

Hambolu, M.O. (2000). *Studies in the Late Stone Age archaeology of the southwest Chad Basin*, Unpublished Ph.D. thesis, University of Maiduguri, Maiduguri.

Hanotte, O., Bradley, D.G., Ochieng, J.W., Verjee, Y., Hill, E.W. and Rege, J.E.O. (2002). African pastoralism: genetic imprints of origins and migrations. *Science* **296**: 336-339.

Hanotte, O., Tawah, C.L., Bradley, D.G., Okomo, M., Verjee, Y., Ochieng, J.W. and Rege, J.E.O. (2000). Geographic distribution and frequency of a taurine (*Bos taurus*) and an indicine (*Bos indicus*) Y specific allele amongst sub-saharan African cattle breeds. *Molecular Ecology* **9**: 387-396.

Happold, D.C.D. (1987). *The mammals of Nigeria*, Clarendon Press, Oxford.

Harlan, J.R. (1971). Agricultural origins: centers and noncenters. *Science* **174**: 468-474.

Harris, D.R. (1996). *The origins and spread of agriculture and pastoralism in Eurasia*, UCL Press, London.

Harrison, D.L. and Bates, P.J.J. (1991). *The mammals of Arabia* (2nd edition), Kent, Harrison Zoological Museum.

Hassan, F. (1985). Radiocarbon chronology of Neolithic and Predynastic sites in Upper Egypt and the Delta. *African Archaeological Review* **3**: 95-116.

Hassan, F.A. (1996). Abrupt Holocene climatic events in Africa. In Pwiti, G. and Soper, R. (eds), *Aspects of African archaeology. Papers from the 10th congress of the PanAfrican Association for Prehistory and Related Studies*, University of Zimbabwe publications, Harare, pp. 83-89.

Hassan, F.A. (2000). Climate and cattle in North Africa: a first approximation. In Blench, R.M. and MacDonald, K.C. (eds), *The origins and development of African livestock: archaeology, genetics, linguistics and ethnograpy*, UCL Press, London, pp. 61-86.

Hassan, F.A. (2002). Palaeoclimate, food and culture change in Africa: an overview. In Hassan, F.A. (ed.), *Droughts, food and culture. Ecological change and food security in Africa's later prehistory*, Kluwer Academic/Plenum Publishers, New York/Boston/Dordrecht/London/Moscow, pp. 11-26.

Heinzel, H., Fitter, R. and Parslow, J. (1982). *Elseviers gids van de Europese vogels. Alle vogels van Europa, Noord-Afrika en het Midden-Oosten*, Elsevier, Amsterdam and Brussel.

Helmer, D. (2000). Discrimination des genres *Ovis* et *Capra* à l'aide des prémolaires inférieures 3 et 4 et interprétation des âges d'abattage: l'exemple de Dikili Tash (Grèce). *Anthropozoologica* **31**: 29-38.

Hitchcock, R.K. (1999). Introduction: Africa. In Lee, R.B. and Daly, R. (eds), *The Cambridge encyclopedia of hunters and gatherers*, Cambridge University Press, Cambridge, pp. 175-184.

Hoelzmann, P., Gasse, F., Dupont, L.M., Salzmann, U., Staubwasser, M., Leuschner, D.C. and Sirocko, F. (2004). Palaeoenvironmental changes in the arid and subarid belt (Sahara-Sahel-Arabian Peninsula) from 150 kyr to present. In Batterbee, W., Gasse, F. and Stickley, C.E. (eds), *Past climate variability through Europe and Africa*, Springer, Dordrecht, pp. 219-256.

Höhn, A. (2002). Vegetation changes in the Sahel of Burkina Faso (West Africa) – Analysis of charcoal from the Iron Age sites Oursi and Oursi-Village. In Thiébault, S. (ed.), *Charcoal analysis. Methodological approaches, palaeoecological results and wood uses*, BAR International Series 1063, Oxford, pp. 133-139.

Höhn, A., Kahlheber, S. and Hallier-Czerniewicz, M. von (2004). Den frühen Bauern auf der Spur – Siedlungs und Vegetationsgeschichte der Region Oursi (Burkina Faso). In Albert, K.-D., Löhr, D. and Neumann, K. (eds), *Mensch und Natur in Westafrika*, Abschlussbuch des Sonderforschungsbereichs 268, Wiley-VCH Verlag, Weinheim, pp. 221-255.

Holl, A. (1985). Subsistence patterns of the Dhar Tichitt Neolithic, Mauritania. *African Archaeological Review* **3**: 151-162.

Holl, A. (1986). *Economie et société néolithique du Dhar Tichitt (Mauritanie)*, Mémoire n° 69, Editions Recherche sur les Civilisations, Paris.

Holl, A. (1988). *Houlouf I. Archéologie des sociétés protohistoriques du Nord-Cameroun*, BAR International Series 456, Oxford.

Holl, A. (2002). *The land of Houlouf: The genesis of a Chadic polity, 1900 B.C.-A.D.1800*, Memoirs 35, Museum of Anthropology, University of Michigan, Ann Arbor, Michigan.

Holl, A. (2003). *Ethnoarchaeology of Shuwa-Arab settlements*, Lexington books, Lanham.

Holl, A. (2005). Holocene "aquatic" adaptations in north tropical Africa. In Stahl, A.B. (ed.), *African archaeology. A critical introduction*, Blackwell, Malden, pp. 174-186.

Holl, A., Lévy, T., Lechevalier, C. and Bridault, A. (1991). Of men, mounds and cattle: Archaeology and ethnoarchaeology in the Houlouf region (northern Cameroun). *West African Journal of Archaeology* **21**: 7-36.

Hopson, J. (1967). The fisheries of Lake Chad. In Reed, W., Burchard, J., Hopson, A.J., Jenness, J. and Yaro, I. (eds), *Fish and fisheries of northern Nigeria*, Ministry of Agriculture Northern Nigeria, Gaskiya Corporation, Zaria, pp. 189-200.

Hoyo, J. del, Elliott, A. and Sargatal, J. (1992). *Handbook of the birds of the world, Vol. 1*, Lynx Edicions, Barcelona.

Insoll, T. (1995). A cache of hippopotamus ivory at Gao, Mali; and a hypothesis of its use. *Antiquity* **69**: 327-336.

Insoll, T. (1996). *Islam, archaeology and history. Gao Region (Mali) ca. 900-1250*, BAR International Series 647, Oxford.

Insoll, T. (1999). *The archaeology of Islam*, Blackwell, Oxford.

Insoll, T. (2000). *Urbanism, archaeology and trade: further observations on the Gao region (Mali): the 1996 fieldseason results*, BAR International Series 829, Oxford.

Iverson, J.B. (1992). *A revised checklist with distribution maps of the turtles of the world*, Privately Printed, Richmond, Indiana.

Jansen, T., Forster, P., Levine, M.A., Oelke, H., Hurles, M., Renfrew, C., Weber, J. and Olek, K. (2002). Mitochondrial DNA and the origins of the domestic horse. *Proceedings of the National Academy of Sciences of the United States of America* **99**(16): 10905-10910.

Jemkur, J.F. (2004). The beginnings of iron metallurgy in West Africa. In Bocoum, H. (ed.), *The origins of iron metallurgy in Africa. New light on its antiquity in West and Central Africa*, Unesco publishing, Paris, pp. 33-42.

Jesse, F., Kröpelin, S., Lange, M., Pöllath, N. and Berke, H. (2004). On the periphery of Kerma – The Handessi horizon in Wadi Hariq, Northwestern Sudan. *Journal of African Archaeology* **2**(2): 123-164.

Johnson, M. (1970). The cowrie currencies of West Africa (Part I). *Journal of African History* **11**: 17-49.

Jolly, D., Harrison, S.P., Damnati, B. and Bonnefille, R. (1998). Simulated climate and biomes of Africa during the late Quaternary: comparison with pollen and lake status data. *Quaternary Science Reviews* **17**: 629-657.

Jones, K.S. (1997). Refuse disposal and mound formation: an ethnoarchaeological approach in northern Cameroon. *Nyame Akuma* **48**: 11-19.

Jousse, H. (2004a). *Impact des variations environnementales sur la structure des communautés mammaliennes et l'anthropisation des milieux: exemple des faunes holocènes du Sahara occidental*, Documents des Laboratoires de Géologie de Lyon 160, U.F.R. des Sciences de la Terre, Université de Claude-Bernard, Lyon 1.

Jousse, H. (2004b). A new contribution to the history of pastoralism in West Africa. *Journal of African Archaeology* **2**(2): 187-201.

Jousse, H. (2006). What is the impact of Holocene climatic changes on human societies? Analysis of Neolithic populations dietary customs. *Quaternary Science Reviews* **151**: 63-73.

Jousse, H. and Chenal-Vélardé, I. (2001-2002). Nouvelles données sur la faune mammalienne de Kobadi (Mali) au Néolithique: implications paléoéconomiques et paléoenvironnementales. *Préhistoire Anthropologie méditerranéennes* **10-11**: 145-158.

Kahlheber, S. (1999). Indications for agroforestery. Archaeobotanical remains of crops and woody plants from medieval Saouga, Burkina Faso. In van der Veen, M. (ed.), *The exploitation of plant resources in ancient Africa*, Kluwer Academic/Plenum Publishers, New York, pp. 89-100.

Kahlheber, S. (2003). *Perlhirse und Baobab – Archäobotanische Untersuchungen im Norden Burkina Fasos*. Ph.D. thesis, Johann Wolfgang Goethe-University, Frankfurt am Main. http://publikationen.ub.uni-frankfurt.de/volltexte/2005/561/

Kahlheber, S. (submitted). Archaeobotanical remains of Early Iron Age sites in the Bama deltaic complex of NE Nigeria. *Journal of African Archaeology*.

Kahlheber, S., Albert, K.-D. and Höhn, A. (2001). A contribution to the palaeoenvironment of the archaeological site Oursi in North Burkina Faso. In Kahlheber, S. and Neumann, K. (eds), Man and environment in the West African Sahel – an interdisciplinary approach. *Berichte des Sonderforschungsbereichs 268* **17**: 145-159.

Kahlheber, S. and Neumann, K. (in press). The development of plant cultivation in semi-arid West Africa. In Denham, T., Iriarte, J. and Vrydaghs, L. (eds), *Rethinking agriculture: archaeological and ethnoarchaeological perspectives*, UCL Press, London.

Keepax, C.A. (1981). Avian egg-shell from archaeological sites. *Journal of Archaeological Science* **8**: 315-335.

Kingdon, J. (1971). *East African mammals. An atlas of evolution in Africa, Vol. 1*, Academic Press, London and New York.

Kingdon, J. (1974). *East African mammals. An atlas of evolution in Africa, Vol. 2 A-B*, Academic Press, London and New York.

Kingdon, J. (1977). *East African mammals. An atlas of evolution in Africa, Vol. 3 A*, Academic Press, London/New York/San Fransisco.

Kingdon, J. (1979). *East African mammals. An atlas of evolution in Africa, Vol. 3 B*, Academic Press, London/New York/San Fransisco.

Kingdon, J. (1982). *East African mammals. An atlas of evolution in Africa, Vol. 3 C-D*, Academic Press, London/New York/San Fransisco.

Kingdon, J. (1997). *The Kingdon field guide to African mammals*, Academic Press Limited, London.

Klee, M. and Zach, B. (1999). The exploitation of wild and domestic food plants at settlement mounds in North-East Nigeria (1800 cal BC to today). In van der Veen, M. (ed.), *The exploitation of plant resources in ancient Africa*, Kluwer Academic/Plenum, New York, pp. 81-88.

Klee, M., Zach, B. and Neumann, K. (2000). Four thousand years of plant exploitation in the Chad Basin of northeast Nigeria, part I: The archaeobotany of Kursakata. *Vegetation History and Archaeobotany* **9**: 223-237.

Klee, M., Zach, B. and Stika, H.-P. (2004). Four thousand years of plant exploitation in the Lake Chad Basin (Nigeria), part III: Plant impressions in potsherds from the Final Stone Age Gajiganna Culture. *Vegetation History and Archaeobotany* **13**: 131-142.

Klein, R.G. and Cruz-Uribe, K. (1984). *The analysis of animal bones from archaeological sites,* Prehistoric Archaeology and Ecology Series, The University of Chicago Press, Chicago.

Kottusch, R. (1999). *Die Knochenartefakte der endsteinzeitlichen Gajiganna-Kultur Nordost-Nigerias*, Unpublished M.A. thesis, Johann Wolfgang Goethe-University, Frankfurt am Main.

Krings, M. (2004). Fishing for Naira. In Krings, M. and Platte, E. (eds), *Living with the lake: perspectives on history, culture and economy of Lake Chad,* Studien zur Kulturkunde 121, Rüdiger Köppe Verlag, Köln, pp. 204-226.

Krings, T.F. (1980). *Kulturgeographischer Wandel in der Kontaktzone von Nomaden und Bauern im Sahel von Obervolta. Am Beispiel des Oudalan (Nordost-Obervolta)*, Hamburger Geographische Studien 36, Ferdinand Hirt, Hamburg.

Krzyzaniak, L. (1992). The later prehistory of the Upper (main) Nile: comments on the current state of research. In Klees, F. and Kuper, R. (eds), *New light on the northeast African past. Current prehistoric research,* Africa Praehistoria 5, Köln, pp. 238-248.

Kühltrunk, P. (2000). *Taphonomische und typologische Keramikanalyse des steinzeitlichen Fundplatzes Tin-Akoff im Norden Burkina Fasos*, Unpublished M.A. thesis, Johann Wolfgang Goethe-University, Frankfurt am Main.

Kühltrunk, P. (in prep.). Céramiques de l'Age de la Pierre récent du Nord Burkinabé.

Lagercrantz, S. (1950). *Contributions to the ethnology of Africa,* Studia Ethnographica Upsaliensia 1, Håkan Ohlssons, Lund.

Lambrecht, C.L. (1964). Aspects of evolution and ecology of tsetse flies and trypanosomiasis in prehistoric African environments. *Journal of African History* **5**: 1-24.

Lambrecht, S. (1997). *Archeozoölogische studie van het site Mege (1e millenium BC – 1983 AD). Aanduidingen voor de eerste veeteelt in het ZW-Tchaadbekken*, Unpublished M.A. thesis, Catholic University Leuven, Leuven.

Lamotte, M. and Xavier, F. (1981). Amphibiens. In Durand, J.-R. and Lévêque, C. (eds), *Flore et faune aquatiques de l'Afrique Sahelo-Soudanienne, Tome II*, ORSTOM, Paris, pp. 773-816.

Law, R. (1980). *The horse in West African history: the role of the horse in the societies of pre-colonial West Africa*, Oxford University Press, for International Africa Institute, London.

Law, R. (1995). The horse in pre-colonial West-Africa. In Pezzoli, G. (ed.), *Cavalieri dell'Africa*, Centro studi archaeologia Africana, Milano, pp. 175-184.

Lawson, R.M. (1963). The economic organisation of the *Egeria radiata* fishing industry on the River Volta. *Proceedings of the Malacological Society London* **35**: 273-287.

Lebeuf, J.-P. (1962). *Archéologie tchadienne. Les Sao du Cameroun et du Tchad*, Hermann, Paris.

Lebeuf, J.-P. (1969). *Carte archéologique des abords du Lac Tchad (Cameroun, Nigeria, Tchad)*, Centre National de Recherche Scientifique, Paris.

Lebeuf, J.-P. (1976). *Etudes Kotoko*, Mouton, Paris.

Lebeuf, J.-P., Lebeuf, A.M.D., Treinen-Claustre, F. and Courtin, J. (1980). *Le gisement Sao de Mdaga (Tchad). Fouilles 1960-1968*, Société d'ethnographie, Paris.

Leeuw, P.N. de, Lesslie, A. and Tuley, P. (1972). *The land resources of North East Nigeria, Vol. 4, Present and potential landuse*, Land Resource Study 9, Surbiton.

Le Houérou, H.N. (1989). *The grazing land ecosystems of the African Sahel*, Ecological Studies 75, Springer, Berlin/Heidelberg/New York.

Lepetit, P. (1989). Cypraeidae of the Mediterranean and Western Africa. *Xenophora* **46**: 4-8.

Le Quellec, J.L. (1999). Répartition de la grande faune sauvage dans le nord de l'Afrique durant l'Holocène. *L'Anthropologie* **103**(1): 161-176.

Lernia, S. di (2002). Dry climatic events and cultural trajectories: adjusting Middle Holocene pastoral economy of the Libyan Sahara. In Hassan, F.A. (ed.), *Droughts, food and culture. Ecological change and food security in Africa's later Prehistory*, Kluwer Academic/Plenum Publishers, New York/Boston/Dordrecht/London/ Moscow pp. 225-250.

Lesur, J. (2004). *Exploitation de la diversité faunique et début du pastoralisme dans la Corne de L'Afrique à l'Holocène*, unpublished Ph.D. thesis, Université de Paris I/Panthéon – Sorbonne, Paris.

Lesur, J. and Langlois, O. (2002). Une communauté d'"Agro-chasseurs" au XIXe siècle dans la Haute Bénoué: analyse du matériel archéozoologique extrait du site Hossere-Djaba. *Actes du colloque Méga-Tchad, « Ressources vivrières et choix alimentaires dans le bassin du lac Tchad », Nanterre, 20-22 novembre 2002.* htttp://www.Uni-bayreuth.de/afrikanistik/mega-tchad/ Table/Colloque2002/Lesur.pdf

Lévêque, C. (1972). Mollusques benthiques du Lac Tchad: écologie, étude des peuplements et estimation des biomasses. *Cahiers ORSTOM. Série Hydrobiologie* **6**: 3-45.

Lévêque, C. (1990). Relict tropical fish fauna in central Sahara. *Ichthyological Exploration of Freshwaters* **1**: 39-48.

Lévêque, C. and Paugy, D. (2003). Protopteridae. In Paugy, D., Lévêque, C. and Teugels, G. (eds), *Poissons d'eaux douces et saumâtres de l'Afrique de l'Ouest. The fresh and brackish water fishes of West Africa, Vol. 1*, Collection Faune et Flore tropicales 40, Editions de l'IRD, Paris, pp. 97-100.

Lévêque, C. and Quensière, J. (1988). Les peuplements ichtyologiques des lacs peu profonds. In Lévêque, C., Bruton, M.N. and Ssentongo, G.W. (eds), *Biology and ecology of African freshwater fishes*, ORSTOM, Paris, pp. 303-324.

Levtzion, N. and Hopkins, J. (1981). *Corpus of early Arabic sources for West African history*, Cambridge University Press, Cambridge.

Lewicki, T. (1974). *West African food in the Middle Ages*, Cambridge University Press, London.

Lézine, A.M. and Casanova, J. (1989). Pollen and hydrological evidence for the interpretation of past climates in tropical West Africa during the Holocene. *Quaternary Science Reviews* **8**: 45-55.

Linares de Sapir, O. (1971). Shell middens of lower Casamance and problems of Diola protohistory. *West African Journal of Archaeology* **1**: 23-54.

Linseele, V. (2003). Cultural identity and the consumption of dogs in western Africa. In O'Day, S.J., Van Neer, W. and Ervynck, A. (eds), *Behaviour behind bones. The zooarchaeology of ritual, religion, status and identity*, Oxbow Books, Oxford, pp. 318-326.

Linseele, V. (2004). Size and size change of the African aurochs. *Journal of African Archaeology* **2**(2): 165-185.

Linseele, V. (2005). Humans and animals in the southern Lake Chad area during the 2nd millennium AD. An archaeozoological case study from Ngala, Borno State, Nigeria. *Proceedings of the Mega Chad Conference "Man and the Lake" Maiduguri, Borno State, Nigeria, 2-10 December 2003,* pp. 21-40.

Linseele, V. (in press). The animal remains. In Petit, L., Hallier, M., Pelzer, C. (eds), *Oursi Hu-beero.*

Loftus, R. and Cunningham, P. (2000). Molecular genetic analysis of african zeboid populations. In Blench, R.M. and MacDonald, K.C. (eds), *The origins and development of African livestock: archaeology, genetics, linguistics and ethnography*, UCL Press, London, pp. 251-258.

Loonbeek, L. (2001). *Archeozoölogische studie van een aantal sites uit het 4e en 3e millenium BP in Noordoost-Nigeria*, Unpublished M.A. thesis, Catholic University Leuven, Leuven.

Lyman, R.L. (1994). *Vertebrate taphonomy*, Cambridge Manuals in Archaeology, Cambridge University Press, Cambridge.

MacDonald, K.C. (1992). The domestic chicken (*Gallus gallus*) in Sub-Saharan Africa: a background to its introduction and its osteological differentiation from indigenous fowls (Numidinae and *Francolinus* sp.). *Journal of Archaeological Science* **19**: 303-318.

MacDonald, K.C. (1994). *Socio-economic diversity and the origins of cultural complexity along the Middle Niger (2000 BC to AD 300)*, Unpublished Ph.D. thesis, University of Cambridge, Cambridge.

MacDonald, K.C. (1995). Analysis of the mammalian, avian and reptilian remains. In McIntosh, S.K. (ed.), *Excavations at Jenné-Jeno, Hambarketolo, and Kaniana (Inland Niger Delta), the 1981 Season*, Anthropology 20, University of California Press, Berkeley and Los Angeles, pp. 291-318.

MacDonald, K.C. (1996). The Windé Koroji Complex: Evidence for the peopling of the eastern Inland Niger Delta (2100-500 BC). *Préhistoire Anthropologie méditerranéennes* **5**: 147-165.

MacDonald, K.C. (1997a). Korounkorokalé revisited: The pays Mande and the West African microlithic technocomplex. *African Archaeological Review* **14**: 161-200.

MacDonald, K.C. (1997b). The Late Stone Age and Neolithic cultures of West Africa and the Sahara. In Vogel, J.O. (ed.), *Encyclopedia of precolonial Africa: archaeology, history, languages, cultures and environments*, Altamira Press, Walnut Creek/London/New Delhi, pp. 394-408.

MacDonald, K.C. (1999). Invisible pastoralists: an inquiry into the origins of nomadic pastoralism in the West African Sahel. In Gosden, C. and Hather, J. (eds), *The prehistory of food. Appetites for change*, Routledge, London, pp. 147-165.

MacDonald, K.C. and MacDonald, R.H. (2000). The origins and development of domesticated animals in arid West-Africa. In Blench, R.M. and MacDonald, K.C. (eds), *The origins and development of African livestock: archaeology, genetics, linguistics and ethnography*, UCL Press, London, pp. 127-162.

MacDonald, K.C. and Van Neer, W. (1993). An initial report on the fauna of Akumbu (Mali). Appendix to Togola, T., *Archaeological investigations of Iron Age sites in the Méma region, Mali (West Africa)*, unpublished Ph.D. thesis, Rice University, Houston.

MacDonald, K.C. and Van Neer, W. (1994). Specialised fishing peoples in the later Holocene of the Méma Region (Mali). In Van Neer, W. (ed.), *Fish exploitation in the past, Proceedings of the 7th meeting of the ICAZ Fish Remains Working Group*, Annales du Musée royal de l'Afrique centrale, Sciences Zoologiques 274, Tervuren, pp. 243-251.

MacDonald, K.C., Togola, T., MacDonald, R.H. and Capezza, C. (1994). Douentza, Mali. *Past* **17**: 12-14.

MacDonald, R.H. and MacDonald, K.C. (1996). A preliminary report on the faunal remains recovered from Gao Ancien and Gao Saney (1993 season). In Insoll, T. (ed.), *Islam, archaeology and history: Gao Region (Mali) c. AD 900-1250,* BAR International Series 647, Oxford, pp. 124-126.

MacEachern, S. (1996). Iron Age beginnings north of the Mandara Mountains, Cameroon and Nigeria. In Pwiti, G. and Soper, R. (eds), *Aspects of African archaeology. Papers from the 10th congress of the PanAfrican Association for Prehistory and Related Studies*, University of Zimbabwe publications, Harare, pp. 489-495.

MacEachern, S., Bourges, C. and Reeves, M. (2001). Early horse remains from Northern Cameroon. *Antiquity* **75**(1): 62-67.

MacHugh, D.E., Shriver, M.D., Loftus, R.T., Cunningham, P. and Bradley, D.G. (1997). Microsatellite DNA variation and the evolution, domestication and phylogeography of taurine and zebu cattle (*Bos taurus* and *B. indicus*). *Genetics* **146**: 1071-1086.

MacLean, R. and Insoll, T. (1999). The social context of food in Iron Age Gao, Mali. *World Archaeology* **31**: 78-92.

Magnavita, C. (2002). Recent archaeological finds of domesticated *Sorghum bicolor* in the Lake Chad region. *Nyame Akuma* **57**: 14-20.

Magnavita, C. (2003). *Studien zur endsteinzeitlichen und früheisenzeitlichen Besiedlung im südwestlichen Tschadbecken (1300 BC-700 AD)*, unpublished Ph.D. thesis, Johann Wolfgang Goethe-University, Frankfurt am Main.

Magnavita, C. (in press). Ancient humped cattle in Africa: a view from the Chad Basin. *African Archaeological Review.*

Magnavita, C., Kahlheber, S. and Eichhorn, B. (2004). The rise of organisational complexity in mid-first millennium BC Chad Basin. *Antiquity* **78**(301). http://antiquity.ac.uk/ProjGall/magnavita/

Magnavita, C. and Magnavita, S. (2001). New evidence for proto-urban settlements in the Lake Chad area. *Nyame Akuma* **55**: 46-50

Magnavita, S., Hallier, M., Pelzer, C., Kahlheber, S. and Linseele, V. (2002). Nobles, guerriers, paysans. Une nécropole de l'Age de Fer et son emplacement dans l'Oudalan pré- et protohistorique. *Beiträge zur allgemeinen und vergleichenden Archäologie* **22**: 21-64.

Maley, J. (1973). Les variations climatiques dans le bassin du Tchad durant le dernier millénaire: essai d'interprétation climatique de l'Holocène africain. *Comptes rendus de l'Académie des Sciences, Série D* **276**: 1673-1675.

Maley, J. (1981). Etudes palynologiques dans le bassin du Tchad et paléoclimatologie de l'Afrique nord-tropicale de 30000 ans à l'époque actuelle. *Palaeoecology of Africa* **13**: 45-52.

Maley, J. (1996) The African rain-forest – main characteristics of changes in vegetation and climate change from the Upper Cretaceous to the Quaternary. *Proceedings of the Royal Society of Edinburgh B* **104**: 31-73.

Manning, K. and MacDonald, K.C. (2005). Analyse des restes d'animaux collectés à Dia. In Bedaux, R., Polet, J. Sanogo, K. and Schmidt, A. (eds), *Recherches archéologiques à Dia dans le Delta intérieur du Niger (Mali): bilan des saisons de fouilles 1998-2003.* Mededelingen van het Rijksmuseum voor Volkenkunde, Leiden no. 33, pp. 363-385

Marchi, S. (1998). Sub-Saharan rock art. *TRACCE* 10. http://rupestre.net/tracce/subsaha.html

Marliac, A. (1991). *De la préhistoire à l'histoire au Cameroun septentrional* (2 vols.), Collections Etudes et Thèses, Editions de l'ORSTOM, Paris.

Marshall, F. (1990). Cattle herds and caprine flocks. In Robertshaw, P., *Early pastoralists of south-western Kenya,* British Institute in Eastern Africa Memoir 11, Nairobi, pp. 205-241.

Marshall, F. (1998). Early food production in Africa. In Bar-Yosef, O. (ed.), The transition to agriculture in the Old World. *Review of Archaeology* **19**(2): 47-58.

Marshall, F. (2000). The origins and spread of domestic animals in East Africa. In Blench, R.M. and MacDonald, K.C. (eds), *The origins and development of African livestock: archaeology, genetics, linguistics and ethnography*, UCL Press, London, pp. 191-221.

Marshall, F. and Hildebrand, E. (2002). Cattle before crops: the beginnings of food production in Africa. *Journal of World Prehistory* **16**(2): 99-143.

Marshall, F. and Stewart, K. (1994). Hunting, fishing and herding pastoralists of western Kenya: the fauna from Gogo Falls. *Archaeozoologia* **7**(1): 7-27.

Masonen, P. (1997). Trans-Saharan trade and the West African discovery of the Mediterranean. In Sabour, M. and Vikør, K.S. (eds), *Ethnic encounter and cultural change. Papers from the third Nordic conference on Middle Eastern Studies*, Nordic Research on the Middle East 3, Hurst and Co Publishers, London, pp. 116-142.

Mbida, C.M., Van Neer, W., Doutrelepont, H. and Vrydaghs, L. (2000). Evidence for banana cultivation and animal husbandry during the first millennium BC in the forest of southern Cameroon. *Journal of Archaeological Science* **27**: 151-162.

Mbida, C.M., Doutrelepont, H., Vrydaghs, L., Swennen, R.L., Swennen, R.J., Beeckman, H., de Langhe, E. and de Maret, P. (2001). First archaeological evidence of banana plant cultivation in central Africa during the third millennium before present. *Vegetation History and Archaeobotany* **10**: 1-6.

McIntosh, R.J. (1993). The Pulse Model: genesis and accommodation of specialization in the Middle-Niger. *Journal of African History* **34**: 181-220.

McIntosh, R.J. (1997). Agricultural beginnings in sub-Saharan Africa. In Vogel, J.O. (ed.), *Encyclopedia of precolonial Africa: archaeology, history, languages, cultures and environments*, Altamira Press, Walnut Creek/London/New Delhi, pp. 409-418.

McIntosh, R.J. and McIntosh, S.K. (1995). Stratigraphy, features and chronology. In McIntosh, S.K. (ed.), *Excavations at Jenné-Jeno, Hambarketolo, and Kaniana (Inland Niger Delta), the 1981 Season*, Anthropology 20, University of California Press, Berkeley and Los Angeles, pp. 27-129.

McIntosh, S.K. (1994). Changing perceptions of West Africa's past: archaeological research since 1988. *Journal of Archaeological Research* **2**(2): 165-198.

McIntosh, S.K. (1995). *Excavations at Jenné-Jeno, Hambarketolo, and Kaniana (Inland Niger Delta), the*

1981 Season, Anthropology 20, University of California Press, Berkeley and Los Angeles.

McIntosh, S.K. (1999). A tale from two floodplains: comparative perspectives on the emergence of complex societies and urbanism in the Middle Niger and Senegal Valleys. In Sinclair, P. (ed.), *Proceedings of the Second World Archaeological Congress Intercongress*, Mombasa. http://www.arkeologi.uu.se/afr/projects/BOOK/Mcintosh/mcintosh.htm

McIntosh, S.K. (2001). West African Neolithic. In Peregrine, P.N. and Ember, M. (eds), *Encyclopedia of prehistory, Vol. 1, Africa*, Kluwer Academic/Plenum, New York/Boston/Dordrecht/London/Moscow, pp. 323-338.

McIntosh, S.K. and Bocoum, H. (2000). New perspectives on Sincu Bara, a first millenium site in the Senegal Valley. *African Archaeological Review* **17**: 1-43.

Meadow, R.H. (1980). Animal bones: problems for the archaeologist together with some possible solutions. *Paléorient* **6**: 65-77.

Meadow, R.H. (1993). Animal domestication in the Middle East: a revised view from the Eastern margin. In Possehl, G. (ed.), *Harappan civilization*, Oxford & IBH, New Delhi, pp. 295-320.

Meadow, R.H. (1999). The use of size index scaling techniques for research on archaeozoological collections from the Middle East. In Becker, C., Manhart, H., Peters, J. and Schibler, J, (eds), *Historia animalium ex ossibus: Beiträge zur Paläoanatomie, Archäologie, Ägyptologie, Ethnology und Geschichte der Tiermedizin,* Festschrift für A. von den Driesch zum 65. Geburtstag, Internationale Archäologie, Studia Honaria 8, Leidorf, Rahden/Westfalen, pp. 285-300.

Meghen, C., MacHugh, D.E. and Bradley, D.G. (1994). Genetic characterization and West African cattle. *World Animal Review* **78**: 59-66.

Meghen, C., MacHugh, D.E., Sauveroche, B., Kana, G. and Bradley, D.G. (2000). Characterization of the Kuri cattle of Lake Chad using molecular genetic techniques. In Blench, R.M. and MacDonald, K.C. (eds), *The origins and development of African livestock: archaeology, genetics, linguistics and ethnography*, UCL Press, London, pp. 259-268.

Mertens, R. (1942). Die Familie der Warane (Varanidae). *Abhandlungen der Senckenbergischen naturforschenden Gesellschaft* **462, 465, 466**: 1-391.

Milner, N. (2000). The marine and freshwater mollusks. In Insoll, T. (ed.), *Urbanism, archaeology and trade: further observations on the Gao region (Mali): the 1996 fieldseason results*, BAR International Series 829, Oxford, pp. 36-38.

Mones, H. (1992). The conquest of North Africa and Berber resistance. In Hrbek, I. (ed.), *Africa from the seventh to eleventh century*, General history of Africa, Abridged Edition, Unesco Publishing, Paris, James Currey, London, California University Press, Berkeley, pp. 118-129.

Mongin, P. and Plouzeau, M. (1984). Guinea-fowl. In Mason, I.L. (ed.), *Evolution of domesticated animals*, Longman, London, pp. 322-325.

Morczinek, I. (1995). *Diatomeen aus dem Mare d'Oursi. Ein Beitrag zur holozänen Paläoökologie des westafrikanischen Sahel*, unpublished B.Sc. thesis, Johann Wolfgang Goethe-University, Frankfurt am Main.

Moreno-García, M. (2003). Hunting practices and consumption patterns in rural communities in the Rif mountains (Morocco) – some ethno-zoological notes. In O'Day, S.J., Van Neer, W. and Ervynck, A. (eds), *Behaviour behind bones. The zooarchaeology of ritual, religion, status and identity*, Oxbow Books, Oxford, pp. 327-334.

Muzzolini, A. (1995). *Les images rupestres du Sahara*, Collection: Préhistoire du Sahara 1, Edité par l'auteur, Toulouse.

Muzzolini, A. (2000). Livestock in Saharan rock art. In Blench, R.M. and MacDonald, K.C. (eds), *The origins and development of African livestock: archaeology, genetics, linguistics and ethnography*, UCL Press, London, pp. 87-110.

Neiland, A.E. (1992). *Artisanal fisheries of the Chad Basin in Africa. Overview of research and annotated bibliography 1920–1990*, CIFA Occasional Paper 18. http://www.fao.org/documents/show_cdr.asp?url_file=/DOCREP/005/U7989E/U7989E00.HTM

Neumann, K. (2003). The late emergence of agriculture in Sub-Saharan Africa: archaeobotanical evidence and ecological considerations. In Neumann, K., Butler, A. and Kahlheber, S. (eds), *Food, fuel and fields*, Progress in African Archaeobotany, Africa Praehistoria 15, Heinrich-Barth-Institut, Köln, pp. 71-92.

Neumann, K. (2005). The romance of farming: plant cultivation and domestication in Africa. In Stahl, A.B. (ed.), *African archaeology. A critical introduction*, Blackwell, Malden, pp. 249-275.

Neumann, K., Breunig, P. and Kahlheber, S. (2000). Early food production in the Sahel of Burkina Faso. *Berichte des Sonderforschungsbereichs 268* **14**: 327-334.

Neumann, K., Hahn-Hadjali, K. and Salzmann, U. (2004). Die Savannen der Sudanzone in Westafrika – natürlich oder menschengemacht? In Albert, K.-D., Löhr, D. and Neumann, K. (eds), *Mensch und Natur in Westafrika*, Abschlussbuch des Sonderforschungs-bereichs 268, Wiley-VCH Verlag, Weinheim, pp. 39-68.

Neumann, K., Kahlheber, S. and Uebel, D. (1998). Remains of woody plants from Saouga, a medieval West African village. *Vegetation History and Archaeobotany* **7**: 57-77.

Newman, J.L. (2000). Africa south from the Sahara. In Kiple, K.F. and Ornelas, K.C. (eds), *The Cambridge world history of food* (2 vols.), Cambridge University Press, Cambridge, pp. 1330-1339.

Nicholson, S.E. (1980). Saharan climate in historic times. In Williams, M.A.J. and Faure, H. (eds), *The Sahara and the Nile. Quaternary environments and prehistoric occupation in northern Africa*, A.A. Balkema, Rotterdam, pp. 173-200.

Nicolotti, M. and Guérin, C. (1992). Le zébu (*Bos indicus*) dans l'Egypte ancienne. *Archaeozoologia* **5**(1): 87-108.

Ntiamoa-Baidu, Y. (1997). *Wildlife and food security in Africa*, FAO Conservation Guide 33. http://www.fao.org/documents/show_cdr.asp?url_file=/docrep/w7540e/w7540e00.htm

O'Connor, T. (2000). *The archaeology of animal bones*, Sutton Publishing, Stroud.

Olaosebikan, B.D. and Raji, A. (1998). *Field guide to Nigerian freshwater fishes,* Federal College of Freshwater Fisheries Technology, New Bussa, Nigeria.

Osborn, D.J. and Helmy, I. (1980). *The contemporary land mammals of Egypt (including Sinai)*, Fieldiana (Zoology) 5, The Field Museum of Natural History, Chicago.

Paris, F. (1984). *La région d'In Gall – Tegidda N Tesemt (Niger): Programme archéologique d'urgence 1977-1981. III. Les Sépultures du Néolithique Final à l'Islam*, Etudes Nigériennes 50, Institut de Recherches en Sciences Humaines, Niamey.

Paris, F. (1997). Les inhumations de *Bos* au Sahara méridional au Néolithique. *Archaeozoologia* **9**: 113-122.

Paris, F. (2000). African livestock remains from Saharan mortuary contexts. In Blench, R.M. and MacDonald, K.C. (eds), *The origins and development of African livestock: archaeology, genetics, linguistics and ethnography*, UCL Press, London, pp. 111-126.

Paugy, D., Lévêque, C. and Teugels, G. (2003), *Poissons d'eaux douces et saumâtres de l'Afrique de l'Ouest. The fresh and brackish water fishes of West Africa* (2 vols.), Faune et Flore tropicales 40, Editions de l'IRD, Paris.

Payne, W.J.A. (1964). The origin of domestic cattle in Africa. *Empire Journal of Experimental Agriculture* **32**: 97-113.

Payne, S. (1969). A metrical distinction between sheep and goat metacarpals. In Ucko, P. and Dimbleby, G.W. (eds), *The domestication of plants and animals*, Gerald Duckworth and Co., London, pp. 295-305.

Payne, S. (1972). Partial recovery and sample bias: the results of some sieving experiments. In Higgs, E.S. (ed.), *Papers in economic prehistory*, Cambridge University Press, Cambridge, pp. 49-64.

Payne, S. (1985). Morphological distinctions between the mandibular teeth of young sheep, *Ovis* and goats, *Capra*. *Journal of Archaeological Science* **12**: 139-147.

Payne, S. and Munson, P.J. (1985). Ruby and how many squirrels? The destruction of bones by dogs. In Fieller, N.R.J., Gilbertson, D.D. and Ralph, N.G.A. (eds), *Palaeobiological investigations. Research design, methods and data analysis*, BAR International Series 266, Oxford, pp. 31-39.

Pélissier, P. (1980). L'arbre dans les paysages agraires de l'Afrique noire. *Cahiers ORSTOM* **17**(3-4): 131-136.

Pelzer, C., Müller, J. and Albert, K.-D. (2004). Die Nomadisierung des Sahel. Siedlungsgeschichte, Klima und Vegetation in der Sahelzone Burkina Fasos in historischer zeit. In Albert, K.-D., Löhr, D. and Neumann, K. (eds), *Mensch und Natur in Westafrika*, Abschlussbuch des Sonderforschungsbereichs 268, Wiley-VCH Verlag, Weinheim, pp. 256-288.

Peters, J. (1986a). *Osteomorphology and osteometry of the appendicular skeleton of African buffalo,* Syncerus caffer *(Sparrman, 1779) and cattle,* Bos primigenius *f. taurus (Bojanus, 1827*), Occasional papers 1, Laboratorium voor Paleontologie, Ghent University, Ghent.

Peters, J. (1986b). *Osteomorphology and osteometry of the appendicular skeleton of Grant's gazelle,* Gazella granti *(Brooke, 1872), Bohor Reedbuck,* Redunca redunca *(Pallas, 1767) and bushbuck,* Tragelaphus scriptus *(Pallas, 1766*), Occasional papers 2, Laboratorium voor Paleontologie, Ghent University, Ghent.

Peters, J. (1986c). *Bijdrage tot de archeozoölogie van Soedan en Egypte*, Unpublished Ph.D. thesis, Ghent University, Ghent.

Peters, J. (1988). Osteomorphological features of the appendicular skeleton of African buffalo, *Syncerus caffer* (Sparrman, 1779) and of domestic cattle, *Bos primigenius* f. taurus (Bojanus, 1827). *Zeitschrift für Säugetierkunde* **53**: 108-123.

Peters, J. (1995). Mesolithic subsistence between the 5th and 6th Nile cataract: the archaeofaunas from Abu Dharbein, El Damer and Aneibis (Sudan). In Haaland, R. and Abdul-Magid, A. (eds), *Aqualithic sites along the River Nile and Atbara, Sudan*, Alma Mater Forlag, Bergen, pp. 178-224.

Peters, J., Helmer, D., Driesch, A. von den and Saña Segui, M. (1999). Early animal husbandry in the Northern Levant. *Paléorient* **25**(2): 27-47.

Peters, J. and Pöllath, N. (2003). Holocene faunas of the eastern Sahara: zoogeographical and palaeoecological aspects. In Roel, C. and Plug, I. (ed.), *The future from the past*, Oxbow Books, Oxford, pp. 34-50.

Petter, F. (1959). Eléments d'une révision des lièvres africains du sous-genre *Lepus*. *Mammalia* **23**(1): 41-67.

Philippart, J.-Cl. and Ruwet, J.-Cl. (1982). Ecology and distribution of tilapias. In Pullin, R.S.V. and Lowe-McConnell, R.H. (eds), The biology and culture of tilapias. *ICLARM Conference Proceedings* **7**: 15-60.

Phillipson, D.W. (1985). *African Archaeology*, Cambridge University Press, Cambridge.

Pilsbry, H.A. and Bequaert, J. (1927). The aquatic mollusks of the Belgian Congo, with a geographical and ecological account of Congo malacology. *Bulletin of the American Museum of Natural History* **53**: 69-602.

Plug, I. (1997). Early Iron Age buffalo hunters on the Kadzi River, Zimbabwe. *African Archaeological Review* **14**(2): 85-105.

Plug, I. and Badenhorst, S. (2001). *Distribution of macromammals in southern Africa over the past 30000 years as reflected in animal remains from archaeological sites*, Transvaal Museum Monograph 12, Transvaal Museum, Pretoria.

Poll, M. (1939). Les Polyptères. *Cercle Zoologique Congolais* **16**(4): 108-113.

Poll, M. (1971). *Révision des* Synodontis *africains (Famille Mochocidae)*, Annales Sciences Zoologiques 191, Musée royal de l'Afrique centrale, Tervuren

Poplin, F. (1976). Remarques théoriques et pratiques sur les unités utilisées dans les études d'ostéologie quantitative, particulièrement en archéologie préhistorique. In Barral, L. (ed.), *Union Internationale des Sciences Préhistoriques et Protohistoriques, XIe Congrès*, Union Internationale des Sciences Préhistoriques et Protohistoriques, Conseil International de la Philosophie et des Sciences Humaines de l'UNESCO, Nice, pp. 124-141.

Porter, V. (1991). *Cattle. A handbook to the breeds of the world*, Christopher Helm/A & C Black, London.

Prummel, W. and Frisch, H.-J. (1986). A guide for the distinction of species, sex and body side in bones of sheep and goat. *Journal of Archaeological Science* **13**: 567-577.

Quéval, R., Petit, J.P., Tacher, G., Provost, A. and Pagot, J. (1971). Le kouri: race bovine du lac Chad I. Introduction générale à son étude zootechnique et biochémique: origines et écologie de la race. *Revue d'Elevage et de Médecine vétérinaire des Pays Tropicaux* **24**(4): 667-687.

Raimbault, M. and Dutour, O. (1989). Les nouvelles données du site néolithique de Kobadi dans le Sahel Malien: la mission 1989. *Traveaux de LAPMO, 1989*: 175-183.

Raimbault, M., Guérin, C. and Faure, M. (1987). Les vertébrés du gisement néolithique de Kobadi (Mali). *Archaeozoologia* **1**: 219-238.

Raimbault, M. and Sanogo, K. (1991). Les données de la fouille sur la butte de Mouyssam II (KNT 2), campagnes de 1985 et 1986. In Raimbault, M. and Sanogo, K. (eds), *Recherches archéologiques au Mali*, Éditions Karthala, Paris, pp. 301-323.

Redding, R.W. (1984). Theoretical determinants of a herder's decisions: modeling variation in the sheep/goat ratio. In Clutton-Brock, J. and Grigson, C. (eds), *Animals and archaeology: 3. Early herders and their flocks*, BAR International Series 202, Oxford, pp. 223-241.

Redmond, P.M. (1978). Notes on the trade in dried fish in the Central Savanna 1800-1930. *Paper for the Seminar on Economic History of he Central Savanna of West Africa. Kano, 5-10 Jan. 1976.*

Reed, W., Burchard, J., Hopson, A.J., Jenness, J. and Yaro, I. (1967). *Fish and fisheries of northern Nigeria*, Ministry of Agriculture Northern Nigeria, Gaskiya Corporation, Zaria.

Reid, A. (1996). Cattle herds and the redistribution of cattle resources. *World Archaeology* **28**: 43-57.

Reimer, P.J., Baillie, M.G.L., Bard, E., Bayliss, A., Beck, J.W., Bertrand, C., Blackwell, P.G., Buck, C.E., Burr, G.,

Cutler, K.B., Damon, P.E., Edwards, R.L., Fairbanks, R.G., Friedrich, M., Guilderson, T.P., Hughen, K.A., Kromer, B., McCormac, F.G., Manning, S., Bronk Ramsey, C., Reimer, R.W., Remmele, S., Southon, J.R., Stuiver, M., Talamo, S., Taylor, F.W., van der Plicht, J. and Weyhenmeyer, C.E. (2004), IntCal04 Terrestrial radiocarbon age calibration, 26 – 0 ka BP. *Radiocarbon* **46**: 1029-1058.

Reitz, E.J. and Wing, E.S. (1999). *Zooarchaeology*, Cambridge Manuals in Archaeology, Cambridge University Press, Cambridge.

Rivallain, J. and Van Neer, W. (1983). Les fouilles de Koyom (Sud du Tchad). Etude du matériel archéologique et faunique. *L'Anthropologie* **87**(2): 221-239.

Rivallain, J. and Van Neer, W. (1984). Inventaire du matériel archéologique et faunique de Koyom, Sud du Tchad. *L'Anthropologie* **88**(3): 441-448.

Robertshaw, P. (1990). *Early pastoralists of south-western Kenya*, British Institute in Eastern Africa Memoir 11, Nairobi.

Robertshaw, P. (1999). Archaeology of African hunters and gatherers. In Lee, R.B. and Daly, R. (eds), *The Cambridge encyclopedia of hunters and gatherers*, Cambridge University Press, Cambridge, pp. 185-189.

Robertshaw, P. and Collett, D. (1983). The identification of pastoral peoples in the archaeological record: an example from East Africa. *World Archaeology* **15**: 67-78.

Rödel, M.-O. (1996). *Amphibien der westafrikanischen Savanne*, Edition Chimaira, Frankfurt am Main.

Rohlfs, G. (1868). *Reise durch Nord-Afrika vom Mittelländischen Meere bis zum Busen von Guinea 1865 bis 1867. 1. Hälfte: von Tripoli nach Kuka (Fesan, Sahara, Bornu)*, Ergänzungsh. No. 25 zu Petermann's "Geographischen Mittheilungen", Gotha.

Rohlfs, G. (1872). *Reise durch Nord-Afrika vom Mittelländischen Meere bis zum Busen von Guinea 1865 bis 1867. 2. Hälfte: von Kuka nach Lagos (Bornu, Bautschi, Saria, Nupe, Yoruba)*, Ergnänzungsh. No. 34 zu Petermann's "Geographischen Mittheilungen", Gotha.

Rosevear, D.R. (1969). *The rodents of West Africa*, Trustees of the British Museum (Natural History), London.

Rosevear, D.R. (1974). *The carnivores of West Africa*, Trustees of the British Museum (Natural History), London.

Roth, H.H. and Douglas-Hamilton, I. (1991). Distribution and status of elephants in West Africa. *Mammalia* **55**(4): 489-527.

Rowley-Conwy, P. (1988). The camel in the Nile valley: new radiocarbon accelerator (AMS) dates from Qasr Ibrim. *Journal of Egyptian Archaeology* **74**: 245-248.

Rupp, N. (2004). *Land ohne Steine. Die Rohmaterialversorgung in Nordost-Nigeria von der Endsteinzeit bis zur Eisenzeit*, unpublished Ph.D. thesis, Johann Wolfgang Goethe-University, Frankfurt am Main.

Salzmann, U. (1996). Pollenanalytical studies in NE-Nigeria: preliminary results from the Manga grasslands and Lake Tilla, Biu Plateau. *Berichte des Sonderforschungsbereichs 268* **8**: 321-333.

Salzmann, U. (1999). Zur holozänen Vegetations- und Klimaentwicklung der westafrikanischen Savannen. Paläoökologische Untersuchungen in der Sahel- und Sudanzone NO-Nigerias. *Berichte des Sonderforschungsbereichs 268* **13**: 1-144.

Salzmann, U. (2000). Are modern savannas degraded forests? – A Holocene pollen record from the Sudanian vegetation zone of NE Nigeria. *Vegetation History and Archaeobotany* **9**: 1-15.

Salzmann, U. and Waller, M. (1998). The Holocene vegetational history of the Nigerian Sahel based on multiple pollen profiles. *Review of Palaeobotany and Palynology* **100**: 39-72.

Schnelle, H. (1971). *Die traditionelle Jagd Westafrikas. Analyse ihrer wirtschaftlichen Bedeutung für anbautreibende Gruppen*, Ph.D. thesis, Institut für Völkerkunde der Universität zu Göttingen, Klaus Renner Verlag, München.

Schramm, M. (1967). Morphological differences of some goat and sheep bones. *Roczniki Wyczsej Szkoly Roliczej w Pozaniu* **36**: 107-133.

Schultze, A. (1968). *The sultanate of Bornu, translated from the German of A. Schultze with additions and appendices by P.A. Benton*, Frank Cass, London.

Serle, W., Morel, G.J. and Hartwig, W. (1977). *A field guide to the birds of West Africa*, Collins, London.

Servant, M. and Servant-Vildary, S. (1980). L'environnement quaternaire du bassin du lac Tchad. In Williams, M.A.J. and Faure, H. (eds), *The Sahara and the Nile*, A.A. Balkema, Rotterdam, pp. 133-162.

Shareika, N. (2004). Bush for beasts. Lake Chad from the nomadic Woodabe point of view. In Krings, M. and

Platte, E. (eds), *Living with the lake: perspectives on cultures, economies and histories of Lake Chad*, Studien zur Kulturkunde 21, Rüdiger Köppe Verlag, Köln, pp. 171-190.

Shaw, T. (1981). The Late Stone Age in West Africa and the beginnings of African food production. In Roubet, C., Hugot, H.J. and Souville, G., (eds), *Préhistoire africaine, Mélanges offerts au doyen Lionel Balout*, Recherche sur les grandes civilisations 6, Editions ADPF, Paris, pp. 213-235.

Sheil, C.A. (1999). Osteology and skeletal development of *Pyxicephalus adspersus* (Anura: Ranidae: Raninae). *Journal of Morphology* **240**: 49-75.

Sherratt, A. (1981). Plough and pastoralism: aspects of the secondary products revolution. In Hodder, I., Isaac, G. and Hammond, N. (eds), *Pattern of the past. Studies in honour of David Clark*, Cambridge University Press, Cambridge, pp. 261-305

Sherratt, A. (1983). The secondary exploitation of animals in the Old-World. *World Archaeology* **15**: 90-104.

Shinnie, P.L. (1992). Ghana. Radiocarbon dates from Asante sites. *Nyame Akuma* **38**: 28-29.

Shinnie, P.L. and Kense, F.J. (1989). *Archaeology of Gonja, Ghana: excavations at Daboya*, University of Calgary Press, Calgary.

Silver, I.A. (1963). The ageing of domestic animals. In Brothwell, D. and Higgs, E. (eds), *Science in Archaeology*, Thames and Hudson, London, pp. 250-268.

Simoons, F.J. (1971). The antiquity of dairying in Asia and Africa. *Geographical Review* **61**: 431-439.

Simoons, F.J. (1981). Dogs as human food in northwest Africa. *Appetite: Journal for Intake Research* **2**: 253-266.

Sinclair, P., Shaw, T. and Andah, B. (1993). Introduction. In Shaw, T., Sinclair, P., Andah, B. and Okpoko, A. (eds), *The archaeology of Africa. Food, metals and towns*, Routledge, London and New York, pp. 1-31.

Smith, A.B. (1974). Preliminary report of excavations at Karkarichinkat Nord and Karkarichinkat Sud, Tilemsi Valley, Republic of Mali, Spring 1972. *West African Journal of Archaeology* **4**: 33-55.

Smith, A.B. (1975). A note on the flora and fauna from the post-palaeolithic sites of Karkarichinkat Nord and Sud. *West African Journal of Archaeology* **5**: 201-204.

Smith, A.B. (1980). Domesticated cattle in the Sahara and their introduction into West Africa. In Williams, M.A.J. and Faure, H. (eds), *The Sahara and the Nile. Quaternary environments and prehistoric occupation in northern Africa*, A.A. Balkema, Rotterdam, pp. 489-501.

Smith, A.B. (1984a). Environmental limitations on prehistoric pastoralism in Africa. *African Archaeological Review* **2**: 99-111.

Smith, A.B. (1984b). Origins of the Neolithic in the Sahara. In Clark, J.D. and Brandt, S.A. (eds), *From hunters to farmers. The causes and consequences of food-production in Africa*, University of California Press, Berkeley, pp. 88-92.

Smith, A.B. (1992). *Pastoralism in Africa. Origins and development ecology*, Hurts and Company, London.

Smith, A.B. (2000). The origins of the domesticated animals of southern Africa. In Blench, R.M. and MacDonald, K.C. (eds), *The origins and development of African livestock: archaeology, genetics, linguistics and ethnography*, UCL Press, London, pp. 222-238.

Smith, S.E. (1980). The environmental adaptation of nomads in the West African Sahel: a key to understanding prehistoric pastoralists. In Williams, M.A.J. and Faure, H. (eds), *The Sahara and the Nile. Quaternary environments and prehistoric occupation in northern Africa*, A.A. Balkema, Rotterdam, pp. 467-487.

Sobania, N. (1988). Fishermen herders: subsistence, survival and cultural change in northern Kenya. *Journal of African History* **29**(1): 41-56.

Spinage, C.A. (1986). *The natural history of antelopes*, Cromm Helm, London.

Stahl, A.B. (1985). Reinvestigation of Kintampo 6 rock shelter, Ghana: implications for the nature of culture change. *African Archaeological Review* **3**: 117-150.

Stahl, A.B. (1993). Intensification in the West African Late Stone Age: a view from central Ghana. In Shaw, T., Sinclair, P., Andah, B. and Okpoko, A. (eds), *The archaeology of Africa. Food, metals and towns*, Routledge, London and New York, pp. 261-273.

Stangroome, C. (2000). The faunal remains from Gadei (excluding fish). In Insoll, T. (ed.), *Urbanism, archaeology and trade: further observations on the Gao region (Mali): the 1996 fieldseason results*, BAR International Series 829, Oxford, pp. 56-61.

Starkey, P. (2000). The history of working animals in Africa. In Blench, R.M. and MacDonald, K.C. (eds), *The origins and development of African livestock:*

archaeology, genetics, linguistics and ethnography, UCL Press, London, pp. 478-502.

Stenning, D.J. (1959). *Savannah nomads: a study of the Wodaabe pastoral Fulani of western Bornu Province, Northern Region, Nigeria*, Oxford University Press, London.

Stiner, M.C., Munro, N.D., Surovell, T.A., Tchernov, E., Bar-Yosef, O. (1999). Paleolithic population growth pulses evidenced by small animal exploitation. *Science* **283**: 190-194.

Street-Perrott, F.A., Holmes, J.A., Waller, M.P., Allen, M.J., Barber, N.G.H., Fothergill, P.A., Harkness, D.D., Ivanovich, M., Kroon, D. and Perrott, R.A. (2000). Drought and dust deposition in the West African Sahel: a 5500-year record from Kajemarum Oasis, northeastern Nigeria. *The Holocene* **10**(3): 293-302.

Street-Perrott, F.A., and Perrott, R.A. (1993). Holocene vegetation, lake levels and climate of Africa. In Wright, H.E., Kutzbach, J.E., Webb T., Ruddiman, W.F., Street-Perrott, F.A. and Bartlein, J. (eds), *Global climates since the last glacial maximum*, University of Minnesota Press, Minneapolis, pp. 318-356.

Sturm, H.-J. (1997). Kulturbaumlandschaften in der Provinz Boulgou (Burkina Faso). *Berichte des Sonderforschungsbereichs 268* **9**: 169-188.

Sturm, H.-J., Franke, I., Kirscht, H., Ataholo, M. and Skorupinski, T. (1996). Natural environment and land use in the Chad Basin, NE-Nigeria. Preliminary results of an interdisciplinary research. *Berichte des Sonderforschungsbereichs 268* **8**: 335-348.

Sundström, L. (1972). *Ecology and symbiosis: Niger water folk*, Studia Ethnographica Upsaliensia 35, Uppsala.

Sutton, J.E.G. (1974). The Aquatic Civilization of Middle Africa. *Journal of African History* **15**: 527-546.

Sutton, J.E.G. (1985). The antiquity of horse and asses in West Africa: a correction. *Oxford Journal of Archaeology* **4**: 117-118.

Talbot, M.R. (1980). Environmental responses to climatic change in the West African Sahel over the past 20000 years. In Williams, M.A.J. and Faure, H. (eds), *The Sahara and the Nile. Quaternary environments and prehistoric occupation in northern Africa*, A.A. Balkema, Rotterdam, pp. 37-62.

Teugels, G. (2003). Clariidae. In Paugy, D., Lévêque, C. and Teugels, G. (eds), *Poissons d'eaux douces et saumâtres de l'Afrique de l'Ouest. The fresh and brackish water fishes of West Africa* (2 vols.), Faune et Flore tropicales 40, Editions de l'IRD, Paris, pp. 144-173.

Tewe, O.O., Ajayi, S.S. and Faturoti, E.O. (1984). Giant rat and cane rat. In Mason, I.L. (ed.), *Evolution of domesticated animals*, Longman, London, pp. 291-293.

Thiemeyer, H. (1992). On the age of the Bama Ridge: a new 14C-record from Konduga area, Borno State, NE Nigeria. *Zeitschrift für Geomorphologie* **36**(1): 113-118.

Thiemeyer, H. (2004). Environmental and climatic history of Lake Chad during the Holocene. In Krings, M. and Platte, E. (eds), *Living with the lake: perspectives on history, culture and economy of Lake Chad*, Studien zur Kulturkunde 121, Rüdiger Köppe Verlag, Köln, pp. 41-51.

Thilmans, G. and Ravisé, E. (1980). *Protohistoire du Sénégal. Recherches archéologiques, Vol. 2, Sintiou-Bara et les sites du fleuve*, Mémoires de l'Institut Français d'Afrique Noire 91, IFAN, Dakar.

Thomas, H. (1980). Faune. In Lebeuf, J.-P., Lebeuf, A.M.D., Treinen-Claustre, F. and Courtin, J., *Le gisement Sao de Mdaga (Tchad). Fouilles 1960-1968*, Société d'ethnographie, Paris, pp. 189-192.

Tobor, J.G. (1974). *A contribution to the study of* Lates niloticus, *Nile perch, in Lake Chad*, Occasional Papers 17, Federal Department of Fisheries, Lagos, Nigeria.

Togola, T. (1996). Iron Age occupation in the Méma region, Mali. *African Archaeological Review* **13**: 91-110.

Togola, T. and Raimbault, M. (1991). Les missions d'inventaire dans le Méma, Karéri et Farimaké. In Raimbault, M. and Sanogo, K. (eds), *Les sites protohistoriques de la zone lacustre*, ACCT-Karthala, Paris, pp. 81-98.

Troy, C.S., MacHugh, D.E., Bailey, J.F., Magee, D.A., Loftus, R.T., Cunningham, P., Chamberlain, A.T., Sykes, B.C. and Bradley, D.G. (2000). Genetic evidence for Near-Eastern origins of European cattle. *Nature* **410**: 1088-1091.

Tschierschke, K. (1998). Statistische Analyse und Interpretation langjäriger Niederschlags- und Temperaturdaten von Klimastationen im Tschadseegebiet. *Berichte des Sonderforschungsbereichs 268* **10**: 11-140.

Uerpmann, H.-P. (1973). Animal bone finds and economic archaeology: a critical study of 'osteo-archaeological' method. *World Archaeology* **4**(3): 307-322.

Uerpmann, H.-P. (1991). *Equus africanus* in Arabia. In: Meadow, R.H. and Uerpmann, H.-P. (eds), *Equids in the Ancient World II. Beihefte zum Tübinger Atlas des Vorderen Orients, Reihe A (Naturwissenschaften)* 19(2), pp. 12-33.

Uerpmann, H.-P. (1993). Proposal for a separate nomenclature of domestic animals. In Clason, A., Payne, S. and Uerpmann, H.-P. (eds), *Skeletons in her cupboard, Festschrift for Juliet Clutton-Brock*, Oxbow Monograph 34, Oxbow books, Oxford, pp. 239-241.

Uerpmann, M. and Uerpmann, H.-P. (1994). Animal bones. In Højlund, F. and Andersen, H.H. (eds), *Qala'at al Bahrein, Vol. 1, The northern city wall and the Islamic fortress*, Jutland Archaeological Society Publications 30, pp. 417-454.

Urban, E.K., Fry, C.H. and Keith, S. (1986). *The birds of Africa, Vol. 2*, Academic Press, New York and London.

Van Damme, D. (1984). *The freshwater mollusca of northern Africa: Distribution, biogeography and palaeoecology*, Dr. W. Junk, Dordrecht.

Van De Velde, M. (1991). *L'élevage d'aulacodes au Zaire*, Publication du Service agricole, Administration générale de la Coopération au Développement, Bruxelles.

Van Neer, W. (1986). Some notes on the fish remains from Wadi Kubbaniya (Upper Egypt, Late Palaeolithic). In Brinkhuizen, D.C. and Clason, A.T. (eds), *Fish and Archaeology*, BAR International Series 294, Oxford, pp. 103-113.

Van Neer, W. (1989a). *Contribution to the archaeozoology of Central Africa*, Annales Sciences Zoologiques 259, Musée royal de l'Afrique centrale, Tervuren.

Van Neer, W. (1989b). Recent and fossil fish from the Sahara and their palaeohydrological meaning. *Palaeoecology of Africa* **20**: 1-18.

Van Neer, W. (1989c). Fishing along the prehistoric Nile. In Krzyzaniak, L. and Kubasiewicz, M. (eds), *Late Prehistory of the Nile Basin and the Sahara*, Poznan Archaeological Museum, Poznan, pp. 49-56.

Van Neer, W. (1995). Analysis of the fish remains. In McIntosh, S.K. (ed.), *Excavations at Jenné-Jeno, Hambarketolo, and Kaniana (Inland Niger Delta), the 1981 season*, Anthropology 20, University of California Press, Berkeley and Los Angeles, pp. 291-318.

Van Neer, W. (1997). Etudes des ossements d'animaux de l'amas coquiller de Ndiamon-Badat (Delta du Saloum). *Fouille et dégradations dans les îles du Saloum* **3**: 15-21.

Van Neer, W. (2000). Domestic animals from archaeological sites in central and west-central Africa. In Blench, R.M. and MacDonald, K.C. (eds), *The origins and development of African livestock: archaeology, genetics, linguistics and ethnography*, UCL Press, London, pp. 163-190.

Van Neer, W. (2002a). Food security in western and central Africa during the late Holocene: the role of domestic stock keeping, hunting and fishing. In Hassan, F.A. (ed.), Droughts, food and culture. Ecological change and food security in Africa's later Prehistory. Kluwer Academic/Plenum Publishers, New York/Boston/Dordrecht/ London/Moscow, pp. 251-274.

Van Neer, W. (2002b). Le matériel faunique. In Midant-Reynes, B. and Buchez, N. (eds), *Adaïma. 1. Economie et habitat*, Fouilles de l'Institut français d'archéologie orientale 45, Institut français d'archéologie orientale, Cairo, pp. 521-565.

Van Neer, W. (2004). Evolution of prehistoric fishing in the Nile Valley. *Journal of African Archaeology* **2**(2): 251-269.

Van Neer, W. and Bocoum, H. (1991). Etude archéozoologique de Tulel-Fobo, site protohistorique (IVe-Xe siècle) de la moyenne vallée du Fleuve Sénégal (République du Sénegal). *Archaeozoologia* **4**(1): 93-114.

Van Neer, W. and Gayet, M. (1988). Etude des poissons en provenance des sites holocènes du bassin de Taoudenni-Araouane (Mali). *Bulletin du Muséum National d'Histoire Naturelle Paris 4e sér., 10, section C* **4**: 343-383.

Van Neer, W., Lernau, O., Friedman, R., Mumford, G., Poblome, J. and Waelkens, M. (2004a). Fish remains from archaeological sites as indicators of former trade connections in the Eastern Mediterranean. *Paléorient* **30**(1): 101-148.

Van Neer, W. and Lesur, J. (2004). The ancient fish fauna from Asa Koma (Djibouti) and modern osteometric data on Tilapiini and two *Clarias* catfish species. *Documenta Archaeobiologiae* **2**: 141-160.

Van Neer, W., Linseele, V. and Friedman, R. (2004b). Animal burials and food offerings at the elite cemetery HK6 of Hierakonpolis. In Hendrickx, S., Friedman, R., Cialowicz, K.M. and Chlodnicki, M. (eds), *Egypt at its origins. Studies in memory of Barbara Adams, Proceedings of the International Conference "Origin of the State. Predynastic and Early Dynastic Egypt", Krakow, 28th August – 1st September 2002*, Orientalia

Lovaniensia Analecta 138, Peeters Publishing, Leuven, pp. 67-130.

Vansina, J. (1999). Population movements and the emergence of new socio-political forms in Africa. In Ogot, B.A. (ed.), *Africa from the sixteenth to the eighteenth century*, General history of Africa, Abridged Edition, Unesco Publishing, Paris, James Currey, London, California University Press, Berkeley, pp. 25-38.

Vélardé, I. (1995). La répartition des ossements animaux comme indicateur des modes de rejet et de l'origine des couches: le cas d'une concession à Hamdallahi (Mali, 19ème S.). *Anthropozoologica* **21**: 207-216.

Vermeersch, P.M., Van Peer, P., Moeyersons, J. and Van Neer, W. (1994). Sodmein Cave Site, Red Sea Mountains (Egypt). *Sahara* **6**: 32-40.

Vermeersch, P.M., Van Peer, P., Moeyersons, J. and Van Neer, W. (1996). Neolithic occupation of the Sodmein Area, Red Sea Mountains, Egypt. In Pwiti, G. and Soper, R. (eds), *Aspects of African Archaeology. Papers from the 10th congress of the PanAfrican Association for Prehistory and Related Studies*, University of Zimbabwe publications, Harare, pp. 411-419.

Vernet, R. (1993). *Préhistoire de la Mauritanie*, Centre Culturel Français A. de Saint Exupéry, Sepia, Nouakchott.

Vernet, R. (2002). Climate during the late Holocene in the Sahara and the Sahel: evolution and consequences on human settlement. In Hassan, F.A. (ed.), *Droughts, food and culture. Ecological change and food security in Africa's later Prehistory*, Kluwer Academic/Plenum Publishers, New York/Boston/Dordrecht/London/Moscow, pp. 47-63.

Vernet, R. and Aumassip, G. (1991). *Le Sahara et ses marges. Paleoenvironnements et occupation humaine à l'Holocène. Inventaire des datations 14C (1ère édition jusqu'en 1989)*, Université de Niamey, Niamey, CNRS, Paris, G 0848.

Vigne, J.-D. (1991). The meat and offal weight (MOW) method and the relative proportion of ovicaprines in some ancient meat diets in the northwestern Mediterranean. *Revista di Studi Liguri* **57**: 21-47.

Vigne, J.-D., Guilane, J., Debue, K., Haye, L. and Gerard, P. (2004). Early taming of the cat in Cyprus. *Science* **304**: 259.

Vilà, C., Leonard, J.A. and Beja-Pereira, A. (2006). Genetic documentation of horse and donkey domestication. In Zeder, M., Bradley, D.G., Emshwiller, E. and Smith, B.D. (eds.), *Documenting Domestication – New Genetic and Archaeological Paradigms*, University of California Press, Berkeley, pp. 342-353.

Villiers, A. (1958). *Tortues et crocodiles de l'Afrique Noire Française. Initiations africaines*, Institut Français d'Afrique Noire, Dakar.

Vivian, B.C. (1996). Ghana. Recent excavations of Adansemanso. *Nyame Akuma* **46**: 37-42.

Vogelsang, R. (2000). Archäologische Forschungen in der Sahel-Region Burkina Fasos – Ergebnisse der Grabungskampagnen 1994, 1995, 1996. *Beiträge zur allgemeinen und vergleichenden Archäologie* **20**: 173-203.

Vogelsang, R., Albert, K.-D. and Kahlheber, S. (1999). Le sable savant: les cordons dunaires sahéliens au Burkina Faso comme archive archéologique et paléoécologique du Holocène. *Sahara* **11**: 51-68.

Walker, R. (1985). *A guide to the post-cranial bones of East African mammals*, Hylochoerus Press, Norwich, England.

Wasylikowa, K., Harlan, J.R., Evans, J., Wendorf, F., Schild, R., Close, A.E., Krolik, H. and Housley, R.A. (1993). Examination of botanical remains from early neolithic houses at Nabta Playa, Western Desert, Egypt, with special reference to sorghum grains. In Shaw, T., Sinclair, P., Andah, B. and Okpoko, A. (eds), *The archaeology of Africa. Food, metals and towns*, Routledge, London and New York, pp. 155-164.

Welbourne, R.G. (1975). Tautse Iron Age site: its yield of bones. *Botswana Notes and Records* **7**: 1-16.

Welcomme, R.L. (1988). *International introductions of inland aquatic species*, FAO Fisheries Technical Paper 294, FAO, Rome.

Welcomme, R.L. and de Merona, B. (1988). Fish communities of rivers. In Lévêque, C., Bruton, M.N. and Ssentongo, G.W. (eds), *Biology and ecology of African freshwater fishes*, ORSTOM, Paris, pp. 251-276.

Wendt, K.P. (2005). *Beiträge zur Entwicklung der prähistorischen Keramik des inneren Tschadbeckens in Nordost-Nigeria,* Ph.D. thesis, Johann Wolfgang Goethe-University, Frankfurt am Main.

Wheeler, A. and Jones, A.K.G. (1989). *Fishes*, Cambridge University Press, Cambridge.

White, F. (1983). *Vegetation of Africa. A descriptive memoir to accompany the Unesco/AETFAT/UNSO vegetation map of Africa*, Natural Resources Research Report 20, Unesco, Paris.

Wiesmüller, B. (2001). *Die Entwicklung der Keramik von 3000 BP bis zu Gegenwart in den Tonebenen südlich des Tschadsees*, Ph.D. thesis, Johann Wolfgang Goethe-University, Frankfurt am Main. http://publikationen.stub.uni-frankfurt.de/volltexte/2003/337/

Wilkinson, K. and Stevens, C. (2003). *Environmental archaeology. Approaches, techniques and applications*, Tempus Publishing Ltd., Gloucestershire.

Williamson, K. (2000). Did chickens go west? In Blench, R.M. and MacDonald, K.C. (eds), *The origins and development of African livestock: archaeology, genetics, linguistics and ethnography*, UCL Press, London, pp. 368-448.

Wilson, R.T. (1984). *The camel*, Longman, London/New York.

Woelfle, E. (1967). *Vergleichend morphologische Untersuchungen an Einzelknochen des postcranialen Skelettes in Mitteleuropa vorkommender Enten, Halbgänse and Säger*, Unpublished Ph.D. thesis, Ludwig-Maximilians-University, Munich.

Yellen, J.E. (1998). Barbed bone points: tradition and continuity in Saharan and sub-Saharan Africa. *African Archaeological Review* **15**: 173-198.

Zach, B., Kirscht, H., Löhr, D., Neumann, K. and Platte, E. (1996). Masakwa dry season cropping in the Chad Basin. *Berichte des Sonderforschungsbereichs 268* **8**: 349-356.

Zimmermann, H. (1994). Sacrifice and business. A comparative study of ritual and commercial cattle slaughtering in Tenkodogo, Burkina Faso. *Berichte des Sonderforschungsbereichs 268* **4**: 131-156.

Zohary, D. and Hopf, M. (2000). *Domestication of plants in the Old World. The origin and spread of cultivated plants in West Asia, Europe and the Nile Valley*, Oxford University Press, New York

Zvelebil, M. (1992). Hunting in farming societies – the prehistoric perspective. In Grant, A. (ed.), Animals and their products in trade and exchange. *Anthropozoologica* **16**: 7-18.

Appendix A: Radiocarbon dates of the studied localities and sites cited from the literature

locality	site number	lab code	dated material	14C age bp	cal. age (2σ)	reference
Dori	BF 94/40	UtC-5162*	charcoal	6990 +/- 50	5990-5750 BC	Vogelsang, 2000; Vogelsang *et al.*,1999
Dori	BF 94/40	UtC-3950	charcoal	1916 +/- 55	40 BC-AD 230	Vogelsang, 2000; Vogelsang *et al.*,1999
Dori	BF 94/40	UtC-5161	charcoal	1378 +/- 33	AD 600-690	Vogelsang, 2000; Vogelsang *et al.*,1999
Dori	BF 94/96	UtC-5154*	charcoal	9230 +/- 50	8600-8300 BC	Vogelsang, 2000; Vogelsang *et al.*,1999
Dori	BF 94/96	UtC-3949	charcoal	2950 +/- 40	1300-1020 BC	Vogelsang, 2000; Vogelsang *et al.*,1999
Dori	BF 94/96	UtC-5155	charcoal	2316 +/- 49	530-340 BC/330-200 BC	Vogelsang, 2000; Vogelsang *et al.*,1999
Dori	BF 94/96	UtC-5157	charcoal	1852 +/ 49	AD 50-260/AD 290-320	Vogelsang, 2000; Vogelsang *et al.*,1999
Dori	BF 94/96	UtC-5156	charcoal	698 +/- 43	AD 1220-1330/AD 1340-1400	Vogelsang, 2000; Vogelsang *et al.*,1999
Tin Akof	BF 94/133	KN-4777	charcoal	3479 +/- 45	1920-1680 BC	Breunig and Neumann, 2002a
Tin Akof	BF 94/133	UtC-6466	charcoal	3413 +/- 37	1880-1840 BC/1820-1790 BC/1780-1610 BC	Breunig and Neumann, 2002a
Tin Akof	BF 94/133	KN-4776	charcoal	3380 +/- 100	1920-1440 BC	Breunig and Neumann, 2002a
Tin Akof	BF 94/133	UtC-4906	*Pennisetum americanum*	2840 +/- 49	1200-1170 BC/1160 -890 BC/880-840 BC	Breunig and Neumann, 2002a
Corcoba	BF 97/5	UtC-7282	charcoal	3722 +/- 33	2210-2020 BC	Breunig and Neumann, 2002a
Oursi	BF 94/45	UtC-7281	charcoal	2931 +/- 32	1260-1020 BC	von Czerniewicz, 2004, Table 2
Oursi	BF 94/45	KN-4971	charcoal	2167 +/ 45	380-90 BC	von Czerniewicz, 2004, Table 2
Oursi	BF 94/45	KN-4785	charcoal	1907 +/- 46	AD 0-230	von Czerniewicz, 2004, Table 2
Oursi	BF 94/45	Erl-3109	charcoal	1895 +/- 41	AD 20-230	von Czerniewicz, 2004, Table 2
Oursi	BF 94/45	UtC-7353	charcoal	1891 +/- 33	AD 50-220	von Czerniewicz, 2004, Table 2
Oursi	BF 94/45	UtC-8513	charcoal	1864 +/- 42	AD 50-250	von Czerniewicz, 2004, Table 2
Oursi	BF 94/45	Erl-3126	charcoal	1845 +/- 41	AD 70-260	von Czerniewicz, 2004, Table 2
Oursi	BF 94/45	Erl-2650	charcoal	1829 +/- 52	AD 60-340	von Czerniewicz, 2004, Table 2
Oursi village	BF 97/13	UtC-7354	charcoal	1587 +/- 35	AD 400-560	von Czerniewicz, 2004, Table 16
Oursi village	BF 97/13	KI-4551	charcoal	1130 +/- 35	AD 780-790/AD 800-990	von Czerniewicz, 2004, Table 16
Oursi village	BF 97/13	KI-4550	charcoal	1070 +/- 30	AD 890-1030	von Czerniewicz, 2004, Table 16
Oursi village	BF 97/13	KI-4549	charcoal	960 +/- 25	AD 1020-1160	von Czerniewicz, 2004, Table 16
Oursi village	BF 97/13	KI-4862	charcoal	860 +/- 40	AD 1050-1080/AD 1150-1230	von Czerniewicz, 2004, Table 16
Oursi hu-beero	BF 97/30	KL 4860	charcoal	1092 +/- 43	AD 860-1030	Petit, pers. comm.
Oursi hu-beero	BF 97/30	KL 4861	charcoal	960 +/- 30	AD 1020-1160	Petit, pers. comm.
Kissi 22	BF 96/22	UtC-6467	*Vitex* sp. seed	1882 +/- 34	AD 50-230	Breunig and Neumann, 2002a
Kissi 22-B		KI-4487	charcoal	1240 +/- 25	AD 680-870	S. Magnavita, pers. comm.
Kissi 40	BF 97/31	KIA-10613	charcoal	1097 +/- 29	AD 880-1020	S. Magnavita, pers. comm.
Kissi 40	BF 97/31	KI-4885	charcoal	1030 +/- 25	AD 900-920/AD 970-1040	S. Magnavita, pers. comm.
Saouga	BF 94/120	KN-4783	charcoal	964 +/ 35	AD 1010-1170	von Czerniewicz, 2004, Table 29
Saouga	BF 94/120	KI-4552	charcoal	1150 +/- 30	AD 770-980	von Czerniewicz, 2004, Table 29
Saouga	BF 94/120	KN-4784	charcoal	1139 +/- 40	AD 770-990	von Czerniewicz, 2004, Table 29
Saouga	BF 94/120	KN-4775	charcoal	1136 +/- 44	AD 770-990	von Czerniewicz, 2004, Table 29
Saouga	BF 95/7	KI-4362	charcoal	1230 +/- 35	AD 680-890	von Czerniewicz, 2004, Table 30
Saouga	BF 95/7	UtC-6465	*Vitellaria paradoxa*	1197 +/- 29	AD 710-750/AD 760-900/AD 920-940	von Czerniewicz, 2004, Table 30
Saouga	BF 95/7	KN-4940	charcoal	1135 +/- 36	AD 780-990	von Czerniewicz, 2004, Table 30
Saouga	BF 95/7	KN-4939	charcoal	720 +/- 36	AD 1220-1310/AD 1360-1390	von Czerniewicz, 2004, Table 30
Saouga	BF 95/7	KN-4939	charcoal	657 +/- 60	AD 1260-1420	von Czerniewicz, 2004, Table 30

*rejected or disputed date

Table A.1: Radiocarbon dates of the studied sites in northern Burkina Faso

locality	site number	labcode	dated material	14C age bp	cal. age (2σ)	reference
Gajiganna A	NA 90/5A	UtC-2795	charcoal	2960 +/- 50	1370-1340 BC/1320-1010 BC	Breunig and Neumann, 2002a
Gajiganna A	NA 90/5A	UtC-2329	charcoal	2930 +/- 60	1320-970 BC/ 960-930 BC	Breunig and Neumann, 2002a
Gajiganna A	NA 90/5A	KIA-603	human bone	2180 +/- 80	400-40 BC	Breunig and Neumann, 2002a
Gajiganna A	NA 90/5A	KN-4674	cattle bone	2141 +/- 66	380-30 BC	Breunig and Neumann, 2002a
Gajiganna BI	NA 90/5BI	UtC-2330	charcoal	3150 +/- 70	1610-1260 BC	Breunig and Neumann, 2002a
Gajiganna BI	NA 90/5BI	UtC-2332	charcoal	3140 +/- 110	1700-1050 BC	Breunig and Neumann, 2002a
Gajiganna BI	NA 90/5BI	KI-3605	charcoal	3040 +/- 120	1550-900 BC	Breunig and Neumann, 2002a
Gajiganna BI	NA 90/5BI	KN-4675	cattle bone	2880 +/- 165	1500-750 BC	Breunig and Neumann, 2002a
Gajiganna BI	NA 90/5BI	UtC-2331	charcoal	2740 +/- 50	770-410 BC	Breunig and Neumann, 2002a
Gajiganna BII	NA 90/BII	UtC-2797	charcoal	2750 +/- 70	1090-790 BC	Breunig and Neumann, 2002a
Gajiganna BII	NA 90/BII	UtC-2796	charcoal	2730 +/- 50	1000-800 BC	Breunig and Neumann, 2002a
Gajiganna C	NA 90/5C	UtC-2799	charcoal	3070 +/- 70	1500-1120 BC	Breunig and Neumann, 2002a
Gajiganna D	NA 90/5D	UtC-2801	charcoal	3360 +/- 80	1880-1490 BC/1480-1460 BC	Breunig and Neumann, 2002a; Breunig *et al.*, in prep.
Gilgila	NA 99/65	Ki-4745	charcoal	2910 +/- 70	1320-910 BC	C. Magnavita, pers. comm.
Gilgila	NA 99/65	Erl-3105	charcoal	2859 +/- 40	1190-1170 BC/1160-910 BC	C. Magnavita, pers. comm.
	NA 91/1A	-	-	-	-	-
	NA 93/10ST1	UtC-3513	charcoal	2470 +/- 70	780-400 BC	Breunig and Neumann, 2002a
	NA 93/36	UtC-3514	charcoal	3120 +/- 70	1530-1190 BC	Breunig and Neumann, 2002a
	NA 93/42	UtC-3151	charcoal	3690 +/- 120	2500-1750 BC	Breunig and Neumann, 2002a
	NA 93/42	UtC-8297	charcoal	3489 +/- 34	1900-1730 BC/1720-1690 BC	Breunig and Neumann, 2002a
Dumge	NA 97/18	-	-	-	-	-
Labe	NA 97/24	UtC-6964	charcoal	3275 +/- 33	1630-1450 BC	Breunig and Neumann, 2002a; Breunig *et al.*, in prep.
Labe	NA 97/24	UtC-6781	charcoal	3223 +/- 33	1610-1570 BC/1540-1420 BC	Breunig and Neumann, 2002a; Breunig *et al.*, in prep.
Zilum	NA 97/37	Erl-4836	palm fruit	2548 +/- 41	810-530 BC	C. Magnavita, pers. comm.
Zilum	NA 97/37	Erl-4835	charcoal	2537 +/- 46	810-510 BC	C. Magnavita, pers. comm.
Zilum	NA 97/37	Kn-5457	charcoal	2510 +/- 40	800-500 BC/440-420 BC	C. Magnavita, pers. comm.
Zilum	NA 97/37	UtC-13071	bean seed	2508 +/- 33	790-520 BC	C. Magnavita, pers. comm.
Zilum	NA 97/37	Erl-3130	charcoal	2483 +/- 41	780-480 BC/470-410 BC	C. Magnavita, pers. comm.
Zilum	NA 97/37	Ki-4959	charcoal	2480 +/- 25	770-500 BC/440-410 BC	C. Magnavita, pers. comm.
Zilum	NA 97/37	Erl-3312	*Pennisetum* seeds	2359 +/- 42	740-680 BC/670-640 BC/550-360 BC	C. Magnavita, pers. comm.
Kariari C	NA 95/1	UtC-5158	charcoal	3146 +/- 38	1500-1360 BC/1350-1310 BC	Breunig and Neumann, 2002a
Kariari C	NA 95/1	UtC-5159	charcoal	3076 +/- 43	1440-1250 BC/1240-1210 BC	Breunig and Neumann, 2002a
Kariari C	NA 95/1	KI-4073	charcoal	2780 +/- 95	1220-790 BC	Breunig and Neumann, 2002a
Kelumeri	NA 96/45	N-794	charcoal	3960 +/- 160	2900-2000 BC	Connah, 1981, 84
Kelumeri	NA 96/45	N-794	charcoal	2960 +/- 160	1550-800 BC	Breunig and Neumann, 2002a
Kelumeri	NA 96/45	KI-4206	charcoal	2940 +/- 125	1450-800 BC	Breunig and Neumann, 2002a
Kelumeri	NA 96/45	UtC-5746	charcoal	2920 +/- 35	1260-1230 BC/1220-1000 BC	Breunig and Neumann, 2002a
Kelumeri	NA 96/45	N-795	charcoal	2880 +/- 140	1450-800 BC	Connah, 1981, 84
Kelumeri	NA 96/45	KI-4204	charcoal	2780 +/- 95	1220-790 BC	Breunig and Neumann, 2002a

Table A.2: Radiocarbon dates of the studied sites in the southern Lake Chad area

locality	site number	labcode	dated material	14C age bp	cal. age (2σ)	reference
Kelumeri	NA 96/45	UtC-5747	charcoal	2773 +/- 38	1010-830 BC	Breunig and Neumann, 2002a
Kelumeri	NA 96/45	N-796	charcoal	2590 +/- 170	1200-350 BC	Connah, 1981, 84
Bukarkurari	NA 97/33	UtC-6779	charcoal	3205 +/- 43	1610-1570 BC/1560-1400 BC	Breunig and Neumann, 2002a; Breunig *et al.*, in prep.
Bukarkurari	NA 97/33	UtC-6841	charcoal	2991 +/- 46	1390-1050 BC	Breunig *et al.*, in prep.
Bukarkurari	NA 97/33	KI-4206	charcoal	2940 +/- 125	1450-800 BC	Breunig and Neumann, 2002a
Galaga	NA 92/2C	UtC-2802	charcoal	230 +/- 40	AD 1520-1590/AD 1620-1700/AD/1720-1820/AD 1910-1960	Breunig *et al.*, in prep.
Galaga	NA 92/2C	KI-3756	charcoal	180 +/- 36	AD 1640-1710/AD 1720-1820/AD 1830-1880/AD 1910-1960	Breunig *et al.*, in prep.
Galaga	NA 92/2C	KI-3757	charcoal	180 +/- 45	AD 1640-1890/AD 1910-1960	Breunig *et al.*, in prep.
Labe Kanuri	NA 97/26	Erl-3106	charcoal	1944 +/ 37	40 BC-AD 130	C. Magnavita, pers. comm.; Breunig *et al.*, in prep.
Labe Kanuri	NA 97/26	Erl-5141	charcoal	1614 +/- 44	AD 330-550	C. Magnavita, pers. comm.; Breunig *et al.*, in prep.
Elkido North	NA 99/75	Ki-4743	charcoal	1770 +/- 60	AD 120-410	C. Magnavita, pers. comm.; Breunig *et al.*, in prep.
Elkido North	NA 99/75	Ki-4742	*Sorghum* seeds	1660 +/- 35	AD 250-300/AD310-450/AD 480-540	C. Magnavita, 2002; Breunig *et al.*, in prep.
Dorota	NA 97/13	Erl-2971	charcoal	1797 +/- 38	AD 120-340	C. Magnavita, 2002; Breunig *et al.*, in prep.
Dorota	NA 97/13	Ki-4730	charcoal	1610 +/- 30	AD 390-540	C. Magnavita, 2002; Breunig *et al.*, in prep.
Dorota	NA 97/13	Erl-3104	charcoal	1559 +/- 35	AD 410-580	C. Magnavita, 2002; Breunig *et al.*, in prep.
Dorota	NA 97/13	Erl-3103	*Sorghum* seeds	1546 +/- 35	AD 420-600	C. Magnavita, 2002; Breunig *et al.*, in prep.
Dorota	NA 97/13	Erl-3127*	charcoal	92 +/- 34	AD 1680-1740/AD1800-1940	C. Magnavita, pers. comm.
Kursakata	NA 93/46	N-480	charcoal	2880 +/- 140	1450-800 BC	Connah, 1981, 91
Kursakata	NA 93/46	UtC-3517	charcoal	2860 +/- 60	1206-1230 BC/1120-890 BC	Gronenborn, 1998; Wiesmüller 2001, 34-36
Kursakata	NA 93/46	Utc-6476	charcoal	2615 +/- 36	850-750 BC/690-660 BC	Gronenborn, 1998; Wiesmüller 2001, 34-36
Kursakata	NA 93/46	UtC-5452	*Pennisetum* seeds	2430 +/- 70	770-390 BC	Gronenborn, 1998; Wiesmüller 2001, 34-36
Kursakata	NA 93/46	UtC-5453	*Pennisetum* seeds	2290 +/- 70	750-650 BC/550-150 BC	Gronenborn, 1998; Wiesmüller 2001, 34-36
Kursakata	NA 93/46	UtC-3516	charcoal	2250 +/- 80	510-50 BC	Gronenborn, 1998; Wiesmüller 2001, 34-36
Kursakata	NA 93/46	UtC-3518	charcoal	1920 +/- 50	40 BC-220 AD	Gronenborn, 1998; Wiesmüller 2001, 34-36
Mege	NA 94/7	KN-4811*	charcoal	3834 +/- 270	3100-1500 BC	Gronenborn, 1998; Wiesmüller, 2001, 36-38
Mege	NA 94/7	UtC-4204	charcoal	2659 +/- 36	900 BC-790 BC	Gronenborn, 1998; Wiesmüller, 2001, 36-38
Mege	NA 94/7	UtC-4934	charcoal	2570 +/- 50	830-530 BC	Gronenborn, 1998; Wiesmüller, 2001, 36-38
Mege	NA 94/7	UtC-6472	charcoal	2481 +/- 42	770-480 BC/470-410 BC	Gronenborn, 1998; Wiesmüller, 2001, 36-38
Mege	NA 94/7	UtC-4933	charcoal	2425 +/- 34	750-680 BC/670-630 BC/600-400 BC	Gronenborn, 1998; Wiesmüller, 2001, 36-38
Mege	NA 94/7	UtC-8507	charcoal	1382 +/- 41	AD 570-710/AD 740-770	Wiesmüller, 2001, 36-38
Mege	NA 94/7	UtC-6471	charcoal	1310 +/- 60	AD 630-880	Gronenborn, 1998; Wiesmüller, 2001, 36-38
Mege	NA 94/7	UtC-4935	charcoal	411 +/- 36	AD 1420-1530/AD 1570-1630	Gronenborn, 1998; Wiesmüller, 2001, 36-38
Mege	NA 94/7	UtC-8508	charcoal	399 +/- 33	AD 1430-1530/AD 1550-1630	Wiesmüller, 2001, 36-38
Mege	NA 94/7	KN-4812	charcoal	365 +/- 66	AD 1430-1650	Gronenborn, 1998; Wiesmüller, 2001, 36-38
Ngala	NA 93/45	UtC-5150	charcoal	2584 +/- 34	820-740 BC/690-660 BC/650-590 BC/580-560 BC	Gronenborn, 1998
Ngala	NA 93/45	UtC-5148	charcoal	2565 +/- 39	810-730 BC/690-660 BC/650-540 BC	Gronenborn, 1998
Ngala	NA 93/45	UtC-5146	charcoal	2099 +/- 40	350 BC-310 BC/AD 210-0	Gronenborn, 1998
Ngala	NA 93/45	UtC-5149	charcoal	1630 +/- 39	AD 330-540	Gronenborn, 1998
Ngala	NA 93/45	UtC-5147	charcoal	1490 +/- 33	AD 440-490/AD 530-650	Gronenborn, 1998

*rejected or disputed date

Table A.2 cont. 1

locality	site number	labcode	dated material	14C age bp	cal. age (2σ)	reference
Ngala	NA 93/45	UtC-5151	charcoal	1467 +/- 33	AD 540-650	Gronenborn, 1998
Ngala	NA 93/45	UtC-5152	charcoal	1165 +/- 33	AD 770-970	Gronenborn, 1998
Blé A		RT 1668b	charcoal	1095 +/- 45	AD 810-850/AD 860-1030	Holl, 2002, 76
Blé A		RT 1668a	charcoal	1010 +/- 45	AD 890-920/AD 960-1160	Holl, 2002, 76
Blé A		RT-1666	charcoal	775 +/- 40	AD 1180-1290	Holl, 2002, 81
Blé A		RT-1667	charcoal	755 +/- 45	AD 1180-1300/AD 1360-1380	Holl, 2002, 80
Blé B		RT-1670	charcoal	560 +/- 50	AD 1290-1440	Holl, 2002, 97
Blé C		RT-1671	charcoal	1080 +/- 50	AD 810-1040	Holl, 2002, 105
Blé C		RT-1672	charcoal	915 +/- 50	AD 1020-1220	Holl, 2002, 105
Blé C		RT-1673	charcoal	905 +/- 45	AD 1020-1220	Holl, 2002, 102
Blé E		RT-1680*	charcoal	1355 +/- 45	AD 600-780	Holl, 2002, 125
Blé E		RT-1681	charcoal	992 +/- 40	AD 980-1160	Holl, 2002, 120-121
Blé E		RT-1679	charcoal	850 +/- 50	AD 1040-1270	Holl, 2002, 128

*rejected or disputed date

Table A.2 cont. 2

site	labcode	dated material	14C age bp	cal. age (2σ)	reference
Adansemanso	Gd-10365	charcoal	1360 +/- 160	AD 350-1050	Vivian, 1996
Adansemanso	Gd-6537	charcoal	1110 +/- 110	AD 680-1060/AD 1070-1160	Shinnie, 1992; Vivian, 1996
Adansemanso	Gd-6541	charcoal	1050 +/- 100	AD 770-1220	Shinnie, 1992; Vivian, 1996
Adansemanso	Gd-6545	charcoal	740 +/- 70	AD 1210-1300/AD 1360-1390	Shinnie, 1992; Vivian, 1996
Adansemanso	Gd-5996	charcoal	680 +/- 50	AD 1250-1410	Shinnie, 1992; Vivian, 1996
Adansemanso	Gd-11232	charcoal	620 +/- 50	AD 1280-1410	Vivian, 1996
Adansemanso	Gd-9533	charcoal	580 +/- 160	AD 1000-1700	Vivian, 1996
Adansemanso	Gd-11233	charcoal	540 +/- 50	AD 1290-1370/AD 1380-1450	Vivian, 1996
Adansemanso	Gd-10366	charcoal	530 +/- 150	AD 1150-1700	Vivian, 1996
Adansemanso	Gd-11230	charcoal	510 +/- 60	AD 1200-1490	Vivian, 1996
Adansemanso	Gd-10364	charcoal	490 +/-80	AD 1290-1530/AD 1550-1640	Vivian, 1996
Adansemanso	Gd-10360	charcoal	430 +/- 90	AD 1300-1360/AD 1380-1660	Vivian, 1996
Adansemanso	Gd-6540	charcoal	310 +/- 70	AD 1400-1850/AD 1900-2000	Shinnie, 1992; Vivian, 1996
Adansemanso	Gd-7735	charcoal	310 +/- 30	AD 1480-1650	Vivian, 1996
Adansemanso	Gd-10363	charcoal	100 +/- 150	AD 1490-1960	Vivian, 1996
Adansemanso	Gd-10362	charcoal	modern	-	Vivian, 1996
Adrar Bous	UCLA-1658	cattle bone	5760 +/- 500	5700-3500 BC	Carter and Clark, 1976
Adrar Bous	N-870	cattle bone	4910 +/- 135	4000-3350 BC	Carter and Clark, 1976
Aissa Dugjé	TO-7515	bone horse 1	1310 +/- 60	AD 630-880	MacEachern *et al.*, 2001
Aissa Dugjé	TO-7516	bone horse 3	950 +/- 50	AD 990-1210	MacEachern *et al.*, 2001
Aissa Dugjé	TO-7517	bone horse 5	680 +/- 50	AD 1250-1410	MacEachern *et al.*, 2001
Ajere	-	-	-	-	-
Akumbu	Beta 40446	charcoal	1630 +/- 60	AD 250-300/AD 310-560	Togola, 1996
Amachita	BDY 226	charcoal	-	AD 1650-1965	Holl, 2002, 67
Anyinam	Gd-5998	charcoal	550 +/- 50	AD 1290-1450	Shinnie, 1992
Anyinam	Gd-6546	charcoal	170 +/- 70	AD 1030-1960	Shinnie, 1992
Arouane (MN27)	Pa373/GrA-22576	mammal tooth	5065 +/- 45	3970-3760 BC	Jousse, 2004b
Arouane (MN35)	Pa376/GrA-22581	mammal tooth	4910 +/- 45	3790-3630 BC	Jousse, 2004b
Arouane (MN10)	Pa-1024	plant temper	4885 +/- 200	4300-3000 BC	Jousse, 2004b
Arouane (MN6)	Pa375/GrA-22580	mammal tooth	4255 +/- 45	3010-2980 BC/2940-2830 BC/2820-2670 BC	Jousse, 2004b
Arouane (MN24)	Pa374/GrA-22579	mammal tooth	3940 +/- 45	2570-2510 BC/2500-2290 BC	Jousse, 2004b
Asa Koma	Gif-7404	charcoal	3440 +/- 90 BP	1980-1520 BC	Guérin and Faure, 1996
Birnin Gazargamo	N-481	charcoal	330 +/- 105	AD 1400-2000	Connah, 1981, 216
Birimi	Beta 099377 AMS	charcoal	3830 +/- 100	2600-1950 BC	Casey *et al.*, 1997
Birimi	Beta 104756 AMS	charcoal	3710 + 120 -130	-	Casey *et al.*, 1997
Birimi	Beta 099308	charcoal	3520 + 160 -130	-	Casey *et al.*, 1997
Birimi	Beta 099306	charcoal	3460 +/- 100	2050-1500 BC	Casey *et al.*, 1997
Blabli	TO-1128*	ovicaprine bone	6960 +/- 200	6250-5500 BC	David and Sterner, 1989
Blabli	TO-1127*	ovicaprine bone	4390 +/- 220	3700-2400 BC	David and Sterner, 1989
Blé Mound D	RT-1677	charcoal	1395 +/- 55	AD 540-720/AD 740-770	Holl, 2002, 116
Blé Mound D	RT-1676	charcoal	1165 +/- 50	AD 710-750/AD 760-990	Holl, 2002, 117
Blé Mound D	RT-1678	charcoal	1050 +/- 45	AD 880-1050/AD 1090-1120	Holl, 2002, 115
Blé Mound D	RT-1675	charcoal	1010 +/- 50	AD 820-920/AD 840-1160	Holl, 2002, 117
Blé Mound D	RT-1674	charcoal	740 +/- 50	AD 1180-1320/AD 1350-1390	Holl, 2002, 119
Borno 38 = Kelumeri	see Table A.2	see Table A.2	see Table A.2	see Table A.2	see Table A.2
Boû Khzâmâ	Pa-2153	charcoal	3765 +/- 35	2290-2120 BC/2100-2040 BC	Jousse, 2004b
Chami	GIF 2492	n.d.	3500 +/- 120	BC 2150-1500	Vernet and Aumassip, 1989
Chin Tafidet	Pa-1054	human bone	3910 +/- 150	BC 2900-950	Paris, 1997
Chin Tafidet	Pa-292	cattle bone	3325 +/- 260	BC 2300-900	Paris, 1997

*rejected or disputed date

Table A.3: Radiocarbon dates of sites cited from the literature

site	labcode	dated material	14C age bp	cal. age (2σ)	reference
Cubalel	-	-	-	-	-
Daima	I-2945	charcoal	2520+/- 110	900-350 BC	Connah, 1981, 114
Daima	I-2372	charcoal	2400 +/- 95	800-350 BC/300-200 BC	Connah, 1981, 114
Daima	I-2371	animal bone	1500 +/- 670	1000 BC-AD 1800	Connah, 1981, 147
Daima	I-2370	animal bone	1470 +/- 270	100 BC-AD 1200	Connah, 1981, 147
Daima	I-2943	animal bone	1320 +/- 190	AD 250-1200	Connah, 1981, 147
Daima	I-2368	charcoal	1140 +/- 90	AD 670-1040	Connah, 1981, 164
Daima	I-2369	charcoal	970 +/- 90	AD 890-1260	Connah, 1981, 164
Daima	I-3181	charcoal	890 +/- 90	AD 990-1280	Connah, 1981, 164
Deguesse	Ly-4177	charcoal	3350 +/- 270	2500-900 BC	Holl, 2002, 36
Deguesse	Ly-4178	charcoal	1870 +/- 180	400 BC-AD 550	Holl, 2002, 37
Deguesse	Ly-4176	charcoal	1275 +/- 220	AD 300-1250	Holl, 2002, 37
Dhar Tichitt (site 38)	n.d.	bone and ashes	3850 +/- 250	3000-1600 BC	Holl, 1985, 1986, Table 36
Dhar Tichitt (site 38)	Dak-52	ashes	3776 +/- 120	2600-1850 BC	Holl, 1985, 1986, Table 36
Dhar Tichitt (site 38)	MC-427	bone	3490 +/- 50	1940-1680 BC	Holl, 1985, 1986, Table 36
Dhar Tichitt (site 38)	Dak-203	n.d.	3122 +/- 120	1700-1000 BC	Holl, 1985, 1986, Table 36
Dhar Tichitt (site 38)	I-3565	charcoal	3100 +/- 105	1650-1050 BC	Holl, 1985, 1986, Table 36
Dhar Tichitt (site 38)	Dak-187	n.d.	2975 +/- 110	1450-900 BC	Holl, 1985, 1986, Table 36
Dhar Tichitt (site 38)	Dak-190	n.d.	2760 +/- 160	1400-500 BC	Holl, 1985, 1986, Table 36
Dhar Tichitt (site 38)	Gif-6083	charcoal	2430 +/- 80	780-390 BC	Holl, 1985
Dhar Tichitt (site 30)	I-3564	charcoal	3205 +/- 105	1750-1200 BC	Holl, 1985, 1986, Table 36
Dhar Tichitt (site 30)	GX-1888	charcoal	2885 +/- 140	1450-800 BC	Holl, 1985, 1986, Table 36
Dhar Tichitt (site 30)	GX-1324	charcoal	2700 +/- 115	1250-500 BC	Holl, 1985, 1986, Table 36
Dhar Tichitt (site 11)	I-3819	charcoal	2170 +/- 105	410 BC-AD 60	Holl, 1985, 1986, Table 36
Dhar Tichitt (site 12)	GX-1325	charcoal	2780 +/- 140	1400-550 BC	Holl, 1985, 1986, Table 36
Dhar Tichitt (site 17)	GX-1890	shell	3700 +/- 130	2500-1750 BC	Holl, 1985, 1986, Table 36
Dhar Tichitt (site 3)	GX-1326	charcoal	2600 +/- 105	1000-400 BC	Holl, 1985, 1986, Table 36
Dhar Tichitt (site 4)	I-3566	charcoal	2330 +/- 105	800-150 BC	Holl, 1985, 1986, Table 36
Dhar Tichitt (site 45)	GX-1421	bone	3465 +/- 160	2250-1400 BC	Holl, 1985, 1986, Table 36
Dhar Tichitt (site 45)	GX-1889	charcoal	3425 +/- 130	2150-1400 BC	Holl, 1985, 1986, Table 36
Dhar Tichitt (site 45)	I-3561	charcoal	3350 +/- 110	1920-1410 BC	Holl, 1985, 1986, Table 36
Dhar Tichitt (site 45)	GX-1323	charcoal	3205 +/- 95	1740-1710 BC/1700-1260 BC	Holl, 1985, 1986, Table 36
Dhar Tichitt (site 45)	I-3562	charcoal	3190 +/- 110	1750-1100 BC	Holl, 1985, 1986, Table 36
Dhar Tichitt (site 45)	I-3563	charcoal	2950 +/- 100	1420-910 BC	Holl, 1985, 1986, Table 36
Dhar Tichitt (site 46)	Gif-2884	bone and ashes	3830 +/- 120	2660-1900 BC	Holl, 1985, 1986, Table 36
Dhar Tichitt (site 46)	Gif-4110	n.d.	2610 +/- 110	1000-400 BC	Holl, 1985, 1986, Table 36
Dia-Shoma	GX-25786-AMS	charcoal	2550 +/- 50	820-510 BC	Bedaux *et al.*, 2001
Dia-Shoma	GX-25785-AMS	charcoal	2470 +/- 40	770-410 BC	Bedaux *et al.*, 2001
Dia-Shoma	GX-27050-AMS	charcoal	2450 +/-30	760-680 BC/670-610 BC/600-400 BC	Bedaux *et al.*, 2001
Dia-Shoma	GX-27055	charcoal	2470 +/- 70	780-400 BC	Bedaux *et al.*, 2001
Dia-Shoma	GX-27051	charcoal	2220 +/- 100	550 BC-AD 50	Bedaux *et al.*, 2001
Dia-Shoma	GX-27053	charcoal	2070 +/- 100	370 BC-AD 130	Bedaux *et al.*, 2001
Dia-Shoma	GX-27054	charcoal	1990 +/- 50	160-130 BC/120 BC-AD 130	Bedaux *et al.*, 2001
Dia-Shoma	GX-25637-AMS	charcoal	1200 +/- 40	AD 680-900/AD 910-950	Bedaux *et al.*, 2001
Dia-Shoma	GX-25634-LS	charcoal	310 +/- 90	AD 1400-2000	Bedaux *et al.*, 2001
Dia-Shoma	GX-25633-LS	charcoal	"less than 1000 years"		Bedaux *et al.*, 2001
Dufuna	KI-3587	wood dugout	7670 +/- 110	6800-6250 BC	Breunig, 1996
Dufuna	KN-4683	wood dugout	7264 +/- 55	6230-6020 BC	Breunig, 1996
Erg Inc Sakane (MK42)	Gif 5818	charcoal	4710 +/- 120	3750-3050 BC	Jousse, 2004b

Table A.3 cont. 1

site	labcode	dated material	14C age bp	cal. age (2σ)	reference
Erg Ine Sakane (AZ21)	Gif-5442	mammal bone	3990 +/- 130	2900-2100 BC	Jousse, 2004b
Erg Ine Sakane (AZ21)	Gif-5227	charcoal	3750 +/- 100	2500-1900 BC	Jousse, 2004b
Erg Ine Sakane (AZ21)	Gif-5441	human bone	3600 +/- 180	2500-1500 BC	Jousse, 2004b
Esiease	Gd-6542	n.d.	450 +/- 80	AD 1310-1360/1380-1650	Shinnie, 1992
Esiease	Gd-5997	n.d.	230 +/- 50	AD 130-1760/AD 1720-1820/AD 1830-1880/AD 1910-	Shinnie, 1992
Esiease	Gd-6543	n.d.	190 +/- 60	AD 1630-1960	Shinnie, 1992
Faboura	Mc-1390	*Anadara* shell	1940 +/- 80	170 BC-AD 250	Descamps *et al.*, 1977
Faboura	Mc-2073	shell	1315 +/- 80	AD 580-900	Descamps *et al.*, 1977
Gadei	GX-22811*	old wood	3510 +/- 150	1800-1000 BC	Insoll, 2000, 18
Gadei	GX-22812	charcoal	1195 +/- 205	AD 400-1250	Insoll, 2000, 18
Gadei	GX-22810	charcoal	515 +/- 70	AD 1280-1520/AD 1600-1620	Insoll, 2000, 18
Gao Ancien	GFX-21657	charcoal	1430 +/- 90	AD 420-780	Insoll, 1996, 41
Gao Ancien	GX-22808*	charcoal	1375 +/- 75	AD 560-870	Insoll, 2000, 7
Gao Ancien	GX-21656	charcoal	1160 +/- 75	AD 680-1020	Insoll, 1996, 41
Gao Ancien	GX-20623	wood	1105 +/- 165	AD 600-1300	Insoll, 1996, 41
Gao Ancien	GX-22809*	wood	<100		Insoll, 2000, 7
Gao Saney	-	-	-	-	-
Goray	Hv 12297	n.d.	1595 +/- 450	800 BC-AD 1300	Marliac, 1991, Table 4
Goray	Gif 5463	n.d.	1180 +/- 120	AD 600-1150	Marliac, 1991, Table 4
Goray	Gif 5251	n.d.	1030 +/- 80	AD 810-1210	Marliac, 1991, Table 4
Goray	Hv 13964	n.d.	945 +/- 75	AD 970-1260	Marliac, 1991, Table 4
Goray	Orsay	n.d.	920 +/- 120	AD 870-1300	Marliac, 1991, Table 4
Goray	Gif 5250	n.d.	920 +/- 80	AD 980-1270	Marliac, 1991, Table 4
Goray	Hv 13965	n.d.	910 +/- 180	AD 700-1450	Marliac, 1991, Table 4
Goray	Orsay	n.d.	870 +/- 110	AD 960-1310	Marliac, 1991, Table 4
Goray	Orsay	n.d.	870 +/- 80	AD 1020-1080	Marliac, 1991, Table 4
Goray	Orsay	n.d.	750 +/- 200	AD 800-1650	Marliac, 1991, Table 4
Goray	Orsay	n.d.	745 +/- 150	AD 980-1450	Marliac, 1991, Table 4
Goray	Orsay	n.d.	610 +/- 75	AD 1270-1440	Marliac, 1991, Table 4
Goray	Orsay	n.d.	470 +/- 105	AD 1290-1650	Marliac, 1991, Table 4
Goray	Gif 4990*	n.d.	<90		Marliac, 1991, Table 4
Hamadallahi	-	-	-	-	-
Hamei	Hv 16373	charcoal	845 +/- 55	AD 1040-1280	Holl, 2002, 51
Hamei	Hv 16374	charcoal	85 +/- 65	AD 1670-1780/AD 1790-1960	Holl, 2002, 52
Hassi-el-Abiod	UQ368	shell	8450 +/- 60	5050-4500 BC	Vernet and Aumassip, 1989, 28
Hassi-el-Abiod	Gif5495	nile perch bone	6970 +/- 130	9150-9000 BC/8950-8550 BC	Vernet and Aumassip, 1989, 27
Hassi-el-Abiod	Gif6462	shell	5920 +/- 100	6070-5620 BC	Vernet and Aumassip, 1989, 27
Hosséré-Djaba	Ly-11462	n.d.	85 +/- 60	AD 1670-1780/AD 1790-1960	Lesur and Langlois, 2002
Hosséré-Djaba	LY-11463	n.d.	65 +/- 55	AD 1670-1780/AD 1800-1940/1950-1980	Lesur and Langlois, 2002
Houlouf	Hv 16381	charcoal	1290 +/- 50	AD 650-870	Holl, 2002, 152
Houlouf	Hv 16382*	charcoal	1030 +/- 65	AD 670-1170	Holl, 2002, 149
Houlouf	OBDY 200	charcoal	1030 +/- 55	AD 890-1160	Holl, 2002, 155
Houlouf	OBDY 191	charcoal	760 +/- 50	AD 1160-1300/AD 1360-1380	Holl, 2002, 157
Houlouf	OBDY 192	charcoal	680 +/- 135	AD 1250-1470	Holl, 2002, 165
Houlouf	Hv 16380	charcoal	675 +/- 50	AD 1360-1410	Holl, 2002, 165
Houlouf	Gif 6106	charcoal	BP < 100		Holl, 2002, 181
Houlouf	Gif 6107	charcoal	BP < 100		Holl, 2002, 181
Houlouf	Gif 6105	charcoal	BP >100		Holl, 2002, 181
Jenné-Jeno	RL1622	n.d.	2090 +/- 110	400 BC-AD 150	McIntosh and McIntosh, 1995

*rejected or disputed date

Table A.3 cont. 2

site	labcode	dated material	14C age bp	cal. age (2σ)	reference
Jenné-Jeno	RL1620	n.d.	2060 +/- 110	400 BC-AD 200	McIntosh and McIntosh, 1995
Jenné-Jeno	RL1619	n.d.	2050 +/- 120	400 BC-AD 250	McIntosh and McIntosh, 1995
Jenné-Jeno	RL1621	n.d.	1910 +/- 110	200 BC-AD 400	McIntosh and McIntosh, 1995
Jenné-Jeno	RL1574	n.d.	1860 +/- 120	400 BC-AD 150	McIntosh and McIntosh, 1995
Jenné-Jeno	RL1581	n.d.	1800 +/- 120	200 BC-AD 400	McIntosh and McIntosh, 1995
Jenné-Jeno	RL1576	n.d.	1790 +/- 120	50 BC-AD 550	McIntosh and McIntosh, 1995
Jenné-Jeno	RL1580	n.d.	1750 +/- 100	AD 60-540	McIntosh and McIntosh, 1995
Jenné-Jeno	RL1573	n.d.	1590 +/- 110	AD 220-660	McIntosh and McIntosh, 1995
Jenné-Jeno	RL1618	n.d.	1300 +/- 110	AD 550-980	McIntosh and McIntosh, 1995
Jenné-Jeno	RL1578	n.d.	1310 +/- 110	AD 540-980	McIntosh and McIntosh, 1995
Jenné-Jeno	RL1809	n.d.	1310 +/- 110	AD 540-980	McIntosh and McIntosh, 1995
Jenné-Jeno	RL1575	n.d.	1310 +/- 120	AD 450-1000	McIntosh and McIntosh, 1995
Jenné-Jeno	RL1577	n.d.	1220 +/- 110	AD 630-1030	McIntosh and McIntosh, 1995
Jenné-Jeno	RL1579	n.d.	1210 +/- 110	AD 630-1030	McIntosh and McIntosh, 1995
Jenné-Jeno	RL1571	n.d.	1060 +/- 110	AD 700-1250	McIntosh and McIntosh, 1995
Jenné-Jeno	RL1808	n.d.	970 +/- 110	AD 860-1280	McIntosh and McIntosh, 1995
Jenné-Jeno	RL1807	n.d.	940 +/- 110	AD 880-1280	McIntosh and McIntosh, 1995
Jenné-Jeno	RL1617	n.d.	790 +/- 100	AD 1020-1330/AD 1340-1400	McIntosh and McIntosh, 1995
Jenné-Jeno	RL1616	n.d.	550 +/- 100	AD 1260-1530/AD 1570-1630	McIntosh and McIntosh, 1995
Karkarichinkat Sud	N-1395	charcoal	3960 +/- 160	2900-2000 BC	Smith, 1974
Karkarichinkat Sud	N-1394	charcoal	3640 +/- 100	2300-1700	Smith, 1974
Karkarichinkat Nord	N-1396	charcoal	3950 +/- 90	2900-2800/2750-2100 BC	Smith, 1974
Karkarichinkat Nord	N-1397	charcoal	3710 +/- 85	2450-1850 BC	Smith, 1974
Karkarichinkat Nord	N-1398	charcoal	3620 +/- 80	2200-1750 BC	Smith, 1974
Kerma (dog inhumations)	n.d.	n.d.	3650 +/- 60	2200-1880 BC	Chaix, pers. comm. in Guérin and Faure, 1996
Kerma (dog inhumations)	n.d.	n.d.	3035 +/- 170	1440 BC-1050 BC	Chaix, pers. comm. in Guérin and Faure, 1996
Kerma (dog inhumations)	n.d.	n.d.	2825 +/- 115	1400-750 BC	Chaix, pers. comm. in Guérin and Faure, 1996
Kerma (sheep imhumations)	n.d.	n.d.	n.d.	n.d.	n.d.
Kintampo (K6)	UCR-1692	palm nut	6100 +/- 250	5600-4400 BC	Stahl, 1985
Kintampo (K6)	UCR-1691	palm nut	3700 +/- 90	2500-1750 BC	Stahl, 1985
Kintampo (K6)	UCR-1693	palm nut	3605 +/- 100	2300-1650 BC	Stahl, 1985
Kintampo (K6)	Birm-29	n.d.	3570 +/- 84	2150-1730 BC/1720-1690 BC	Flight, 1968
Kintampo (K6)	UCR-1692	palm nut	3550 +/- 127	2300-1500 BC	Stahl, 1985
Kintampo (K6)	UCR-1690	palm nut	3495 +/- 100	2150-1500 BC	Stahl, 1985
Kissi 3	UtC-5670	wood	1495 +/- 45	AD 430-650	Magnavita *et al.*, 2002
Kissi 3	KIA 8548	wool	1456 +/- 38	AD 540-600	Magnavita *et al.*, 2002
Kissi 3	UtC-5671	wood	1393 +/- 33	AD 595-680	Magnavita *et al.*, 2002
Kissi 3B	UtC-9322	charcoal	1250 +/- 40	AD 670-880	Magnavita *et al.*, 2002
Kissi-49	KIA-12488	plant temper	2294 +/- 36	410-340 BC/300-200 BC	S. Magnavita, pers. comm.
Kobadi	Pa-221	burned hippo bone	3335 +/- 100	1890-1420 BC	Raimbault *et al.*, 1987
Kobadi	Pa-916	charcoal	3305 +/- 80	1770-1410 BC	Jousse and Chenal-Vélardé, 2001-2002
Kobadi	Pa-222	burned hippo bone	2880 +/- 120	1400-800 BC	Raimbault *et al.*, 1987
Kobadi	Dak-56	animal bone	2663 +/- 145	1200-400 BC	Raimbault and Dutour, 1989
Kobadi	Pa-223	human bone	2415 +/- 120	850-200 BC	Raimbault *et al.*, 1987
Kolima-Sud	GX-19814-AMS	cattle tooth	3084 +/- 73	1500-1120 BC	MacDonald and Van Neer, 1994
Konduga	Kn-4300	charcoal	6340 +/-250	5800-4700 BC	Thiemeyer, 1992
Korounkorakolé	GX-18376	charcoal	5185 +/- 95	4260-3770 BC	MacDonald, 1997a
Korounkorakolé	GX-18516	charcoal	1020 +/- 105	770-1230 AD	MacDonald, 1997a
Koyom	-	-	-	-	-

Table A.3 cont. 3

site	labcode	dated material	14C age bp	cal. age (2σ)	reference
Krenak	Hv 16376	charcoal	2110 +/- 90	380 BC-AD 60	Holl, 2002, 44
Krenak	Hv 16377	charcoal	1825 +/- 50	AD 70-340	Holl, 2002, 43
Krenak	Hv 16379	charcoal	1720 +/- 65	AD 130-440/AD 490-530	Holl, 2002, 43
Krenak	Hv 16378	charcoal	1630 +/- 60	AD 250-300/AD 310-560	Holl, 2002, 43
Krenak-Sao	RT-1683	charcoal	1160 +/- 45	AD 720-740/AD 770-990	Holl, 2002, 136
Krenak-Sao	RT-1684	charcoal	1120 +/- 45	AD 780-790	Holl, 2002, 135
Kursakata	see Table A.2	see Table A.2	see Table A.2	see Table A.2	see Table A.2
Maadaga	-	-	-	-	-
Maadi	Beta-2805	grain	5010 +/- 50	3950-3690 BC	Hassan, 1985
Maadi	Beta-2804	wood	4730 +/- 60	3640-3370 BC	Hassan, 1985
Madaf	Hv 16375	charcoal	1205 +/- 60	AD 670-970	Holl, 2002, 63
Madaf	Gif 7241	charcoal	1050 +/- 70	AD 810-1160	Holl, 2002, 62
Mandara	Beta-12177*	charcoal	3480 +/- 70	1980-1620 BC	Shinnie and Kense, 1989, 286
Mandara	Beta-12178	charcoal	140 +/- 80	AD 1640-1960	Shinnie and Kense, 1989, 286
Mandara	Beta-12719	charcoal	1100 +/- 260	AD 400-1400	Shinnie and Kense, 1989, 286
Mankuma	Beta-12180	charcoal	400 +/- 90	AD 1320-1350/AD 1390-1650	Shinnie and Kense, 1989, 286
Mankuma	Beta-12181	charcoal	120 +/- 60	AD 1660-1950	Shinnie and Kense, 1989, 286
Mankuma	Beta-12182	charcoal	170 +/- 60	AD 1640-1960	Shinnie and Kense, 1989, 286
Mdaga	Gif 742	n.d.	2375 +/- 150	850-100 BC	Lebeuf *et al.*, 1980, 16
Mdaga	Dak 10	n.d.	2150 +/- 135	550 BC-AD 150	Lebeuf *et al.*, 1980, 16
Mdaga	Dak 26	n.d.	1383 +/- 125	AD 400-950	Lebeuf *et al.*, 1980, 16
Mdaga	Gif 741	n.d.	1260 +/- 100	AD 610-990	Lebeuf *et al.*, 1980, 16
Mdaga	Dak 28	n.d.	981 +/- 117	AD 800-1270	Lebeuf *et al.*, 1980, 16
Mdaga	Gif 740	n.d.	910 +/- 100	AD 890-920/ AD 960-1290	Lebeuf *et al.*, 1980, 16
Mdaga	Dak 11	n.d.	794 +/- 120	AD 1010-1410	Lebeuf *et al.*, 1980, 16
Mdaga	Gif 1171	n.d.	750 +/- 95	AD 1040-1090/AD 1120-1410	Lebeuf *et al.*, 1980, 16
Mdaga	Gif 1172	n.d.	680 +/- 95	AD 1160-1440	Lebeuf *et al.*, 1980, 16
Mdaga	Gif 429	n.d.	330 +/- 120	AD 1400-2000	Lebeuf *et al.*, 1980, 16
Mdaga	Dak 27	n.d.	284 +/- 109	AD 1400-2000	Lebeuf *et al.*, 1980, 16
Mdaga	Gif 1367	n.d.	170 +/- 90	AD 1520-1600/AD 1620-1960	Lebeuf *et al.*, 1980, 16
Mdaga	Gif 428*	n.d.	contemporary		Lebeuf *et al.*, 1980, 16
Mishiskwa	RT-1685	charcoal	635 +/- 45	AD 1280-1410	Holl, 2002, 55
Mishiskwa	BDY 228	charcoal	510 +/- 85	AD 1280-1530/AD1570-1630	Holl, 2002, 55
Mongossi	Hv 12292*	n.d.	1905 +/- 255	600 BC-AD 700	Marliac, 1991, Table 5
Mongossi	Hv 12293	n.d.	915 +/- 85	AD 980-1270	Marliac, 1991, Table 5
Mongossi	Orsay*	n.d.	715 +/- 85	AD 1150-1420	Marliac, 1991, Table 5
Mongossi	Gif 5843*	n.d.	560 +/- 70	AD 1280-1450	Marliac, 1991, Table 5
Mongossi	Orsay	n.d.	450 +/- 95	AD 1300-1370/AD 1300-1650	Marliac, 1991, Table 5
Mongossi	Orsay	n.d.	440 +/- 60	AD 1390-1640	Marliac, 1991, Table 5
Mongossi	Hv 12294*	n.d.	395 +/- 600	out of range	Marliac, 1991, Table 5
Mongossi	Gif 5842*	n.d.	350 +/- 80	AD 1400-1700/AD1750-1800	Marliac, 1991, Table 5
Mouyssam II (KNT 2)	n.d.	charcoal	1715 +/- 60	AD 130-440	Raimbault and Sanogo, 1991
Mouyssam II (KNT 2)	n.d.	charcoal	1660 +/- 80	AD 100-650	Raimbault and Sanogo, 1991
Mouyssam II (KNT 2)	n.d.	charcoal	1650 +/- 120	AD 210-580	Raimbault and Sanogo, 1991
Mouyssam II (KNT 2)	n.d.	charcoal	1640 +/- 70	AD 240-570	Raimbault and Sanogo, 1991
Mouyssam II (KNT 2)	n.d.	charcoal	1580 +/- 75	AD 260-280/AD 330-640	Raimbault and Sanogo, 1991
Mouyssam II (KNT 2)	n.d.	charcoal	1510 +/- 60	AD 420-650	Raimbault and Sanogo, 1991
Mouyssam II (KNT 2)	n.d.	temper	1500 +/- 70	AD 420-660	Raimbault and Sanogo, 1991
Mouyssam II (KNT 2)	n.d.	temper	1490 +/- 70	AD 420-660	Raimbault and Sanogo, 1991

*rejected or disputed date

Table A.3 cont. 4

site	labcode	dated material	14C age bp	cal. age (2σ)	reference
Mouyssam II (KNT 2)	n.d.	charcoal	1430 +/- 70	AD 430-710/AD 740-770	Raimbault and Sanogo, 1991
Mouyssam II (KNT 2)	n.d.	charcoal?	1390 +/- 70	AD 530-780	Raimbault and Sanogo, 1991
Mouyssam II (KNT 2)	n.d.	charcoal	1370 +/- 60	AD 560-780	Raimbault and Sanogo, 1991
Ndiamon-Badat	Beta-85886	*Anadara* shell	2370 +/- 50	BC 750-680/BC 670-630/BC 600-360	Ba *et al.*, 1997
Ndiamon-Badat	Pa-001	*Anadara* shell	1555 +/- 80	AD 340-650	Ba *et al.*, 1997
Ndufu (NA 93/49)	UtC-5153	charcoal	2617 +/- 35	840-750 BC/690-660 BC	Gronenborn, 1998; Wiesmüller, 2001, 38-40
Ndufu (NA 93/49)	UtC-6475	charcoal	2609 +/- 39	850-740 BC/690-600 BC/640-590 BC	Gronenborn, 1998; Wiesmüller, 2001, 38-40
Ndufu (NA 93/49)	UtC-5302	charcoal	2278 +/- 33	410-340 BC/310-200 BC	Gronenborn, 1998; Wiesmüller, 2001, 38-40
Ndufu (NA 93/49)	UtC-5301	charcoal	2088 +/- 34	200-20 BC/10-0 BC	Gronenborn, 1998; Wiesmüller, 2001, 38-40
Ndufu (NA 93/49)	UtC-5298	charcoal	1961 +/- 49	AD 230-440/AD 480-530	Gronenborn, 1998; Wiesmüller, 2001, 38-40
Ndufu (NA 93/49)	UtC-5299	charcoal	1638 +/- 38	AD 260-280/AD 330-540	Gronenborn, 1998; Wiesmüller, 2001, 38-40
Ndufu (NA 93/49)	UtC-6473	charcoal	1599 +/- 45	AD 350-570	Gronenborn, 1998; Wiesmüller, 2001, 38-40
Ndufu (NA 93/49)	UtC-5300	charcoal	1577 +/- 30	AD 410-560	Gronenborn, 1998; Wiesmüller, 2001, 38-40
Ndufu (NA 93/49)	UtC-6474	charcoal	1536 +/- 39	AD 420-610	Gronenborn, 1998; Wiesmüller, 2001, 38-40
Ntereso	SR-52	charred wood	3580 +/- 130	BC 2300-1500	Davies, 1980
Ntereso	SR-61	organic fill	3190 +/- 120	BC 1750-1100	Davies, 1980
Nyanga	Beta-12184	charcoal	180 +/- 60	AD 1640-1960	Shinnie and Kense, 1989, 286
Nyanga	Beta-12183	charcoal	160 +/- 60	AD 1650-1960	Shinnie and Kense, 1989, 286
Ogo	Ly-2159*	n.d.	1910 +/- 210	400 BC-AD 550	Evin *et al.*, 1983
Ogo	Gif-4530	n.d.	1020 +/- 90	AD 800-1220	Evin *et al.*, 1983
Ogo	Gif-4529	n.d.	910 +/- 90	AD 900-1280	Evin *et al.*, 1983
Ogo	Ly-2034	n.d.	790 +/- 100	AD 1020-1330/AD 1340-1400	Evin *et al.*, 1983
Péntènga	KI-3351	charcoal	7590 +/- 90	6610-6240 BC	Frank *et al.*, 2001
Péntènga	UtC-1993	charcoal	7480 +/- 60	6440-6230 BC	Frank *et al.*, 2001
Péntènga	UtC-1992	charcoal	7140 +/- 60	6210-6140 BC/6110-5880 BC	Frank *et al.*, 2001
Péntènga	KI-3352	charcoal	7000 +/- 100	6060-5700 BC	Frank *et al.*, 2001
Péntènga	KI-3474	charcoal	6990 +/- 160	6250-5600 BC	Frank *et al.*, 2001
Péntènga	KN-5187	charcoal	4825 +/- 40	3700-3520 BC	Frank *et al.*, 2001
Péntènga	UtC-2317	charcoal	2590 +/- 50	800-720 BC/700-530 BC	Frank *et al.*, 2001
Péntènga	UtC-2314	charcoal	2580 +/- 40	820-740 BC/690-660 BC/650-540 BC	Frank *et al.*, 2001
Péntènga	KI-3470	charcoal	1650 +/- 90	AD 170-200/AD 210-610	Frank *et al.*, 2001
Péntènga	UtC-2316	charcoal	1640 +/- 50	AD 250-300/AD 320-550	Frank *et al.*, 2001
Péntènga	KI-3350	charcoal	890 +/- 38	AD 1030-1220	Frank *et al.*, 2001
Péntènga	UtC-2315*	charcoal	recent	-	Frank *et al.*, 2001
Qasr Ibrim	OxA 1062	barley straw	2800 +/- 80	1200-800 BC	Rowley-Conwy, 1988
Qasr Ibrim	OxA1061	camel dung pellets	2690 +/- 90	1150-500 BC	Rowley-Conwy, 1988
Qasr Ibrim	OxA 1060	camel mandible	2470 +/- 160	1000-150 BC	Rowley-Conwy, 1988
Rop	I-I,460	human bone	1975 +/- 120	400 BC-AD 350	Fagg, 1965, 1972
Siouré	n.d.	n.d.	n.d.	n.d.	n.d.
Salak	Hv 13966	n.d.	1280 +/- 125	AD 540-102	Marliac, 1991, Table 2
Salak	Hv 13967	n.d.	1165 +/- 90	AD 980-1280	Marliac, 1991, Table 2
Salak	Gif 4567	n.d.	1080 +/- 90	AD 710-750/AD 760-1160	Marliac, 1991, Table 2
Salak	Gif 4568	n.d.	910 +/- 90	5540-5200 BC	Marliac, 1991, Table 2
Salak	Gif 3932	n.d.	210 +/- 80	AD 1490-1960	Marliac, 1991, Table 2
Sodmein	Lv-2086	charcoal	7090 +/- 80	6100-5750 BC	Vermeersch *et al.*, 1994, 1996
Sodmein	Lv-2083	charcoal	6940 +/- 100	6010-5650 BC	Vermeersch *et al.*, 1994, 1996
Sodmein	Lv-2085	charcoal	6390 +/- 90	5840-5040 BC	Vermeersch *et al.*, 1994, 1996
Sodmein	Lv-2082	charcoal	6320 +/- 100	6010-5650 BC	Vermeersch *et al.*, 1994, 1996
Sororo	-	-	-	-	-

*rejected or disputed date

Table A.3 cont. 5

site	labcode	dated material	14C age bp	cal. age (2σ)	reference
Tegdaoust	Dak96	charcoal	1261 +/- 121	AD 570-1020	Vernet and Aumassip, 1989, 27
Tegdaoust	Dak193	charcoal	1233 +/- 100	AD 640-1020	Vernet and Aumassip, 1989, 27
Tegdaoust	Dak195	charcoal	1170 +/- 100	AD 660-1030	Vernet and Aumassip, 1989, 27
Tegdaoust	UCLA696	charcoal	1150 +/- 80	AD 680-1020	Vernet and Aumassip, 1989, 27
Tegdaoust	Dak63	charcoal	1143 +/- 124	AD 650-1160	Vernet and Aumassip, 1989, 27
Tegdaoust	Dak195	charcoal	1080 +/- 110	AD 680-1170	Vernet and Aumassip, 1989, 27
Tegdaoust	Dak94	charcoal	1066 +/- 120	AD 650-1250	Vernet and Aumassip, 1989, 27
Tegdaoust	Dak173	charcoal	1031 +/- 100	AD 770-1220	Vernet and Aumassip, 1989, 27
Tegdaoust	Dak172	charcoal	1017 +/- 100	AD 770-1220	Vernet and Aumassip, 1989, 27
Tegdaoust	Dak97	charcoal	1003 +/- 19	A 980-1050/AD 1100-1120	Vernet and Aumassip, 1989, 27
Tegdaoust	Dak93	charcoal	881 +/- 18	AD 1050-1090/AD 1120-1140/AD 1150-1220	Vernet and Aumassip, 1989, 27
Tegdaoust	Dak95	charcoal	831 +/- 117	AD 980-1330/AD 1340-1400	Vernet and Aumassip, 1989, 27
Tiabel Goudiodé	Pa 434	shell	3980 +/- 60	2850-2800 BC/2700-2250 BC	Togola and Raimbault, 1991
Toguéré Doupwil	WA-103	charcoal	990 +/- 70	AD 890-1120	Bedaux *et al.*, 1978
Toguéré Doupwil	WA-110	charcoal	935 +/- 110	AD 890-1280	Bedaux *et al.*, 1978
Toguéré Doupwil	GrN-7944	charcoal	850 +/- 45	AD 1040-110/AD 1110-1270	Bedaux *et al.*, 1978
Toguéré Doupwil	GrN-7943	charcoal	800 +/ 45	AD 1150-1290	Bedaux *et al.*, 1978
Toguéré Doupwil	GrN-8553	charcoal	675 +/- 45	AD 1260-1400	Bedaux *et al.*, 1978
Toguéré Doupwil	GrN-8118	charcoal	650 +/- 45	AD 1270-1400	Bedaux *et al.*, 1978
Toguéré Doupwil	GrN-8552	charcoal	580 +/- 25	AD 1300-1370/AD 1380-1420	Bedaux *et al.*, 1978
Toguéré Doupwil	GX-3768	human bone	545 +/- 45	AD 1300-1370/AD 1380-1450	Bedaux *et al.*, 1978
Toguéré Doupwil	GX-3767	charcoal	470 +/- 100	AD 1290-1650	Bedaux *et al.*, 1978
Toguéré Galia	GrN-8554	charcoal	1005 +/- 40	AD 900-920/AD 970-1160	Bedaux *et al.*, 1978
Toguéré Galia	GrN-8555	charcoal	930 +/- 50	AD 1020-1220	Bedaux *et al.*, 1978
Toguéré Galia	GrN-7945	charcoal	800 +/- 80	AD 1030-1300/AD1360-1390	Bedaux *et al.*, 1978
Tongo Maaré Diabel	GX-21728	charcoal	1105 +/- 80	AD 690-750/AD 760-1050/AD 1090-1150	MacDonald and MacDonald, 2000
Tulel Fobo	n.d.	n.d.	n.d.	n.d.	n.d.
West Tessalit (MN 51)	Pa-0181	n.d.	4320 +/- 100	3350-2600 BC	Jousse, 2004b
West Tessalit (MN 51)	Pa-0172	n.d.	3920 +/- 100	2850-2800 BC/2750-2050 BC	Jousse, 2004b
Windé Koroji Ouest	GX-19990	charcoal	3635 +/- 90	2300-1700 BC	MacDonald, 1996
Windé Koroji Ouest	GX-19234	charcoal	3115 +/- 195	1900-800 BC	MacDonald, 1996
Windé Koroji Sud-Ouest	GX-19254-AMS	warthog tooth	3319 +/- 57	1750-1490 BC/1480-1460 BC	MacDonald, 1996
Yau	N-476	charcoal	1100 +/- 110	AD 680-1160	Connah, 1981, 205
Yau	N-478	charcoal	1020 +/- 90	AD 800-1220	Connah, 1981, 205
Yau	N-479	charcoal	990 +/- 110	AD 800-1270	Connah, 1981, 205
Yau	N-477	charcoal	675 +/- 110	AD 1050-1090/AD 1150-1450	Connah, 1981, 205
Zampia	GX-19232-AMS	charcoal	2607 +/- 86	950-400 BC	MacDonald, 1994, 238, 1996

Table A.3 cont. 6

Appendix B: Faunal weights by studied locality

	hand and 10 mm													AB								
	E5	E6	E7	E8	F5	F6	F7	F8	G5	G6	G7	G8	**Total**	E5	E6	E7	E8	F6	F7	G6	G8	**Total**
Surface	37	25	16	5	96	137	16	8	32	-	31	11	**414**	-	2	-	-	-	-	-	-	**2**
-10	8	16	14	28	42	101	10	34	28	34	4	13	**331**	-	-	-	-	-	-	-	-	**0**
-20	15	74	73	75	57	56	48	39	65	71	76	21	**670**	-	-	-	-	-	-	-	-	**0**
-30	44	18	170	51	31	44	27	91	25	22	93	43	**659**	-	-	-	-	-	-	-	-	**0**
-40	13	56	15	47	42	42	48	66	37	26	61	31	**483**	-	-	-	-	-	-	-	-	**0**
-50	26	28	35	22	39	51	19	13	39	28	92	85	**476**	-	-	-	-	-	-	-	-	**0**
-60	39	36	39	21	22	21	37	65	-	21	22	28	**353**	-	-	-	-	-	-	-	-	**0**
-70	-	-	35	34	-	-	35	-	-	-	-	13	**117**	-	-	-	-	-	-	-	-	**0**
-80	-	-	-	-	-	-	3	-	-	6	-	20	**29**	-	1	-	-	-	-	-	1	**2**
-90	39	58	47	12	60	19	18	58	-	47	41	21	**420**	-	-	-	-	-	1	-	1	**2**
-100	25	54	74	11	45	66	9	32	-	12	66	67	**460**	-	6	-	-	-	1	-	3	**10**
-110	5	11	21	35	-	45	11	9	-	26	30	14	**206**	-	-	-	-	-	1	-	1	**2**
-120	20	6	15	25	-	56	13	36	-	24	28	50	**273**	-	1	-	-	-	1	-	1	**3**
-130	15	4	24	25	-	15	15	29	-	37	18	3	**185**	-	1	-	-	-	2	2	1	**6**
-140	10	27	13	34	-	27	7	21	-	8	56	32	**235**	1	2	-	-	-	3	-	1	**7**
-150	6	7	19	53	-	21	23	21	-	19	30	12	**211**	-	3	-	-	-	1	-	2	**6**
-160	-	29	37	20	-	21	8	25	-	25	13	23	**201**	-	2	-	-	-	1	-	1	**4**
-170	-	20	22	22	-	11	11	12	-	16	27	41	**181**	-	1	-	-	-	1	-	1	**3**
-180	-	4	43	77	-	15	21	16	-	13	12	4	**205**	-	1	-	-	1	3	-	1	**5**
-190	-	4	15	9	-	7	37	75	-	15	7	29	**196**	-	1	1	-	-	-	-	1	**3**
-200	-	14	19	14	-	14	14	25	-	32	26	30	**188**	-	1	-	-	-	1	-	6	**7**
-210	-	6	8	22	-	25	5	16	-	49	27	5	**163**	-	8	-	-	-	4	-	4	**16**
-220	-	3	19	6	-	99	5	3	-	40	34	-	**210**	-	3	-	-	-	4	-	26	**34**
-230	-	11	91	72	-	34	11	75	-	27	17	15	**352**	-	1	-	-	-	3	-	3	**7**
-240	-	28	25	21	-	80	6	15	-	34	11	7	**226**	-	1	-	-	-	-	-	-	**1**
-250	-	12	22	19	-	37	12	12	-	32	87	26	**258**	-	2	-	-	-	4	-	3	**8**
-260	-	9	10	30	-	43	16	11	-	14	13	24	**170**	-	-	-	-	-	4	-	3	**7**
-270	-	14	31	7	-	28	22	21	-	33	13	15	**183**	-	2	-	-	-	1	-	2	**5**
-280	-	56	21	20	-	20	16	10	-	17	26	14	**199**	-	1	-	-	-	1	-	1	**3**
-290	-	35	29	18	-	13	20	36	-	30	25	5	**211**	-	-	-	-	-	3	-	1	**4**
-300	-	50	18	14	-	29	35	41	-	56	21	-	**264**	-	1	-	-	-	1	-	2	**4**
-310	-	17	140	26	-	25	29	23	-	26	44	25	**354**	-	-	-	-	-	1	-	1	**1**
-320	-	23	43	29	-	17	37	46	-	39	37	11	**281**	-	12	-	-	-	1	-	2	**15**
-330	-	94	101	24	-	61	46	47	-	136	32	18	**560**	-	-	-	-	-	6	-	1	**7**
-340	-	33	124	44	-	28	44	32	-	19	58	23	**403**	-	13	-	-	-	9	-	9	**31**
-350	-	17	40	57	-	11	2	23	-	6	27	25	**208**	-	2	-	-	-	2	-	-	**4**
-360	-	-	14	32	-	1	-	6	-	7	10	-	**70**	-	2	-	-	-	2	-	1	**5**
-370	-	6	26	10	-	5	7	4	-	3	5	3	**70**	-	2	-	-	-	2	-	-	**4**
-380	-	10	1	10	-	14	7	7	-	3	3	5	**61**	-	1	-	-	-	1	-	1	**3**
-390	-	1	1	5	-	2	10	4	-	1	5	1	**28**	-	1	-	-	-	1	-	-	**2**
Total EIA	**302**	**913**	**1510**	**1084**	**434**	**1339**	**757**	**1106**	**226**	**1054**	**1225**	**812**	**10761**	**1**	**63**	**1**	**0**	**1**	**63**	**2**	**75**	**206**
-400	-	1	4	4	-	1	1	3	-	-	4	6	**23**	-	1	-	-	-	1	-	1	**3**
-410	-	-	1	1	-	1	6	2	-	5	20	5	**40**	-	-	-	-	-	-	-	-	**0**
-420	-	7	5	5	-	5	2	1	-	1	7	8	**41**	-	-	-	-	-	1	-	-	**1**
-430	-	-	4	1	-	3	-	5	-	3	2	-	**18**	-	-	-	-	-	-	-	1	**1**
-440	-	-	3	2	-	2	-	1	-	-	-	3	**10**	-	-	-	-	-	-	-	-	**0**
-450	-	2	-	1	-	3	3	1	-	1	2	1	**13**	-	-	-	-	-	-	-	3	**3**
-460	-	-	1	2	-	2	25	-	-	6	2	2	**41**	-	-	-	-	-	-	-	2	**2**
-470	-	-	-	-	-	-	-	-	-	-	-	-	**0**	-	-	-	-	-	-	-	1	**1**
-480	-	-	-	-	-	-	-	-	-	-	-	-	**0**	-	-	-	1	-	-	-	-	**1**
-490	-	-	-	-	-	-	-	-	-	-	-	-	**0**	-	-	-	-	-	-	-	1	**1**
-500	-	-	-	-	-	-	-	1	-	-	1	-	**2**	-	-	-	1	-	-	-	1	**2**
-510	-	-	/1	-	-	-	/1	-	-	-	-	10	**12**	-	-	/-	1	-	/-	-	1	**2**
-520	-	-	/-	-	-	-	/-	-	-	1	1	-	**2**	-	-	/-	-	-	/-	-	-	**0**
-530	-	-	/-	-	-	-	/-	1	-	-	-	-	**1**	-	-	/-	-	-	/-	-	-	**0**
-540	-	-	/-	-	-	-	/-	1	-	-	-	-	**1**	-	-	/-	-	-	/-	-	-	**0**
-550	-	-	/-	-	-	-	/-	-	-	-	-	-	**0**	-	-	/-	1	-	/-	-	-	**1**
-560	-	-	/-	-	-	-	/-	-	-	-	-	-	**0**	-	-	/-	-	-	/-	-	1	**1**
-570	-	-	/-	-	-	-	/-	-	-	-	-	-	**0**	-	-	/-	-	-	/-	-	-	**0**
-580	-	-	/-	-	-	-	/-	-	-	-	-	-	**0**	-	-	/-	-	-	/-	-	-	**0**
-590	-	-	/-	-	-	-	/-	-	-	-	-	-	**0**	-	-	/-	-	-	/-	-	-	**0**
-600	-	-	/-	-	-	-	/-	-	-	-	-	-	**0**	-	-	/-	-	-	/-	-	-	**0**
Total LSA	**-**	**9**	**18**	**15**	**-**	**17**	**36**	**15**	**-**	**17**	**39**	**34**	**202**	**-**	**1**	**0**	**4**	**0**	**2**	**0**	**11**	**18**

/: only half of square excavated; shaded cells indicate maximal excavated depth square

Table B.1: Faunal weights (g) by square and depth (cm) of Oursi (BF 94/45)

	hand and 10 mm										AB
	K5	K6	K7	L5	L6	L7	M5	M6	M7	**Total**	L6
Surface	73	77	43	51	7	94	65	62	124	**596**	1
-10	110	72	16	43	21	182	125	109	151	**829**	3
-20	223	81	50	38	51	163	85	114	151	**956**	18
-30	60	114	44	58	53	232	67	120	63	**811**	11
-40	24	75	60	102	82	67	34	67	82	**593**	20
-50	77	137	93	65	106	145	35	90	212	**960**	48
-60	67	111	74	55	36	196	49	142	164	**894**	-
-70	84	146	108	48	89	184	39	74	106	**878**	-
-80	85	235	148	50	90	209	46	110	241	**1214**	-
-90	161	603	187	112	124	408	31	140	228	**1994**	14
-100	107	392	141	157	98	290	52	159	286	**1682**	8
-110	218	384	294	164	205	296	-	80	95	**1736**	15
-120	164	517	190	160	429	462	-	164	177	**2263**	20
-130	160	383	100	235	250	405	-	228	185	**1946**	9
-140	193	269	81	255	262	387	-	528	234	**2209**	14
-150	130	152	76	176	201	183	-	276	80	**1274**	1
-160	84	353	34	370	143	237	-	235	87	**1543**	1
-170	70	93	88	147	163	197	-	338	195	**1291**	1
-180	88	98	102	123	250	13	-	154	176	**1004**	1
-190	80	57	76	39	109	54	-	393	76	**884**	6
-200	50	29	100	37	53	57	-	167	56	**549**	1
-210	68	60	101	-	26	45	-	49	64	**413**	-
-220	9	69	113	-	16	58	-	58	61	**384**	-
-230	34	201	55	-	4	159	-	93	45	**591**	-
-240	45	13	66	-	16	65	-	80	19	**304**	-
Total LIA	**2464**	**4721**	**2440**	**2485**	**2883**	**4788**	**628**	**4030**	**3358**	**27797**	**191**
-250	18	34	40	-	5	19	-	19	12	**147**	1
-260	-	54	97	-	8	69	-	-	360	**588**	-
-270	-	119	93	-	13	80	-	6	22	**333**	-
-280	-	176	210	-	27	112	-	11	39	**575**	-
-290	-	266	270	-	78	93	-	54	68	**829**	-
-300	-	129	355	-	159	113	-	10	214	**980**	2
-310	-	-	193	-	100	149	-	47	116	**605**	-
-320	-	-	133	-	140	435	-	174	131	**1013**	-
-330	-	-	51	-	287	293	-	195	299	**1125**	1
-340	-	-	62	-	504	96	-	329	293	**1284**	-
-350	-	-	48	-	318	54	-	396	193	**1009**	1
-360	-	-	-	-	150	113	-	413	76	**752**	-
-370	-	-	-	-	94	84	-	142	78	**398**	-
-380	-	-	-	-	65	58	-	110	66	**299**	-
-390	-	-	-	-	71	83	-	125	80	**359**	-
-400	-	-	-	-	72	116	-	157	61	**406**	1
-410	-	-	-	-	91	48	-	114	90	**343**	-
-420	-	-	-	-	86	51	-	144	134	**415**	-
-430	-	-	-	-	7	26	-	10	15	**58**	-
-440	-	-	-	-	6	1	-	1	-	**8**	-
-450	-	-	-	-	5	-	-	-	-	**5**	-
-460	-	-	-	-	-	-	-	1	15	**16**	-
-470	-	-	-	-	10	3	-	1	-	**14**	-
-480	-	-	-	-	10	-	-	-	-	**10**	-
-490	-	-	-	-	4	17	-	-	20	**41**	-
-500	-	-	-	-	-	4	-	1	35	**40**	-
-510	-	-	-	-	4	-	-	-	-	**4**	-
-520	-	-	-	-	1	1	-	-	5	**7**	-
-530	-	-	-	-	-	16	-	1	-	**17**	-
Total MIA	**18**	**778**	**1552**	**-**	**2317**	**2134**	**-**	**2461**	**2422**	**11682**	**5**

shaded cells indicate maximal excavated depth square

Table B.2: Faunal weights (g) by square and depth (cm) of Oursi village (BF 97/13)

	hand and 10 mm										AB
	K5	K6	K7	L5	L6	L7	M5	M6	M7	**Total**	L6
-540	-	-	-	-	-	3	-	1	2	**6**	-
-550	-	-	-	-	1	3	-	-	-	**4**	1
-560	-	-	-	-	-	2	-	1	3	**6**	-
-570	-	-	-	-	-	5	-	2	2	**9**	-
-580	-	-	-	-	3	1	-	2	6	**12**	-
-590	-	-	-	-	23	-	-	3	6	**32**	-
-600	-	-	-	-	11	6	-	6	4	**27**	1
-610	-	-	-	-	7	-	-	7	4	**18**	-
-620	-	-	-	-	4	-	-	14	7	**25**	-
-630	-	-	-	-	3	-	-	-	11	**14**	-
-640	-	-	-	-	11	-	-	1	5	**17**	-
-650	-	-	-	-	2	-	-	14	31	**47**	1
-660	-	-	-	-	10	-	-	2	88	**100**	-
-670	-	-	-	-	1	-	-	1	41	**43**	-
-680	-	-	-	-	2	-	-	6	-	**8**	-
-690	-	-	-	-	1	-	-	7	4	**12**	-
-700	-	-	-	-	3	-	-	7	4	**14**	1
-710	-	-	-	-	3	-	-	4	-	**7**	-
-720	-	-	-	-	1	-	-	8	-	**9**	-
-730	-	-	-	-	30	-	-	8	-	**38**	-
-740	-	-	-	-	4	-	-	7	-	**11**	-
-750	-	-	-	-	2	-	-	16	-	**18**	1
-760	-	-	-	-	1	-	-	10	-	**11**	-
-770	-	-	-	-	3	-	-	14	-	**17**	-
-780	-	-	-	-	13	-	-	-	-	**13**	-
-790	-	-	-	-	2	-	-	18	-	**20**	-
-800	-	-	-	-	5	-	-	-	-	**5**	1
-810	-	-	-	-	13	-	-	9	-	**22**	-
Total EIA	**-**	**-**	**-**	**-**	**157**	**20**	**-**	**168**	**218**	**563**	**5**

shaded cells indicate maximal excavated depth square

Table B.2 cont.

	hand 2.5 and 1 mm		hand 2.5 and 1 mm		hand 2.5 and 1 mm
-50	49	-165	22	-280	7
-55	44	-170	42	-285	54
-60	39	-175	-	-290	45
-65	62	-180	67	-295	115
-70	70	-185	19	-300	132
-75	35	-190	17	-305	118
-80	21	-195	44	-310	26
-85	27	-200	34	-315	48
-90	27	-205	28	-320	-
-95	47	-210	29	-325	-
-100	31	-215	27	-330	17
-105	40	-220	75	-335	2
-110	24	-225	96	-338	15
-115	51	-230	43	-340	6
-120	5	-235	33	-345	5
-125	24	-240	31	-350	4
-130	61	-245	19	-355	-
-135	8	-250	14	-360	4
-140	24	-255	18	-365	-
-145	26	-260	18	-370	5
-150	36	-265	9	-375	-
-155	37	-270	21	-380	7
-160	22	-275	-	**Total**	**2126**

Table B.3: Faunal weights (g) by depth (cm) of Kissi 22 (BF 96/22)

	hand, 2.5 and 1 mm							
	E5	F4	F5	F6	G4	G5	Feats.	**Total**
+20	-	20	6	-	-	6	-	**32**
+10	-	-	2	-	-	1	-	**3**
Surface	-	45	11	-	-	24	-	**80**
-10	-	4	8	-	1	4	-	**17**
-20	-	2	-	-	6	14	-	**22**
-30	-	20	12	-	6	3	-	**41**
-40	-	7	5	-	-	3	-	**15**
-50	-	3	1	-	8	5	-	**17**
-60	-	13	7	-	44	4	-	**68**
-70	-	3	23	-	1	1	-	**28**
-80	-	5	21	-	2	7	-	**35**
-90	-	2	1	-	1	18	-	**21**
-100	-	9	11	-	-	2	47	**23**
-110	-	4	4	-	4	2	-	**14**
-120	-	-	6	-	7	1	-	**14**
-130	-	3	2	-	3	2	-	**11**
-140	-	51	7	-	14	15	4	**87**
-150	-	45	8	-	24	19	97	**96**
-160	-	7	8	-	4	6	92	**25**
-170	-	-	30	-	17	84	-	**131**
-180	-	36	20	-	62	31	20	**149**
-190	-	79	10	-	54	7	69	**149**
-200	-	21	5	10	85	8	37	**128**
-210	-	55	44	27	68	2	-	**196**
-220	-	-	52	10	145	9	100	**215**
-230	-	4	9	-	64	19	164	**96**
-240	-	15	13	-	148	1	-	**176**
-250	-	-	3	-	13	-	-	**16**
-260	12	-	-	-	-	2	-	**15**
-270	-	4	37	-	3	2	-	**46**
-280	-	4	13	-	4	12	-	**34**
-290	-	9	24	-	3	6	-	**42**
-300	-	9	6	-	29	5	54	**50**
-310	-	2	-	-	14	3	-	**19**
-320	-	3	2	-	6	9	-	**20**
-330	-	6	3	-	4	-	-	**13**
-340	-	7	12	-	18	-	-	**37**
-350	-	-	3	-	1	9	-	**14**
-360	-	-	2	-	1	3	-	**6**
-370	-	-	-	-	-	9	-	**9**
-380	-	-	4	-	-	6	-	**10**
-390	-	-	1	-	-	-	-	**1**
-400	-	-	5	-	-	1	-	**6**
-410	-	-	-	-	1	-	-	**1**
-420	-	-	2	-	-	-	-	**2**
-430	-	-	1	-	-	-	-	**1**
-440	-	-	1	-	-	-	-	**1**
-450	-	-	-	-	-	-	-	**0**
Total	**12**	**431**	**425**	**47**	**865**	**336**	**684**	**2117**

Feats.=features=hearths and pits at the site

Table B.4: Faunal weights (g) by square and depth (cm) of Kissi 40 (BF 97/31)

	hand and 10 mm									
	E5	E6	E7	F5	F6	F7	G5	G6	G7	**Total**
-10	37	54	74	-	24	34	12	20	32	**287**
-20	62	68	111	81	40	46	47	34	26	**514**
-30	40	50	149	46	37	58	27	42	10	**460**
-40	70	62	119	63	67	26	38	20	108	**572**
-50	72	76	26	37	57	20	50	23	62	**421**
-60	65	52	3	58	29	19	59	40	25	**350**
-70	157	34	7	52	23	28	84	68	9	**462**
-80	32	34	28	17	-	52	53	25	35	**275**
-90	64	19	54	45	37	505	9	104	159	**996**
-100	34	21	43	22	33	134	57	48	99	**491**
-110	49	16	66	12	86	29	67	82	-	**407**
-120	32	16	96	164	50	24	99	81	-	**564**
-130	86	57	49	132	73	53	153	75	-	**679**
-140	56	68	44	83	37	96	211	88	51	**733**
-150	39	131	45	162	173	93	97	55	-	**795**
-160	28	40	45	105	72	76	77	37	-	**479**
-170	42	55	39	140	52	108	37	38	-	**509**
-180	97	1	32	26	28	35	45	46	-	**311**
-190	97	62	78	75	106	60	22	49	-	**549**
-200	58	63	70	26	39	-	28	44	-	**327**
-210	91	55	112	25	91	-	20	24	-	**418**
-220	34	46	89	42	26	-	31	90	-	**358**
-230	22	40	51	104	35	-	19	49	-	**320**
-240	85	56	104	62	48	-	20	-	-	**375**
-250	9	22	30	35	11	-	-	56	-	**162**
-260	20	7	29	27	114	-	29	63	-	**289**
-270	29	57	15	21	584	-	63	33	-	**802**
-280	88	-	50	78	34	-	32	17	-	**299**
-290	15	-	-	29	39	-	91	18	-	**193**
-300	41	39	18	42	15	-	21	20	-	**195**
-310	-	43	14	12	12	-	32	29	-	**143**
-320	153	141	12	48	23	-	16	16	-	**409**
-330	83	14	22	-	29	-	26	94	-	**268**
-340	8	60	74	46	-	62	14	20	-	**284**
-350	22	8	74	61	20	-	36	45	-	**266**
-360	2	28	1	23	64	74	28	54	-	**276**
-370	20	28	63	34	53	-	70	72	-	**339**
-380	2	79	44	78	113	-	69	53	-	**437**
-390	-	87	173	88	154	-	49	83	-	**634**
-400	28	11	-	6	23	25	12	66	-	**171**
-410	15	20	-	20	10	-	20	48	-	**134**
-420	17	32	-	54	121	-	44	15	-	**283**
-430	29	45	-	80	118	-	87	118	-	**477**
-440	61	30	-	19	1	-	80	57	-	**249**
-450	33	44	-	22	25	-	24	-	-	**148**
-460	94	65	-	33	42	-	48	9	-	**290**
-470	50	31	-	19	27	-	43	32	-	**202**
-480	47	39	-	42	70	-	32	82	-	**312**
-490	36	46	-	64	117	-	68	63	-	**394**
-500	35	24	-	56	64	-	11	34	-	**224**
-510	31	12	-	35	22	-	16	30	-	**146**
-520	14	-	-	13	10	-	19	11	-	**68**
-530	3	-	-	7	9	-	5	17	-	**41**
-540	1	-	-	7	4	-	4	5	-	**21**
-550	6	-	-	-	15	-	4	4	-	**29**
-560	13	-	-	4	-	-	9	-	-	**26**
-570	8	-	-	1	-	-	2	-	-	**10**
Total	**2464**	**2188**	**2153**	**2682**	**3203**	**1657**	**2466**	**2445**	**616**	**19873**

shaded cells indicate maximal excavated depth square

Table B.5: Faunal weights (g) by square and depth (cm) of Saouga 94/120 (BF 94/120)

	hand and 10 mm						
	E8	E9	F8	F9	G8	G9	**Total**
-120	18	19	39	14	43	62	**195**
-130	63	68	28	21	57	54	**291**
-140	48	35	82	43	36	53	**297**
-150	51	45	86	61	14	24	**281**
-160	73	95	13	18	5	10	**213**
-170	11	53	6	10	24	8	**111**
-180	9	12	52	77	46	21	**218**
-190	19	3	4	17	29	4	**77**
-200	2	8	12	5	26	8	**61**
-210	9	5	3	6	7	6	**35**
-220	1	7	9	5	1	21	**44**
-230	-	12	-	5	1	13	**31**
-240	-	22	86	29	64	49	**251**
-250	-	61	52	-	96	66	**275**
-260	-	83	231	1	143	61	**520**
-270	-	193	55	72	63	110	**493**
-280	-	12	46	65	54	89	**265**
-290	-	50	59	82	56	88	**335**
-300	31	-	27	23	53	77	**211**
-310	-	87	-	-	40	53	**180**
-320	-	58	-	39	81	57	**236**
-330	-	19	-	36	111	52	**217**
-340	-	28	-	17	27	12	**84**
-350	-	10	-	17	35	9	**72**
-360	-	11	-	40	32	34	**117**
-370	-	14	-	35	36	37	**121**
-380	-	3	-	58	36	98	**195**
-390	-	27	-	-	10	98	**135**
-400	-	28	-	31	18	22	**97**
Total	**336**	**1067**	**890**	**826**	**1244**	**1297**	**5659**

shaded cells indicate maximal excavated depth square

Table B.6: Faunal weights (g) by square and depth (cm) of Saouga 95/7 (BF 95/7)

	hand and 5 mm									
	A1	A3	A3	B1	B2	B3	C1	C2	C3	**Total**
-10	-	-	-	-	-	-	-	-	-	**0**
-20	-	-	-	-	-	-	-	-	-	**0**
-30	-	-	-	-	-	-	-	-	-	**0**
-40	-	-	-	-	-	-	-	-	-	**0**
-50	-	-	-	-	-	-	-	-	-	**0**
-60	-	-	-	-	-	-	-	1	-	**1**
-70	3	-	-	-	-	-	222	-	-	**224**
-80	11	-	-	4	-	-	-	2	-	**17**
-90	-	-	4	12	1	-	-	-	-	**18**
-100	4	-	-	7	59	-	-	-	15	**84**
-110	3	-	-	-	1	2	-	23	-	**29**
-120	9	-	-	-	-	-	2	-	-	**11**
-130	-	-	-	6	-	-	-	-	-	**6**
-140	-	-	-	-	-	-	1	-	9	**11**
-150	-	-	-	9	-	-	-	-	-	**9**
-160	-	-	-	-	16	-	-	-	-	**16**
-170	-	-	-	-	-	-	-	-	-	**0**
-180	-	-	-	-	-	-	-	-	-	**0**
-190	-	-	-	5	-	-	-	-	-	**5**
-200	-	-	-	-	-	-	-	-	-	**0**
-210	-	-	-	15	-	-	-	-	-	**15**
-220	-	-	-	4	-	-	-	-	-	**4**
-230	1	-	-	7	-	-	-	-	-	**8**
-240	2	-	-	-	-	-	-	-	-	**2**
-250	-	-	-	-	-	-	-	-	-	**0**
Total	**31**	**0**	**4**	**70**	**77**	**2**	**225**	**27**	**24**	**461**

Table B.8: Faunal weights (g) by square and depth (cm) of Gajiganna C (NA 90/5C)

	hand and 5 mm									
	A1	A2	A3	B1	B2	B3	C1	C2	C3	**Total**
Surface	-	-	-	-	-	4	-	-	-	**4**
-10	3	10	4	-	1	8	-	-	-	**27**
-20	-	-	-	51	-	-	47	20	-	**118**
-30	-	-	1	3	13	-	-	-	-	**17**
-40	1	17	-	-	3	-	2	-	-	**23**
-50	2	-	5	-	-	-	-	-	-	**8**
-60	-	-	-	-	-	-	-	-	-	**0**
-70	3	1	-	-	-	-	-	-	6	**10**
-80	19	-	1	-	-	-	-	-	-	**20**
Total	**27**	**29**	**11**	**54**	**17**	**12**	**49**	**20**	**7**	**225**

Table B.7: Faunal weights (g) by square and depth (cm) of Gajiganna D (NA 90/5D)

	hand and 5 mm										
	A1	A2	A3	B1	B2	B3	C1	C2	C1/C2	C3	**Total**
Surface	-	-	-	-	-	-	-	-	-	-	**0**
-20	-	-	-	-	-	-	-	-	-	-	**0**
-30	-	-	-	-	-	-	-	-	-	-	**0**
-40	-	-	-	-	-	-	-	-	-	-	**0**
-50	-	-	-	-	-	-	-	-	-	-	**0**
-60	-	-	-	-	-	-	-	-	-	-	**0**
-70	-	-	-	-	-	-	-	-	-	-	**0**
-80	-	-	-	-	-	-	-	-	-	-	**0**
-90	-	-	-	-	-	-	-	-	-	-	**0**
-100	-	1	-	-	-	8	-	-	-	-	**9**
-110	-	4	2	-	-	12	-	-	-	11	**28**
-120	152	135	67	75	75	47	50	5	-	11	**618**
-130	42	90	42	45	111	29	10	4	-	7	**380**
-140	11	33	38	59	6	19	11	23	-	25	**225**
-150	-	5	15	5	34	5	16	41	-	24	**146**
-160	27	3	67	86	173	213	107	193	-	8	**876**
-170	52	60	85	127	196	69	68	115	-	26	**797**
-180	227	244	358	210	192	19	830	18	4056*	2	**6153**
-190	108	249	63	-	9	3	43	-	-	3	**477**
-200	-	9	23	14	47	6	51	-	-	4	**153**
-210	-	23	-	-	-	23	3	-	-	-	**49**
-220	30	-	-	3	14	-	57	38	-	2	**145**
-240	55	3	11	49	8	59	94	-	-	38	**317**
-260	23	17	-	112	-	10	18	-	-	-	**181**
-280	171	74	31	61	-	-	-		-	-	**336**
-300	47	96	5	42	145	-	-	-	-	-	**335**
-320	638	114	-	239	-	-	382	7	-	-	**1380**
Total	**1583**	**1159**	**805**	**1127**	**1012**	**519**	**1742**	**444**	**4056**	**159**	**12605**

*-160 to 180 cm

Table B.9: Faunal weights (g) by square and depth (cm) of Gajiganna BII (NA 90/5II)

	hand 5 mm
1	3
2	2
3	-
4	172
5	43
Total	**221**

Table B.10: Faunal weights (g) by strat. layer of NA 91/1A

	hand 5 mm
-10	-
-20	-
-30	105
-40	319
-50	396
-60	30
-70	-
-80	-
-90	-
-100	-
-110	-
-120	-
-130	-
-140	-
-150	-
-160	-
-170	-
-180	49
-190	15
-200	26
-210	9
-220	21
-230	2
Total	**972**

Table B.11: Faunal weights (g) by depth (cm) of NA 93/42

	hand 5 mm
-10	-
-20	-
-30	-
-40	-
-50	-
-60	-
-70	13
-80	3
-90	9
-100	-
-110	-
-120	-
-130	-
-140	-
-150	-
-160	-
-170	-
-180	191
-190	47
-200	82
-210	25
-220	-
Total	**371**

Table B.12: Faunal weights (g) by depth (cm) of Labe (NA 97/24)

	hand 5 mm
-10	30
-20	33
-30	48
-40	152
-50	102
-60	-
-70	10
-80	3
-90	3
-100	3
-110	22
-120	4
-130	5
-140	-
-150	1
-160	-
-170	-
-180	-
-190	-
-200	-
-210	-
-220	-
-230	-
-240	-
-250	-
Total	**417**

Table B.13: Faunal weights (g) by depth (cm) of NA 93/36

	hand and 5 mm						
	A1	A2	A3	B1	B2	B3	**Total**
-10	2	-	-	-	-	-	**2**
-20	17	-	8	9	-	3	**37**
-30	29	7	35	3	5	-	**78**
-40	21	47	30	37	379	-	**514**
-50	78	29	42	53	145	23	**371**
-60	73	72	185	345	93	54	**821**
-70	20	132	187	43	106	-	**489**
Total Gaj IIc	**240**	**286**	**486**	**491**	**728**	**80**	**2312**
-80	59	859	199	21	45	91	**1275**
-90	50	35	45	13	195	90	**427**
-100	43	146	52	288	130	37	**695**
-110	19	10	11	7	26	8	**81**
-120	12	31	8	14	19	17	**101**
-130	6	-	4	10	73	7	**100**
Total Gaj IIb	**189**	**1081**	**318**	**353**	**489**	**250**	**2679**
-140	-	-	-	17	-	341	**358**
-150	-	-	-	-	-	-	**0**
-160	-	-	-	23	-	-	**23**
-170	-	-	-	-	-	-	**0**
Total Gaj IIa	**0**	**0**	**0**	**40**	**0**	**341**	**381**

Table B.14: Faunal weights (g) by square and depth (cm) of Gilgila (NA 99/65)

	hand 5 mm
1/2	116
2	6
3	562
4	241
Total	**925**

Table B.15: Faunal weights (g) by strat. layer of NA 93/10

	hand and 5 mm												
	Zilum 1	Zilum 3	Zilum 4	Zilum 5	Zilum 5 (pit 1)	Zilum 5 (pit 2)	Zilum 5 - all	Zilum 6	Zilum 6 (pit 1)	Zilum 6 (pit 2)	Zilum 6 (pit 3)	Zilum 6 - all	**Total**
Surface	-	-	-	34	-	-	34	21	-	-	-	21	90
0-20	-	-	-	-	-	-	-	14	-	-	-	14	14
-10	-	15	-	63	-	-	63	121	-	-	-	121	263
-20	-	-	-	221	-	-	221	159	-	-	-	159	600
-20/-30	-	52	-	-	-	-	-	-	-	-	-	0	52
-30	-	-	-	92	-	-	92	87	-	-	-	87	272
-40	-	-	-	218	-	-	218	131	-	-	-	131	567
-50	-	4	-	83	207	-	290	305	-	-	-	305	682
-60	-	-	-	543	142	-	685	308	-	-	-	308	1537
-70	4	-	-	74	359	-	432	146	-	-	-	146	656
-80	-	2	-	23	-	-	23	129	-	-	-	129	177
-90	-	-	-	106	-	-	106	128	307	17	-	452	663
-100	-	-	-	13	15	-	27	331	-	-	20	351	391
-110	-	-	-	-	56	-	56	168	-	-	-	168	224
-120	-	-	-	65	-	104	170	51	-	-	-	51	286
-130	-	-	12	2	-	14	16	32	54	25	5	116	147
-140	-	-	14	-	-	-	0	4	57	-	-	61	75
-150	-	-	29	-	-	-	0	11	-	-	-	11	40
-150/-160	-	-	-	-	-	-	0	-	292	-	-	292	292
-160	-	-	29	-	-	-	0	18	-	-	-	18	47
-170	-	-	52	-	-	-	0	118	-	-	-	118	170
-180	-	-	12	-	-	-	0	-	-	-	-	0	12
-190	-	-	-	-	-	-	0	-	-	-	-	0	0
-200	-	-	-	-	-	-	0	-	-	-	-	0	0
-210	-	-	-	-	-	-	0	-	-	-	-	0	0
-220	-	-	8	-	-	-	0	-	-	-	-	0	8
-230	-	-	-	-	-	-	0	-	-	-	-	0	0
-240	-	-	-	-	-	-	0	-	-	-	-	0	0
-250	-	-	-	-	-	-	0	-	-	-	-	0	0
-260	-	-	-	-	-	-	0	-	-	-	-	0	0
Total	**4**	**73**	**156**	**1537**	**779**	**119**	**2435**	**2282**	**711**	**42**	**25**	**3059**	**7264**

shaded cells indicate maximal excavated depth trench

Table B.16: Faunal weights (g) by square and depth (cm) of the studied trenches at Zilum (NA 97/37)

	hand and 5 mm									
	A1	A2	A3	B1	B2	B3	C1	C2	C3	**Total**
-10	22	-	23	-	13	19	7	10	7	**101**
-20	230	13	37	5	18	106	32	99	41	**579**
-30	-	-	28	33	46	49	32	72	267	**526**
-40	6	6	155	72	43	137	35	40	-	**494**
-50	16	262	41	66	83	40	21	23	22	**575**
-60	33	43	54	43	63	69	-	11	31	**347**
-70	40	9	25	5	9	43	33	16	61	**240**
-80	32	47	179	16	58	62	53	-	-	**448**
-90	4	47	10	51	33	10	79	14	10	**260**
-100	4	50	19	3	13	9	37	17	18	**170**
-110	7	6	9	-	22	7	6	-	7	**64**
-120	-	-	8	8	-	-	9	-	-	**25**
-130	-	-	-	12	4	-	4	-	-	**20**
-140	5	-	-	9	-	-	9	-	-	**24**
-150	20	-	-	12	-	-	5	-	-	**36**
-160	4	-	-	32	31	-	24	-	-	**91**
-170	34	-	-	27	35	-	95	-	-	**191**
-180	64	-	-	101	83	-	32	-	-	**279**
-190	18	-	-	240	178	-	3	-	-	**438**
-200	172	-	-	89	37	-	17	-	-	**315**
Total Gaj IIa/b	**709**	**483**	**587**	**822**	**771**	**552**	**532**	**302**	**465**	**5222**
-210	34	-	-	24	19	-	15	-	-	**93**
-220	-	-	-	16	-	-	78	-	-	**94**
-230	-	-	-	6	86	-	-	-	-	**92**
-240	68	-	-	56	32	-	25	-	-	**181**
-250	82	-	-	65	78	-	84	-	-	**309**
-260	60	-	-	33	148	-	22	-	-	**264**
-270	106	-	-	137	125	-	113	-	-	**481**
-280	141	-	-	26	51	-	48	-	-	**266**
-290	51	-	-	-	74	-	-	-	-	**125**
-300	-	-	-	20	40	-	11	-	-	**70**
-310	109	-	-	361	35	-	331	-	-	**836**
-320	148	-	-	1368	1914	-	2003	-	-	**5432**
-330	166	-	-	462	501	-	905	-	-	**2035**
-340	5	-	-	39	33	-	-	-	-	**76**
-350	12	-	-	67	7	-	18	-	-	**104**
-360	19	-	-	67	-	-	54	-	-	**141**
-370	-	-	-	3	-	-	1	-	-	**4**
Total Gaj I	**1001**	**-**	**-**	**2750**	**3142**	**-**	**3709**	**-**	**-**	**10601**

Table B.17: Faunal weights (g) by square and depth (cm) of Bukarkurari (NA 97/33)

	hand 5 mm
-10	-
-20	-
-30	-
-40	-
-50	-
-60	-
-70	21
-80	44
-90	69
-100	121
-110	115
-120	-
-130	99
-140	228
-150	361
-160	328
-170	634
-180	118
-190	21
-200	166
-210	709
-220	279
-230	143
-240	369
-250	379
-260	906
-270	163
-280	-
-290	9
-300	74
-310	30
-320	6
Total	**5392**

Table B.18: Faunal weights (g) by depth (cm) of Kelumeri (NA 96/45)

	hand and 5 mm																		
	Surface-40	-50	-60	-70	-80	-90	-100	-110	-120	-130	-140	-150	-160	-170	-180	-190	-200	-210	**Total**
A1	-	-	-	-	-	2	60	3	2	-	20	46	10	29	16	22	4	3	**217**
A2	7	-	-	-	1	-	4	22	3	8	21	44	1	38	19	38	54	23	**282**
A3	1	-	-	-	-	1	4	26	-	4	29	40	19	26	21	25	6	4	**204**
A4	1	-	-	-	11	-	-	-	1	19	15	4	65	25	9	6	-	-	**154**
A5	-	-	-	-	-	5	-	2	-	-	3	41	29	41	16	19	5	-	**161**
A6	-	-	-	1	-	1	-	1	12	2	22	-	33	26	5	-	-	-	**101**
A7	-	-	-	-	-	-	7	4	-	-	4	16	27	-	12	-	-	-	**71**
A8	-	-	-	2	-	-	-	-	2	-	72	15	16	8	7	-	-	-	**122**
A9	-	-	-	6	10	-	49	6	6	10	25	23	18	39	20	1	-	-	**213**
A10	1	-	1	-	-	-	8	56	24	132	79	39	46	33	18	-	-	-	**436**
B1	-	-	-	-	2	14	1	-	23	10	19	80	58	5	19	17	34	6	**290**
B2	-	-	-	-	3	5	16	18	5	12	60	56	20	7	-	3	52	23	**277**
B3	-	-	-	-	4	8	1	7	2	12	42	49	13	11	10	16	32	-	**206**
B4	-	-	-	-	15	30	-	4	19	-	38	14	9	20	73	9	5	-	**235**
B5	1	-	-	-	-	8	-	8	-	4	38	28	12	5	-	-	-	-	**103**
B6	-	-	-	-	4	-	-	13	-	-	28	18	30	18	5	-	-	-	**117**
B7	4	-	1	-	-	-	7	-	8	34	16	26	-	25	10	8	-	-	**139**
B8	-	-	-	2	2	4	3	17	20	28	10	5	47	127	35	6	12	-	**318**
B9	-	-	-	1	-	6	3	22	37	34	21	33	18	9	9	-	-	-	**194**
B10	-	6	-	-	-	2	12	11	-	23	10	47	94	33	1	12	-	-	**250**
C1	-	-	-	-	-	12	14	13	8	6	82	79	43	11	17	-	6	-	**291**
C2	4	-	-	2	17	8	8	65	3	9	52	12	19	10	10	7	10	-	**238**
C3	-	-	-	-	3	13	9	4	13	38	64	34	17	-	48	19	30	2	**294**
C4	2	-	-	-	11	6	17	22	46	29	62	61	107	-	28	5	-	-	**396**
C5	-	-	-	-	5	3	9	43	8	7	118	32	148	418	138	20	-	-	**949**
C6	4	4	-	2	7	4	15	75	48	258	302	131	61	168	18	17	6	-	**1118**
C7	4	-	1	21	3	8	39	18	17	95	57	9	34	44	57	28	1	-	**432**
C8	-	-	2	4	-	3	74	51	16	7	79	28	38	36	35	12	-	1	**386**
C9	-	-	3	-	-	-	3	11	27	5	46	9	6	64	7	-	-	-	**180**
C10	-	1	-	5	-	2	42	112	37	38	45	48	-	18	4	14	-	-	**365**
Total	**28**	**10**	**8**	**45**	**98**	**142**	**403**	**633**	**387**	**824**	**1478**	**1066**	**1037**	**1294**	**666**	**304**	**255**	**62**	**8738**

Table B.19: Faunal weights (g) by square and depth (cm) of Kariari C (NA 95/1)

	hand and 5 mm		
	A1	A2	Total
-10	2	7	**9**
-20	17	43	**59**
-30	64	16	**80**
-40	46	52	**97**
-50	19	26	**45**
-60	13	39	**52**
-70	23	14	**37**
-80	12	36	**47**
-90	26	57	**83**
-100	12	34	**46**
-110	12	8	**21**
-120	58	16	**73**
-130	22	10	**32**
-140	26	17	**42**
-150	22	10	**31**
-160	14	32	**47**
-170	11	22	**33**
-180	24	27	**51**
-190	11	5	**16**
-200	10	23	**34**
-210	11	-	**11**
-200	-	2	**2**
-210	-	-	**0**
-220	-	-	**0**
-230	-	-	**0**
Total	**455**	**494**	**949**

Table B.20: Faunal weights (g) by square and depth (cm) of Labe Kanuri (NA 97/26)

	hand and 5 mm			
	A1	A1+ A2	A2	Total
Surface	3	-	3	**5**
-20	-	-	-	**0**
-30	-	-	-	**0**
-40	-	-	-	**0**
-50	-	-	-	**0**
-60	-	-	-	**0**
-70	15	-	3	**17**
-80	16	-	9	**25**
-90	14	-	1	**15**
-100	11	-	11	**22**
-110	11	-	12	**23**
-120	7	-	12	**18**
-130	42	-	-	**42**
-140	23	-	26	**49**
-150	48	-	22	**70**
-160	-	-	-	**0**
-170	-	-	-	**0**
-180	-	-	-	**0**
-190	-	86	-	**86**
-200	-	-	-	**0**
-210	-	-	-	**0**
-220	-	41	-	**41**
-230	24	-	47	**72**
-240	47	-	-	**47**
-250	20	-	16	**37**
-260	91	-	8	**98**
-270	46	-	24	**71**
-280	31	-	-	**31**
-290	-	-	40	**40**
-300	23	-	140	**163**
-310	12	-	-	**12**
-320	-	-	-	**0**
-330	-	-	80	**80**
Total	**485**	**127**	**452**	**1064**

Table B.21: Faunal weights (g) by square and depth (cm) of Elkido North (NA 99/75)

	hand and 5 mm				
	A1	A2	B1	B2	Total
-10	-	-	-	4	**4**
-20	-	-	-	-	**0**
-30	-	-	-	-	**0**
-40	-	-	-	-	**0**
-50	5	4	15	4	**27**
-60	6	7	7	9	**29**
-70	10	16	6	30	**61**
-80	23	24	12	18	**77**
-90	30	17	19	43	**108**
-100	16	99	9	20	**144**
-110	15	10	20	19	**63**
-120	12	32	21	48	**112**
-130	45	-	69	3	**118**
-140	5	15	8	42	**70**
-150	17	12	87	12	**128**
-160	-	-	8	-	**8**
-170	5	-	-	-	**5**
-180	-	-	2	-	**2**
-190	-	-	5	-	**5**
-200	-	-	3	-	**3**
-210	-	-	64	-	**64**
-220	-	-	7	-	**7**
-230	-	-	-	-	**0**
Total	**189**	**236**	**360**	**250**	**1035**

Table B.22: Faunal weights (g) by square and depth of Dorota (NA 97/13)

	hand and 5 mm													
	-20	-40	-60	-80	-100	-120	-140	-160	-180	-200	-220	-240	-260	**Total**
A1	161	473	385	739	1018	758	653	514	155	158	132	5	13	**5164**
A2	161	177	516	1177	428	442	278	441	58	151	78	-	-	**3907**
A3	209	304	298	576	323	148	327	675	662	199	25	-	-	**3746**
A4	204	492	583	1130	529	633	335	234	75	489	39	-	-	**4743**
A5	150	400	565	845	485	470	173	299	403	442	6	6	-	**4244**
B1	-	235	548	600	765	424	722	310	191	107	10	-	83	**3995**
B2	217	235	331	589	297	311	294	154	174	135	9	-	-	**2746**
B3	-	268	675	155	429	452	317	188	166	507	22	-	-	**3179**
B4	208	550	200	751	604	488	78	221	186	265	125	7	-	**3683**
B5	87	507	313	619	761	634	129	178	336	404	10	-	-	**3978**
C1	181	136	439	197	480	214	470	203	167	156	31	190	99	**2963**
C2	77	70	206	300	287	554	276	154	392	274	35	-	-	**2625**
C3	91	169	210	692	527	217	293	196	124	401	2	-	-	**2922**
C4	193	526	707	678	474	631	87	197	539	239	15	-	-	**4286**
C5	66	398	281	627	443	702	62	176	444	215	-	42	-	**3456**
D1	123	102	267	271	664	267	584	111	181	123	8	30	11	**2742**
D2	141	93	155	592	600	151	115	122	375	478	61	-	-	**2883**
D3	150	19	673	539	207	676	175	67	205	372	7	-	-	**3090**
D4	211	278	509	763	464	782	121	172	451	249	5	-	-	**4005**
D5	10	270	349	882	1541	120	154	90	153	212	-	3	-	**3784**
E1	-	89	337	312	575	551	391	47	167	200	-	5	-	**2674**
E2	117	-	274	264	379	395	162	71	39	50	-	-	-	**1751**
E3	152	108	585	165	283	357	166	44	208	182	3	-	-	**2253**
E4	208	241	279	776	400	582	105	109	440	98	38	-	-	**3276**
E5	280	245	395	419	767	441	109	67	241	184	65	-	-	**3213**
F1	-	-	755	-	-	-	-	-	-	-	-	-	-	**755**
F2	187	-	-	-	-	-	-	-	-	-	-	-	-	**187**
F5	164	-	-	-	-	-	-	-	-	-	-	-	-	**164**
F6	-	-	-	277	-	-	-	-	-	-	-	-	-	**277**
G2	96	-	502	-	-	-	-	-	-	-	-	-	-	**598**
I1	-	-	224	-	-	-	-	-	-	-	-	-	-	**224**
I2	-	289	-	-	-	-	-	-	-	-	-	-	-	**289**
K2	-	-	224	-	-	-	-	-	-	-	-	-	-	**224**
K4	-	-	364	-	-	-	-	-	-	-	-	-	-	**364**
Total	**3844**	**6674**	**12149**	**14935**	**13730**	**11400**	**6576**	**5040**	**6532**	**6290**	**726**	**288**	**206**	88390

Table B.23: Faunal weights (g) by square and depth (cm) of Galaga (NA 92/2C)

	hand 5 mm	AB		hand 5 mm	AB		hand 5 mm	AB
-10	113	-	-250	739	3	-480	162	1
-20	128	-	-260	343	2	-490	70	4
-30	38	-	-270	407	-	-500	56	1
-40	72	1	-280	589	-	-510	7	-
-50	185		-290	1441	2	-520	4	-
-60	198	-	-300	214	-	-530	-	-
-70	343	-	-310	307	-	-540	-	-
-80	196	-	-320	333	1	-550	-	-
-90	231	-	-330	272	-	-560	-	-
-100	174	-	-340	672	-	-570	-	-
-110	241	-	-350	470	-	-580	-	-
-120	198	-	-360	118	-	-590	-	-
-130	157	1	-370	122	-	-600	-	-
-140	235	16	-380	188	-	-610	-	-
-150	155	-	-390	105	-	-620	-	-
-160	232	-	-400	87	-	-630	-	-
-170	70	-	**Total IA**	**19535**	**40**	-640	-	-
-180	503	-	-410	103	-	-650	-	-
-190	295	1	-420	168	-	-660	-	-
-200	1824	3	-430	170	1	-670	-	-
-210	4309	9	-440	290	2	-680	-	-
-220	943	-	-450	207	-	-690	-	-
-230	1546	1	-460	135	-	-700	-	-
-240	742	-	-470	137	-	**Total LSA**	**1509**	**9**

Table B.24: Faunal weights (g) by depth (cm) of Kursakata (NA 93/46)

	hand and 5 mm									
	A1	A2	A3	B1	B2	B3	C1	C2	C3	**Total**
Surface	2	-	-	-	-	-	2	-	-	**4**
-10	84	41	7	59	60	28	33	34	8	**353**
-20	40	41	47	69	45	69	71	83	115	**580**
-30	60	85	91	112	191	79	56	54	80	**809**
-40	51	108	104	17	45	5	56	18	66	**471**
-50	52	19	23	25	7	40	13	3	9	**191**
-60	1	9	5	9	29	4	12	6	166	**242**
-70	21	21	-	-	35	9	17	26	25	**154**
-80	21	25	34	36	14	-	10	15	13	**169**
-90	26	14	36	28	9	48	18	11	11	**200**
-100	47	59	43	146	103	75	25	57	191	**745**
-110	96	150	71	115	165	114	41	125	251	**1127**
-120	98	361	36	57	-	36	10	84	189	**871**
-130	53	68	65	168	110	65	21	87	185	**821**
Total phase IV	**653**	**999**	**561**	**842**	**811**	**571**	**384**	**602**	**1312**	**6735**
-140	27	162	42	23	45	25	-	20	34	**378**
-150	12	2	137	20	69	29	48	51	55	**422**
-160	75	583	-	23	205	105	29	24	75	**1118**
-170	16	146	235	19	262	-	20	67	37	**801**
-180	8	84	197	86	205	286	7	1	3	**878**
-190	15	90	220	14	157	198	17	8	3	**722**
-200	10	33	191	2	250	97	8	13	3	**606**
-210	31	81	154	14	130	95	1	1	-	**506**
-220	28	59	26	22	27	62	16	17	32	**289**
-230	87	9	44	11	275	37	11	57	154	**685**
-240	56	65	16	59	315	65	8	34	43	**661**
Total phase III	**365**	**1314**	**1262**	**293**	**1939**	**999**	**164**	**294**	**437**	**7067**
-250	3	59	17	21	27	46	-	30	74	**276**
-260	18	16	11	18	16	28	-	44	49	**199**
-270	3	29	39	19	35	63	7	-	176	**370**
-280	29	67	87	9	34	120	118	31	40	**534**
-290	208	90	137	179	26	254	66	127	300	**1387**
-300	445	246	152	-	309	197	280	363	196	**2189**
-310	381	373	235	198	115	226	144	253	22	**1946**
-320	682	403	190	175	187	283	329	89	30	**2368**
-330	179	553	345	711	1617	70	113	128	45	**3759**
-340	654	29	260	454	265	97	47	77	83	**1966**
-350	552	7	113	83	82	155	64	133	92	**1281**
Total phase II	**3153**	**1871**	**1586**	**1867**	**2713**	**1538**	**1169**	**1273**	**1107**	**16275**

Table B.25: Faunal weights (g) by square and depth (cm) of Mege (NA 94/7)

	hand and 5 mm									
	A1	A2	A3	B1	B2	B3	C1	C2	C3	**Total**
-360	7	2	1	3	1	2	10	20	107	**154**
-370	4	2	7	1	3	3	113	26	157	**315**
-380	5	-	-	2	28	-	13	205	1600	**1853**
-390	62	3	36	-	25	-	8	43	-	**177**
-400	30	3	1	1	2	-	8	6	-	**51**
-410	6	2	3	-	2	7	5	-	10	**34**
-420	12	43	-	3	96	-	6	31	4	**195**
-430	16	-	-	5	121	6	32	46	2	**228**
-440	26	2	9	56	54	-	65	7	4	**223**
-450	1	2	-	2	7	-	3	51	-	**66**
-460	18	41	-	5	3	-	28	6	-	**101**
-470	9	60	-	-	13	-	2	17	2	**103**
-480	327	186	-	121	75	4	2	2	-	**715**
-490	1206	664	82	810	194	47	408	11	-	**3422**
-500	957	722	929	609	577	199	254	150	22	**4420**
-510	791	1258	281	805	539	-	249	161	9	**4093**
-520	1006	284	513	6	220	95	486	204	77	**2891**
-530	925	826	658	523	1130	129	877	456	58	**5583**
-540	1974	2062	250	1001	2016	165	602	903	47	**9020**
-550	1265	440	-	628	814	-	233	577	-	**3957**
-560	747	773	549	996	1224	147	269	413	-	**5117**
-570	359	847	401	174	150	1610	35	295	1220	**5090**
-580	295	611	690	141	85	389	151	268	462	**3092**
-590	-	128	1185	55	51	587	9	171	280	**2466**
-600	-	19	332	30	206	-	234	1206	70	**2097**
-610	141	9	496	-	595	167	760	1396	93	**3656**
-620	674	512	200	338	1224	143	2274	597	1035	**6998**
-630	751	2590	285	5356	6481	1926	1910	574	3285	**23158**
-640	1378	1814	-	1499	270	3195	59	3020	919	**12154**
-650	1051	950	1282	1645	2202	1616	1742	395	722	**11605**
-660	1038	2870	2444	874	683	2270	74	173	940	**11366**
-670	665	171	2032	5	-	2625	-	-	1381	**6878**
-680	-	-	731	-	-	375	-	-	701	**1807**
Total phase I	**15745**	**17896**	**13396**	**15692**	**19091**	**15705**	**10920**	**11430**	**13205**	**133078**

Table B.25 cont.

	hand and 5 mm									
	A1	A2	A3	B1	B2	B3	C1	C2	C3	**Total**
surface	9	20	-	5	33	-	-	-	-	**67**
-10	-	-	-	-	27	-	-	-	-	**27**
-20	-	43	-	-	-	9	-	12	-	**63**
-30	40	-	20	51	-	112	48	43	12	**326**
-40	-	49	102	100	105	80	47	15	-	**499**
-50	22	43	152	127	101	94	35	-	72	**646**
-30 to -60	-	-	-	-	-	-	-	52	-	**52**
-60	12	55	717	13	94	384	-	-	118	**1393**
Total phase IV	**84**	**209**	**990**	**296**	**361**	**679**	**130**	**122**	**202**	**3073**
-70	60	7	505	44	48	94	28	11	347	**1144**
-80	95	23	134	116	-	232	81	224	27	**932**
-90	42	85	309	125	168	250	73	18	138	**1209**
-100	42	128	76	241	233	99	27	4	48	**899**
Total phase IIIb	**239**	**243**	**1025**	**526**	**449**	**676**	**210**	**257**	**561**	**4184**
-110	92	27	-	114	34	118	14	10	128	**536**
-120	27	8	22	28	2	22	10	10	-	**129**
-130	13	1	29	20	1	115	12	30	19	**240**
-140	40	10	3	72	-	17	9	12	35	**198**
-150	29	54	19	98	13	46	-	14	6	**279**
-160	40	65	35	156	16	31	112	40	-	**494**
-170	27	18	156	48	25	-	66	63	21	**423**
-180	43	95	74	36	40	97	88	69	41	**582**
-190	55	95	203	55	54	66	71	126	15	**739**
Total phase IIIa	**366**	**372**	**540**	**627**	**184**	**513**	**381**	**373**	**264**	**3620**
-200	136	280	57	113	55	16	43	102	54	**856**
-210	29	49	39	83	9	34	35	29	56	**363**
-220	149	17	45	17	56	94	17	34	48	**477**
-230	8	10	32	36	21	39	16	21	-	**182**
-240	7	8	21	37	17	32	23	23	107	**276**
-250	13	11	59	41	15	-	37	43	61	**280**
-260	26	77	242	38	29	62	25	49	45	**593**
-270	27	121	104	23	29	39	72	65	64	**543**
-280	78	61	83	43	68	68	82	40	83	**607**
-290	50	70	53	36	62	54	79	69	107	**580**
-300	35	36	279	33	21	103	23	24	44	**597**
-310	79	92	67	65	46	101	14	16	22	**502**
-320	140	69	57	28	56	27	6	42	9	**432**
-330	5	9	68	72	-	48	12	-	29	**242**
-340	-	91	-	32	4	21	23	-	43	**213**
-350	-	21	71	23	47	44	129	9	-	**343**
-360	33	31	96	24	68	-	3	82	40	**376**
Total phase II	**815**	**1052**	**1373**	**744**	**604**	**781**	**638**	**646**	**809**	**7463**
-370	57	142	25	26	28	44	83	120	28	**553**
-380	106	102	22	16	41	107	14	138	53	**600**
-390	81	49	-	50	68	93	45	42	8	**435**
-400	92	31	-	51	59	90	11	15	70	**419**
-410	68	142	21	37	21	20	19	35	28	**393**
-420	73	81	57	33	32	30	33	43	43	**424**
-430	68	58	111	17	65	34	57	39	37	**485**
-440	95	-	64	42	57	17	63	64	123	**525**
-450	36	175	55	34	58	75	56	30	88	**607**
-460	55	109	4	24	63	4	46	51	45	**402**
-470	61	14	1	10	49	-	12	36	28	**212**
-480	9	3	-	5	5	-	8	6	13	**49**
-490	21	-	-	-	-	-	31	-	12	**64**
-500	2	1	-	-	-	-	-	-	18	**21**
-510	-	-	-	-	-	-	-	-	-	**0**
-520	-	-	-	-	-	-	-	-	8	**8**
-530	-	-	-	-	-	-	-	-	47	**47**
Total Phase I	**825**	**908**	**360**	**345**	**545**	**515**	**476**	**619**	**650**	**5244**

Table B.26: Faunal weights (g) by square and depth (cm) of Ngala (NA 93/45)

Appendic C: Tables Chapter 4. Description of the faunal remains by taxon

		M.G.	F.G.				T.G.		F.B.						
		Cypraea moneta/annulus	*Pila wernei*	*Lanistes varicus*	*Cleopatra bulimoides*	*Lymnaea natalensis*	*Limicolaria* sp.	Gastropod	*Chambardia* sp./*Spathopsis* sp.	*Mutela dubia*	*Eupera ferruginea*	Nile oyster (*Etheria elliptica*)	Bivalve	Unidentified mollusc	**Total**
BF 97/5	EIA	-	-	-	-	-	-	-	-	-	-	-	2	-	**2**
BF 94/45	LSA	-	-	-	-	-	-	1	-	-	-	-	9	1	**11**
BF 94/45	EIA	-	-	-	-	-	-	47	6	-	-	-	162	11	**226**
BF 94/45	AB EIA	-	-	-	-	-	-	9	1	-	-	-	-	-	**10**
BF 97/13	LIA	-	-	-	-	-	-	-	-	-	-	-	1	-	**1**
BF 97/30	LIA	-	-	-	-	-	-	1	11	-	-	-	-	-	**12**
BF 96/22	MIA	-	-	-	-	-	-	10	-	-	-	-	23	-	**33**
BF 97/31	LIA	-	-	-	-	-	-	3	3	-	-	-	20	1	**27**
BF 94/120	LIA	3	-	5	-	-	-	31	55	-	1	2	324	-	**421**
BF 95/7	LIA	4	-	-	-	-	-	12	12	-	-	1	89	-	**118**
NA 90/5C	Gaj IIa	-	1	-	-	-	-	-	-	-	-	-	-	-	**1**
NA 90/5A	Gaj IIb	-	50*	-	-	-	3	-	4	-	-	-	-	-	**57**
NA 90/5BI	Gaj IIa/b	-	36	-	-	-	-	-	-	-	-	-	-	-	**36**
NA 90/5BII	Gaj IIc	-	20*	-	-	-	-	27	-	-	-	-	3	3	**53**
NA 93/42	Gaj I	-	1	-	-	-	-	3	-	-	-	-	-	-	**4**
NA 97/24	Gaj I	-	1	-	-	-	-	-	-	-	-	-	-	-	**1**
NA 93/36	Gaj IIa	-	2	-	-	-	-	4	-	-	-	-	-	-	**6**
NA 93/10	Gaj IIc	-	-	-	-	-	-	-	-	-	-	-	1	-	**1**
NA 97/37	Gaj III	-	11	-	-	-	-	8	-	-	-	-	-	-	**19**
NA 97/33	Gaj I	-	29*	-	-	-	-	-	-	-	-	-	9	1	**39**
NA 97/33	Gaj IIa/b	-	40*	-	-	-	1	-	2	-	-	-	10	1	**54**
NA 96/45	Gaj IIa/b	-	1	-	-	-	-	1	5	-	-	-	9	5	**21**
NA 97/26	IA	-	4	-	-	1	34	33	-	-	-	-	-	8	**80**
NA 99/75	IA	-	7	-	1	-	1	9	-	-	-	-	2	-	**20**
NA 97/13	IA	-	-	-	-	-	-	4	2	-	-	-	3	-	**9**
NA 92/2C	subrec	-	5*	-	-	-	-	9	-	-	-	-	5	2	**21**
NA 93/46	LSA	-	1	-	-	-	-	-	-	-	-	-	-	-	**1**
NA 93/46	EIA	-	-	-	-	-	-	6	-	-	-	-	2	24	**32**
NA 94/7	II	-	-	-	-	-	-	3	-	-	-	-	1	1	**5**
NA 94/7	III	-	1	-	-	-	-	5	-	-	-	-	2	5	**13**
NA 94/7	IV	-	-	-	-	-	1	3	-	-	-	-	7	1	**12**
NA 93/45	I	-	2	-	-	-	-	7	-	-	-	-	-	-	**9**
NA 93/45	II	-	-	-	-	-	-	6	-	-	-	-	-	-	**6**
NA 93/45	IIIa	-	-	-	-	-	-	3	-	-	-	-	-	-	**3**
NA 93/45	IIIb	-	-	-	-	-	-	2	-	-	-	-	-	-	**2**
NA 93/45	IV	-	-	-	-	-	-	4	-	-	-	-	-	-	**4**
Blé A		-	8	-	-	-	-	-	-	-	-	-	2	-	**10**
Blé B		-	2	-	-	-	1	-	-	-	-	-	4	-	**7**
Blé C		-	6*	-	-	-	-	-	1	-	-	1	2	-	**10**
Blé E		-	13	-	-	-	1	-	7	1	-	1	13	-	**36**
All sites		**7**	**257**	**5**	**1**	**1**	**42**	**251**	**109**	**1**	**1**	**5**	**705**	**64**	**1433**

*with *Pila wernei* opercula

M.G.=marine gastropod, F.G.= freshwater gastropod, T.G.=terrestrial gastropod, F.B.=freshwater bivalve

Table C.1: Mollusc taxa by site and phase

	postrostrale	frontale	parietale	parasphenoideum	palatopterygoideum	coronoides	palat./cor.	keratohyale	interoperculare	intertemporale	operculare	cranial rib	costae	unidentified	**Total**
BF 97/5 EIA	-	-	-	1	-	-	5	-	-	-	-	1	-	2	**9**
BF 94/45 EIA	2	1	4	22	1	8	194	11	-	5	4	26	-	3	**281**
BF 94/45 AB EIA	1	-	2	4	7	8	6	5	3	5	2	4	-	-	**47**
BF 97/13 EIA	-	1	-	-	1	7	-	-	-	-	-	-	1	-	**10**
BF 97/13 MIA	-	-	-	-	-	1	-	1	-	-	-	-	-	-	**2**
BF 97/13 LIA	-	1	-	-	1	-	-	1	-	-	-	-	-	-	**3**
BF 97/30 LIA	-	-	-	-	1	-	-	1	-	-	-	1	-	-	**3**
BF 96/22 MIA	-	-	1	3	4	8	-	7	-	1	-	2	1	-	**27**
BF 97/31 LIA	-	1	1	1	6	6	-	4	-	-	1	-	-	-	**20**
BF 94/120 LIA	-	-	2	1	11	29	8	5	-	-	-	5	1	-	**62**
BF 95/7 LIA	-	-	1	5	7	11	2	2	2	1	-	2	-	-	**33**
NA 97/37 Gai III	2	2	1	2	8	17	-	4	-	1	-	3	-	-	**40**
NA 97/33 Gaj I	-	-	-	-	-	-	1	-	-	-	-	-	-	-	**1**
NA 97/26 IA	-	-	1	6	7	22	-	3	2	-	-	2	-	-	**43**
NA 99/75 IA	1	-	-	2	2	6	1	-	-	-	1	-	-	-	**13**
NA 92/2C subrec	-	-	-	-	1	1	-	2	-	-	-	-	-	-	**4**
NA 93/46 EIA	-	-	-	-	2	2	-	2	-	-	-	1	-	-	**7**
NA 93/45 I	-	-	-	1	1	4	-	-	-	-	-	-	-	1	**7**
NA 93/45 II	1	-	-	3	3	4	1	2	-	-	1	1	-	1	**17**
NA 93/45 IIIa	-	-	-	1	3	1	-	-	1	-	-	-	-	-	**6**
NA 93/45 IIIb	-	-	-	1	2	2	-	1	-	-	-	-	-	-	**6**
NA 93/45 IV	-	-	-	1	2	1	-	1	-	-	-	-	-	-	**5**
Blé A	-	-	-	1	2	-	10	-	-	-	-	3	-	-	**16**
Blé B	-	-	-	1	-	-	3	-	-	-	-	2	-	-	**6**
Blé C	-	-	-	1	-	-	3	-	-	-	-	-	-	-	**4**
Blé E	-	-	-	1	-	-	6	-	-	-	-	4	-	-	**11**
All sites	**7**	**6**	**13**	**58**	**72**	**138**	**240**	**52**	**8**	**13**	**9**	**57**	**3**	**7**	**683**

		10-20	20-30	30-40	40-50	50-60	60-70	>80
BF 97/5	EIA	-	1	-	2	3	2	-
BF 94/45	EIA	17	37	70	36	43	34	15
BF 94/45	AB EIA	2	9	11	4	9	1	-
BF 97/13	EIA	-	1	2	2	1	2	-
BF 97/13	MIA	-	-	1	-	-	1	-
BF 97/13	LIA	-	-	1	-	2	-	-
BF 97/30	LIA	-	-	-	-	-	2	1
BF 96/22	MIA	-	-	4	12	5	2	-
BF 97/31	LIA	-	1	3	5	4	3	-
BF 94/120	LIA	-	5	14	12	11	5	2
BF 95/7	LIA	-	1	6	2	8	8	2
NA 97/37	Gaj III	-	-	1	4	8	8	17
NA 97/33	Gaj I	-	-	-	-	-	-	-
NA 97/26	IA	1	8	16	3	9	4	-
NA 99/75	IA	-	-	-	1	4	2	3
NA 92/2C	subrec	-	1	-	1	1	1	-
NA 93/46	EIA	-	-	1	-	1	-	3
NA 93/45	I	-	-	-	1	1	3	2
NA 93/45	II	-	-	1	1	-	6	6
NA 93/45	IIIa	-	-	-	1	-	3	2
NA 93/45	IIIb	-	-	-	-	4	1	1
NA 93/45	IV	-	-	-	-	-	3	2
Blé A		-	-	1	2	4	6	1
Blé B		-	-	-	-	-	5	1
Blé C		-	-	-	-	-	-	3
Blé E		-	-	-	-	2	8	1
All sites		**20**	**62**	**119**	**70**	**103**	**97**	**59**

Table C.2a-b: Skeletal element distribution and SL (cm) reconstructions of lungfish by site and phase

		neurocranium	articulare	dentale	maxillare	palatinum	praemaxillare	dentale/maxillare/praemaxillare	hyomandibulare	branchiostegale	operculare	praeoperculare	suboperculare	vertebrae	cleithrum	scapula	supracleithrale	spinna pinnae dorsalis	squamae	**Total**
BF 97/5	LSA	-	-	-	-	-	-	-	-	-	-	-	-	1	-	-	-	-	5	**6**
BF 97/5	EIA	-	-	-	-	-	-	-	-	-	-	-	-	2	-	-	-	-	-	**2**
BF 94/45	AB EIA	-	-	-	-	-	-	-	-	-	-	-	-	1	-	-	-	1	-	**2**
NA 90/5A	Gaj IIb	-	-	-	-	-	-	-	-	-	-	-	-	1	-	-	-	-	-	**1**
NA 90/5BI	Gaj IIa/b	-	-	-	-	-	-	-	-	-	-	-	-	27	-	-	-	-	1	**28**
NA 93/42	Gaj I	-	-	-	-	-	-	-	-	-	-	-	-	13	-	-	-	-	-	**13**
NA 97/18	Gaj I	-	-	-	-	-	-	-	-	-	-	-	-	15	-	-	-	-	-	**15**
NA 95/1	Gaj IIa/b	-	-	-	-	-	-	-	-	-	-	-	-	4	2	-	-	-	-	**6**
NA 99/75	IA	-	-	1	-	-	-	-	-	-	-	-	-	-	-	-	-	-	-	**1**
NA 93/46	LSA	11	-	-	-	-	-	-	-	-	-	-	-	27	2	-	-	5	77	**122**
NA 93/46	AB LSA	2	-	-	-	-	-	-	-	-	-	-	-	1	-	-	-	1	49	**53**
NA 93/46	EIA	300	9	67	8	1	1	3	1	-	1	-	1	1084	41	-	4	160	1819	**3500**
NA 93/46	AB EIA	6	-	1	1	-	-	-	-	-	-	-	-	8	-	-	-	40	401	**457**
NA 94/7	I	11	-	-	-	-	-	-	-	-	-	-	-	39	2	-	-	2	3	**57**
NA 94/7	II	1	-	-	-	-	-	-	-	-	-	-	-	3	1	-	-	1	-	**6**
NA 94/7	III	-	-	-	-	-	-	-	-	-	-	-	-	1	-	-	-	-	-	**1**
NA 93/45	I	16	-	-	-	-	-	-	-	1	-	1	-	7	2	1	-	-	1	**29**
NA 93/45	II	10	1	1	-	-	-	-	-	-	-	-	-	46	2	-	-	1	3	**64**
NA 93/45	IIIa	5	-	-	-	-	-	-	-	-	-	-	-	15	-	-	-	-	-	**20**
NA 93/45	IIIb	1	-	-	-	-	-	-	-	-	-	-	-	1	-	-	-	-	-	**2**
NA 93/45	IV	1	-	-	-	-	-	-	-	-	-	-	-	5	1	-	1	-	-	**8**
Blé B		1	-	-	-	-	-	-	-	-	-	-	-	-	-	-	-	-	-	**1**
Blé C		1	-	-	-	-	-	-	-	-	-	-	-	-	-	-	-	-	-	**1**
Blé E		-	-	-	-	-	-	-	-	-	-	-	-	12	2	-	-	1	-	**15**
All sites		**366**	**10**	**70**	**9**	**1**	**1**	**3**	**1**	**1**	**1**	**1**	**1**	**1313**	**55**	**1**	**5**	**212**	**2359**	**4410**

		<10	10-20	20-30	30-40	40-50	50-60	>60
BF 95/7	LSA	-	-	-	1	-	-	-
BF 95/7	EIA	-	-	2	-	-	-	-
BF 94/45	AB EIA	-	1	1	-	-	-	-
NA 90/5BI	Gaj IIa/b	-	-	-	8	5	13	-
NA 93/42	Gaj I	-	-	-	9	1	-	-
NA 97/18	Gaj I	-	-	-	5	5	2	-
NA 95/1	Gaj IIa/b	-	-	-	-	6	-	-
NA 99/75	IA	-	-	-	-	-	-	1
NA 93/46	LSA	1	-	3	4	8	17	-
NA 93/46	AB LSA	-	1	5	3	-	-	-
NA 93/46	EIA	-	3	47	268	726	173	5
NA 94/7	I	-	-	-	-	-	2	2
NA 94/7	II	-	-	-	-	-	-	3
NA 94/7	III	-	-	-	-	-	-	1
NA 93/45	I	-	-	-	1	4	6	2
NA 93/45	II	-	-	-	3	21	28	-
NA 93/45	IIIa	-	-	-	-	6	8	3
NA 93/45	IIIb	-	-	-	-	1	1	-
NA 93/45	IV	-	-	-	-	5	2	-
Blé C		-	-	-	-	1	-	-
Blé E		-	-	1	-	5	9	-
All sites		**1**	**5**	**59**	**302**	**794**	**261**	**17**

Table C.3a-b: Skeletal element distribution and SL (cm) reconstructions of *Polypterus* sp. by site and phase

		neurocranium	dentale	maxillare	palatinum	praemaxillare	quadratum	keratohyale	epihyale	hyomandibulare	operculare	praeoperculare	vertebrae praecaudales	vertebrae caudales	vertebrae	costae	cleithrum	scapula	supracleithrale	**Total**
BF 97/5	LSA	-	-	-	-	-	-	-	-	-	-	-	-	-	3	-	-	-	-	**3**
NA 90/5A	Gaj IIb	-	-	-	-	-	-	-	-	-	-	-	1	-	-	-	-	-	-	**1**
NA 90/5BI	Gaj IIa/b	-	-	-	-	2	-	-	-	-	3	-	5	-	25	1	-	-	-	**36**
NA 93/42	Gaj I	-	-	-	-	-	-	-	-	-	-	-	-	-	11	-	-	-	-	**11**
NA 97/37	Gaj III	1	-	-	-	-	-	-	-	-	-	-	6	3	3	1	-	-	1	**15**
NA 95/1	Gaj IIa/b	-	-	-	-	-	-	-	-	-	-	-	1	-	-	-	-	-	-	**1**
NA 99/75	IA	-	-	-	-	-	-	-	-	-	-	-	2	-	-	-	-	-	-	**2**
NA 92/2C	subrec	-	-	-	-	-	-	-	-	-	-	-	1	2	-	1	-	-	-	**4**
NA 93/46	LSA	4	-	-	-	-	-	-	-	-	1	1	3	4	-	2	1	-	-	**16**
NA 93/46	EIA	23	17	5	1	4	-	1	3	3	30	4	185	110	118	34	18	1	-	**557**
NA 93/46	AB EIA	-	-	-	-	1	-	-	-	-	-	-	2	-	-	-	-	-	-	**3**
NA 94/7	I	-	-	-	-	-	-	-	-	-	-	-	12	11	2	-	-	-	-	**25**
NA 94/7	II	1	-	-	-	-	-	-	-	-	-	-	8	13	10	-	-	-	-	**32**
NA 94/7	III	-	-	-	-	-	-	-	-	-	-	-	2	3	-	-	-	-	-	**5**
NA 94/7	IV	-	-	-	-	-	-	-	-	-	-	-	-	5	3	-	-	-	-	**8**
NA 93/45	I	4	1	-	-	-	-	-	-	-	2	1	6	2	3	1	3	-	1	**24**
NA 93/45	II	3	-	1	-	-	2	1	-	-	1	1	13	1	1	2	-	-	-	**26**
NA 93/45	IIIa	6	-	-	-	-	-	1	1	-	1	-	4	-	-	-	-	-	1	**14**
NA 93/45	IIIb	2	-	-	-	-	-	-	-	-	2	1	3	-	3	-	1	-	-	**12**
NA 93/45	IV	2	-	-	-	-	-	-	-	-	-	-	4	2	4	-	1	-	-	**13**
Blé A		1	-	-	-	-	-	-	-	-	-	-	-	1	-	-	-	-	-	**2**
Blé B		-	-	-	-	-	-	-	-	-	3	-	-	-	-	-	1	-	-	**4**
Blé C		-	-	-	-	-	1	-	-	-	4	-	15	3	-	1	1	-	-	**25**
Blé E		1	1	-	-	-	-	1	-	-	16	1	11	-	-	4	9	-	-	**44**
All sites		**48**	**19**	**6**	**1**	**7**	**3**	**4**	**4**	**3**	**63**	**9**	**284**	**160**	**186**	**47**	**35**	**1**	**3**	**883**

		10-20	20-30	30-40	40-50	50-60	60-70	>70
BF 97/5	LSA	-	-	-	1	-	-	-
NA 90/5BI	Gaj IIa/b	-	-	-	-	12	21	3
NA 97/37	Gaj III	-	1	-	5	1	3	1
NA 95/1	Gaj IIa/b	-	-	-	-	-	-	1
NA 99/75	IA	-	-	-	1	-	1	-
NA 92/2C	subrec	-	-	-	-	-	1	1
NA 93/46	LSA	-	-	1	3	1	1	-
NA 93/46	EIA	1	20	49	89	94	57	24
NA 93/46	AB EIA	-	1	-	-	-	-	-
NA 94/7	I	-	-	-	-	-	5	18
NA 94/7	II	-	-	-	-	-	7	14
NA 94/7	III	-	-	-	-	-	-	4
NA 94/7	IV	-	-	-	-	-	-	5
NA 93/45	I	-	-	5	6	3	2	5
NA 93/45	II	-	-	2	-	-	4	1
NA 93/45	IIIa	-	-	-	-	-	1	3
NA 93/45	IIIb	-	-	-	-	-	2	8
NA 93/45	IV	-	-	-	4	-	-	5
Blé A		-	-	-	-	-	1	-
Blé B		-	-	2	1	-	-	-
Blé C		-	-	-	13	5	3	4
Blé E		-	-	4	10	14	1	2
All sites		**1**	**22**	**63**	**133**	**130**	**110**	**99**

Table C4.a-b: Skeletal element distribution and SL (cm) reconstructions of *Heterotis niloticus* by site and phase

		quadratum	vertebrae praecaudales	vertebrae caudales	Vertebrae	**Total**
BF 97/5	LSA	-	-	1	-	**1**
NA 93/46	LSA	-	-	1	-	**1**
NA 93/46	AB LSA	-	-	-	10	**10**
NA 93/46	EIA	-	5	6	-	**11**
NA 93/46	AB LSA	-	-	2	-	**2**
NA 94/7	II	-	3	1	-	**4**
Blé C		1	-	1	-	**2**
All sites		**1**	**8**	**12**	**10**	**31**

		<10	10-20	20-30	30-40	40-50	50-60	60-70	70-80
BF 97/5	LSA	-	-	-	-	1	-	-	-
NA 93/46	LSA	-	-	-	-	1	-	-	-
NA 93/46	AB LSA	1	1	-	-	-	-	-	-
NA 93/46	EIA	-	-	-	4	3	1	-	-
NA 93/46	AB EIA	10	-	-	-	-	-	-	-
NA 94/7		-	-	-	-	-	-	-	4
Blé C		-	-	-	-	-	-	1	1
Total		**11**	**1**	**0**	**4**	**5**	**1**	**1**	**5**

Table C.5a-b: Skeletal element distribution and SL (cm) reconstructions of mormyrid by site and phase

		neurocranium	angulare	articulare	dentale	dentes	quadratum	keratohyale	epihyale	hyomandibulare	interoperculare	operculare	praeoperculare	urohyale	vertebrae	cleithrum	supracleithrale	**Total**
BF 97/5	LSA	1	-	1	-	-	-	-	-	-	-	-	-	-	22	-	-	**24**
BF 97/30	LIA	-	-	-	-	-	-	-	-	-	-	-	-	-	1	-	-	**1**
NA 90/5A	Gaj IIb	-	-	-	-	-	-	-	-	-	-	-	-	-	2	-	-	**2**
NA 93/42	Gaj I	-	-	-	-	-	-	-	-	-	-	-	-	-	2	-	-	**2**
NA 97/37	Gaj III	-	-	-	-	-	-	1	-	-	-	-	-	-	-	-	-	**1**
NA 95/1	Gaj IIa/b	-	-	-	-	-	-	-	-	-	-	-	-	-	1	-	-	**1**
NA 92/2C	subrec	-	-	-	-	-	-	-	-	-	-	-	-	-	1	-	-	**1**
NA 93/46	LSA	1	-	1	-	-	1	-	-	-	-	-	-	-	29	-	-	**32**
NA 93/46	EIA	11	2	1	-	1	5	-	-	2	3	1	-	1	170	6	5	**208**
NA 93/46	AB EIA	-	-	-	-	-	-	-	-	-	-	-	-	-	1	-	-	**1**
NA 94/7	I	-	-	-	-	-	-	-	-	-	-	-	-	-	10	-	-	**10**
NA 94/7	II	1	-	-	-	-	-	-	-	-	-	-	-	-	18	-	-	**19**
NA 94/7	III	-	-	-	-	-	-	-	-	-	-	-	-	-	1	-	-	**1**
NA 93/45	I	-	-	-	-	-	-	1	1	-	-	-	1	-	12	1	2	**18**
NA 93/45	II	-	1	-	-	-	-	1	-	-	-	-	-	-	18	3	-	**23**
NA 93/45	IIIa	-	-	-	-	-	-	-	1	-	1	-	-	1	6	-	-	**9**
NA 93/45	IIIb	-	-	-	-	-	-	-	-	-	-	-	-	-	5	-	-	**5**
NA 93/45	IV	1	-	-	-	-	-	-	-	-	1	-	-	-	1	1	-	**4**
Blé B		-	-	-	-	-	-	-	-	-	-	-	-	-	3	-	-	**3**
Blé C		-	-	-	-	-	1	-	-	-	-	-	-	-	4	-	-	**5**
Blé E		-	-	-	1	-	-	1	-	-	-	-	-	-	3	-	-	**5**
All sites		**15**	**3**	**3**	**1**	**1**	**7**	**4**	**2**	**2**	**5**	**1**	**1**	**2**	**310**	**11**	**7**	**375**

		20-30	30-40	40-50	50-60	60-70	70-80	80-90	90-100	100-120	120-140	140-160	>160
BF 97/5	LSA	-	-	-	-	1	5	4	2	2	1	-	-
BF 97/30	LIA	-	-	-	-	-	-	-	-	1	-	-	-
NA 90/5A	Gaj IIb	-	-	-	-	-	-	-	-	-	-	1	-
NA 97/37	Gaj III	-	-	-	-	-	-	-	-	-	-	-	1
NA 92/2C	subrec	-	-	-	-	-	-	-	-	1	-	-	-
NA 93/46	LSA	-	-	-	-	-	-	1	3	1	14	1	4
NA 93/46	EIA	3	-	2	8	24	14	8	43	39	23	7	10
NA 94/7	I	-	-	-	-	-	-	-	-	10	-	-	-
NA 94/7	II	-	-	-	-	-	-	-	-	19	-	-	-
NA 94/7	III	-	-	-	-	-	-	-	-	1	-	-	-
NA 93/45	I	-	-	1	-	-	-	-	4	2	9	-	-
NA 93/45	II	-	1	-	-	-	2	-	2	3	6	2	11
NA 93/45	IIIa	-	-	-	-	-	-	-	1	2	5	-	-
NA 93/45	IIIb	-	-	-	-	-	-	-	-	2	2	-	-
NA 93/45	IV	-	-	-	1	-	-	-	-	2	1	-	1
Blé B		-	-	-	-	-	-	-	-	3	-	-	-
Blé C		-	-	-	-	-	-	-	-	-	4	1	-
Blé E		-	-	-	-	-	-	-	3	1	1	-	-
All sites		**3**	**1**	**3**	**9**	**25**	**21**	**13**	**58**	**89**	**66**	**12**	**27**

Table C.6a-b: Skeletal element distribution and SL (cm) reconstructions of *Gymnarchus niloticus* by site and phase

	operculare	vertebrae praecaudales	vertebrae caudales	vertebrae	basipterygium	supracleithrale	**Total**
NA 92/2C subrec	-	6	1	-	-	-	**7**
NA 93/46 LSA	-	1	-	-	-	-	**1**
NA 93/46 EIA	-	5	8	2	-	-	**15**
NA 93/46 AB EIA	-	2	-	-	-	-	**2**
Blé B	-	-	6	-	-	1	**7**
Blé C	-	1	3	-	-	-	**4**
Blé E	1	-	13	-	1	-	**15**
All sites	**1**	**15**	**31**	**2**	**1**	**1**	**51**

	<10	10-20	20-30	30-40	40-50	50-60	60-70
NA 92/2C subrec	-	-	-	-	-	1	4
NA 93/46 LSA	-	-	-	-	-	1	-
NA 93/46 EIA	-	-	1	-	2	4	5
NA 93/46 AB EIA	1	1	-	-	-	-	-
Blé B	-	-	-	2	1	-	-
Blé C	-	-	-	-	4	-	-
Blé E	-	-	-	-	1	14	-
All sites	**1**	**1**	**1**	**2**	**8**	**20**	**9**

Table C.7a-b: Skeletal element distribution and SL (cm) reconstructions of Distichodontidae or Citharinidae by site and phase

	neurocranium	operculare	vertebrae praecaudales	vertebrae caudales	costae	basipterygium	spina pinnae pectoralis	pterygiophorus	**Total**
NA 93/46 EIA	-	-	1	2	-	-	-	-	**3**
NA 93/45 I	1	-	2	-	-	-	-	-	**3**
Blé A	-	-	1	-	-	-	-	-	**1**
Blé B	-	1	1	-	1	-	-	2	**5**
Blé C	-	-	2	7	2	1	4	2	**18**
Blé E	-	-	4	5	5	-	-	3	**17**
All sites	**1**	**1**	**11**	**14**	**8**	**1**	**4**	**7**	**47**

	40-50	50-60	60-70
NA 93/46 EIA	1	1	-
NA 93/45 I	-	1	-
Blé A	-	-	1
Blé B	-	3	1
Blé C	2	3	7
Blé E	1	3	4
All sites	**4**	**11**	**13**

Table C.8a-b: Skeletal element distribution and SL (cm) reconstructions of cyprinid by site and phase

		neurocranium	epihyale	hyomandibulare	operculare	apparatus Weberianus	vertebrae praecaudales	vertebrae caudales	vertebrae	**Total**
NA 93/46	EIA	1	-	1	-	-	-	2	1	**5**
NA 93/45	I	-	1	-	1	-	-	-	-	**2**
NA 93/45	IIIa	1	-	-	-	-	-	-	-	**1**
NA 93/45	IV	1	-	-	-	1	1	-	-	**3**
Blé C		-	-	-	-	1	-	-	-	**1**
All sites		**3**	**1**	**1**	**1**	**1**	**1**	**2**	**1**	**11**

		20-30	30-40	40-50	50-60	60-70	70-80
NA 93/46	EIA	1	1	-	-	-	-
NA 93/45	I	-	-	1	1	-	-
NA 93/45	IIIa	-	-	-	1	-	-
NA 93/45	IV	-	-	1	1	-	-
Blé C		-	-	-	-	-	1
All sites		**1**	**1**	**2**	**3**	**0**	**1**

Table C.9a-b: Skeletal element distribution and SL (cm) reconstructions of *Bagrus bayad* by site and phase

		neurocranium	articulare	keratohyale	hyomandibulare	operculare	vertebrae praecaudales	vertebrae caudales	cleithrum	coracoideum	spina pinnae pectoralis	spina pinnae dorsalis	pterygiophorus dorsalis	**Total**
BF 97/5	LSA	17	1	-	-	-	1	-	7	-	5	3	-	**34**
NA 93/46	EIA	5	-	-	-	-	-	1	3	1	-	1	2	**13**
NA 94/7	I	-	-	-	-	-	-	-	5	-	2	-	-	**7**
NA 94/7	II	-	-	-	-	-	-	-	3	-	-	-	-	**3**
NA 94/7	III	-	-	-	-	-	-	-	-	-	-	1	-	**1**
NA 94/7	IV	-	-	-	-	-	-	-	-	-	-	1	-	**1**
NA 93/45	I	1	-	-	-	-	-	-	1	-	-	-	-	**2**
NA 93/45	IIIa	-	-	-	-	-	-	-	-	-	-	1	-	**1**
NA 93/45	IIIb	-	-	-	-	-	-	-	1	-	-	-	-	**1**
NA 93/45	IV	-	-	1	-	-	-	-	-	-	-	-	-	**1**
Blé B		1	-	-	1	-	-	-	-	-	-	-	-	**2**
Blé C		1	-	-	-	1	-	-	-	-	1	-	-	**3**
Blé E		1	-	-	1	-	-	-	3	-	1	-	-	**6**
All sites		**26**	**1**	**1**	**2**	**1**	**1**	**1**	**23**	**1**	**9**	**7**	**2**	**75**

		10-20	20-30	30-40	40-50	>50
BF 97/5	LSA	2	7	4	-	-
NA 93/46	EIA	-	-	1	1	4
NA 94/7	I	-	1	-	1	2
NA 94/7	II	-	-	-	3	-
NA 94/7	III	-	1	-	-	-
NA 94/7	IV	-	-	-	1	-
NA 93/45	I	-	1	1	-	-
NA 93/45	IIIa	-	-	-	1	-
NA 93/45	IIIb	-	-	-	1	-
NA 93/45	IV	-	-	-	1	-
Blé B		-	-	2	-	-
Blé C		-	-	-	1	2
Blé E		-	-	4	-	2

Table C.10a-b: Skeletal element distribution and SL (cm) reconstructions of *Auchenoglanis* sp. by site and phase

		H	G	A	C	Clariidae																							
		spina pinnae pectoralis	vomer	vomer	spina pinnae pectoralis	vomer	neurocranium	mandibulare	maxillare	mandibulare/maxillare	palatinum	praemaxillare	quadratum	branchiostegale	epi-, hypo- and keratohyale	hyomandibulare	operculare	urohyale	keratobranchiale	apparatus Weberianus	vertebrae praecaudales	vertebrae caudales	vertebrae	cleithrum	coracoideum	cleithrum/coracoideum	spina pinnae pectoralis	unidentified	**Total**
BF 97/5	LSA	-	-	-	-	-	6	2	-	-	-	-	-	-	1	-	-	-	-	-	-	-	-	1	-	-	1	-	**11**
BF 97/5	EIA	-	-	-	1	-	105	9	-	-	1	-	1	2	9	2	1	1	-	6	40	128	-	25	1	-	15	-	**346**
BF 95/45	LSA	-	-	-	-	-	15	-	-	-	-	-	-	-	-	-	-	-	-	-	1	3	1	1	-	-	-	-	**21**
BF 95/45	AB LSA	-	-	-	-	-	10	-	-	-	-	-	-	-	-	-	-	-	-	-	1	-	-	-	-	-	-	-	**11**
BF 95/45	EIA	-	2	-	38	1	621	31	-	-	3	-	11	6	6	3	6	5	1	2	31	22	13	95	12	-	2	-	**871**
BF 95/45	AB EIA	3	-	-	-	-	68	8	-	-	3	-	4	2	4	2	3	3	-	2	29	12	9	17	8	-	7	-	**181**
BF 97/13	EIA	-	-	-	2	-	30	1	-	-	-	-	-	-	-	1	-	-	-	-	1	4	-	2	-	-	1	-	**40**
BF 97/13	AB EIA	-	-	-	-	-	1	-	-	-	-	-	-	-	-	-	-	-	-	-	1	2	1	-	-	-	1	-	**6**
BF 97/13	MIA	-	-	-	4	-	11	3	-	-	-	-	-	-	-	-	-	-	-	-	2	5	5	6	2	-	1	-	**35**
BF 97/13	LIA	-	-	-	7	-	17	6	1	-	-	-	1	1	2	-	-	2	-	-	-	2	-	6	2	-	1	-	**41**
BF 97/13	AB LIA	-	-	-	1	-	1	1	-	-	-	-	-	2	-	-	-	-	-	-	2	2	1	-	1	-	-	-	**10**
BF 97/30	LIA	-	-	-	12	-	39	3	1	-	-	-	-	-	7	1	-	2	-	-	-	1	1	7	1	-	3	-	**66**
BF 96/22	MIA	-	-	-	2	-	38	2	-	-	-	-	1	2	-	-	1	1	-	-	5	2	-	1	1	-	1	-	**55**
BF 97/31	LIA	-	-	-	4	-	47	6	-	-	2	-	3	-	2	1	-	1	-	-	2	3	4	14	3	-	5	-	**93**
BF 94/120	LIA	-	-	-	47	-	99	31	-	-	2	1	4	3	4	3	-	-	-	6	19	29	21	23	4	-	-	1	**250**
BF 95/7	LIA	-	-	-	10	-	40	4	-	-	2	-	-	6	-	1	-	-	-	1	4	6	5	15	-	-	4	-	**88**
NA 90/5A	Gaj IIb	-	1	9	14	-	206	38	-	-	2	-	3	1	23	5	1	-	-	-	11	8	1	61	1	-	-	-	**361**
NA 90/5BI	Gaj IIa/b	-	1	-	12	-	431	53	-	-	-	-	11	7	43	1	2	1	-	16	28	22	-	91	11	5	10	1	**733**
NA 90/5BII	Gaj IIc	-	-	-	2	-	2	-	-	-	-	-	-	-	1	-	-	-	-	-	-	-	-	1	-	-	-	-	**4**
NA 93/42	Gaj I	-	-	-	-	-	27	4	-	-	-	-	-	-	3	-	-	-	-	-	1	2	-	1	-	-	2	-	**40**
NA 97/18	Gaj I	-	-	-	1	-	2	-	-	-	-	-	1	-	1	-	-	-	-	-	1	3	-	1	-	-	-	-	**9**
NA 93/36	Gaj IIa	-	-	-	-	-	5	-	-	-	-	-	-	-	-	-	-	-	-	-	-	-	-	-	-	-	1	-	**6**
NA 99/65	Gaj IIb	-	-	-	3	-	27	4	-	-	-	-	1	-	-	-	-	-	-	-	6	3	-	14	-	-	-	-	**55**
NA 99/65	Gaj IIc	-	-	-	-	-	2	-	-	-	-	-	1	-	-	-	-	-	-	-	-	1	1	-	-	-	-	-	**5**
NA 97/37	Gaj III	-	1	1	6	1	71	14	-	-	-	-	1	-	6	1	-	-	-	1	5	7	1	17	21	-	-	-	**146**
NA 97/33	Gaj I	-	-	-	-	-	2	1	-	-	-	-	-	-	-	-	-	-	-	-	-	-	-	-	-	-	-	-	**3**
NA 97/33	Gaj IIa/b	-	-	-	2	-	2	-	-	-	-	-	-	-	-	-	-	-	-	-	-	-	-	-	1	-	-	-	**3**
NA 96/45	Gaj IIa/b	-	-	-	3	-	5	7	-	-	-	-	-	-	1	-	1	-	-	3	-	1	-	14	1	-	-	-	**33**
NA 95/1	Gaj IIa/b	-	-	-	3	-	10	1	-	-	-	-	-	-	-	-	-	-	-	-	6	1	-	2	-	-	2	-	**22**
NA 97/26	IA	-	-	-	1	-	4	2	-	-	-	-	-	-	-	-	-	-	-	1	-	2	-	1	-	-	-	-	**10**
NA 99/75	IA	-	-	-	4	-	24	5	-	-	-	-	1	-	1	-	-	1	-	1	1	2	-	4	2	-	1	-	**43**
NA 92/2C	subrec	-	-	-	-	-	14	2	-	-	-	-	1	-	1	-	2	-	-	1	6	4	-	3	3	-	-	-	**37**
NA 93/46	LSA	-	2	-	22	-	217	27	-	-	6	-	3	4	9	3	3	3	-	8	20	16	6	37	19	-	10	-	**391**
NA 93/46	AB LSA	-	-	-	-	1	34	2	-	-	1	-	1	1	-	-	1	-	-	-	3	5	2	2	1	-	2	-	**56**
NA 93/46	EIA	1	26	4	424	21	6828	470	5	1	82	13	103	20	238	54	72	44	-	96	198	222	62	953	224	-	217	1	**9924**
NA 93/46	AB EIA	-	2	2	2	1	149	6	3	-	3	-	1	2	1	-	2	-	-	1	12	17	7	15	12	-	7	-	**239**
NA 94/7	I	-	-	-	-	13	1161	59	-	-	1	-	7	2	21	-	4	-	-	-	20	26	-	160	4	-	56	-	**1534**
NA 94/7	II	-	-	-	-	5	290	14	-	-	2	-	3	-	4	-	-	-	-	-	6	2	-	35	3	-	11	-	**375**
NA 94/7	III	1	2	-	-	4	255	15	-	-	3	-	6	3	16	-	2	-	-	-	10	25	-	28	5	-	11	-	**383**
NA 94/7	IV	-	-	-	-	1	218	12	-	-	-	-	5	1	4	1	6	-	-	-	4	21	-	26	3	-	7	-	**309**
NA 93/45	I	1	-	-	10	-	132	11	-	-	-	-	2	-	11	5	1	2	-	3	-	4	-	21	11	-	5	-	**208**
NA 93/45	II	-	1	-	18	1	439	24	-	-	-	-	9	-	14	11	4	4	-	15	24	-	10	35	22	-	5	-	**617**
NA 93/45	IIIa	-	-	-	11	-	161	10	-	-	2	-	6	-	10	3	5	1	-	2	7	1	1	30	11	-	1	-	**251**
NA 93/45	IIIb	-	-	-	8	-	168	5	-	-	-	-	-	-	2	1	2	-	-	3	4	3	1	21	9	-	2	-	**221**
NA 93/45	IV	-	-	1	9	-	176	16	-	-	-	-	3	-	7	3	-	1	-	6	4	3	-	37	7	-	3	-	**266**
Blé A		-	-	-	2	-	18	4	-	-	-	-	-	-	2	-	-	-	-	-	-	-	-	13	1	-	9	-	**47**
Blé B		-	-	-	1	-	72	5	-	-	-	-	-	-	3	-	-	-	-	-	5	4	-	18	3	-	2	-	**112**
Blé C		-	-	-	-	-	53	6	-	-	-	-	-	-	6	-	-	-	-	-	2	1	-	18	-	-	4	-	**90**
Blé E		-	-	-	-	-	83	14	-	-	1	-	1	-	14	-	1	1	-	-	6	7	-	25	4	-	3	-	**160**
All sites		**6**	**38**	**17**	**686**	**49**	**12484**	**938**	**10**	**1**	**116**	**14**	**195**	**65**	**477**	**102**	**120**	**73**	**1**	**174**	**528**	**634**	**153**	**1897**	**414**	**5**	**413**	**3**	**18818**

H=*Heterobranchus* sp., G=*Clarias gariepinus*, A=*Clarias anguillaris*, C=*Clarias sp.*

Table C.11a: Skeletal element distribution of clariid catfish by site and phase

		<10	10-20	20-30	30-40	40-50	50-60	60-70	70-80	80-90	90-100	100-110	110-120	120-130
BF 97/5	LSA	-	-	1	2	-	-	-	-	-	-	-	-	-
BF 97/5	EIA	-	7	43	72	44	4	3	3	-	-	-	-	-
BF 95/45	LSA	-	4	1	1	-	-	-	-	-	-	-	-	-
BF 95/45	AB LSA	-	-	1	-	-	-	-	-	-	-	-	-	-
BF 95/45	EIA	-	7	17	81	64	20	9	21	5	2	-	-	-
BF 95/45	AB EIA	-	3	24	31	26	7	3	-	-	-	-	-	-
BF 97/13	EIA	-	1	1	5	-	1	-	-	-	-	-	-	-
BF 97/13	AB EIA	-	2	1	-	-	-	-	-	-	-	-	-	-
BF 97/13	MIA	-	-	-	5	4	2	-	-	-	1	-	-	-
BF 97/13	LIA	-	-	-	10	6	1	-	-	-	-	-	-	-
BF 97/13	AB LIA	-	-	2	2	2	-	-	-	-	-	-	-	-
BF 97/30	LIA	-	-	8	19	5	4	1	1	-	-	-	-	-
BF 96/22	MIA	-	-	2	8	2	2	1	-	-	-	-	-	-
BF 97/31	LIA	-	-	6	10	7	2	2	-	-	-	-	-	-
BF 94/120	LIA	1	25	44	54	31	11	3	-	-	1	-	-	-
BF 95/7	LIA	-	4	17	7	4	2	2	-	-	-	-	-	-
NA 90/5A	Gaj IIb	-	-	12	51	62	15	9	2	-	-	-	-	-
NA 90/5BI	Gaj IIa/b	-	-	6	46	88	31	5	-	-	-	-	-	-
NA 90/5BII	Gaj IIc	-	-	-	4	1	-	-	-	-	-	-	-	-
NA 93/42	Gaj I	-	-	3	3	-	-	-	-	-	-	-	-	-
NA 97/18	Gaj I	-	-	-	3	-	-	-	-	-	-	-	-	-
NA 99/65	Gaj IIb	-	-	-	-	-	1	1	-	-	-	-	-	-
NA 99/65	Gaj IIc	-	3	5	3	3	4	3	-	-	-	-	-	-
NA 97/37	Gaj III	-	-	2	14	21	14	3	3	-	-	-	-	-
NA 97/33	Gaj I	-	-	-	-	-	-	1	-	-	-	-	-	-
NA 97/33	Gaj IIa/b	-	-	-	-	-	1	-	-	-	-	-	-	-
NA 96/45	Gaj IIa/b	-	-	-	1	8	6	6	10	-	-	-	-	-
NA 95/1	Gaj IIa/b	-	-	-	4	4	5	-	-	-	-	-	-	-
NA 97/26	IA	-	-	1	3	1	1	1	-	-	-	-	-	-
NA 99/75	IA	-	2	4	5	2	4	-	-	-	-	-	-	-
NA 92/2C	subrec	-	-	4	6	6	-	2	-	3	-	-	-	-
NA 93/46	LSA	-	16	37	66	39	15	9	2	2	-	-	-	-
NA 93/46	AB LSA	-	4	4	-	-	-	-	-	-	-	-	-	-
NA 93/46	EIA	-	115	526	1082	661	180	93	52	12	1	-	-	-
NA 93/46	AB EIA	4	24	35	18	3	-	-	3	-	-	-	-	-
NA 94/7	I	-	-	1	7	19	5	2	5	5	3	-	-	-
NA 94/7	II	-	-	1	4	15	3	9	9	14	15	-	-	2
NA 94/7	III	-	-	2	3	10	8	7	11	6	2	-	-	-
NA 94/7	IV	-	-	-	12	57	32	29	12	9	6	2	-	-
NA 93/45	I	-	1	3	38	49	7	5	1	1	-	-	-	-
NA 93/45	II	-	-	12	62	78	35	25	23	8	1	-	-	-
NA 93/45	IIIa	-	-	4	32	47	15	13	2	8	-	-	2	-
NA 93/45	IIIb	-	-	4	6	23	10	14	7	5	-	-	-	-
NA 93/45	IV	1	-	10	20	29	10	17	9	5	1	-	-	-
Blé A		-	-	5	14	12	1	1	-	1	1	1	-	-
Blé B		-	-	1	10	12	11	3	5	2	-	-	-	-
Blé C		-	-	-	7	15	3	5	4	1	1	-	-	-
Blé E		-	-	3	21	36	13	5	6	1	-	-	-	-
All sites		**6**	**218**	**847**	**1852**	**1496**	**484**	**290**	**191**	**88**	**35**	**3**	**2**	**2**

Table C.11b: SL (cm) reconstructions of clariid catfish by site and phase

		basioccipitale	neurocranium	angulare	operculare	apparatus Weberianus	vertebrae praecaudales	vertebrae caudales	cleithrum	coracoideum	spina pinnae pectoralis	spina pinnae dorsalis	pterygiophorus dorsalis	**Total**
BF 97/5	LSA	-	9	-	-	-	-	-	1	-	8	2	-	**20**
BF 94/45	AB LSA	-	-	-	-	-	-	-	-	-	1	-	-	**1**
BF 94/45	EIA	-	2	-	-	-	1	1	-	-	2	1	-	**7**
BF 94/45	AB EIA	-	-	-	-	-	-	2	-	-	1	-	-	**3**
BF 96/22	MIA	-	-	-	-	-	-	-	-	-	1	-	-	**1**
BF 97/31	LIA	-	-	-	-	-	-	-	-	-	1	-	-	**1**
BF 94/120	LIA	-	1	-	-	-	-	-	1	-	3	4	-	**9**
BF 95/7	LIA	-	-	-	-	-	-	-	2	1	5	-	-	**8**
NA 96/45	Gaj IIa/b	-	-	-	-	-	-	-	-	-	1	-	-	**1**
NA 95/1	Gaj IIa/b	-	-	-	-	-	-	-	-	-	1	-	-	**1**
NA 92/2C	subrec	-	4	-	1	-	-	2	10	4	3	2	1	**27**
NA 93/46	EIA	-	14	-	-	-	-	1	6	2	6	7	-	**36**
NA 94/7	I	-	-	-	-	-	-	-	-	-	3	2	-	**5**
NA 94/7	II	-	1	-	-	-	-	-	-	-	1	-	-	**2**
NA 94/7	III	-	-	-	-	-	-	-	-	-	-	1	-	**1**
NA 94/7	IV	-	1	-	-	-	-	-	-	-	-	-	-	**1**
NA 93/45	I	-	1	-	-	-	-	-	4	-	1	-	-	**6**
NA 93/45	II	-	1	-	-	-	-	-	2	-	3	-	-	**6**
NA 93/45	IIIa	-	2	-	-	-	-	-	2	-	1	-	-	**5**
NA 93/45	IIIb	-	3	-	-	-	-	-	1	-	1	-	-	**5**
NA 93/45	IV	-	1	-	1	-	-	-	1	-	1	-	-	**4**
Blé A		-	-	-	-	-	-	-	11	3	13	1	-	**28**
Blé B		-	-	10	-	-	1	1	22	7	22	8	-	**71**
Blé C		1	8	-	-	1	16	-	32	12	27	7	-	**104**
Blé E		-	22	-	1	-	1	2	72	23	44	10	-	**175**
All sites		**1**	**70**	**10**	**3**	**1**	**19**	**9**	**167**	**52**	**150**	**43**	**1**	**528**

		<10	10-20	20-30	30-40	>40
BF 97/5	LSA	1	15	1	-	-
BF 94/45	EIA	4	1	-	-	-
BF 97/31	LIA	1	-	-	-	-
BF 94/120	LIA	5	2	1	-	-
BF 95/7	LIA	2	3	2	-	-
NA 96/45	Gaj IIa/b	1	-	-	-	-
NA 95/1	Gaj IIa/b	1	-	-	-	-
NA 92/2C	subrec	2	13	4	-	-
NA 93/46	EIA	3	11	6	-	-
NA 94/7	I	-	-	1	1	3
NA 94/7	II	-	-	-	-	1
NA 94/7	III	-	-	1	-	-
NA 93/45	I	-	6	-	-	-
NA 93/45	II	1	2	2	1	-
NA 93/45	IIIa	-	3	-	-	-
NA 93/45	IIIb	-	1	1	-	-
NA 93/45	IV	1	-	1	1	-
Blé A		-	4	14	7	-
Blé B		-	2	37	11	-
Blé C		-	4	50	3	1
Blé E		-	19	77	-	-
All sites		**22**	**86**	**198**	**24**	**5**

Table C.12a-b: Skeletal element distribution and SL (cm) reconstructions of *Synodontis* sp. by site and phase

		neurocranium	vertebrae praecaudales	vertebrae caudales	cleithrum	spina pinnae pectoralis	spina pinnae dorsalis	spina	**Total**
BF 97/5	LSA	235	10	59	7	68	1	3	**383**
BF 97/5	EIA	1	-	-	-	-	-	-	**1**
NA 97/18	Gaj I	-	1	-	-	-	-	-	**1**
NA 93/46	EIA	47	-	-	-	3	2	1	**53**
NA 93/45	II	-	-	-	-	1	-	-	**1**
NA 93/45	IIIa	-	-	-	-	1	-	-	**1**
NA 93/45	IV	-	-	-	-	1	-	-	**1**
All sites		**283**	**11**	**59**	**7**	**74**	**3**	**4**	**7**

Table C.13: Skeletal element distribution of unidentified catfish by site and phase

		neurocranium	articulare	dentale	maxillare	palatinum	praemaxillare	quadratum	keratohyale	epihyale	hyomandibulare	operculare	praeoperculare	arcus hyoideus	arcus branchialis	vertebrae praecaudales	vertebrae caudales	vertebrae	cleithrum	supracleithrale	**Total**
BF 97/5	LSA	-	-	1	-	-	1	-	-	-	-	-	-	-	-	2	-	23	-	-	**27**
BF 97/5	EIA	2	1	2	-	-	-	-	-	-	-	-	-	-	-	21	-	4	-	-	**30**
NA 93/46	LSA	1	-	-	-	-	1	-	2	-	-	-	-	-	-	1	-	2	1	-	**8**
NA 93/46	EIA	62	21	46	21	5	13	24	14	5	10	5	-	1	1	21	3	322	23	3	**600**
NA 93/46	AB EIA	-	-	-	1	-	-	-	-	-	-	-	-	-	-	2	-	-	-	-	**3**
NA 94/7	II	-	-	-	1	-	-	-	-	-	-	-	-	-	-	1	-	-	-	-	**2**
NA 94/7	III	-	2	-	-	-	-	-	-	-	-	-	-	-	-	-	-	-	-	-	**2**
NA 93/45	I	1	2	2	1	-	-	-	1	-	-	1	-	-	-	-	-	4	3	-	**15**
NA 93/45	II	1	2	-	1	-	-	1	-	1	1	1	1	1	-	4	-	4	-	1	**19**
NA 93/45	IIIa	-	-	-	-	-	-	-	-	-	-	-	-	-	-	-	-	-	1	-	**1**
NA 93/45	IIIb	1	-	-	-	-	-	-	1	-	-	-	-	-	-	-	-	-	-	-	**2**
NA 93/45	IV	2	1	-	-	-	-	-	-	-	-	-	-	-	-	-	-	-	-	-	**3**
Blé B		-	-	-	-	-	-	-	-	-	-	-	-	-	-	5	-	-	-	-	**5**
Blé E		1	1	-	-	-	-	-	-	-	-	-	-	-	-	2	-	-	-	-	**4**
All sites		**71**	**30**	**51**	**25**	**5**	**15**	**25**	**18**	**6**	**11**	**7**	**1**	**2**	**1**	**59**	**3**	**359**	**28**	**4**	**721**

		10-20	20-30	30-40	>40
BF 97/5	LSA	15	4	7	-
BF 97/5	EIA	3	18	9	-
NA 93/46	LSA	-	1	6	-
NA 93/46	EIA	11	108	279	68
NA 93/46	AB EIA	1	-	2	-
NA 94/7	II	-	-	1	-
NA 94/7	III	-	2	-	-
NA 93/45	I	-	4	9	-
NA 93/45	II	-	-	19	-
NA 93/45	IIIa	-	-	1	-
NA 93/45	IIIb	-	-	1	-
NA 93/45	IV	-	-	1	1
Blé B		-	1	2	-
Blé E		-	2	2	-
All sites		**30**	**140**	**339**	**69**

Table C.14a-b: Skeletal element distribution and SL reconstructions (cm) of *Parachanna obscura* by site and phase

	BF 97/5 LSA	BF 94/45 EIA	BF 97/30 LIA	BF 94/120 LIA	NA 95/1 Gaj IIa/b	NA 92/2C subrec	NA 93/46 LSA	NA 93/46 EIA	NA 94/7 I	NA 94/7 II	NA 94/7 III	NA 94/7 IV	NA 93/45 I	NA 93/45 II	NA 93/45 IIIa	NA 93/45 IIIb	NA 93/45 IV	Blé A	Blé B	Blé C	Blé E	**All sites**
neurocranium	3	-	-	-	-	4	2	38	-	-	-	-	5	5	-	-	2	-	-	3	1	**60**
angulare	-	-	-	-	-	1	-	2	-	-	-	-	-	1	-	-	-	-	1	-	-	**5**
articulare	-	-	-	-	-	1	2	11	-	-	-	-	1	-	1	-	-	-	-	2	2	**20**
angulare + articulare	-	-	-	-	-	-	-	1	-	-	-	-	-	-	-	-	-	-	2	-	-	**3**
dentale	-	-	-	1	-	2	-	30	-	-	-	-	3	4	1	-	-	-	-	-	2	**43**
articulare + dentale	-	-	-	1	-	-	-	-	-	-	-	-	-	-	-	-	-	-	-	-	-	**1**
ectopterygoideum	2	-	-	-	-	-	-	-	-	-	-	-	-	-	-	-	-	-	1	-	-	**1**
maxillare	-	-	-	-	-	-	-	5	-	-	-	-	1	1	2	-	-	-	-	1	3	**13**
metapterygoideum	-	-	-	-	-	-	-	-	-	-	-	-	-	-	1	-	-	-	-	-	-	**1**
dentale/maxillare	-	-	-	-	-	-	-	1	-	-	-	-	-	-	-	-	-	-	-	-	-	**1**
palatinum	-	-	-	-	-	1	1	8	-	-	1	-	-	-	-	-	-	-	-	-	-	**11**
praemaxillare	-	-	-	-	-	1	-	7	-	-	-	-	1	1	-	1	-	1	1	-	6	**19**
quadratum	2	-	-	-	-	2	-	11	-	1	-	1	1	3	2	-	1	1	1	-	-	**24**
branchiostegale	-	-	-	1	-	1	-	12	-	-	-	-	-	2	1	-	-	-	-	-	1	**18**
keratohyale	-	-	-	-	-	-	1	2	-	-	-	-	1	2	-	-	-	1	-	-	1	**8**
epihyale	-	-	-	-	-	-	-	4	-	-	-	-	1	2	-	-	-	-	-	1	2	**10**
hyomandibulare	-	-	-	-	-	1	-	5	-	-	-	-	-	3	1	-	-	-	-	-	4	**14**
hypohyale	1	-	-	-	-	-	-	6	-	-	-	-	1	-	-	-	-	-	-	-	-	**7**
interoperculare	-	-	-	-	-	-	-	-	-	-	-	-	-	-	1	-	-	-	-	-	-	**1**
operculare	-	-	-	-	-	-	3	9	-	-	-	-	2	1	2	-	-	-	-	-	1	**18**
praeoperculare	-	-	-	-	-	2	-	18	-	-	-	-	2	2	2	-	1	-	-	-	10	**37**
suboperculare	-	-	-	-	-	-	-	-	-	-	-	-	-	-	-	-	1	-	-	-	-	**1**
urohyale	-	-	-	-	-	1	-	1	-	-	-	-	-	2	-	-	-	-	1	-	-	**5**
arcus hyoideus	-	-	-	-	-	-	-	1	-	-	-	-	-	-	-	-	-	-	-	-	1	**2**
arcus branchialis	-	-	-	-	-	1	-	-	-	-	-	-	1	1	1	-	-	-	-	-	2	**6**
vertebrae praecaudales	5	1	-	1	-	2	1	34	1	1	1	1	4	10	-	-	4	2	3	2	8	**76**
vertebrae caudales	3	-	1	-	-	4	-	23	1	-	-	-	3	11	2	-	1	-	-	1	6	**53**
vertebrae	5	-	-	-	-	-	2	84	1	1	-	-	1	11	1	1	2	1	1	-	-	**106**
urostylus	-	-	-	-	-	-	2	2	-	-	-	-	-	-	-	-	-	-	-	-	-	**4**
costae	-	-	-	-	-	-	-	17	-	-	-	-	-	-	-	-	-	-	-	1	8	**26**
basipterygium	-	-	-	-	-	-	1	10	2	-	-	1	-	-	-	-	1	-	-	1	6	**22**
spina pinnae ventralis	-	-	-	-	-	2	1	8	-	-	-	-	-	3	4	-	-	-	-	-	-	**18**
cleithrum	1	-	-	-	-	-	-	23	1	-	-	-	4	8	2	3	4	-	7	3	10	**65**
postcleithrale	-	-	-	-	-	-	-	2	-	-	-	-	-	-	-	-	-	-	-	-	-	**2**
scapula	1	-	-	-	-	-	-	2	-	-	-	-	-	-	1	-	-	-	-	-	-	**3**
supracleithrale	-	-	-	-	-	-	1	11	-	-	-	1	3	1	-	-	-	1	-	-	2	**20**
spina pinnae pectoralis	-	-	-	1	-	1	-	-	-	-	-	-	-	-	-	-	-	-	-	-	-	**2**
spina pinnae dorsalis	1	-	-	-	1	2	1	24	-	-	-	-	6	5	3	-	-	-	4	3	17	**66**
pterygiophorus dorsalis	-	-	-	-	1	1	-	13	2	4	1	-	2	2	2	-	1	-	-	1	2	**32**
spina pinnae analis	1	-	-	-	-	2	-	16	-	-	-	-	4	3	-	2	-	-	-	-	-	**27**
pterygiophorus analis	-	-	-	1	-	1	-	6	-	1	-	-	1	1	-	1	-	-	-	-	3	**15**
pterygiophorus	-	-	-	1	-	-	-	-	-	-	-	-	-	-	-	-	-	1	-	-	2	**4**
lepidotrichium	-	-	-	-	-	-	-	-	-	-	-	-	-	-	-	-	-	-	-	-	2	**2**
unidentified	-	-	-	-	-	-	-	4	-	-	-	-	1	-	-	-	-	-	-	-	-	**5**
Total	**25**	**1**	**1**	**7**	**2**	**33**	**18**	**451**	**8**	**8**	**3**	**4**	**49**	**85**	**30**	**8**	**18**	**8**	**22**	**19**	**102**	**877**

Table C.15a: Skeletal element distribution of Nile perch by site and phase

		10-20	20-30	30-40	40-50	50-60	60-70	70-80	80-90	90-100	>100
BF 97/5	LSA	-	-	2	1	7	3	4	-	-	1
BF 94/45	EIA	-	-	1	-	-	-	-	-	-	-
BF 97/30	LIA	-	-	-	-	-	-	-	-	1	-
BF 94/120	LIA	-	3	1	-	-	-	-	1	-	-
NA 95/1	Gaj IIa/b	-	1	-	1	-	-	-	-	-	-
NA 92/2C	subrec	-	-	-	6	4	1	9	3	6	-
NA 93/46	LSA	-	-	2	-	1	6	1	1	1	1
NA 93/46	EIA	1	17	56	71	56	46	35	11	7	3
NA 94/7	I	-	-	1	1	-	3	-	-	-	-
NA 94/7	II	-	-	-	-	-	3	3	-	-	-
NA 94/7	III	-	1	-	-	1	1	-	-	-	-
NA 94/7	IV	-	-	-	1	-	-	-	-	-	-
NA 93/45	I	-	-	2	-	7	12	9	7	-	-
NA 93/45	II	-	1	1	9	17	18	16	3	-	1
NA 93/45	IIIa	-	-	2	8	7	5	-	1	-	3
NA 93/45	IIIb	-	-	-	3	-	2	1	1	-	-
NA 93/45	IV	-	-	3	1	5	3	-	1	-	-
Blé A		-	-	-	-	1	1	2	-	-	1
Blé B		-	3	-	-	3	2	-	-	3	3
Blé C		-	2	1	3	-	4	1	-	2	-
Blé E		-	8	10	17	7	7	11	2	1	2
All sites		**1**	**36**	**82**	**122**	**116**	**117**	**92**	**31**	**21**	**15**

Table C.15b: SL (cm) reconstructions of Nile perch by site and phase

		neurocranium	articulare	dentale	ectopterygoideum	maxillare	palatinum	praemaxillare	quadratum	keratohyale	epihyale	hyomandibulare	operculare	praeoperculare	suboperculare	urohyale	arcus hyoideus	arcus branchialis	vertebrae praecaudales	vertebrae caudales	vertebrae	basipterygium	spina pinnae ventralis	cleithrum	postcleithrale	scapula	supracleithrale	otolith	spina pinnae dorsalis	pterygiophorus dorsalis	spina pinnae analis	pterygiophorus analis	spina	pterygiophorus	unidentified	**Total**
BF 97/5	LSA	1	-	1	-	1	-	-	1	-	-	2	7	-	-	2	-	-	125	91	7	6	-	-	1	1	2	-	84	4	7	11	10	-	-	**364**
BF 97/5	EIA	4	2	1	-	-	-	-	-	-	-	1	-	2	-	-	-	-	189	165	5	2	1	10	-	-	1	2	31	6	13	18	6	-	-	**459**
BF 94/45	LSA	-	-	-	-	-	-	-	-	-	-	-	-	-	-	-	-	-	1	1	-	-	-	-	-	-	-	-	1	1	-	-	-	-	-	**4**
BF 94/45	AB LSA	-	-	-	-	-	-	-	-	-	-	-	-	-	-	-	-	-	1	-	-	-	-	-	-	-	-	-	-	-	-	-	-	-	-	**1**
BF 94/45	EIA	-	-	-	-	-	-	-	-	-	-	2	5	5	-	-	-	-	26	12	-	2	-	9	-	-	2	-	21	-	3	6	-	-	-	**93**
BF 94/45	AB EIA	13	1	-	-	-	1	-	-	-	-	1	5	1	-	-	2	-	62	64	2	3	4	3	-	-	8	-	39	9	12	7	-	-	5	**242**
BF 96/22	MIA	-	-	-	-	-	-	-	-	-	-	1	-	-	-	-	-	-	-	-	-	-	-	-	-	-	-	-	2	-	-	1	-	-	-	**4**
BF 97/31	LIA	-	-	-	-	-	-	-	-	-	-	-	-	-	-	-	-	-	1	-	-	-	-	-	-	-	-	-	-	-	2	2	-	-	-	**5**
BF 94/120	LIA	-	-	-	-	-	-	-	-	-	-	-	-	-	-	-	1	-	-	-	-	-	-	-	-	-	-	-	1	-	-	1	-	-	-	**3**
BF 95/7	LIA	-	-	-	-	-	-	-	-	-	-	-	-	-	-	-	-	-	1	-	-	-	-	-	-	-	-	-	-	-	1	1	1	-	-	**4**
NA 96/45	Gaj IIa/b	-	-	-	-	-	-	-	-	-	-	-	-	-	-	-	-	-	1	-	-	-	-	-	-	-	-	-	1	-	-	1	-	-	-	**3**
NA 92/2C	subrec	-	-	-	-	-	-	-	-	-	-	-	-	-	-	-	-	-	1	-	-	-	-	-	-	-	-	-	-	-	-	-	-	-	-	**1**
NA 93/46	LSA	1	-	-	-	-	-	-	-	-	-	-	-	-	-	-	-	-	1	-	-	1	5	-	-	-	1	-	-	-	1	1	-	-	-	**11**
NA 93/46	AB LSA	-	-	-	-	-	-	-	-	-	-	-	-	-	-	-	-	-	1	-	-	-	1	-	-	-	-	-	-	-	-	-	-	-	-	**2**
NA 93/46	EIA	35	6	2	1	3	2	2	8	2	1	12	24	9	-	2	4	1	184	90	1	18	370	43	11	-	17	-	-	89	48	43	-	-	-	**1028**
NA 93/46	AB EIA	3	-	-	-	-	-	-	-	-	-	-	-	1	-	-	-	-	19	10	-	2	10	3	-	-	-	-	-	4	1	-	-	-	-	**53**
NA 94/7	I	-	-	-	-	-	-	-	-	-	-	-	-	-	-	-	-	-	-	-	-	-	-	-	-	-	-	-	4	-	-	-	-	-	-	**4**
NA 94/7	II	-	-	-	-	-	-	-	-	-	-	1	-	1	-	-	-	-	16	3	-	-	8	1	-	-	-	-	41	2	5	5	-	-	-	**83**
NA 94/7	III	2	-	-	-	-	-	-	-	-	-	-	-	-	-	-	-	-	-	-	-	-	-	-	-	-	-	-	11	1	-	1	-	-	-	**15**
NA 94/7	IV	-	1	-	-	-	-	-	-	-	-	-	-	-	-	-	-	-	-	-	-	-	-	-	-	-	-	-	3	-	1	-	-	-	-	**5**
NA 93/45	I	-	-	-	-	-	-	-	1	-	-	1	2	1	1	-	-	-	1	-	-	1	1	6	-	-	-	-	5	1	1	1	-	-	-	**23**
NA 93/45	II	1	-	-	-	-	-	-	-	1	-	-	2	2	-	-	-	-	1	-	-	-	1	1	1	-	1	-	2	1	-	-	-	-	-	**14**
NA 93/45	IIIa	2	-	-	-	1	-	1	-	-	-	-	-	1	-	-	-	-	1	-	-	-	-	2	1	-	-	-	-	-	1	1	-	-	-	**11**
NA 93/45	IIIb	-	-	-	-	-	-	-	-	-	-	-	1	-	-	-	-	-	-	-	-	-	-	2	-	-	-	-	3	-	-	-	-	-	-	**6**
NA 93/45	IV	-	-	-	-	-	-	-	-	-	-	1	3	-	1	-	-	-	1	-	-	-	-	2	-	-	1	-	-	-	1	-	-	-	-	**10**
Blé A		-	-	-	-	-	-	-	-	-	-	-	-	-	-	-	-	-	-	1	-	-	-	-	-	-	-	-	-	-	-	-	-	-	-	**1**
Blé B		-	-	-	-	-	-	-	-	-	-	-	1	-	-	-	-	-	1	-	-	-	1	1	-	-	-	-	2	-	2	-	-	-	-	**8**
Blé C		-	-	-	-	-	-	-	-	-	-	-	1	4	-	-	-	-	3	2	-	-	2	7	-	-	-	-	5	-	-	1	-	-	-	**25**
Blé E		2	-	-	-	-	-	-	-	-	-	-	10	-	-	1	-	1	3	1	-	4	5	21	-	-	-	-	13	-	-	6	-	1	-	**68**
All sites		**64**	**10**	**4**	**1**	**5**	**3**	**3**	**10**	**3**	**1**	**22**	**61**	**27**	**2**	**5**	**7**	**2**	**640**	**440**	**15**	**39**	**409**	**111**	**14**	**1**	**33**	**2**	**269**	**118**	**99**	**107**	**17**	**1**	**5**	**2550**

Table C.16a: Skeletal element distribution of tilapia by site and phase

		<10	10-20	20-30	30-40	>40
BF 97/5	LSA	1	107	113	46	-
BF 97/5	EIA	3	84	284	2	-
BF 94/45	LSA	-	4	-	-	-
BF 94/45	AB LSA	-	1	-	-	-
BF 94/45	EIA	-	25	45	1	-
BF 94/45	AB EIA	7	73	96	9	-
BF 96/22	MIA	-	3	1	-	-
BF 97/31	LIA	-	3	2	-	-
BF 94/120	LIA	-	2	1	-	-
BF 95/7	LIA	1	1	1	-	-
NA 96/45	Gaj IIa/b	-	1	2	-	-
NA 92/2C	subrec	-	-	1	-	-
NA 93/46	LSA	-	3	5	1	-
NA 93/46	AB LSA	2	-	-	-	-
NA 93/46	EIA	-	200	632	126	7
NA 93/46	AB EIA	22	22	5	1	-
NA 94/7	I	-	1	1	1	-
NA 94/7	II	-	-	11	50	10
NA 94/7	III	-	-	7	1	-
NA 94/7	IV	-	-	7	2	-
NA 93/45	I	-	1	10	12	-
NA 93/45	II	-	1	9	4	-
NA 93/45	IIIa	-	-	8	3	-
NA 93/45	IIIb	-	-	4	2	-
NA 93/45	IV	-	1	2	5	1
Blé A		-	-	-	-	1
Blé B		-	-	5	2	-
Blé C		-	2	21	2	-
Blé E		-	-	58	9	-
All sites		**36**	**535**	**1331**	**279**	**19**

Table C.16b: SL (cm) reconstructions of tilapia by site and phase

		sphenethmoideum	fronto-parietale	exoccipitale	parasphenoideum	angulare	articulare	dentale	pterygoideum	maxillare	praemaxillare	quadratum	cranium	atlas	os sacrum	vertebrae	urostylus	coracoideum	scapula	scapula + coracoideum + clavicula	humerus	radio-ulna	ischio-pubis	ilium	os femoris	tibio-fibula	ossa tarsi	os metatarsale	ossa metapodalia	phalanx	unidentified	**Total**
BF 94/45	LSA	-	-	-	-	-	-	-	-	-	-	-	-	-	-	2	-	-	-	-	-	-	-	-	-	-	-	-	-	-	-	**2**
BF 94/45	AB LSA	-	-	-	-	-	-	-	-	-	-	-	-	-	-	-	-	-	-	-	-	1	-	-	-	1	-	-	-	-	1	**3**
BF 94/45	EIA	-	2	1	-	1	-	6	-	-	-	-	-	-	-	11	5	-	1	-	3	1	2	14	6	6	1	-	-	-	-	**60**
BF 94/45	AB EIA	2	5	1	-	5	3	-	1	2	-	1	-	-	5	12	4	7	8	-	10	5	2	13	8	13	4	-	3	1	31	**145**
BF 97/13	EIA	1	-	-	-	-	-	-	-	-	-	-	-	-	-	-	-	-	-	-	-	2	-	1	-	1	-	-	-	-	-	**5**
BF 97/13	AB EIA	-	-	-	-	-	-	-	-	2	-	-	-	-	1	-	-	-	1	-	2	1	2	-	-	-	-	-	-	5	-	**14**
BF 97/13	MIA	-	-	-	-	-	-	-	-	-	-	-	1	-	-	-	-	-	-	-	1	-	-	-	-	-	-	-	-	-	-	**2**
BF 97/13	AB MIA	-	-	1	-	-	-	-	-	-	-	-	-	-	2	1	-	-	1	-	1	-	-	-	-	1	-	-	2	-	-	**9**
BF 97/13	LIA	-	-	-	-	-	-	-	-	-	-	-	-	-	-	-	1	-	-	-	-	-	-	-	-	-	-	-	-	-	-	**1**
BF 97/13	AB LIA	-	-	-	-	-	-	-	-	-	-	-	-	-	-	-	-	-	-	-	-	-	-	-	1	1	-	-	-	4	7	**13**
BF 97/30	sieve LIA	-	-	-	-	-	-	-	-	-	-	-	-	-	-	-	-	-	-	-	-	-	-	-	-	1	-	-	-	-	-	**1**
BF 96/22	MIA	6	17	7	5	25	5	3	6	25	3	3	12	2	4	38	9	2	8	1	9	3	-	31	12	20	1	-	1	-	2	**260***
BF 97/31	LIA	-	1	-	3	8	-	-	-	5	-	1	-	-	-	4	3	1	1	-	2	2	-	3	5	8	-	-	-	-	-	**47**
BF 94/120	LIA	1	2	-	-	-	-	-	-	8	-	-	-	-	1	3	1	1	1	-	6	1	-	4	4	10	1	-	-	-	-	**44**
BF 95/7	LIA	1	-	-	-	7	-	-	-	3	-	-	1	-	3	2	7	-	1	-	17	2	-	29	9	38	1	-	-	-	6	**127**
NA 97/37	Gaj III	1	-	-	-	2	-	-	-	-	-	-	-	-	-	4	-	-	-	-	1	1	-	1	3	5	-	-	1	-	-	**19**
NA 97/26	IA	1	1	3	-	8	-	-	-	3	-	1	-	-	-	1	2	3	2	-	4	1	-	12	5	7	-	-	-	-	-	**54**
NA 99/75	IA	-	1	1	-	2	-	-	-	3	-	-	-	-	-	-	-	-	-	-	1	-	-	1	2	1	-	-	-	-	-	**12**
NA 97/13	IA	-	-	-	-	1	-	-	-	1	-	-	-	-	-	1	-	-	-	-	1	-	-	2	3	4	-	-	-	-	-	**13**
NA 92/2C	subrec	1	1	1	-	4	-	-	3	7	-	-	1	-	-	3	3	1	-	-	1	-	-	6	1	5	-	1	-	-	-	**39**
NA 93/46	EIA	-	-	-	-	1	-	1	-	-	-	-	-	-	-	1	1	-	1	-	5	1	-	4	3	1	-	-	-	-	-	**19**
NA 94/7	IV	-	1	-	-	-	-	1	-	2	-	-	-	-	-	-	-	-	-	-	-	-	-	-	-	-	-	-	-	-	-	**4**
NA 93/45	I	-	-	-	-	-	-	-	-	-	-	-	-	-	-	-	-	-	-	-	1	-	-	-	1	1	-	-	-	-	-	**3**
NA 93/45	II	1	-	-	-	2	-	-	-	1	-	-	-	-	1	1	2	-	-	-	2	1	-	2	-	5	-	1	-	-	-	**19**
NA 93/45	IIIb	-	-	-	-	-	-	-	-	-	-	-	-	-	-	-	-	-	-	-	-	-	-	-	-	2	-	1	-	-	-	**3**
NA 93/45	IV	-	-	-	-	-	-	-	-	-	-	-	-	-	-	-	-	-	-	-	-	-	-	1	-	1	-	-	-	-	-	**2**
Blé E		-	-	-	-	-	-	-	-	-	-	-	-	-	-	-	-	-	-	-	-	-	-	-	-	1	-	-	-	-	-	**1**
All sites		**15**	**31**	**15**	**8**	**66**	**8**	**11**	**10**	**62**	**3**	**6**	**15**	**2**	**17**	**84**	**38**	**15**	**25**	**1**	**67**	**22**	**6**	**124**	**63**	**133**	**8**	**3**	**7**	**10**	**47**	**921**

*MNI=4

Table C.17: Skeletal element distribution of frogs and toads by site and phase

		T	C				C/D				Trionychidae			
		carapax	carapax	plastron	carapax/plastron	**Total**	carapax	plastron	carapax/plastron	**Total**	humerus	ilium	carapax	**Total**
BF 94/133	LSA	-	-	-	-	**0**	-	-	-	**0**	-	-	1	**1**
BF 97/5	LSA	-	2	-	2	**4**	-	-	-	**0**	-	-	-	**0**
BF 94/45	EIA	-	-	-	-	**0**	-	-	-	**0**	-	-	2	**2**
BF 95/7	LIA	-	-	-	-	**0**	-	-	-	**0**	-	-	1	**1**
NA 97/37	Gaj III	-	-	-	-	**0**	-	-	1	**1**	-	-	-	**0**
NA 97/33	Gaj IIa/b	-	1	-	-	**1**	-	-	-	**0**	-	-	-	**0**
NA 93/46	EIA	3	-	-	-	**0**	-	-	-	**0**	-	-	-	**0**
NA 93/46	EIA	-	-	-	-	**0**	1	1	6	7	-	-	-	**0**
NA 93/45	I	-	-	-	2	**2**	-	-	-	**0**	1	-	-	**1**
NA 93/45	II	1	6	-	5	**11**	-	-	-	**0**	-	-	-	**0**
NA 93/45	IIIa	-	2	-	-	**2**	-	-	-	**0**	-	-	-	**0**
NA 93/45	IIIb	-	1	1	-	**2**	-	-	-	**0**	-	1	-	**1**
NA 93/45	IV	-	-	-	2	**2**	-	-	-	**0**	-	-	-	**0**
Blé B		-	1	-	-	**1**	-	-	-	**0**	-	-	-	**0**
Blé C		-	2	1	-	**3**	-	-	-	**0**	1	-	-	**1**
Blé E		-	2	-	-	**2**	-	-	-	**0**	-	-	-	**0**
All sites		**4**	**17**	**2**	**11**	**30**	**1**	**1**	7	**8**	**2**	**1**	**4**	7

T=*Trionyx triunguis*, C=*Cyclanorbis senegalensis*, D=*Cycloderma aubryi*

Table C.18: Skeletal element distribution of softshell turtle by site and phase

		dentale	maxillare	os sacrum	vertebrae	humerus	ilium	os femoris	tibia	**Total**
BF 94/45	EIA	2	-	-	-	-	-	-	-	**2**
BF 94/45	AB EIA	1	-	-	6	2	-	3	2	**14**
BF 97/13	AB EIA	-	1	-	-	-	-	-	-	**1**
BF 97/30	LIA	-	1	-	-	-	-	-	-	**1**
BF 96/22	MIA	1	-	-	-	-	1	-	-	**2**
BF 97/31	LIA	1	-	1	-	5	-	3	1	**11**
BF 94/120	LIA	-	1	-	-	-	-	-	-	**1**
BF 95/7	LIA	-	-	-	-	-	-	1	-	**1**
NA 99/75	IA	1	-	-	-	-	-	-	-	**1**
NA 93/46	EIA	1	-	-	-	-	-	-	-	**1**
NA 93/46	AB EIA	-	-	-	1	-	-	-	-	**1**
All sites		**7**	**3**	**1**	7	7	**1**	7	**3**	**36**

Table C.19: Skeletal element distribution of agama by site and phase

		cranium	maxillare	mandibulare	vertebrae caudales	vertebrae	coracoideum	scapula	humerus	radius	ulna	pubis	ilium	ischium	os coxae	os femoris	humerus/os femoris	tibia	fibula	carapax	plastron	carapax/plastron	unidentified	**Total**
BF 97/5	LSA	-	-	-	-	-	-	-	-	-	-	-	-	-	-	-	-	-	-	9	3	11	-	**23**
BF 97/5	EIA	-	-	-	-	-	-	-	-	-	-	-	-	-	-	-	-	-	-	4	-	-	-	**4**
BF 94/45	EIA	-	-	-	-	-	-	-	-	-	-	-	-	-	-	1	-	-	-	63	11	11	1	**87**
BF 94/45	AB EIA	-	-	-	-	-	-	-	1	-	-	-	-	-	-	-	-	-	-	-	-	2	-	**3**
BF 97/13	EIA	-	-	-	-	-	-	-	-	-	-	-	-	-	-	-	-	-	-	1	-	-	-	**1**
BF 97/13	LIA	-	-	-	-	-	-	-	-	-	-	-	-	-	-	-	-	-	-	2	-	-	-	**2**
BF 97/30	LIA	-	-	-	-	-	-	-	-	-	-	-	-	-	-	-	-	-	-	1	-	-	-	**1**
BF 96/22	MIA	-	-	-	-	-	-	-	-	-	-	-	-	-	-	-	-	-	-	1	-	-	-	**1**
BF 97/31	LIA	-	-	-	-	-	-	-	-	-	-	-	-	-	-	-	-	-	-	3	-	1	-	**4**
BF 94/120	LIA	-	-	-	-	-	-	-	-	-	-	-	-	-	-	-	-	-	-	3	5	3	-	**11**
BF 95/7	LIA	-	-	-	-	-	-	-	-	-	-	-	-	-	-	-	-	-	-	22	4	3	-	**29**
NA 90/5A	Gaj IIb	-	-	-	-	-	-	-	-	-	-	-	-	-	-	-	-	-	-	19	4	3	-	**26**
NA 90/5BI	Gaj IIa/b	-	-	-	-	-	-	-	-	-	-	-	1	-	-	-	-	-	-	8	-	-	-	**9**
NA 90/5BII	Gaj IIc	-	-	-	-	-	-	-	-	-	-	-	-	-	-	-	-	-	-	2	-	-	-	**2**
NA 93/42	Gaj I	-	-	-	-	-	-	-	-	-	-	-	-	-	-	-	2	-	-	3	-	-	-	**5**
NA 97/18	Gaj I	-	-	-	-	-	-	-	-	-	-	-	-	-	-	-	-	-	-	-	2	4	-	**6**
NA 93/36	Gaj I	-	-	-	-	-	-	-	-	-	-	-	-	-	-	-	-	-	-	2	-	-	-	**2**
NA 99/65	Gaj IIb	-	-	-	-	-	-	-	-	-	-	-	-	-	-	-	-	-	-	1	-	-	-	**1**
NA 97/37	Gaj III	1	-	2	-	2	1	4	4	-	-	2	1	2	-	3	-	2	-	103	22	79	-	**228***
NA 97/33	Gaj IIa/b	-	-	-	-	-	-	-	-	-	-	-	-	-	-	-	-	-	-	3	-	-	-	**3**
NA 95/1	Gaj IIa/b	-	-	-	-	-	-	-	-	-	-	-	-	-	-	-	-	-	-	17	-	2	-	**19**
NA 97/26	IA	-	-	-	-	-	-	-	1	-	-	-	-	-	-	-	-	-	-	3	-	-	-	**4**
NA 99/75	IA	-	-	-	-	-	-	1	-	-	-	-	-	-	-	-	-	-	-	3	7	3	-	**14**
NA 93/46	LSA	-	-	1	-	-	-	-	1	-	-	-	-	-	-	1	-	-	-	7	8	4	1	**23**
NA 93/46	EIA	-	2	1	1	-	5	7	15	2	1	7	8	7	2	11	5	5	1	228	99	362	2	**769**
NA 93/46	AB EIA	-	-	-	-	-	-	-	-	-	-	-	-	-	-	-	-	-	-	1	-	2	-	**3**
NA 94/7	I	-	-	-	-	-	-	-	1	-	-	-	2	-	3	-	-	-	-	48	53	66	-	**173**
NA 94/7	II	-	-	-	-	-	-	-	1	-	-	-	1	-	1	1	-	-	-	43	11	11	-	**69**
NA 94/7	III	-	-	-	-	-	-	-	-	-	-	-	-	-	-	-	-	-	-	1	-	1	-	**2**
NA 93/45	I	-	-	-	-	-	1	-	1	-	-	-	-	1	-	1	-	1	-	24	8	3	-	**40**
NA 93/45	II	-	-	-	-	-	1	2	2	-	-	-	4	-	-	2	-	-	-	78	10	18	-	**117**
NA 93/45	IIIa	-	-	-	-	-	-	-	-	-	-	-	-	-	-	-	-	-	-	20	5	6	-	**31**
NA 93/45	IIIb	-	-	-	-	-	-	-	-	-	-	1	1	-	-	-	-	-	-	43	20	6	-	**71**
NA 93/45	IV	-	-	1	-	-	-	-	-	-	-	-	-	-	1	-	-	-	-	48	14	17	-	**81**
Blé A		-	-	-	-	-	-	-	-	-	-	-	-	-	-	-	-	-	-	12	6	-	-	**18****
Blé B		-	-	-	-	-	-	-	-	-	-	-	-	-	-	-	-	-	-	15	1	-	-	**16**
Blé C		-	-	-	-	-	-	-	-	-	-	-	-	-	-	-	-	-	-	7	-	5	-	**12**
Blé E		-	-	-	-	-	-	-	-	-	-	-	-	-	-	-	-	-	-	25	2	5	-	**32**
All sites		**1**	**2**	**5**	**1**	**2**	**8**	**14**	**27**	**2**	**1**	**10**	**18**	**10**	**7**	**20**	**7**	**8**	**1**	**873**	**295**	**628**	**4**	**1942**

*including 221 elements with MNI=5, **including 15 elements with MNI=2

Table C.20: Skeletal element distribution of Adanson's mud turtle by site and phase

		praefrontale	frontale	parietale	basisphenoideum	angulare	supra-angulare	articulare	coronoid	dentale	pterygoideum	maxillare	palatinum	praemaxillare	quadratum	postfrontorbitale	squamosum	vertebrae	costae	humerus	radius	ulna	ilium	ossa coxae	os femoris	tibia	fibula	ossa metapodalia	phalanx	unidentified	**Total**
BF 94/45	LSA	-	-	-	-	-	-	-	-	-	-	-	-	-	-	-	-	1	-	-	-	-	-	-	-	-	-	-	-	-	**1**
BF 94/45	AB LSA	-	-	-	-	-	-	-	-	-	-	-	-	-	-	-	-	1	-	-	-	-	-	-	-	-	-	-	-	-	**1**
BF 94/45	EIA	1	-	1	-	-	1	-	-	5	-	2	-	1	2	-	-	35	-	1	-	1	1	-	-	-	-	-	-	-	**51**
BF 94/45	AB EIA	-	-	-	-	-	-	-	-	1	-	-	2	-	-	-	-	5	-	-	-	-	-	-	-	-	-	-	1	1	**10**
BF 97/13	EIA	-	-	-	-	-	-	-	-	-	-	-	-	-	-	-	-	1	-	-	-	-	-	-	-	-	-	-	-	-	**1**
BF 97/13	MIA	1	-	-	-	-	-	-	-	-	-	1	-	-	-	-	-	1	-	-	-	-	-	-	-	-	-	-	-	-	**3**
BF 97/13	LIA	-	-	-	-	-	-	1	-	1	-	-	-	-	-	-	-	5	-	1	-	-	-	-	-	-	-	-	-	-	**8**
BF 97/30	LIA	-	-	-	-	-	-	1	-	-	-	-	-	-	-	-	-	1	-	-	-	-	-	-	-	-	-	-	-	-	**2**
BF 96/22	MIA	1	1	1	1	2	-	-	-	5	1	4	1	-	1	1	1	26	-	2	-	-	2	-	3	1	1	2	-	-	**57**
BF 97/31	LIA	-	-	-	-	-	-	-	-	1	1	2	-	2	-	1	1	16	-	-	-	-	1	-	1	-	-	-	-	-	**26**
BF 94/120	LIA	-	1	1	2	-	2	-	-	7	-	-	-	1	3	-	-	50	-	3	-	-	-	-	-	-	1	1	-	-	**71**
BF 95/7	LIA	-	1	1	-	-	1	1	-	5	2	5	1	-	2	-	-	34	1	6	1	1	3	-	2	3	4	-	-	-	**73**
NA 90/5BI	Gaj IIa/b	-	-	-	-	1	-	-	-	-	-	-	-	-	-	-	-	-	-	-	-	-	-	-	-	-	-	-	-	-	**1**
NA 90/5BII	Gaj IIc	-	-	-	-	-	-	-	-	2	-	-	-	-	-	-	-	15	-	3	-	-	-	1	-	-	-	-	-	-	**21**
NA 93/42	Gaj I	-	-	-	-	-	-	-	-	-	-	-	-	-	-	-	-	4	-	-	-	-	-	-	-	-	-	-	-	-	**4**
NA 99/65	Gaj IIb	-	-	-	-	-	-	-	-	-	-	-	-	-	-	-	-	1	-	-	-	-	-	-	-	-	-	-	-	-	**1**
NA 97/37	Gaj III	-	-	-	-	-	-	1	1	-	-	2	-	-	-	-	-	6	-	1	-	-	-	-	1	-	-	-	-	-	**12**
NA 97/33	Gaj I	-	-	-	-	-	-	-	-	-	-	-	-	-	-	-	-	1	-	1	-	-	-	-	-	-	-	-	-	-	**2**
NA 97/26	IA	1	-	-	-	-	-	-	-	1	-	-	-	-	-	-	-	5	-	-	-	-	1	-	1	-	-	-	-	-	**9**
NA 99/75	IA	-	-	-	-	-	-	-	-	-	-	-	-	-	-	-	-	4	-	1	-	-	-	-	-	-	-	-	-	-	**5**
NA 97/13	IA	-	-	-	-	-	-	-	-	-	-	1	-	-	-	-	-	2	-	-	-	-	-	-	-	-	-	-	-	-	**3**
NA 92/2C	subrec	-	-	-	-	-	1	1	-	-	-	-	-	-	-	-	-	24	1	5	-	-	1	-	1	4	-	1	-	-	**39**
		-	-	-	-	-	-	-	-	-	-	-	-	-	1	-	-	3	-	-	-	-	-	-	-	-	-	1	-	-	**5**
NA 93/46	AB EIA	-	-	-	-	-	-	-	-	-	-	-	-	-	-	-	-	1	-	-	-	-	-	-	-	-	-	-	-	-	**1**
NA 94/7	III	-	-	-	-	-	-	-	-	-	-	-	-	-	-	-	-	4	-	-	-	-	-	-	-	-	-	-	-	-	**4**
NA 94/7	IV	-	-	-	-	-	-	-	-	-	-	-	-	-	-	-	-	2	-	-	-	-	-	-	-	-	-	-	-	-	**2**
NA 93/45	I	-	-	-	-	-	1	-	-	3	-	-	-	-	-	-	-	6	-	-	-	-	-	-	-	-	-	-	1	-	**11**
NA 93/45	II	-	-	-	-	-	-	-	-	-	-	-	-	-	1	-	-	9	-	-	-	-	1	-	2	2	-	1	1	-	**17**
NA 93/45	IIIa	-	-	-	-	-	-	-	-	-	-	1	-	-	-	-	-	3	-	-	-	-	1	-	-	-	-	-	-	-	**5**
NA 93/45	IIIb	-	-	-	-	-	-	-	-	-	-	-	-	-	-	-	-	2	-	-	-	-	-	-	-	-	-	-	-	-	**2**
NA 93/45	IV	-	-	-	-	-	-	-	-	1	-	-	-	-	-	-	-	3	-	-	-	-	-	-	-	-	-	-	-	-	**4**
Blé A		-	-	-	-	-	-	-	-	-	-	-	-	-	-	-	-	1	2	-	-	-	-	-	-	-	-	-	-	-	**3**
Blé E		-	-	-	-	-	-	-	-	-	-	-	-	-	-	-	-	2	-	-	-	-	-	-	-	-	-	-	-	-	**2**
All sites		**4**	**3**	**4**	**3**	**3**	**6**	**5**	**1**	**32**	**4**	**18**	**4**	**4**	**10**	**2**	**2**	**274**	**4**	**24**	**1**	**2**	**11**	**1**	**11**	**10**	**6**	**6**	**3**	**1**	**452**

Table C.21: Skeletal element distribution of monitor lizard by site and phase

		vertebrae	costae	**Total**
BF 97/5	LSA	1	-	**1**
BF 94/45	LSA	3	-	**3**
BF 94/45	EIA	8	-	**8**
BF 94/45	AB EIA	11	-	**11**
BF 97/13	EIA	-	1	**1**
BF 97/30	LIA	1	-	**1**
BF 96/22	MIA	4	-	**4**
BF 97/31	LIA	6	-	**6**
BF 94/120	LIA	18	-	**18**
BF 95/7	LIA	15	-	**15**
NA 90/5BI	Gaj IIa/b	22	-	**22**
NA 97/26	IA	14	-	**14**
NA 99/75	IA	1	-	**1**
NA 97/13	IA	1	-	**1**
NA 93/46	LSA	4	-	**4**
NA 93/46	AB LSA	1	-	**1**
NA 93/46	EIA	4	-	**4**
NA 93/45	I	1	-	**1**
NA 93/45	II	1	-	**1**
All sites		**116**	**1**	**117**

Table C.22: Skeletal element distribution of snake by site and phase

		dentale	quadratum	jugale	dentes	vertebrae	costae	humerus	ossa carpi	os femoris	ossa tarsi	ossa metapodalia	phalanx proximalis	phalanx distalis	phalanx	osteoderm	unidentified	**Total**
BF 97/5	LSA	-	-	-	-	1	-	-	-	-	-	-	-	-	-	1	1	**3**
BF 97/30	LIA	1	-	-	-	1	-	-	-	-	-	1	-	1	2	-	-	**6**
BF 94/120	LIA	3	1	3	-	4	-	-	1	1	1	6	3	3	5	17	2	**50**
BF 95/7	LIA	-	-	-	-	-	-	-	-	-	-	-	-	-	-	1	-	**1**
NA 97/37	Gaj III	-	-	-	-	1	1	-	-	-	-	-	-	-	-	2	-	**4**
NA 93/46	EIA	-	-	-	3	-	1	-	-	-	-	-	-	1	-	1	-	**6**
NA 93/45	II	-	-	-	-	2	-	1	-	-	-	-	-	-	-	4	-	**7**
NA 93/45	IIIa	-	-	-	-	-	1	-	-	-	-	-	-	-	-	3	-	**4**
NA 93/45	IIIb	-	-	-	-	-	-	-	-	-	-	-	-	-	-	2	-	**2**
Blé C		-	-	-	-	-	-	-	-	-	-	-	-	-	-	1	-	**1**
All sites		**4**	**1**	**3**	**3**	**9**	**3**	**1**	**1**	**1**	**1**	7	**3**	**5**	7	**32**	**3**	**84**

Table C.23: Skeletal element distribution of crocodile by site and phase

species		**TL (cm)**	
Yellow-billed stork[1,2]	*Mycteria ibis*	100	resident
Openbill stork	*Anastomus lamelligerus*	91	intra-African migrant
Black stork[1,2]	*Ciconia nigra*	97	palearctic migrant
Abdim's stork	*Ciconia abdimii*	76	intra-African migrant
Woolly-necked stork	*Ciconia episcopus*	86	intra-African migrant
White stork[1,2]	*Ciconia ciconia*	117	palearctic migrant
Saddle-billed stork	*Ephippiorhynchus senegalensis*	145	resident
Marabou stork	*Leptoptilus crumeniferus*	127	resident

[1]reference skeleton available, [2]osteology described in Gruber (1990)
TL (Total Length) from Serle *et al.* (1977, 24-25) and Heinzel *et al.* (1982,. 42)

Table C.24: Stork species of present northern Nigeria

		furculum	humerus	ulna	os femoris	tibiotarsus	tarsometatarsus	**Total**	size classes present
NA 90/5C	Gaj IIa	-	1	-	-	-	-	**1**	
NA 90/5A	Gaj IIb	-	-	1	-	1	-	**2**	C, A
NA 92/2C	subrec	1	2	2	1	4	1	**11**	C, A, M
All sites		**1**	**3**	**3**	**1**	**5**	**1**	**14**	

C=*Ciconia ciconia*, A=*Ciconia abdimii*, M=*Mycteria ibis*

Table C.25: Skeletal element distribution of stork by site and phase

humerus	NA 92/2C subrec - A?	**ulna**	NA 90/5A Gaj IIb - C?	NA 92/2C subrec - M?	**os femoris**	NA 92/2C subrec - A?
SC	10.8	Bp	22.2	-	Bd	16
		Did	-	13.3	Dd	15.3

tibiotarsus	NA 92/2C subrec - C?	NA 92/2C subrec - A?	NA 92/2C subrec - A?	**tarsometat.**	NA 92/2C subrec - A?
SC	8.8	6.5	6.5	Bp	13.4
Bd	14.8	10.9	-	SC	5.9
Dd	18.8	12.8	-		

C=*Ciconia ciconia*, A=*Ciconia abdimii*, M=*Mycteria ibis*

Table C.26: Measurements (mm) on stork bones

NA 92/2C subrec	*Threskiornis aethiopica* (n=2)	*Platalea leucordia* (n=3)	*Platalea alba* (n=7)
20.1	19.4-21.1	22.8-23.3	21.4-23.8

shaded cells contain measurements on reference specimens

Table C.27: Distal breadth (mm) Threskiornithidae humerus

	Species		B	N	TL (cm)	
small	African pygmee goose	*Nettapus auritus*	x	x	33	resident
	Common teal[1,2]	*Anas crecca*	x	x	38	palearctic migrant
	Hottentot teal	*Anas hottentota*		x	33	resident
	Garganey[1,2]	*Anas querquedula*	x	x	38	palearctic migrant
medium-sized	Fulvous whistling duck	*Dendrocygna bicolor*	x	x	48	resident
	White-faced whistling duck[1]	*Dendrocygna viduata*	x	x	48	resident
	White-backed duck	*Thallasornis leuconotus*	x	x	43	resident
	European wigeon[1,2]	*Anas penelope*	x	x	46-56	palearctic migrant
	Gadwell[2]	*Anas strepera*	(x)	(x)	51	palearctic migrant
	Cape teal	*Anas capensis*		x	41-48	resident
	Yellow-billed duck	*Anas undulata*		(x)	50	resident
	African black duck	*Anas sparsa*		(x)	48	resident
	Northern pintail[1,2]	*Anas acuta*	x	x	59	palearctic migrant
	Northern shoveler[2]	*Anas clypeata*	x	x	51	palearctic migrant
	Northern porchard[1,2]	*Aythya ferina*	x	x	46	palearctic migrant
	Ferruginous duck[2]	*Aythya nyroca*		x	41	palearctic migrant
	Tufted duck	*Aythya fuligula*	x	x	43	palearctic migrant
large	Egyptian goose[1]	*Alopochen aegyptiacus*	x	x	71	resident
	Spur-winged goose	*Plectropterus gambensis*	x	x	102	intra-African migrant (Lake Chad Feb-May)
	Knob-billed duck	*Sarkidiornis melanotos*	x	x	79	resident

[1]reference skeleton available, [2]osteology described in Woelfle (1967)

B=Burkina Faso, N=Nigeria; TL (Total Length) from Serle *et al.* (1977, 28-34) and Heinzel *et al.* (1982, 50-58)

Table C.28: Duck and goose species of present northern Nigeria and Burkina Faso

		mandibulare	clavicula	coracoideum	scapula	humerus	radius	ulna	os femoris	tibiotarsus	**Total**
NA 93/46	LSA	-	1	-	-	-	1	1	-	-	**3**
NA 93/46	EIA	1	-	5	3	2	2	4	1	1	**19**
NA 94/7	I	-	-	1	-	-	-	1	-	-	**2**
NA 93/45	II	-	-	1	-	-	-	1	-	-	**2**
NA 93/45	IV	-	-	-	-	-	1	-	-	-	**1**
All sites		**1**	**1**	**7**	**3**	**2**	**4**	**7**	**1**	**1**	**27**

Table C.29: Skeletal element distribution of small duck or goose by site and phase

		furculum	mandibulare	atlas	axis	coracoideum	scapula	humerus	radius	ulna	carpometacarpus	os coxae	os femoris	tibiotarsus	tarsometatarsus	**Total**
BF 94/45	EIA	-	-	-	-	2	-	-	-	1	-	-	-	-	-	**3**
BF 94/45	AB EIA	-	-	-	-	1	-	-	-	-	-	-	-	-	-	**1**
BF 94/120	LIA	-	-	-	-	1	-	-	-	-	1	-	-	-	-	**2**
NA 90/5BI	Gaj IIa/b	-	-	-	-	-	-	-	-	-	1	-	-	-	-	**1**
NA 97/37	Gaj III	-	1	-	-	2	-	-	1	2	1	-	1	-	-	**8**
NA 92/2C	subrec	1	1	1	1	2	2	1	2	1	1	1	1	3	2	**20**
NA 94/7	IV	-	-	-	-	1	-	-	-	-	-	-	-	-	-	**1**
NA 93/45	I	-	-	-	-	-	-	-	-	1	-	-	-	-	-	**1**
NA 93/45	II	-	-	-	-	-	-	2	-	-	1	-	-	-	-	**3**
NA 93/45	IIIb	-	-	-	-	-	-	1	-	-	-	-	-	1	-	**2**
NA 93/45	IV	-	-	-	-	-	-	-	1	1	-	-	-	-	-	**2**
All sites		**1**	**2**	**1**	**1**	**9**	**2**	**4**	**4**	**6**	**5**	**1**	**2**	**4**	**2**	**44**

Table C.30: Skeletal element distribution of medium-sized duck or goose by site and phase

		vertebrae cervicales	vertebrae thoracales	synsacrum	sternum	furculum	coracoideum	scapula	humerus	radius	ulna	carpometacarpus	ossa carpi	phalanx proximalis manus	os coxae	os femoris	tibiotarsus	tarsometatarsus	phalanx	**Total**	species present
BF 94/45	LSA	-	-	1	-	-	1	-	2	1	1	1	-	-	-	-	1	-	-	**8**	
BF 94/45	EIA	-	2	-	-	-	1	-	-	1	1	-	-	-	-	-	1	-	-	**6**	A/S
BF 94/45	AB EIA	-	-	-	-	-	1	-	-	-	1	-	-	-	-	-	-	-	-	**2**	
BF 97/13	EIA	-	-	-	-	-	-	1	-	-	-	-	-	-	-	-	-	-	-	**1**	A/S
BF 96/22	EIA	-	-	-	-	-	-	-	1	-	-	-	-	-	-	-	-	-	-	**1**	
BF 97/31	LIA	1	-	-	-	-	-	-	-	-	-	-	-	-	-	-	-	-	-	**1**	
BF 94/120	LIA	-	-	-	-	-	-	-	-	-	-	1	-	-	-	-	-	-	-	**1**	
NA 90/5A	Gaj IIb	-	-	-	-	-	-	2	-	-	1	2	-	-	-	-	-	-	1	**6**	S, P
NA 90/5BII	Gaj IIc	-	-	-	-	-	-	-	-	1	-	-	-	-	-	-	3	1	-	**5**	A/S, P
NA 99/65	Gaj IIb	-	-	1	-	-	-	-	-	-	-	-	-	-	-	-	-	-	-	**1**	
NA 97/37	Gaj III	4	1	-	-	-	4	1	1	-	8	1	-	-	1	-	3	2	2	**28**	A/S
NA 99/75	IA	1	-	-	-	-	-	-	-	-	-	1	1	-	-	1	-	-	2	**6**	P
NA 93/46	EIA	-	-	-	-	-	2	1	-	-	1	-	1	1	-	-	1	-	1	**8**	A/S
NA 94/7	I	-	-	-	-	-	-	1	-	-	-	-	-	-	-	-	-	-	-	**1**	P
NA 94/7	III	-	-	-	-	-	-	1	-	1	-	-	-	-	-	-	-	-	-	**2**	P
NA 93/45	I	3	2	-	-	-	-	-	2	-	-	-	-	-	-	-	-	1	1	**9**	P
NA 93/45	II	1	-	-	-	-	1	-	2	1	2	-	-	-	-	-	1	-	-	**8**	P
NA 93/45	IIIa	-	-	-	-	-	1	-	-	1	-	-	-	-	-	-	-	-	-	**2**	P
NA 93/45	IIIb	-	-	-	-	-	1	1	1	-	-	-	-	-	-	-	-	-	-	**3**	P
NA 93/45	IV	-	-	-	-	-	-	-	-	1	-	-	-	-	-	-	-	1	1	**3**	P
Blé A		-	-	-	-	-	2	-	1	-	1	1	-	-	-	-	1	-	-	**6**	P
Blé B		-	-	-	-	-	-	-	1	-	-	-	-	-	-	-	-	-	-	**1**	P
Blé C		2	-	-	1	1	-	2	2	-	3	1	-	-	-	1	1	-	-	**14**	P
Blé E		-	-	-	1	-	1	1	2	-	-	-	-	-	-	-	-	-	-	**5**	P
All sites		**12**	**5**	**2**	**2**	**1**	**15**	**11**	**15**	**7**	**19**	**8**	**2**	**1**	**1**	**2**	**12**	**5**	**8**	**128**	

A=*Alopochen aegyptiacus*, P=*Plectropterus gambensis*, S=*Sarkidiornis melanotos*

Table C. 31: Skeletal element distribution of large duck or goose by site and phase

corac.	Blé E P	P (n=4)	A (n=4)	NA 92/2C subrec - M
Lm	85.4	68-77	55.2-64.0	-
Bb	-	26.7-27.5	26.4-32.0	22.6

scapula	P (n=1)	Blé E P	Blé C P	A (n=3)	BF 97/13 EIA - A/S	NA 97/37 Gaj III - A/S	NA 93/46 EIA - A/S	NA 92/2C subrec - M
Dic	19.5	(19)	(18)	13.3-15.9	14.6	13.3	11.9	10.2

humerus	P (n=3)	Blé C P	Blé E P	Blé B P	A (n=3)	NA 92/2C subrec - M
Bp	35.7-38.8	(37)	-	-	28.2-31.3	-
SC	-	-	11.9	-	9.1-9.6	6.7
Bd	22.6	-	-	29.6	19.4-21.2	14.7

radius	P (n=1)	NA 90/5BII Gaj IIc - P	A (n=3)	NA 92/2C subrec - M	BF 94/45 EIA - A/S
GL	-	-	118-132	86.0	-
SC	10.3	-	4.1-4.8	3.4	-
Did	21.6	16.5	9.2-9.7	6.9	9.0

ulna	NA 97/37 Gaj III - A/S/P	NA 90/5A Gaj IIb - P	A (n=3)	NA 92/2C subrec - M	BF 94/45 EIA - A/S
GL	-	-	125-139	91	-
Dip	19.8	-	16.8-18.6	8.6	-
Bp	15.7	20.7	13.3-14.7	10.0	-
SC	-	-	6.6-7.9	5.0	-
Did	-	-	13.4-15.0	10.4	14.3

carpomc	NA 90/5A Gaj IIb - P	NA 90/5A Gaj IIb - S	A (n=3)	NA 92/2C subrec - M?	BF 94/45 EIA - A/S
GL	-	-	79.1-83	77.7	-
Bp	24.9	23.0	19.9-21.6	22.9	20.7
Did	-	-	9.2-10.9	9.2	9.4

os fem.	Blé C P	A (n=3)	NA 92/2C subrec - M
Bp	-	14.4-16.4	11.0
Dp	-	14.2-16.4	10.9
Bd	27.1	17.0-18.7	-

tibiotars.	P (n=1)	NA 90/5BII Gaj IIc - A/S	NA 90/5BII Gaj IIc - P	A (n=3)	BF 94/45 EIA - A/S	NA 92/2C subrec - M
GL	-	-	-	136-151	-	90.1
La	-	-	-	130-144	-	88.8
Dip	-	23.5	-	19.9-23.0	-	12.8
SC	-	-	-	6.8-7.3	-	4.6
Bd	19.5	-	17.6	14.1-14.8	12.7	-
Dd	-	-	18.9	14.6-15.2	14.1	11.0

tarsomt	A (n=3)	NA 92/2C subrec - M?	NA 92/2C subrec - M?
GL	81-94	69	-
Bp	16.8-16.7	13.8	-
SC	5.9-6.0	6.3	-
Bd	15.4-16.7	13.5	13.3

M=medium-sized duck or goose, A=*Alopochen aegyptiacus*, P=*Plectropterus gambensis*, S=*Sarkidiornis melanotos*
shaded cells contain measurements on reference specimens; ()=estimate

Table C.32: Measurements (mm) on duck or goose bones

		coracoideum	scapula	radius	carpometacarpus	tarsometatarsus	**Total**
BF 97/13	AB LIA	-	1	-	-	-	**1**
BF 97/30	LIA	1	1	-	-	2	**4**
BF 94/120	LIA	1	1	-	1	-	**3**
BF 95/7	LIA	1	-	-	3	-	**4**
NA 92/2C	subrec	2	-	-	1	-	**3**
NA 94/7	III	1	-	-	-	-	**1**
NA 93/45	I	-	1	-	-	-	**1**
NA 93/45	II	1	-	1	2	1	**5**
NA 93/45	IV	-	-	-	1	-	**1**
Blé B		-	-	-	-	1	**1**
Blé E		-	-	-	-	1	**1**
All sites		**7**	**4**	**1**	**8**	**5**	**25**

Table C.33: Skeletal element distribution of domestic fowl by site and phase

		coracoideum	scapula	sternum	humerus	radius	ulna	carpometacarpus	os coxae	os femoris	tibiotarsus	tarsometatarsus	**Total**
BF 97/13	MIA	-	1	-	-	-	-	-	-	-	-	-	**1**
BF 94/120	LIA	-	-	-	-	1	-	-	-	-	-	-	**1**
BF 95/7	LIA	-	-	-	-	-	-	1	-	-	-	-	**1**
NA 90/5C	Gaj IIa	-	-	-	-	-	1	-	-	-	-	-	**1**
NA 90/5A	Gaj IIb	-	-	-	-	-	-	-	-	-	-	3	**3**
NA 90/5BI	Gaj IIa/b	-	-	-	-	-	-	-	-	-	1	-	**1**
NA 90/5BII	Gaj IIc	4	2	1	3	1	2	1	1	1	9	9	**34**
NA 99/65	Gaj IIb	1	-	-	-	-	-	-	-	-	-	-	**1**
NA 96/45	Gaj IIa/b	-	-	-	-	-	-	-	-	-	1	-	**1**
NA 97/13	IA	1	-	-	-	-	-	-	-	-	-	-	**1**
NA 92/2C	subrec	2	-	-	-	-	-	-	-	-	-	-	**2**
NA 94/7	III	-	1	-	-	-	-	2	-	-	-	-	**3**
All sites		**8**	**4**	**1**	**3**	**2**	**3**	**4**	**1**	**1**	**11**	**12**	**50**

Table C.34: Skeletal element distribution of guineafowl by site and phase

		cranium	mandibulare	quadratum	vertebrae cervicales	vertebrae thoracales	synsacrum	furculum	sternum	coracoideum	scapula	humerus	radius	ulna	ossa carpi	carpometacarpus	os coxae	os femoris	fibula	tibiotarsus	tarsometatarsus	phalanx	**Total**	*Francolinus* sp.	*Gallus gallus* f. domestica	*Numida meleagris*
BF 94/45	EIA	-	-	-	-	-	-	1	1	1	-	1	-	-	-	1	-	2	-	2	2	1	**12**	-	-	-
BF 94/45	AB EIA	-	-	-	-	-	-	-	-	-	-	-	-	-	-	-	-	-	-	-	1	1	**2**	-	-	-
BF 97/13	AB EIA	-	-	-	-	-	-	-	-	-	-	-	-	-	-	-	-	-	-	-	-	1	**1**	-	-	-
BF 97/13	MIA	-	-	-	-	-	-	-	-	-	-	-	-	-	-	-	-	-	1	-	-	-	**1**	-	-	1
BF 97/13	AB MIA	-	-	-	-	-	-	-	-	-	-	-	-	-	-	-	-	-	-	-	-	2	**2**	-	-	-
BF 97/13	LIA	-	-	-	-	-	-	-	-	-	-	1	-	-	-	-	1	-	-	-	-	-	**2**	-	-	-
BF 97/13	AB LIA	-	-	-	-	-	-	-	-	-	-	-	-	1	1	-	-	-	-	-	-	6	**8**	-	1	-
BF 97/30	LIA	-	1	-	-	1	-	-	-	3	2	1	2	2	-	1	-	3	-	4	3	1	**24**	-	4	-
BF 96/22	MIA	-	2	-	-	-	-	-	-	1	-	-	-	-	-	-	1	1	-	-	1	6	**12**	-	-	-
BF 97/31	LIA	-	-	-	-	-	-	-	2	2	4	2	1	-	-	-	-	2	-	4	5	-	**22**	-	-	-
BF 94/120	LIA	3	-	2	2	3	-	2	-	12	1	8	2	10	-	3	-	7	1	17	15	15	**103**	-	3	1
BF 95/7	LIA	-	2	-	-	-	-	-	3	3	1	5	7	10	-	2	2	-	-	9	12	-	**56**	-	4	1
NA 99/65	Gaj IIb	-	-	1	-	-	1	-	-	-	-	-	-	-	-	-	-	-	-	-	-	-	**2**	-	-	1
NA 95/1	Gaj IIa/b	-	-	-	-	-	-	-	-	-	-	-	-	-	-	-	-	1	-	1	-	-	**2**	-	-	-
NA 99/75	IA	-	-	-	-	-	-	-	-	1	-	-	-	-	-	-	-	-	-	2	-	-	**3**	-	-	-
NA 97/13	IA	-	-	-	-	-	1	-	-	-	-	-	1	-	-	-	-	-	-	1	2	-	**5**	-	-	1
NA 92/2C	subrec	-	-	-	1	-	-	1	2	1	1	6	-	4	-	-	-	5	-	1	7	1	**30**	2	3	2
NA 94/7	III	-	-	-	-	-	-	-	-	-	-	-	-	1	-	-	-	2	-	-	-	-	**3**	-	1	-
NA 94/7	IV	-	-	-	-	-	-	-	-	-	-	-	-	-	-	-	-	-	-	-	2	-	**2**	-	-	-
NA 93/45	I	-	-	-	-	-	-	-	1	1	3	-	-	2	-	-	-	3	-	5	4	-	**19**	-	1	-
NA 93/45	II	-	-	-	-	-	-	1	1	1	2	7	1	5	-	-	-	7	-	8	2	-	**35**	-	5	-
NA 93/45	IIIa	-	-	-	-	-	-	-	-	-	1	-	-	1	-	-	-	1	-	2	-	-	**5**	-	-	-
NA 93/45	IIIb	-	-	-	-	-	-	-	-	-	-	1	-	-	-	1	-	-	-	1	-	-	**3**	-	-	-
NA 93/45	IV	-	-	-	-	-	-	-	1	1	-	-	-	-	-	-	-	1	-	5	-	-	**8**	-	1	-
Blé A		-	-	-	-	-	-	-	-	1	-	-	-	-	-	-	-	1	-	1	-	-	**3**	-	-	-
Blé B		-	-	-	-	-	-	-	-	1	-	-	-	-	-	-	-	-	-	-	-	-	**1**	-	1	-
Blé C		-	-	-	-	-	-	-	-	-	-	-	-	-	-	-	-	-	-	-	1	-	**1**	-	-	-
Blé E		-	-	-	-	-	-	-	1	-	-	1	-	-	-	-	-	-	-	-	1	-	**3**	-	1	-
All sites		**3**	**5**	**3**	**3**	**4**	**2**	**5**	**12**	**29**	**15**	**33**	**14**	**36**	**1**	**8**	**4**	**36**	**2**	**63**	**58**	**34**	**370**	**2**	**25**	**7**

Table C.35: Skeletal element distribution of large galliform by site and phase

	NA 92/2C	BF 94/120	Blé B	BF 95/7	NA 93/45	NA 97/13	NA 92/2C	BF 94/120	NA 90/5BII	NA 92/2C
coracoideum	subrec - F?	LIA - G		LIA - G	II - G	IA - N	subrec - N	LIA	Gaj IIc - N	subrec - G
GL	44.9	42.9	-	-	-	-	-	-	-	-
Lm	42.3	43.6	45.4	-	-	-	-	-	-	-
Bb	12.3	9.4	-	14.4	13.9	12.3	12.0	-	-	-
BF	10.2	8.7	-	11.2	11.6	10.5	9.6	11.3	11.2	8.6

	BF 97/13	NA 90/5BII	BF 97/13	BF 94/120	NA 93/45	BF 97/30
scapula	MIA - N	Gaj IIc -N	LIA - G	LIA - G	I - G	LIA - G
Dic	(13)	12.9	12.2	11.9	10.5	10.0

	NA 92/2C	NA 92/2C	NA 92/2C	Blé E	BF 94/120	BF 97/13	BF 95/7	NA 92/2C	NA 90/5BII	BF 95/7	BF 94/120
humerus	subrec	subrec	subrec		LIA	LIA	LIA	subrec	Gaj IIc - N	LIA	LIA
GL	76.7	67.6	62.7	59.2	-	-	-	-	-	-	-
Bp	20.6	17.6	15.8	16.1	21.4	18.7	15.8	-	-	-	-
SC	7.0	6.4	5.6	5.1	-	-	-	5.5	-	-	-
Bd	16.0	13.2	12.2	11.5	-	-	-	12.5	16.2	13.5	13.4

	BF 97/30	BF 94/120	BF 95/7	BF 94/120	BF 94/120
humerus	LIA	LIA	LIA	LIA	LIA
Bd	(13)	12.0	12.0	11.7	9.9

	NA 93/45	BF 94/120	NA 90/5BII	BF 95/7	BF 94/120	BF 95/7	BF 95/7
radius	III - G	LIA - N	Gaj IIc - N	LIA	LIA	LIA	LIA
HL	52.1	-	-	-	-	-	-
SC	2.5	-	-	-	-	-	-
Bd	5.2	7.3	6.5	6.1	5.3	5.3	4.7

	BF 95/7	NA 92/2C	BF 94/120	BF 94/120	NA 90/5BII	NA 92/2C	BF 94/120	BF 97/13	BF 94/120	BF 95/7	BF 94/120
ulna	LIA	subrec	LIA	LIA	Gaj IIc - N	subrec	LIA	LIA	LIA	LIA	LIA
GL	71.7	61.0	-	-	-	-	-	-	-	-	-
Dip	11.5	7.5	11.3	8.2	-	-	-	-	-	-	-
Bp	8.7	9.6	7.9	6.5	-	-	-	-	-	-	-
SC	4.2	3.8	3.9	-	4.4	4.1	3.9	-	-	-	-
Did	8.6	8.1	-	-	-	-	-	8.5	8.5	7.9	7.3

	NA 92/2C	NA 94/7	BF 94/120	BF 95/7	BF 95/7	NA 93/45	NA 93/45	NA 93/45	BF 94/120	BF 95/7	NA 94/7
carpometacarpus	subrec - G	III - N	LIA - G	LIA - G	LIA - G	II - G	IV - G	II - G	LIA	LIA - G	III - N
GL	41.6	41.1	40.5	38.3	(36)	34.7	33.8	33.7	27.0	-	-
L	-	-	37.2	-	-	-	-	-	25.7	-	-
Bp	12.7	-	-	11.0	11.0	11.0	10.0	11.0	8.1	10.4	-
Did	7.7	12.3	7.7	7.3	7.2	7.2	7.1	8.0	4.7	7.6	11.7

F=*Francolinus* sp., G=*Gallus gallus* f. domestica, N=*Numida meleagris*; ()=estimate

Table C.36: Measurements on large galliform bones

	BF 95/7	NA 90/5BII	BF 95/7		NA 92/2C	NA 94/7	NA 93/45	NA 90/5BII	NA 92/2C	NA 94/7	NA 92/2C
os coxae	LIA	Gaj IIc - N	LIA	**os fem.**	subrec	III	II	Gaj IIc - N	subrec	III	subrec
DiA	7.6	7.1	6.7	GL	77.1	-	-	-	-	-	-
				Lm	73.5	-	-	-	-	-	-
				Bp	15.4	17.4	13.0	-	-	-	-
				Dp	10.8	-	8.7	-	-	-	-
				SC	6.6	-	-	11.0	5.6	-	-
				Bd	12.2	-	-	-	-	14.5	13.7
				Dd	12.1	-	-	-	-	-	12.2

	BF 94/120	NA 90/5BII	NA 90/5BII	NA 90/5BII	NA 90/5BII	BF 94/120	NA 90/5BII	NA 92/2C	NA 90/5BII	BF 94/120	NA 94/7
tibiotarsus	LIA	Gaj IIc - N	Gaj IIc - N	Gaj IIc - N	Gaj IIc - N	LIA	Gaj IIc - N	subrec	Gaj IIc - N	LIA	IV
Bp	14.4	-	-	-	-	-	-	-	-	-	-
SC	-	10.4	10	-	-	-	-	-	-	-	-
Dd	-	-	10.9	12.1	11.8	11.3	11.3	11.3	11.1	10.9	10.5
Bd	-	-	10.9	12.1	11.2	-	-	11.2	11.6	-	-

	NA 94/7	BF 97/30	NA 93/45	BF 94/120	94/45	BF 94/120	BF 94/120	Blé E	BF 94/120	BF 94/120	BF 94/120
tibiotarsus	IV	LIA	I	LIA	EIA	LIA	LIA		LIA	LIA	LIA
Dd	10.1	-	-	-	-	-	-	-	-	-	-
Bd	-	11.6	10.0	9.8	9.5	9.2	8.8	8.7	8.5	8.4	8.3

	BF 95/7	94/45	BF 94/120
tibiotarsus	LIA	EIA	LIA
Bd	(8)	7.9	7.5

	NA 92/2C	Blé E	NA 92/2C	NA 92/2C	NA 93/45	NA 90/5BII	NA 90/5BII	BF 94/120	NA 97/13	NA 90/5A	NA 90/5BII
tarsometatarsus	subrec	G	subrec - F	subrec	I	Gaj IIc - N	Gaj IIc - N	LIA	IA	Gaj IIb - F	Gaj IIc - N
GL	70.0	68.8	64.8	59.3	-	-	-	-	-	-	-
Bp	12.3	11.4	11.0	10.5	13.3	12.9	12.6	(10)	9.9	7.4	-
SC	5.5	6.0	5.6	5.2	-	-	-	-	-	-	5.7
Bd	-	(10)	10.6	9.4	-	-	-	-	-	-	-

	NA 90/5A	BF 95/7	BF 95/7	BF 94/45	BF 94/120	NA 90/5A	BF 94/120	BF 94/120
tarsometatarsus	Gaj IIb - N	LIA	LIA	LIA	LIA	Gaj IIb	LIA	LIA
Bd	13.1	(13)	12.0	12.0	11.9	10.9	9.3	8.8

F=*Francolinus* sp., G=*Gallus gallus* f. domestica, N=*Numida meleagris*; ()=estimate

Table C.36 cont.

coracoideum		**scapula**		**humerus**		**radius**		**ulna**	
BF	8.4	Dic	12.4	Bp	20.5	Bd	6.7	Bp	8.9
				Bd	15.4			Did	9.3
carpometac.		**os femoris**		**tibiotarsus**		**tarsomet.**			
Bp	11.7	Bp	17.0	Bd	11.6	Bp	13.0		
Did	7.4	Bd	15.0			Bd	13.3		

Table C.37: Measurements (mm) on a recent guineafowl specimen from northern Burkina Faso (sex unknown). Standard for LSI calculations

		coracoideum	humerus	ulna	**Total**
BF 94/45	LSA	-	1	-	**1**
BF 97/30	LIA	-	-	1	**1**
BF 94/120	LIA	1	-	-	**1**
BF 95/7	LIA	1	1	-	**2**
All sites		**2**	**2**	**1**	**5**

Table C.38: Skeletal distribution of pigeon or dove by site and phase

site	phase	n° of bones	remarks
BF 94/45	EIA	2	child
BF 97/30	LIA	1	permanent tooth
BF 96/22	MIA	8	goes with known adult burial
BF 97/31	LIA	1	permanent tooth
BF 95/7	LIA	8	adult
NA 93/36	Gaj IIa	4	adult
NA 99/65	Gaj IIa-IIc	9	adult
NA 95/1	Gaj IIa/b	15	baby
NA 97/26	IA	3	child?
NA 99/75	IA	52	goes with known adult burials
NA 97/13	IA	8	child
NA 93/46	EIA	1	deciduous tooth
NA 93/45	I	35	baby
NA 93/45	II	11	baby; associated with known adult burial?
NA 93/45	IIIa	1	permanent tooth
Blé E		1	goes with known adult burials

Table C.39: Human remains by site and phase

		mandibulare	dentes	humerus	tibia	**Total**
BF 94/45	AB LSA	1	-	-	-	**1**
BF 94/45	EIA	1	-	-	-	**1**
BF 94/45	AB EIA	1	-	1	1	**3**
BF 96/22	MIA	3	-	-	-	**3**
BF 94/120	LIA	-	1	-	-	**1**
BF 95/7	LIA	4	-	-	-	**4**
Total		**10**	**1**	**1**	**1**	**13**

Table C.40: Skeletal element distribution of African hedgehog by site and phase

	BF 94/45 EIA	BF 94/45 AB EIA	BF 97/13 MIA	BF 97/13 LIA	BF 97/13 AB LIA	BF 97/30 LIA	BF 96/22 MIA	BF 97/31 LIA	BF 94/120 LIA	BF 95/7 LIA	NA 97/26 IA	NA 97/13 IA	NA 92/2C subrec	NA 93/45 II	**All sites**
cranium	-	-	-	-	-	-	1	1	-	-	-	-	1	-	**3**
maxillare	-	1	-	-	1	-	1	1	-	2	-	-	-	-	**6**
mandibulare	-	-	1	-	-	-	-	2	5	1	-	-	-	-	**9**
dentes	-	-	-	-	-	-	-	-	1	-	-	-	1	-	**2**
atlas	-	-	-	-	-	-	-	-	-	1	-	-	-	-	**1**
vertebrae cervicales	-	-	-	-	-	-	-	-	1	-	-	-	-	-	**1**
vertebrae thoracales	-	-	-	-	-	-	-	-	1	-	-	-	1	-	**2**
vertebrae lumbales	-	-	-	-	-	-	-	-	1	-	1	-	-	-	**2**
sternum	-	-	-	-	-	-	1	-	-	-	-	-	-	-	**1**
scapula	-	-	-	-	-	1	-	-	1	-	-	-	2	-	**4**
humerus	-	-	-	1	-	1	-	-	7	3	-	-	1	1	**14**
radius	2	-	-	-	-	-	1	-	2	-	-	-	-	-	**5**
ulna	1	1	-	-	1	1	-	1	2	2	-	1	-	-	**10**
os metacarpale II	-	-	-	-	-	-	-	-	1	-	-	-	1	-	**2**
os metacarpale III	-	-	-	1	-	-	1	-	-	-	-	-	-	-	**2**
os metacarpale IV	1	-	-	1	-	-	-	-	-	-	-	-	-	-	**2**
os coxae	3	-	1	-	-	1	-	4	5	-	-	-	-	-	**14**
os femoris	-	-	1	1	-	2	-	2	4	2	-	-	-	-	**12**
tibia	2	-	1	-	-	-	1	1	1	1	-	-	-	-	**7**
talus	-	-	-	-	-	-	1	-	-	-	-	-	-	-	**1**
calcaneus	2	-	-	1	-	-	1	1	4	-	-	-	-	-	**9**
ossa tarsi	-	-	-	-	-	-	-	-	1	-	-	-	-	-	**1**
os metatarsale II	-	-	-	1	-	-	3	1	1	2	-	-	-	-	**8**
os metatarsale III	-	-	-	1	-	-	1	-	2	-	-	-	-	-	**4**
os metatarsale IV	-	-	-	-	-	-	1	1	-	2	-	-	-	-	**4**
os metatarsale V	1	-	-	-	-	-	-	1	1	2	-	-	-	-	**5**
os metatarsale	1	-	-	-	-	-	-	-	-	-	-	-	-	-	**1**
ossa metapodalia	-	-	-	-	-	-	2	1	5	-	-	-	-	-	**8**
phalanx proximalis	2	-	-	-	-	-	-	-	2	1	-	-	-	-	**5**
phalanx media	-	-	-	-	-	-	-	-	1	-	-	-	-	-	**1**
phalanx	-	-	-	-	-	-	-	-	1	-	-	-	-	-	**1**
Total	**15**	**2**	**4**	**7**	**2**	**6**	**15**	**17**	**50**	**19**	**1**	**1**	**7**	**1**	**147**

Table C.41: Skeletal element distribution of hare by site and phase

	BF 97/31	BF 95/7		BF 96/22		BF 97/31	BF 94/120					
mand.	LIA	LIA	**max.**	MIA	**scapula**	LIA	LIA					
2	15.9	14.9	9	15.7	LG	9.8	8.9					
3	32.2	-	11	5.7	BG	7.8	6.7					
4	16.5	-										
	BF 95/7	NA 93/45	BF 95/7	NA 92/2C	BF 97/13	BF 94/120	BF 94/120		BF 94/120	BF 94/120	BF 94/45	BF 94/45
hum.	LIA	II	LIA	subrec	LIA	LIA	LIA	**radius**	LIA	LIA	EIA	EIA
Bd	10.7	9.6	9.2	9.2	8.8	8.4	8.2	Bp	6.8	6.3	6.1	-
								Bd	-	-	-	6.4
	BF 94/120	BF 97/13	NA 92/2C		BF 96/22		BF 94/45		BF 97/30	BF 95/7	BF 97/30	
os mc II	LIA	LIA	subrec	**os mc III**	MIA	**os mc IV**	EIA	**os fem.**	LIA	LIA	LIA	
GL	18.7	18.2	18.2	GL	22.6	GL	17.5	SD	7.4	-	-	
Bd	-	3.3	3.1	Bd	3.0	Bd	3.0	Bd	12.8	12.3	12.2	
	BF 94/45	BF 97/13	BF 95/7		BF 96/22		BF 96/22	BF 97/13	BF 94/120	BF 94/120	BF 94/120	BF 94/45
tibia	EIA	MIA	LIA	**talus**	MIA	**calc.**	MIA	LIA	LIA	LIA	LIA	EIA
Bp	14.1	-	-	GL	11.5	GL	23.3	23.1	22.5	20.8	20.6	-
Bd	-	10.7	10.7			GB	6.9	8.0	8.0	7.5	6.6	8.3
	BF 96/22	BF 96/22	BF 95/7	BF 97/31	BF 96/22	BF 95/7		BF 95/7		BF 94/120	BF 95/7	
os mt II	MIA	MIA	LIA	LIA	MIA	LIA	**os mt IV**	LIA	**os mt V**	LIA	LIA	
GL	36.6	36.3	35.4	35.0	33.9	33.3	GL	35.9	GL	29.9	-	
Bd	4.2	4.3	4.4	4.0	4.3	4.2	Bd	3.9	Bd	3.7	3.3	

Table C.42: Measurements (mm) on hare bones

		E		G	T	D	A			M																																
		mandibulare	dentes mandibulare	mandibulare	mandibulare	cranium	maxillare	maxillare	mandibulare	cranium	maxillare	mandibulare	dentes	cranium	maxillare	mandibulare	dentes mandibulare	maxillare/mandibulare	dentes	atlas	axis	vertebrae cervicales	vertebrae thoracales	vertebrae lumbales	os sacrum	vertebrae caudales	vertebrae	costae	scapula	humerus	radius	ulna	radius + ulna	os coxae	os femoris	tibia	fibula	talus	calcaneus	ossa metapodalia	phalanx distalis	unidentified	**Total**
BF 97/5	LSA	-	-	-	-	-	-	-	-	-	-	-	-	-	-	-	-	-	1	-	-	-	-	-	-	-	-	-	-	-	-	-	-	-	2	-	-	-	-	-	-	-	**3**
BF 97/5	EIA	-	-	-	-	-	-	-	-	-	-	-	-	-	-	-	-	-	-	-	-	-	-	-	-	-	1	-	-	-	-	-	-	-	-	-	-	-	-	-	-	-	**1**
BF 94/45	LSA	-	-	-	-	-	-	-	1	-	-	-	-	-	-	-	-	-	-	-	-	-	-	-	-	-	-	-	-	1	-	-	-	3	2	1	-	-	1	-	-	-	**8**
BF 94/45	AB LSA	-	-	-	-	-	-	-	1	-	-	-	-	-	-	-	-	-	-	-	-	1	-	-	-	1	-	-	-	-	-	-	-	-	-	1	-	-	-	-	-	-	**3**
BF 94/45	EIA	-	-	1	-	-	-	-	-	-	-	2	-	1	-	4	-	-	3	-	-	-	-	-	-	1	1	-	-	4	1	-	-	9	13	10	-	-	-	1	-	1	**49**
BF 94/45	AB EIA	-	1	-	-	-	-	1	-	-	7	-	-	1	2	1	-	-	5	2	1	6	2	3	4	8	3	-	1	2	-	5	-	10	13	8	-	1	1	2	-	-	**81**
BF 97/13	EIA	-	-	-	-	-	-	-	-	-	-	3	-	-	-	1	-	-	-	-	-	-	1	1	-	4	10	-	-	2	1	-	-	1	10	7	-	-	-	-	-	-	**38**
BF 97/13	AB EIA	-	-	-	-	-	-	-	-	-	-	1	-	-	-	-	-	-	-	-	-	2	-	2	1	3	-	-	-	2	-	-	-	1	2	1	-	-	-	-	4	-	**18**
BF 97/13	MIA	-	-	-	-	-	-	-	-	-	-	1	-	-	-	-	-	-	-	-	-	-	-	-	-	-	-	-	-	-	-	-	-	3	2	-	-	-	-	-	-	-	**5**
BF 97/13	AB MIA	-	-	-	-	-	-	-	-	-	-	-	-	-	-	-	-	-	-	-	-	-	-	-	1	2	-	-	1	-	-	-	-	-	-	-	-	1	-	-	-	-	**5**
BF 97/13	LIA	-	-	-	-	-	-	-	-	-	-	-	-	-	-	1	-	-	-	-	-	-	-	-	-	-	1	-	-	-	-	-	-	-	2	-	-	-	-	1	-	-	**5**
BF 97/13	AB LIA	-	-	-	-	-	-	-	-	-	-	-	-	-	-	-	-	-	1	-	-	-	-	-	-	2	-	-	-	-	-	-	-	-	1	-	-	-	-	-	1	-	**5**
BF 97/30	LIA	1	-	-	-	-	-	-	-	1	-	8	5	2	1	5	-	4	12	-	2	-	-	1	1	-	54	28	3	16	5	4	-	7	28	23	1	1	2	17	-	23	**240**
BF 96/22	MIA	-	-	-	2	-	-	-	-	-	-	3	-	-	1	1	-	-	4	1	-	1	-	-	-	-	1	-	-	2	-	-	-	4	8	2	-	-	-	1	-	-	**26**
BF 97/31	LIA	-	-	-	-	-	-	-	1	-	-	4	-	-	-	-	-	-	6	1	-	-	-	-	-	1	3	-	1	5	-	1	-	3	13	5	-	-	-	-	-	-	**39**
BF 94/120	LIA	1	-	-	-	1	2	-	2	-	1	1	2	1	1	6	-	-	8	-	1	3	-	-	-	1	7	-	4	3	1	1	-	9	28	10	-	-	1	1	-	8	**94**
BF 95/7	LIA	-	-	-	-	-	-	-	2	-	-	-	-	-	5	8	-	-	1	-	-	-	-	-	-	-	12	1	-	1	1	2	-	6	19	15	-	-	-	4	-	1	**76**
NA 90/5A	Gaj IIb	-	-	-	-	-	-	-	2	-	-	-	-	-	-	1	-	-	-	-	-	-	-	-	-	-	-	-	-	-	-	-	-	-	2	1	-	-	-	-	-	-	**4**
NA 90/5BII	Gaj IIa/b	1	-	-	-	-	-	-	-	-	-	-	-	-	-	-	-	-	-	-	-	-	-	-	-	-	-	-	-	-	-	-	-	-	1	-	-	-	-	-	-	-	**1**
NA 99/65	Gaj IIb	-	-	-	-	-	-	-	2	-	-	-	-	-	-	-	-	-	-	-	-	-	-	-	-	-	-	-	-	-	-	-	-	-	2	2	-	-	-	-	-	-	**4**
NA 99/65	Gaj IIc	-	-	-	-	-	-	-	-	-	-	-	-	-	-	-	-	-	-	-	-	1	-	-	-	-	-	-	-	-	-	-	-	-	-	-	-	-	-	-	-	-	**1**
NA 97/37	Gaj III	-	-	-	-	-	-	-	-	-	-	-	-	-	-	-	-	-	-	-	-	-	-	-	-	-	-	-	-	-	-	-	-	1	4	2	-	-	-	-	-	-	**7**
NA 95/1	Gaj IIa/b	-	-	-	-	-	-	-	-	-	-	-	-	-	-	-	-	-	-	-	-	-	-	-	-	-	-	-	-	-	-	-	-	-	2	-	-	-	-	-	-	-	**2**
NA 97/26	IA	-	-	-	-	-	-	-	4	-	-	1	-	-	-	5	-	-	-	-	-	-	-	-	-	-	-	-	-	-	-	-	-	2	5	3	-	-	-	-	-	-	**15**
NA 99/75	IA	-	-	-	-	-	-	-	1	-	-	-	-	-	-	2	-	-	-	-	-	-	-	-	-	-	-	-	-	2	-	-	-	5	5	2	-	-	-	-	-	-	**16**
NA 97/13	IA	-	-	-	-	-	-	-	-	-	2	5	-	1	-	2	-	-	-	-	1	-	-	-	-	-	-	-	1	4	-	-	1	8	10	4	-	-	-	-	-	-	**32**
NA 92/2C	subrec	-	-	-	-	-	-	-	-	-	-	1	-	-	-	2	-	-	-	-	-	-	-	-	-	-	-	-	-	-	-	-	-	-	-	-	-	-	-	-	-	-	**2**
NA 93/46	EIA	-	-	-	-	-	-	1	12	-	-	4	-	-	1	4	1	-	6	1	-	-	-	-	-	1	-	-	-	6	-	-	-	9	17	7	-	-	-	-	-	-	**53**
NA 93/46	AB EIA	-	-	-	-	-	-	-	-	-	-	-	-	-	-	-	1	-	1	-	-	-	-	-	-	-	-	-	-	-	-	-	-	-	2	1	-	-	-	-	-	-	**5**
NA 97/7	II	-	-	-	-	-	-	-	-	-	-	1	-	-	-	-	-	-	1	-	-	-	-	-	-	-	-	-	-	-	-	-	-	1	-	-	-	-	-	-	-	-	**2**
NA 97/7	III	-	-	-	-	-	-	-	-	-	-	-	-	-	-	-	-	-	-	-	-	-	-	-	-	-	-	-	-	-	-	-	-	-	3	2	-	-	-	-	-	-	**5**
NA 97/7	IV	-	-	-	-	-	-	-	-	-	1	3	-	-	-	-	-	-	-	-	-	-	-	-	-	-	-	-	-	-	-	-	-	5	2	2	-	-	-	-	-	-	**9**
NA 93/45	I	-	-	-	-	-	-	-	2	1	1	7	-	-	-	-	-	-	-	-	-	-	-	-	-	-	-	-	-	2	-	-	-	12	11	16	-	-	-	-	-	-	**41**
NA 93/45	II	-	-	-	-	-	-	1	4	-	-	1	-	-	-	3	-	-	2	-	-	-	-	-	-	-	-	-	-	-	-	-	-	3	5	6	-	-	-	-	-	-	**19**
NA 93/45	IIIa	-	-	-	-	-	-	-	1	-	-	1	-	-	-	-	-	-	-	-	-	-	-	-	-	-	-	-	-	1	-	-	-	2	-	-	-	-	-	-	-	-	**3**
NA 93/45	IIIb	-	-	-	-	-	-	-	-	-	-	-	-	-	-	-	-	-	-	-	-	-	-	-	-	-	-	-	-	-	-	-	-	-	2	-	-	-	-	-	-	-	**2**
NA 93/45	IV	-	-	-	-	-	-	-	-	-	-	-	-	1	-	1	-	-	-	-	-	-	-	-	-	-	-	-	-	-	-	-	-	3	-	-	-	-	-	-	-	-	**5**
All sites		**3**	**1**	**1**	**2**	**1**	**2**	**3**	**35**	**2**	**12**	**47**	**7**	**7**	**11**	**47**	**2**	**4**	**51**	**5**	**5**	**14**	**3**	**7**	**7**	**24**	**93**	**29**	**11**	**53**	**9**	**13**	**1**	**107**	**216**	**131**	**1**	**3**	**5**	**27**	**5**	**33**	**924**

E=*Euxerus erythropus*, G=*Gerbillus* sp., T=*Tatera* sp., D=*Desmodilliscus braueri*, A=*Arvicanthis niloticus*, M=*Mastomys natalensis*

Table C.43: Skeletal element distribution of small rodents by site and phase

		cranium	mandibulare	dentes mandibulare	vertebrae caudales	ulna	os femoris	tibia	fibula	talus	ossa metapodalia	**Total**
BF 94/45	EIA	-	1	-	-	-	-	-	-	-	-	**1**
BF 97/30	LIA	-	-	-	-	-	-	1	-	-	-	**1**
BF 97/31	LIA	1	-	1	-	1	-	-	-	-	-	**3**
BF 94/120	LIA	-	1	-	-	-	-	1	1	-	-	**3**
NA 90/5A	Gaj IIb	-	-	-	-	-	-	1	-	-	-	**1**
NA 97/33	Gaj I	-	-	-	-	-	2	-	-	-	-	**2**
NA 99/75	IA	-	-	-	2	-	1	-	-	1	3	7
NA 92/2C	subrec	-	-	-	-	-	1	-	-	-	-	**1**
All sites		**1**	**2**	**1**	**2**	**1**	**4**	**3**	**1**	**1**	**3**	**19**

Table C.44: Skeletal element distribution of giant pouched rat by site and phase

	NA 99/75		NA 99/75
os fem.	IA	**talus**	IA
Bd	12.2	GL	9.2

	BF 97/30	BF 97/31
tibia	LIA	LIA
Bp	9.4	-
Bd	-	7.0

Table C.45: Measurements (mm) on giant pouched rat bones

	species		**HB (cm)**
Herpestidae	Long-snouted mongoose	*Herpestes naso*	52-59
	Slender mongoose	*Herpestes sanguinea*	26-34
	Banded mongoose	*Mungos mungo*	30-45
	Marsh mongoose	*Atilax paludinosus*	46-64
	White-tailed mongoose	*Ichneumia albicaudi*	47-71
Viverridae	Common genet	*Genetta genetta*	40-55
	Banded mongoose	*Genetta tigrina*	40-55
	Hausa genet	*Genetta thierryi*	38-45
	African civet	*Civettictis civetta*	68-95

HB (Length of head and body) from Kingdon (1997, 238-257, 266-274)

Table C.46: Viverridae and Herpestidae species of present northern Nigeria and northern Burkina Faso

		cranium	mandibulare	dentes	os sacrum	scapula	humerus	ulna	os metacarpale I	os metacarpale V	os metacarpale	os metacarpale + phalanx 1	os femoris	tibia	talus	calcaneus	os metatarsale V	os metatarsale	ossa metapodalia	phalanx proximalis	phalanx media	phalanx distalis	**Total**
BF 97/5	LSA	-	1	-	-	-	-	-	-	-	-	-	-	-	-	-	-	-	-	-	-	-	**1**
BF 94/45	EIA	-	1	-	-	1	-	-	-	1	1	-	-	-	-	-	-	-	-	-	-	-	**4**
BF 97/13	EIA	-	-	-	-	-	-	-	-	-	-	-	-	-	1	1	-	-	-	-	-	-	**2**
BF 97/13	AB MIA	-	-	-	-	-	-	-	-	-	-	-	-	-	-	-	-	-	-	1	-	-	**1**
BF 97/13	LIA	-	-	-	-	-	-	-	-	-	-	-	-	-	-	-	-	-	-	2	-	-	**2**
BF 97/30	LIA	-	-	1	1	-	1	-	2	-	-	-	-	-	-	-	-	-	2	2	1	1	**11***
BF 96/22	MIA	-	-	-	-	1	-	-	-	-	-	-	-	-	-	-	-	-	-	-	-	-	**1**
BF 97/31	LIA	-	-	-	-	1	-	-	-	-	-	-	-	-	-	-	-	-	-	-	-	-	**1**
BF 94/120	LIA	-	3	-	-	1	1	-	-	-	-	-	-	-	-	1	-	-	-	1	-	-	**7**
BF 95/7	LIA	-	-	-	-	-	-	1	-	-	-	-	-	-	-	-	1	-	-	-	-	-	**2**
NA 90/5A	Gaj IIb	-	-	-	-	-	3	-	-	-	1	1	-	3	-	-	-	2	-	-	-	-	**10**
NA 90/5BI	Gaj IIa/b	-	-	-	-	-	-	-	-	-	-	-	1	-	-	-	-	-	-	-	-	-	**1**
NA 90/5BII	Gaj IIc	-	1	-	-	-	3	-	-	-	-	-	-	1	-	-	-	-	-	-	-	-	**5**
NA 93/42	Gaj I	-	-	-	-	-	1	-	-	-	-	-	-	-	-	-	-	-	-	-	-	-	**1**
NA 97/37	Gaj III	-	-	-	-	-	-	-	-	-	-	-	-	1	-	1	-	-	-	-	-	-	**2**
NA 97/26	IA	-	-	-	-	-	1	-	-	-	-	-	1	-	-	-	-	-	-	-	-	-	**2**
NA 99/75	IA	-	-	-	-	-	1	-	-	-	-	-	-	-	-	-	-	-	-	-	-	-	**1**
NA 92/2C	subrec	1	-	-	-	-	-	-	-	-	-	-	-	-	-	-	-	-	-	-	-	-	**1**
NA 94/7	IV	-	-	-	-	-	1	-	-	-	-	-	-	-	-	-	-	-	-	-	-	-	**1**
All sites		**1**	**6**	**1**	**1**	**4**	**12**	**1**	**2**	**1**	**2**	**1**	**2**	**5**	**1**	**3**	**1**	**2**	**2**	**6**	**1**	**1**	**45**

*Including 7 bones with MNI=1

Table C.47: Skeletal element distribution of Viverridae or Herpestidae by site and phase

	BF 94/45 EIA	BF 97/13 EIA	BF 97/13 MIA	BF 97/13 LIA	BF 97/13 AB LIA	BF 97/30 LIA	BF 96/22 MIA	BF 97/31 LIA	BF 94/120 LIA	BF 95/7 LIA	NA 90/5A Gaj IIb	NA 90/5BI Gaj IIa/b	NA 99/65 Gaj IIc	NA 97/37 Gaj III	NA 97/26 IA	NA 99/75 IA	NA 97/13 IA	NA 92/2C subrec	NA 94/7 I	NA 93/45 I	**All sites**
cranium	-	-	-	1	-	3	-	1	2	2	-	-	-	-	-	-	-	-	-	-	**9**
maxillare	1	-	1	1	-	1	-	1	2	8	-	-	1	-	-	-	1	-	-	-	**17**
dentes maxillare	-	-	-	-	-	-	3	1	9	15	-	-	-	-	-	-	-	-	-	-	**28**
mandibulare	-	-	1	-	-	6	-	3	5	14	-	-	-	-	1	1	-	-	-	1	**32**
dentes mandibulare	-	-	2	-	-	-	-	1	4	11	-	-	-	-	-	-	-	-	-	-	**18**
dentes	-	-	1	-	-	1	-	4	13	17	-	-	-	-	-	-	-	-	-	-	**36**
atlas	-	-	-	-	-	-	-	-	1	1	-	-	-	-	-	-	-	-	-	-	**2**
axis	-	-	-	-	-	-	-	-	1	5	-	-	-	-	-	-	-	-	-	-	**6**
vertebrae cervicales	-	-	-	1	-	-	-	-	2	1	-	-	-	-	-	-	-	-	-	1	**5**
vertebrae thoracales	-	-	-	-	-	-	-	-	-	1	-	-	-	-	-	-	1	-	-	-	**2**
vertebrae lumbales	-	-	-	-	-	1	-	-	-	-	-	-	-	-	-	-	-	1	-	-	**2**
vertebrae caudales	-	-	-	-	-	-	-	-	-	1	-	-	-	-	-	-	-	-	-	-	**1**
scapula	-	-	-	-	-	-	-	-	-	-	2	-	-	-	-	-	-	-	-	-	**2**
humerus	-	-	-	-	-	1	-	-	-	-	-	-	-	-	-	-	-	-	-	-	**1**
radius	-	-	1	3	-	5	-	-	4	1	-	-	-	-	-	-	-	-	-	-	**14**
ulna	-	-	2	1	-	1	-	-	7	5	-	1	-	1	-	-	-	-	1	-	**19**
ossa carpi	-	-	1	1	1	-	-	-	1	2	-	-	-	-	-	-	-	-	-	-	**6**
os metacarpale I	-	-	-	1	-	1	-	1	-	1	-	-	-	-	-	-	-	-	-	-	**4**
os metacarpale II	-	-	1	1	-	-	-	-	2	6	-	-	-	1	-	-	-	-	-	-	**11**
os metacarpale III	-	-	1	-	-	3	-	-	3	5	-	-	-	-	-	-	-	-	-	1	**13**
os metacarpale IV	-	-	1	-	-	-	-	-	3	3	-	-	-	-	-	-	-	-	-	-	**7**
os metacarpale V	-	-	2	2	-	1	1	-	9	6	-	-	-	-	-	-	-	-	-	-	**21**
os metacarpale	-	-	-	-	-	-	-	-	-	1	-	-	-	-	-	-	-	-	-	-	**1**
os coxae	-	-	-	-	-	1	-	-	-	-	-	-	-	-	-	-	-	-	-	-	**1**
os femoris	-	-	-	-	-	1	-	-	-	-	2	1	-	1	-	-	-	-	-	-	**5**
patella	-	-	-	-	-	1	-	-	1	1	-	-	-	-	-	-	-	1	-	-	**4**
tibia	-	-	-	-	-	-	-	-	3	2	-	1	-	-	-	-	-	-	-	-	**6**
fibula	-	-	-	1	-	-	-	-	2	1	-	-	-	-	-	-	-	-	-	-	**4**
calcaneus	-	-	-	-	-	2	-	-	4	-	1	-	-	-	-	-	-	-	-	-	**7**
talus	-	-	1	1	-	6	-	-	-	1	-	-	-	-	-	-	-	-	-	-	**9**
os intermedioradiale	-	-	-	1	-	1	-	-	3	-	-	-	-	-	-	-	-	-	-	-	**5**
ossa tarsi	-	-	1	-	-	1	-	-	4	3	-	-	-	-	-	-	-	-	-	-	**9**
os metatarsale I	-	-	-	-	-	-	-	-	1	-	-	-	-	-	-	-	-	-	-	-	**1**
os metatarsale II	-	-	-	-	-	-	1	1	2	1	-	-	-	-	-	-	-	-	-	-	**5**
os metatarsale III	-	1	1	-	-	3	-	-	4	6	-	-	-	-	-	-	-	-	-	-	**15**
os metatarsale IV	-	-	-	2	-	-	-	-	3	6	-	-	-	-	-	-	-	-	-	-	**11**
os metatarsale V	-	-	-	1	-	-	1	1	6	6	-	-	-	-	-	-	-	-	-	-	**15**
ossa sesamoidea	-	-	-	-	-	2	-	-	-	-	-	-	-	-	-	-	-	-	-	-	**2**
phalanx proximalis	-	-	10	4	2	7	4	6	16	23	1	-	-	-	-	-	-	-	-	-	**73**
phalanx media	-	-	1	3	-	4	3	3	7	16	-	-	-	-	-	-	-	-	-	-	**37**
phalanx distalis	-	-	-	-	-	1	-	-	4	4	-	-	-	-	-	-	-	-	-	-	**9**
Total	**1**	**1**	**28**	**25**	**3**	**54**	**13**	**23**	**128**	**176**	**6**	**3**	**1**	**3**	**1**	**1**	**2**	**2**	**1**	**3**	**475**

Table C.48: Skeletal element distribution of domestic dog by site and phase

mand.	BF 97/31 LIA	BF 97/30 LIA	BF 94/120 LIA	BF 97/30 LIA	BF 95/7 LIA	BF 95/7 LIA	BF 97/30 LIA	BF 94/120 LIA	BF 97/30 LIA	BF 95/7 LIA	BF 95/7 LIA
7	74.5	67.0	-	-	-	-	-	-	-	-	-
8	70.4	63.3	66.7	65.0	55.4	-	-	-	-	-	-
9	66.0	57.3	-	58.7	50.5	58.0	-	-	-	-	-
10	32.3	27.1	32.0	30.3	28.4	35.2	28.9	-	-	-	-
11	39.1	37.0	35.3	36.0	29.8	-	-	32.7	-	-	-
12	33.2	30.0	30.1	30.1	25.1	33.3	-	27.4	35	34.4	30.3
14	19.9	16.4	20.2	19.2	17.5	17.8	16.6	-	-	19.8	-
18	-	-	-	-	-	-	-	-	-	-	-
19	22.0	18.6	-	18.6	15.4	18.2	21.0	-	-	18.4	-
20	-	14.3	17.2	19.0	12.2	16.4	-	14.6	17.1	14.2	16.9
B M1	-	-	-	-	-	7.4	-	-	-	-	-
L M2/ B M2	7.2/-	-	-	-	-	7.2/-	-	8.1/5.5	-	-	-

	BF 95/7	BF 95/7	BF 94/120		BF 97/31		BF 95/7		BF 95/7	BF 95/7	BF 95/7
mand.	LIA	LIA	LIA	**M2**	LIA	**atlas**	LIA	**axis**	LIA	LIA	LIA
14	19.2	-	-	L	7.5	GL	24.5	LCDe	46.6	41.6	40.2
18	-	41.3	-		BF 97/13	BFcr	34.5	LAPa	45.4	41.2	38.8
19	17.7	-	17.9	**M2**	MIA	LAd	13.3	BFcr	27.7	23.9	25.6
20	-	16.6	13.4	L	7.9	H	24	BPacd	28.2	23.3	26.1
L M2/ B M2	-	-	7.3/5.4					BPcr	31.9	-	-
								SBV	20.9	-	20.3
								BFcd	17.0	14.4	13.2
								H	40	30	35

	NA 90/5A		BF 97/13	BF 97/13	BF 97/13			
scapula	Gaj IIb	**radius**	LIA	MIA	LIA	Ad (n=1)	Au (n=1)	Lu (n=4)
GLP	22.8	Bp	16.2	13.3	-	12	14.5	14-16.4
		Bd	-	-	16.8			

	BF 94/120	BF 97/13	NA 94/7	BF 94/120	BF 94/120	BF 94/120		BF 97/13	BF 95/7	BF 97/30
ulna	LIA	MIA	I	LIA	LIA	LIA	**os mc I**	LIA	LIA	LIA
DPA	21.1	20.2	19.2	-	-	-	GL	20.3	18.4	17.7
BPC	-	13	-	14.0	13.4	11.4	Bd	4.4	5.2	5.0
SDO	-	-	14.9	-	-	-				
BPL	-	-	11.8	-	-	-				

	BF 95/7	BF 95/7	BF 97/13	BF 95/7	BF 95/7		BF 94/120	BF 95/7	BF 95/7
os mc II	LIA	LIA	LIA	LIA	LIA	**os mc III**	LIA	LIA	LIA
GL	54.2	50.0	49.5	49.0	48.8	GL	62.0	61.8	54.9
Bd	7.7	7.7	8.4	6.8	7.7	Bd	7.8	7.1	7.2

	BF 94/120	BF 94/120	BF 97/13	BF 94/120	BF 97/13	BF 94/120		NA 90/5BI		BF 95/7	NA 90/5BI
os mc V	LIA	LIA	LIA	LIA	MIA	LIA	**os fem.**	Gaj IIa/b	**tibia**	LIA	Gaj IIa/b
GL	49.5	48.8	47.6	47.4	43	-	Bd	26.7	Bd	19.9	15.2
Bd	-	7.3	7.9	7.5	6.9	5.3			Dd	-	10.1

Ad=*Canis adustus*, Au=*Canis aureus*, Lu=*Canis lupus* f. familiaris; shaded cells contain measurements on reference specimens (Guérin and Faure, 1996)

Table C.49: Measurements (mm) on domestic dog bones

	BF 97/30	BF 94/120		BF 97/30	BF 97/30	BF 97/30	BF 97/30	BF 95/7			
patella	LIA	LIA	**talus**	LIA	LIA	LIA	LIA	LIA	Ad (n=1)	Au (n=4)	Lu (n=4)
GL	14.3	11.5	GL	24.5	22.8	22.2	22.1	20.7	19	19.3-25	21-27
GB	7.9	9.9									
	BF 94/120	BF 97/30	BF 94/120	BF 94/120	BF 94/120						
calc.	LIA	LIA	LIA	LIA	LIA	Ad (n=1)	Au (n=4)	Lu (n=4)			
GL	36.4	(36)	35.8	33.1	-	31	30.5-40	37.5-45			
GB	13.5	(15)	12.4	12.3	12.7	12.5	12.3-18	15-18.5			
	BF 97/30	BF 94/120	BF 97/13	BF 94/120	BF 94/120		BF 94/120				
tars II-III	LIA	LIA	LIA	LIA	LIA	**os mt I**	LIA				
GB	20.2	19.9	19.6	19.2	15.4	GL	59.2				
						Bd	7.0				
	BF 97/13	BF 95/7	BF 95/7	BF 95/7							
os mt III	MIA	LIA	LIA	LIA							
GL	69.7	66.9	65.9	58.2							
Bd	7.4	7.8	7.5	6.6							
	BF 95/7	BF 95/7	BF 94/120	BF 94/120	BF 95/7						
os mt IV	LIA	LIA	LIA	LIA	LIA	Ad (n=1)	Au (n=4)	Lu (n=5)			
GL	65.8	65.0	62.0	59.1	58.4	60	55-70	66-82.5			
BD	6.9	7.1	6.3	7.0	6.6	6.5	7-7.5	6-8.6			
	BF 95/7	BF 96/22	BF 94/120	BF 95/7							
os mt V	LIA	MIA	LIA	LIA							
GL	61.8	56.4	56.1	53.1							
Bd	6.6	6.4	5.5	5.2							

Ad=*Canis adustus*, Au=*Canis aureus*, Lu=*Canis lupus* f. familiaris; shaded cells contain measurements on reference specimens (Guérin and Faure, 1996)
()=estimate

Table C.49 cont.

		gnawing	etching	coprolite
BF 94/45	EIA	3	5	-
BF 97/13	MIA	1	-	-
BF 97/13	LIA	1	3+1?	-
BF 97/30	LIA	4	-	-
BF 96/22	EIA	5	2	-
BF 97/31	LIA	-	1?	-
BF 94/120	LIA	2	8	1
BF 95/7	LIA	-	1	-
NA 90/5BI	Gaj IIa/b	1	-	min. 3
NA 90/5BII	Gaj IIc	-	-	7
NA 97/33	Gaj IIa/b	1	-	-
NA 95/1	Gaj IIa/b	7	1	-
NA 92/2C	rec	-	1	-
NA 93/46	LSA	1	-	-
NA 94/7	III	1	-	-
NA 93/45	I	2	-	-
NA 93/45	II	1	-	-
Blé C		1	-	-

Table C.50: Indirect evidence for domestic dog by site and phase

Fusion age (months)		7 ph. 1 ph. 2		8-10 os mp dist.		9-10 ulna prox.		11-12 rad. prox. rad. dist. ulna dist.		15 hum. prox. fib. dist.			13-16 tibia dist. calc.		15-18 fib. prox.		18 fem. prox.	
		NF	F	NF	F	NF	F	NF	F	NF	fus	F	NF	F	NF	F	NF	F
BF 97/13	MIA	-	11	-	2	-	1	-	1	-	-	-	-	-	-	-	-	-
BF 97/13	LIA	-	6	-	3	-	-	-	3	-	-	-	-	-	-	1	-	-
BF 97/13	AB LIA	-	3	-	-	-	-	-	-	-	-	-	-	-	-	-	-	-
BF 97/30	LIA	-	11	-	1	-	-	-	2	-	1	-	-	2	-	-	-	-
BF 96/22	MIA	-	7	-	1	-	-	-	-	-	-	-	-	-	-	-	-	-
BF 97/31	LIA	-	9	-	-	-	-	-	-	-	-	-	-	-	-	-	-	-
BF 94/120	LIA	-	23	-	9	-	1	-	2	-	-	2	-	4	-	-	-	-
BF 95/7	LIA	1	36	-	19	1	-	-	2	-	-	1	-	2	-	-	-	-
NA 90/5A	Gaj IIb	-	-	-	-	-	-	-	-	-	-	-	1	-	-	-	-	-
NA 97/37	Gaj III	-	-	1	-	-	-	-	-	-	-	-	-	-	-	-	1	-

NF=not fused, fus=fusing, F=fused

Table C.51: Fusion data of domestic dog bones by site and phase

site		element	cut	chop	worked
BF 97/13	LIA	talus	x		
BF 97/13	LIA	os metacarpale II	x		
BF 95/7	LIA	os metacarpale III prox.			x
BF 95/7	LIA	phalanx proximalis	x		
BF 95/7	LIA	axis	x		
NA 97/37	Gaj III	os femoris prox.		?	?

Table C.52: Butchered or worked domestic dog bones by site and phase

		mandibulare	dentes mandibulare	atlas	axis	scapula	humerus	radius	os metacarpale I	os coxae	os femoris	calcaneus	talus	os metatarsale V	os metatarsale	phalanx proximalis	phalanx media	**Total**
BF 94/45	EIA	-	-	-	-	-	2	1	1	-	-	2	-	2	1	1	-	**10**
BF 94/45	AB EIA	-	-	-	-	-	-	1	-	-	-	-	-	-	-	-	-	**1**
BF 96/22	MIA	-	-	-	-	1	-	-	-	-	-	-	1	-	-	-	-	**2**
BF 94/120	LIA	-	1	-	-	-	1	-	-	-	-	1	-	-	-	1	-	**4**
NA 90/5A	Gaj IIb	-	-	-	-	-	-	-	-	-	-	-	-	-	-	-	1	**1**
NA 90/5BII	Gaj IIc	2	-	-	-	-	1	-	-	1	-	-	-	-	-	-	-	**4**
NA 97/37	Gaj III	-	-	1	-	-	-	-	-	-	-	-	-	-	-	-	-	**1**
NA 92/2C	subrec	-	-	-	1	-	1	-	-	-	1	-	-	-	-	-	-	**3**
NA 94/7	III	-	1	-	-	-	-	-	-	-	-	-	-	-	1	-	-	**2**
All sites		**2**	**2**	**1**	**1**	**1**	**5**	**2**	**1**	**1**	**1**	**3**	**1**	**2**	**2**	**2**	**1**	**28**

Table C.53: Skeletal element distribution of African wild or domestic cat by site and phase

	NA 97/37		BF 96/22		NA 90/5BII	BF 94/45	BF 94/45	NA 92/2C	BF 97/30		BF 94/45	BF 94/45
atlas	Gaj III	**scapula**	MIA	**hum.**	Gaj IIc	EIA	EIA	subrec	LIA	**radius**	AB EIA	EIA
BFcr	17	GLP	(13)	GL	167	-	-	-	-	Bp	8.3	-
BFcd	20	LG	(12)	Dp	22.9	19.5*	16.7	-	-	Bd	-	10.9
GLF	17	BG	9.0	SD	7.9	-	-	-	-			
L	10			Bd	18.7	-	-	17.3	16.5			
H	12											

	BF 94/45		NA 90/5BII		BF 96/22		BF 94/45	BF 94/45		BF 94/45
os mc I	EIA	**os coxae**	Gaj IIc	**talus**	MIA	**calc.**	EIA	EIA	**os mt V**	EIA
GL	10.8	LAR	13.7	GL	15.6	GL	30.1	-	Bd	4.4
Bd	3.9					GB	9.7	9.5		

*unfused; ()=estimate

Table C.54: Measurements on African wild or domestic cat bones

		mandibulare	dentes	axis	radius	ulna	os metacarpale II	os metacarpale III	os metacarpale	os femoris	os metatarsale II	ossa metapodalia	phalanx proximalis	**Total**
BF 97/13	LIA	-	-	-	-	-	1	-	-	-	-	-	-	**1**
BF 97/30	LIA	-	-	-	-	-	-	1	-	-	-	-	4	**5***
NA 90/5C	Gaj IIa	-	-	-	-	-	-	-	-	-	1	1	-	**2**
NA 90/5A	Gaj IIb	-	-	-	1	1	-	-	1	1	-	-	-	**4**
NA 95/1	Gaj IIa/b	-	-	-	1	-	-	-	-	-	-	-	-	**1**
NA 93/45	II	-	1	-	-	1	-	-	-	-	-	-	-	**2**
Blé A		1	-	-	-	-	-	-	-	-	-	-	-	**1**
Blé C		-	-	1	-	-	-	-	-	-	-	-	-	**1**
All sites		**1**	**1**	**1**	**2**	**2**	**1**	**1**	**1**	**1**	**1**	**1**	**4**	**17**

*MNI=1

Table C.55: Skeletal distribution of serval or caracal bones by site and phase

	Blé A		Blé C		NA 90/5A	NA 95/1		BF 97/13		NA 90/5C
mand.		**axis**		**radius**	Gaj IIb	Gaj IIa/b	**os mc II**	LIA	**os mt II**	Gaj IIa
5	26.2	BFcr	21.5	Bp	12.7	-	GL	60.4	GL	71.7
6	9.4/4.7	SBV	16.2	Bd	19.2	18.7	Bd	7.9	Bd	10.1
7	9.5									
9	13.7									
10	11.7									

Table C.56: Measurements (mm) on serval or caracal bones

	BF 94/40 LSA	BF 94/45 EIA	BF 97/13 MIA	BF 97/13 LIA	BF 97/30 LIA	BF 96/22 MIA	BF 97/31 LIA	BF 94/120 LIA	BF 95/7 LIA	NA 99/65 Gaj IIb	NA 99/65 Gaj IIc	NA 97/37 Gaj III	NA 95/1 Gaj IIa/b	NA 97/13 IA	NA 94/7 I	NA 93/45 I	Blé A	**All sites**
cranium	-	-	1	-	-	-	-	3	1	-	-	-	-	-	-	-	-	**5**
maxillare	-	-	1	-	-	-	-	-	4	-	-	-	-	-	-	-	-	**5**
mandibulare	-	-	2	-	1	-	1	1	3	-	-	-	-	-	-	-	-	**8**
dentes	-	2	4	2	6	4	2	19	31	-	1	-	-	-	-	-	-	**71**
vertebrae thoracales	-	-	-	-	-	-	-	-	-	1	-	-	1	-	-	-	-	**2**
vertebrae lumbales	1	-	1	-	-	-	-	-	-	-	-	-	-	-	-	-	-	**2**
vertebrae caudales	-	-	-	-	-	-	-	2	-	-	-	-	-	-	-	-	-	**2**
costae	-	-	1	-	-	-	-	1	-	-	-	-	-	-	-	-	-	**2**
scapula	-	-	-	-	1	-	-	-	-	-	-	-	-	-	-	-	-	**1**
humerus	-	-	-	-	-	-	-	1	-	-	-	-	-	-	-	-	1	**2**
ulna	-	-	-	-	-	-	1	-	-	-	-	-	-	-	-	-	-	**1**
os coxae	-	-	-	-	-	-	-	-	-	-	1	-	2	-	-	-	-	**3**
os femoris	-	-	-	-	-	-	-	-	-	-	-	-	-	-	1	1	-	**2**
tibia	-	-	-	-	-	-	1	-	-	-	-	1	-	-	-	-	-	**2**
patella	-	-	1	-	-	-	-	-	1	-	-	-	1	-	-	-	-	**3**
calcaneus	-	-	-	-	-	-	-	-	1	-	-	-	-	-	-	-	-	**1**
ossa tarsi	-	-	-	1	-	1	-	1	-	-	-	-	-	-	-	-	-	**3**
ossa metapodalia	-	1	-	6	3	1	2	18	35	-	-	-	-	2	-	-	-	**67**
phalanx proximalis	-	-	-	-	-	-	-	-	-	-	2	-	-	-	-	-	-	**2**
phalanx distalis	-	-	-	-	-	-	1	-	-	-	-	-	-	-	-	-	-	**1**
phalanx	-	-	-	-	-	-	-	1	1	-	-	-	-	-	-	-	-	**2**
Total	**1**	**3**	**11**	**9**	**11**	**6**	**8**	**47**	**77**	**1**	**4**	**1**	**4**	**2**	**1**	**1**	**1**	**188**
Vulpes pallida	-	-	1	-	-	-	-	1	2	-	-	-	-	-	-	-	-	**4**
Canis lupus f. familiaris	-	1	28	25	54	13	21	126	175	-	1	3	-	2	1	3	-	**453**
Felis silvestris/F. s. f. catus	-	10	-	-	-	2	-	4	-	-	-	1	-	-	-	-	-	**17**
Felis serval/caracal	-	-	-	1	5	-	-	-	-	-	-	-	1	-	-	-	1	**8**

Table C.57: Skeletal element distribution of medium-sized carnivores by site and phase

fusion age (months)	8-9 hum. dist.		8-10 os mp dist.		15 hum. prox.		13-16 tibia dist. calc.		18 fem. prox.	
	NF	F	NF	F	NF	F	NF	F	NF	F
BF 97/13 LIA	-	-	-	3	-	-	-	-	-	-
BF 97/30 LIA	-	-	1	2	-	-	-	-	-	-
BF 96/22 MIA	-	-	-	1	-	-	-	-	-	-
BF 97/31 LIA	-	-	-	1	-	-	-	-	-	-
BF 94/120 LIA	-	1	-	10	-	-	-	-	-	-
BF 95/7 LIA	-	-	-	28	-	-	-	1	-	-
NA 97/37 Gaj III	-	-	-	-	-	-	1	-	-	-
NA 97/13 IA	-	-	-	2	-	-	-	-	-	-
NA 93/45 I	-	-	-	-	-	-	-	-	1	-
Blé A	-	-	-	-	-	1	-	-	-	-

Table C.58: Fusion data of medium-sized carnivores bones by site and phase

		cranium	dentes mandibulare	dentes	vertebrae thoracales	vertebrae	costae	humerus	radius	ossa carpi	os metacarpale III	os coxae	patella	tibia	fibula	talus	calcaneus	ossa tarsi	ossa metapodalia	phalanx proximalis	phalanx media	phalanx distalis	**Total**
BF 94/45	EIA	-	-	-	-	-	-	-	-	-	-	-	-	-	-	-	-	-	1	-	-	1	**2**
BF 97/13	MIA	-	1	-	-	-	-	-	-	-	-	-	-	-	-	-	-	-	-	-	-	-	**1**
BF 97/13	LIA	-	-	-	-	-	-	-	-	-	-	-	-	-	-	1	-	-	-	-	-	-	**1**
BF 97/30	LIA	-	-	-	-	-	1	-	-	-	-	-	-	-	-	-	-	-	-	-	-	-	**1**
BF 96/22	MIA	-	-	1	-	-	-	-	-	-	-	-	-	-	-	-	-	-	-	-	-	-	**1**
BF 94/120	LIA	1	-	-	-	-	-	-	-	-	1	-	-	-	-	-	-	-	1	-	-	1	**4**
BF 95/7	LIA	-	-	1	-	-	-	-	-	-	-	-	-	-	-	-	1	-	-	-	-	-	**2**
NA 90/5A	Gaj IIb	-	1	2	-	2	-	-	1	1	-	-	1	1	1	1	1	3	1	1	1	1	**19**
NA 90/5BI	Gaj IIa/b	-	-	-	1	-	-	1	-	-	-	-	-	-	-	-	1	-	-	-	-	-	**3**
NA 90/5BII	Gaj IIc	1	-	1	1	-	-	-	-	-	-	2	-	1	-	-	-	-	1	-	1	-	**8**
NA 93/42	Gaj I	-	-	-	-	-	1	-	-	-	-	-	-	-	-	-	-	-	-	-	-	-	**1**
NA 97/33	Gaj I	-	-	-	-	-	1	1	-	-	-	-	-	-	-	-	-	-	-	-	-	-	**2**
NA 96/45	Gaj IIa/b	-	-	-	-	-	1	-	-	-	-	-	-	-	-	-	-	-	-	-	-	-	**1**
NA 95/1	Gaj IIa/b	-	1	-	-	-	-	-	-	-	-	-	-	-	-	-	-	-	1	1	-	-	**3**
NA 92/2C	subrec	-	-	-	-	-	-	1	-	-	-	-	-	-	-	-	-	-	-	-	-	-	**1**
NA 93/46	EIA	1	-	-	-	-	-	-	-	-	-	-	-	-	-	-	-	-	-	-	-	-	**1**
NA 94/7	I	-	-	-	-	-	-	-	-	-	-	-	-	-	-	-	-	-	1	-	-	-	**1**
NA 93/45	IIIa	-	-	2	-	-	-	-	-	-	-	-	-	-	-	-	-	-	-	-	-	-	**2**
Blé A		-	1	-	-	-	-	-	-	-	-	-	-	-	-	-	-	-	-	-	-	-	**1**
Blé E		-	-	-	-	-	-	-	-	-	-	-	-	-	-	-	-	-	1	-	-	-	**1**
All sites		**3**	**4**	**7**	**2**	**2**	**4**	**3**	**1**	**1**	**1**	**2**	**1**	**2**	**1**	**2**	**3**	**3**	**7**	**2**	**2**	**3**	**56**

Table C.59: Skeletal element distribution of common warthog by site and phase

	NA 90/5BI		NA 90/5BII		NA 90/5A
humerus	Gaj IIa/b	**os coxae**	Gaj IIc	**tibia**	Gaj IIb
Bd	47	LA	45.3	Bd	32.5
BT	36				

Table C.60: Measurements (mm) on common warthog bones

	NA 99/65	NA 99/75		NA 99/75		NA 99/75
ph. 1	Gaj IIb	IA	**ph. 2**	IA	**ph. 3**	IA
GLpe	23.1	-	GL	21.1	DLS	(19)
Bp	7.7	-	Bp	9.5	Ld	(17)
Bd	6.6	8.5	SD	6.5	H	11.9
			Bd	7.0	Bp	6.4

()=estimate

Table C.61: Measurements (mm) on forest duiker bones

		mandibulare	dentes mandibulare	scapula	humerus	tibia	talus	os metatarsale III et IV	phalanx proximalis	phalanx media	phalanx distalis	**Total**
BF 94/45	EIA	-	1	-	-	1	-	-	-	-	-	**2**
BF 94/120	LIA	-	-	-	-	1	-	-	1	1	2	**5**
BF 95/7	LIA	-	-	1	-	-	-	-	-	-	-	**1**
NA 90/5A	Gaj IIb	1	-	-	-	-	-	-	1	-	-	**2**
NA 99/65	Gaj IIb	-	-	-	-	-	-	-	-	1	-	**1**
NA 99/65	Gaj IIc	-	-	-	-	-	-	-	-	1	-	**1**
NA 97/33	Gaj IIa/b	-	-	-	-	-	1	-	1	-	-	**2**
NA 95/1	Gaj IIa/b	-	-	2	2	-	4	1	2	1	-	**12**
All sites		**1**	**1**	**3**	**2**	**2**	**5**	**1**	**1**	**1**	**2**	**26**

Table C.62: Skeletal element distribution of bush duiker by site and phase

		maxillare	dentes mandibulare	scapula	humerus	tibia	talus	calcaneus	phalanx proximalis	phalanx media	phalanx distalis	**Total**
BF 94/45	EIA	-	-	-	-	-	-	-	-	-	1	**1**
BF 94/120	LIA	-	-	1	-	-	-	-	1	-	-	**2**
BF 95/7	LIA	-	-	-	-	-	-	-	-	1	-	**1**
NA 90/5BI	Gaj IIa/b	1	1	-	2	2	-	1	-	-	-	**7**
NA 93/46	EIA	-	-	-	-	-	1	-	-	-	-	**1**
All sites		**1**	**1**	**1**	**2**	**2**	**1**	**1**	**1**	**1**	**1**	**12**

Table C.63: Skeletal element distribution of oribi by site and phase

	BF 94/45 EIA	BF 97/13 MIA	BF 97/13 LIA	BF 97/30 LIA	BF 96/22 MIA	BF 97/31 LIA	BF 94/120 LIA	BF 95/7 LIA	NA 90/5C Gaj IIa	NA 90/5BII Gaj IIc	NA 97/33 Gaj I	NA 97/33 Gaj IIa/b	NA 96/45 Gaj IIa/b	NA 95/1 Gaj IIa/b	NA 92/2C subrec	NA 93/46 EIA	NA 94/7 I	NA 94/7 II	NA 94/7 III	NA 94/7 IV	Blé B	**All sites**
processus cornualis	1	-	-	-	-	-	-	-	-	-	-	-	-	-	-	-	-	-	-	-	-	**1**
cranium	-	-	-	-	-	-	-	-	-	-	-	1	-	1	-	-	-	-	-	1	-	**3**
mandibulare	-	-	-	-	-	-	2	-	-	2	-	1	-	1	-	1	-	-	-	-	-	**7**
dentes mandibulare	-	-	-	-	-	-	-	2	-	-	-	-	-	-	-	3	-	-	-	-	-	**5**
dentes	-	-	-	-	-	-	1	-	-	1	-	-	-	-	-	-	-	-	-	-	-	**2**
vertebrae thoracales	-	-	-	-	-	-	-	-	-	-	-	-	-	2	-	-	-	-	-	-	-	**2**
vertebrae lumbales	-	-	-	-	-	-	-	-	-	-	-	1	-	1	-	-	-	-	-	-	-	**2**
os sacrum	-	-	-	-	-	-	-	-	-	-	1	-	-	-	-	-	-	-	-	-	-	**1**
costae	-	-	-	-	-	-	-	-	-	-	-	-	-	2	-	-	-	-	-	-	-	**2**
humerus	3	-	-	-	-	-	-	-	-	-	-	3	1	1	-	-	-	1	-	-	-	**9**
radius	1	-	-	-	-	-	-	-	1	1	-	-	1	2	-	-	-	1	-	-	-	**7**
ulna	-	-	-	-	-	-	1	-	-	-	-	1	-	1	-	-	-	1	-	-	-	**4**
ossa carpi	-	-	-	-	-	-	-	1	-	-	-	-	-	-	-	-	-	-	-	-	-	**1**
os centroquartale	-	-	-	-	-	-	-	-	-	-	-	-	-	1	-	-	-	-	-	-	-	**1**
os metacarpale III et IV	1	-	-	-	-	-	-	-	-	1	-	-	-	-	-	-	-	-	-	-	-	**2**
os coxae	2	-	-	-	-	-	-	-	-	2	-	-	-	1	-	-	-	-	-	-	-	**5**
os femoris	-	-	-	-	-	-	-	-	-	-	-	1	-	-	-	-	-	1	-	-	1	**3**
tibia	-	-	-	-	-	-	-	-	-	-	-	-	-	1	-	-	-	-	1	-	-	**2**
os metatarsale III et IV	-	1	-	-	-	-	-	-	-	1	-	-	-	1	-	-	-	-	1	-	-	**4**
calcaneus	-	-	-	-	-	-	-	-	-	-	-	-	-	1	-	-	-	-	1	-	-	**2**
talus	1	-	-	-	-	-	-	-	-	-	-	-	-	-	-	-	-	-	-	-	-	**1**
phalanx media pedis	-	-	-	-	-	1	-	-	-	-	-	-	-	-	-	-	-	-	-	-	-	**1**
ossa metapodalia	3	-	-	-	-	-	-	-	-	-	-	2	-	-	-	-	-	-	1	-	-	**6**
ossa sesamoidea	2	-	-	-	1	-	-	-	-	-	-	-	-	-	-	-	-	-	-	-	-	**3**
phalanx proximalis	-	-	-	1	-	-	-	3	1	1	-	-	-	1	1	-	1	-	-	-	-	**9**
phalanx media	-	-	-	-	-	1	1	2	-	-	-	-	-	-	-	-	-	-	-	-	-	**4**
phalanx distalis	-	-	1	-	1	-	-	-	-	-	-	-	-	-	-	-	-	-	-	-	-	**2**
Total	**14**	**1**	**1**	**1**	**2**	**2**	**5**	**8**	**2**	**9**	**1**	**10**	**2**	**17**	**1**	**4**	**1**	**4**	**4**	**1**	**1**	**91**
Sylvicapra grimmia	2	-	-	-	-	-	1	1	-	-	-	2	-	12	-	-	-	-	-	-	-	**18**
Ourebia ourebi	1	-	-	-	-	-	2	1	-	-	-	-	-	-	-	1	-	-	-	-	-	**5**

Table C.64: Skeletal element distribution of bush duiker or oribi by site and phase

	NA 97/33		BF 94/120		NA 95/1	BF 94/45	NA 95/1	NA 90/5BI	NA 97/33	NA 90/5BII	NA 90/5BI
os sacr.	Gaj I	**scapula**	LIA - O	**hum.**	Gaj IIa/b - S	EIA	Gaj IIa/b - S	Gaj IIa/b - O	Gaj IIa/b	Gaj III	Gaj IIa/b - O
GB	(60)	GLP	22.2	GL	113	-	-	-	-	-	-
BFcr	26.4	LG	17.8	Bp	24.4	-	-	-	-	-	-
HFcr	10.0	BG	14.2	SD	11.0	11.6	10.4	-	-	-	-
				Bd	21.3	-	22.2	22.8	21.7	20.9	-
				BT	19.1	21.0	18.9	21.7	19.4	18.1	21.0
	NA 95/1		NA 95/1		BF 94/45		BF 94/45	NA 90/5BI	NA 90/5BI		NA 95/1
radius	Gaj IIa/b	**ulna**	Gaj IIa/b	**os mc**	EIA	**tibia**	EIA - S	Gaj IIa/b - O	Gaj IIa/b - O	**os mt**	Gaj IIa/b - S
Bd	16.8	BPC	10.4	Bp	11.6	Bp	32.2	-	-	Bd	17.6
		DPA	19.4	SD	10.2	Bd	-	22.7	22.3		
		SDO	17.0								
		LO	31.3								
	NA 97/33	NA 95/1	NA 95/1	NA 95/1	NA 95/1		NA 95/1	NA 95/1	NA 97/33	NA 90/5A	BF 94/120
talus	Gaj IIa/b - S	Gaj IIa/b - S	Gaj IIa/b - S	Gaj IIa/b - S	Gaj IIa/b - S	**ph. 1**	Gaj IIa/b	Gaj IIa/b	Gaj IIa/b - S	Gaj IIb	LIA - S
GLl	24.3	23.9	23.5	-	-	GLpe	28.0	27.8	26.8	25.5	-
GLm	23.7	22.0	22.2	22	21.8	Bp	8.5	8.5	7.8	6.1	7.5
Dl	13.3	12.2	12.7	13	-	SD	6.7	6.8	6.3	5.0	-
Bd	14.5	14.7	14.3	15	13.2	Bd	7.9	8.2	6.9	5.8	-
	NA 99/65	BF 94/120	NA 95/1	BF 97/31	NA 99/65	BF 95/7	BF 97/31		NA 99/75	BF 94/45	
ph. 2	Gaj IIc -S	LIA	Gaj IIa/b	LIA	Gaj IIb - S	LIA - O	LIA	**ph. 3**	IA	EIA - O	
GL	19.7	18.4	18.3	18.3	18.0	17.9	17.3	DLS	(19)	18.0	
Bp	8.7	7.5	7.7	8.4	7.5	6.8	8.3	Ld	(17)	14.9	
SD	6.6	6.3	5.9	5.5	5.7	5.6	6.3	H	11.9	12.8	
Bd	7.2	6.8	6.4	5.9	6.6	6.0	7.0	Bp	6.4	5.7	

S=*Sylvicapra grimmia*, O=*Ourebia ourebi*; ()=estimate

Table C.65: Measurements (mm) on bush duiker or oribi bones

site			unfused element	fusion age in *Ovis ammon* f. aries
BF 94/45	EIA		2 ossa metapodalia dist.	18-28 months
NA 92/2C	subrec		phalanx proximalis (juv.)	13-16 months
NA 96/45	Gaj IIa/b		radius dist.	36 months
NA 95/1	Gaj IIa/b	S	scapula	6-8 months
NA 95/1	Gaj IIa/b		radius dist.	36 months
NA 95/1	Gaj IIa/b		humerus dist.	10 months

S=*Sylvicapra grimmia*

Table C.66: Unfused bush duiker or oribi bones

		cranium	mandibulare	dentes mandibulare	scapula	humerus	radius	ulna	ossa carpi	os metacarpale III et IV	os coxae	os femoris	patella	tibia	talus	calcaneus	ossa tarsi	os metatarsale III et IV	ossa metapodalia	phalanx proximalis	phalanx media	phalanx distalis	**Total**
BF 97/5	LSA	-	-	1	-	-	-	-	-	-	-	-	-	-	1	-	-	-	-	-	-	-	**2**
BF 94/45	EIA	-	-	-	-	-	-	-	-	1	-	-	-	-	-	-	-	-	-	1	3	-	**5**
BF 97/13	MIA	-	-	-	-	-	-	-	-	-	-	-	-	-	-	-	-	-	-	1	-	-	**1**
BF 97/13	LIA	-	-	-	-	-	-	-	-	-	-	-	-	-	-	-	-	-	-	1	-	-	**1**
BF 97/31	LIA	-	-	-	-	-	-	-	-	-	-	-	-	-	-	-	-	-	-	1	-	-	**1**
NA 90/5C	Gaj IIa	-	-	-	-	1	-	-	-	-	-	-	-	-	-	-	-	-	-	-	-	-	**1**
NA 90/5A	Gaj IIb	-	-	1	-	1	2	-	1	2	-	1	1	3	-	-	1	2	5	6	4	2	**32**
NA 90/5BI	Gaj IIa/b	1	-	1	-	-	1	-	-	1	1	-	-	-	-	-	-	-	-	-	1	-	**6**
NA 90/5BII	Gaj IIc	-	-	-	-	-	-	1	-	-	-	-	-	-	-	-	-	-	-	-	-	-	**1**
NA 99/65	Gaj IIc	-	1	1	-	-	-	-	-	-	-	-	-	-	-	-	-	-	-	-	1	-	**3**
NA 95/1	Gaj IIa/b	-	-	-	1	-	-	-	-	-	-	-	-	-	-	-	-	-	-	-	-	-	**1**
NA 99/75	IA	-	-	-	-	-	-	-	-	-	-	-	-	-	-	-	-	-	-	1	-	-	**1**
NA 93/45	I	-	-	-	-	-	-	-	-	-	-	-	-	-	-	-	-	-	-	-	-	1	**1**
Blé A		-	-	-	-	-	-	-	-	-	-	-	-	-	-	-	-	1	-	-	-	-	**1**
Blé B		-	-	-	-	1	-	-	-	2	-	-	-	-	1	1	-	-	1	-	-	-	**6**
Blé C		-	-	-	-	-	-	-	-	-	-	-	-	-	1	-	-	-	-	3	-	-	**4**
Blé E		-	-	-	-	-	1	-	-	-	-	-	-	-	-	-	-	1	-	-	-	-	**2**
All sites		**1**	**1**	**4**	**1**	**3**	**4**	**1**	**1**	**6**	**1**	**1**	**1**	**3**	**3**	**1**	**1**	**4**	**6**	**13**	**9**	**3**	**69**

Table C.67: Skeletal element distribution of bohor reedbuck by site and phase

	NA 95/1		NA 90/5C		NA 90/5BI		NA 90/5A	NA 90/5BII	BF 94/45	Blé B		NA 90/5A
scapula	Gaj IIa/b	**hum.**	Gaj IIa	**radius**	Gaj IIa/b	**os mc**	Gaj IIb	Gaj IIc	EIA		**tibia**	Gaj IIb
LG	26.2	Bp	39.4	Bd	29.4	Bp	25.2	22.5	-	-	Bd	30.3
BG	23.8					Bd	-	-	23.0	20.6		
SLC	19.5											

	NA 90/5A	Blé E	Blé A
os mt	Gaj IIb		
Bp	22.7	22.0	-
Bd	-	-	24.1

	BF 94/45	NA 90/5A	NA 99/75	Blé C	BF 97/31	Blé C	NA 90/5A	BF 97/13	BF 97/13	Blé C	NA 99/65
ph. 1	EIA	Gaj IIb	IA		LIA		Gaj IIb	LIA	MIA		Gaj IIc
GLpe	47.9	47.8	47.3	46.9	46.0	(44)	43.2	-	-	-	-
Bp	13.4	11.5	13.3	13.4	10.8	(11)	11.2	12	11.4	11.3	11.3
SD	11.2	9.4	10.3	10.9	9.1	9.5	9.0	-	-	-	-
Bd	12.2	10.2	11.8	11.4	9.8	10.6	9.8	-	-	-	-

	NA 90/5BI	BF 94/45	BF 94/45	BF 94/45	NA 97/13		NA 90/5A	NA 93/45
ph. 2	Gaj IIa/b	EIA	EIA	EIA	IA	**ph. 3**	Gaj IIb	I
GL	26.6	25.8	25.2	24.5	22.9	DLS	31.5	31
Bp	11.8	11.1	4.1	10.9	10.2	Ld	-	26
SD	8.9	8.0	8.0	7.9	8.2	H	-	48.2
Bd	9.6	9.0	9.4	9.2	8.9	Bp	-	8.5

()=estimate

Table C.68: Measurements (mm) on bohor reedbuck bones

site		unfused element	fusion age in *Ovis ammon* f. aries
NA 90/5A	Gaj IIb	os femoris prox.	30-36 months
NA 90/5A	Gaj IIb	tibia prox.	36-42 months
NA 90/5A	Gaj IIb	os metapodale III et IV dist.	18-28 months
NA 90/5A	Gaj IIb	2 phalanges proximales	13-16 months
Blé B		calcaneus (juv.)	30-36 months
Blé B		os metapodale III et IV dist.	18-28 months

Table C.69: Unfused bohor reedbuck bones

		processus cornualis	cranium	dentes maxillare	mandibulare	dentes mandibulare	scapula	humerus	radius	ulna	radio-ulna	ossa carpi	os metacarpale III et IV	os coxae	os femoris	patella	tibia	os metatarsale III et IV	calcaneus	talus	ossa metapodalia	ossa sesamoidea	phalanx proximalis	phalanx media	phalanx distalis	**Total**
BF 97/5	LSA	-	-	3	-	-	-	-	-	-	-	-	-	-	-	-	-	-	-	-	-	-	-	-	-	**3**
BF 94/45	EIA	-	-	-	-	-	-	-	-	-	-	-	-	-	-	-	1	-	-	-	-	-	-	-	-	**1**
BF 94/120	LIA	-	-	-	-	-	-	-	-	-	-	-	-	-	1	-	-	-	-	-	-	-	-	-	-	**1**
NA 93/46	EIA	-	-	-	1	-	-	-	-	-	-	-	-	-	-	1	-	-	-	-	-	-	-	1	-	**3**
NA 94/7	I	-	-	5	1	8	1	4	3	-	-	2	2	-	-	-	1	2	5	3	1	-	5	3	9	**55**
NA 94/7	II	-	-	2	-	-	-	-	-	-	-	-	-	-	-	-	-	-	-	1	-	-	-	-	-	**3**
NA 94/7	III	-	-	-	-	-	-	-	1	-	-	-	-	-	-	-	-	-	-	-	-	-	-	-	-	**1**
NA 93/45	I	-	-	-	-	-	-	-	-	-	-	-	1	-	-	-	-	-	1	1	-	-	3	-	1	**7**
NA 93/45	II	-	-	-	-	1	-	-	-	-	-	-	-	-	-	-	-	1	-	-	-	-	2	1	1	**6**
NA 93/45	IIIa	1	-	-	-	-	-	-	-	-	-	-	-	-	-	-	1	1	-	-	-	-	-	1	-	**4**
NA 93/45	IIIb	-	-	2	-	-	-	-	1	-	-	-	2	-	-	-	-	-	-	-	-	-	-	-	-	**5**
NA 93/45	IV	-	-	-	-	-	-	-	-	-	-	-	-	-	1	-	-	-	-	-	-	-	-	-	-	**1**
Blé A		-	-	-	-	-	-	-	-	-	-	-	-	-	-	-	-	-	2	-	-	-	-	1	1	**4**
Blé B		-	-	-	-	-	1	-	1	2	1	1	2	1	-	-	3	5	3	1	-	1	2	4	1	**29**
Blé C		-	2	1	-	-	-	-	-	-	-	-	1	-	1	-	-	-	-	-	-	-	1	1	-	**7**
Blé E		-	1	-	2	-	-	1	1	-	-	-	-	-	-	-	2	-	-	-	-	-	1	1	2	**11**
All sites		**1**	**3**	**13**	**4**	**9**	**2**	**5**	**7**	**2**	**1**	**3**	**8**	**1**	**3**	**1**	**8**	**9**	**11**	**6**	**1**	**1**	**14**	**13**	**15**	**141**

Table C.70: Skeletal element distribution of kob by site and phase

site		unfused element	fusing age in *Ovis ammon* f. aries
NA 93/45	II	os metatarsale III et IV dist.	20-28 months
NA 93/45	II	phalanx proximalis	13-16 months
NA 93/45	IIIb	radius dist.	16 months

Table C.71: Unfused kob bones

	Blé B	NA 94/7		NA 94/7	NA 94/7		NA 94/7	NA 94/7	NA 93/45
scapula		I	**hum.**	I	I	**radius**	I	I	IIIb
GLP	42.8	42.8	Bd	47.0	42.5	Bp	44.0	41.9	-
LG	34.0	-	BT	42.5	39.5	BFp	40.8	40.6	-
BG	33.6	28.3				Bd	-	-	34.7*

	NA 93/45	NA 93/45	Blé C	NA 94/7	NA 93/45	Blé B		NA 93/46
os mc	IIIb	IIIb		I	I		**patella**	EIA
Bp	35.3	-	-	-	-	-	GL	(40)
Bd	-	37.2	33.5	33.5	31.7	30.8	GB	32.2

	Blé E	NA 94/7	BF 94/45	BF 94/45		NA 94/7	NA 93/45	NA 94/7	NA 94/7
tibia		I	EIA	EIA	**talus**	I	I	I	II
Bp	61.4	-	-	-	GLl	44.6	43.2	41.4	41.4
Bd	-	36.1	35.5	35.5	GLm	40.5	40.6	39.4	38.6
					Dl	24.4	24.7	22.5	22.7
					Bd	27.0	26.4	25.3	25.6

	Blé B	NA 94/7		NA 93/45	NA 93/45
calc.		I	**os mt**	I	II
GL	90	87.5	Bp	29.9	-
GB	28	30	Bd	-	33.8*

	Blé B	NA 93/45	NA 93/45	NA 93/45	NA 93/45	Blé B	Blé C	Blé E
ph. 1		I	I	II	I			
GLpe	58.5	56.7	56.5	56*	54.4	54.3	-	-
Bp	16.7	17.4	17.2	16.6	18	16.2	-	-
SD	13.6	14.4	14.0	13.1	14.4	12.6	-	-
Bd	16.2	15.8	14.8	15	15.9	-	17.2	14.0

	Blé A	NA 93/45	Blé B	Blé E	Blé B	NA 94/7	Blé C	NA 93/45	NA 94/7	NA 93/46	NA 94/7
ph. 2		IIIa				I		II	I	EIA	I
GL	36.0	35	33.1	33.0	32.2	31.4	30.8	30.7	-	-	-
Bp	17.7	18.9	15.4	15.7	16.4	15.6	14.7	16.2	16.9	15.5	-
SD	12.3	13.9	11.7	12.8	13.2	12.6	11.2	11.9	11.8	-	-
Bd	13.4	13.9	12.1	13.5	13.1	13.5	11.2	12.7	-	-	13.4

	NA 94/7	NA 93/45	Blé B	Blé E	NA 93/45	Blé E	Blé A	NA 94/7	NA 94/7	NA 94/7
ph. 3	I	II			I			I	I	I
DLS	50	46.6	44.5	43.5	40.7	40.5	40.2	-	-	-
Ld	42.6	38.6	38.2	36.4	35.4	34.6	35.0	-	-	-
H	24.9	28.3	27.9	28.6	25.1	25.2	-	(26)	25.6	-
Bp	12.3	14.5	-	14.6	13.2	13.3	11.7	13.7	14.5	12.7

*unfused; ()=estimate

Table C.72: Measurements (mm) on kob bones

		processus cornualis	cranium	dentes maxillare	mandibulare	dentes mandibulare	dentes	humerus	radius	ulna	ossa carpi	os metacarpale III et IV	os coxae	os femoris	tibia	talus	calcaneus	os centroquartale	os metatarsale III et IV	phalanx proximalis	phalanx media	phalanx distalis	**Total**
BF 94/45	EIA	-	-	1	-	2	-	1	2	-	-	1	3	-	1	3	2	1	1	6	3	2	**29**
BF 97/13	MIA	-	-	-	-	-	-	1	-	-	-	-	-	-	-	-	-	-	-	-	-	-	**1**
BF 97/13	LIA	2	-	-	-	-	-	-	-	-	-	-	-	-	-	-	-	-	-	-	-	-	**2**
BF 96/22	MIA	-	-	-	-	-	-	-	-	-	-	-	-	-	-	-	1	-	-	3	1	1	**6**
BF 97/31	LIA	-	1	-	-	-	-	-	-	2	-	-	-	-	-	-	1	-	-	-	2	-	**6**
BF 94/120	LIA	-	-	6	-	1	1	-	1	-	1	1	-	-	1	2	-	-	1	5	3	2	**25**
BF 95/7	LIA	2	-	-	-	-	-	-	-	-	-	-	-	-	-	-	-	-	-	-	1	1	**4**
NA 97/26	IA	-	-	-	-	-	-	-	-	-	-	-	-	-	-	-	-	-	-	-	-	1	**1**
NA 99/75	IA	-	-	-	-	-	-	-	-	-	-	-	-	-	-	-	-	-	-	-	-	1	**1**
NA 92/2C	subrec	-	-	-	1	-	-	1	1	-	-	1	-	2	-	-	-	-	-	2	-	1	**9**
NA 94/7	III	-	-	1	4	1	-	-	-	-	-	-	-	-	1	-	1	-	-	-	-	-	**8**
NA 93/45	II	-	-	-	-	-	-	-	-	-	-	-	-	-	-	-	-	-	-	-	-	1	**1**
All sites		**4**	**1**	**8**	**5**	**4**	**1**	**3**	**4**	**2**	**1**	**3**	**3**	**2**	**3**	**5**	**5**	**1**	**2**	**11**	**9**	**10**	**93**

Table C.73: Skeletal element distribution of redfronted gazelle by site and phase

mand.	NA 94/7 III	NA 92/2C subrec
7	(50)	-
8	(35)	-
9	17	-
15a	-	29.6
15b	18.7	-
15c	14.7	-

hum.	NA 92/2C subrec	BF 97/13 MIA	BF 94/45 EIA
Bp	34.5	-	-
Bd	-	28.6	-
BT	-	-	25.7

radius	BF 94/120 LIA	BF 94/45 EIA	NA 92/2C subrec	BF 96/22 MIA
Bp	27.2	26.8	-	-
BFp	25.2	25.5	-	-
Bd	-	-	24.0	18.4

ulna	BF 97/31 LIA	BF 97/31 LIA
DPA	22.8	20.9
BPC	14.4	14.5

os mc	NA 92/2C subrec
SD	12.3

os coxae	BF 94/45 EIA
LA	31.6

tibia	BF 94/120 LIA	NA 94/7 III	BF 94/45 EIA
Bd	25.4	25.2	24.4

centroq.	BF 94/45 EIA
GB	24.3

talus	BF 94/45 EIA	BF 94/120 LIA	BF 94/45 EIA	BF 94/45 EIA	BF 94/120 LIA
GLl	31.0	29.8	29.0	28.4	26.8
GLm	27.9	-	26.7	26.7	25.1
Bd	19.5	-	16.6	17.5	15.5
Dl	17.6	16.9	15.8	15.6	14.9

calc.	BF 94/45 EIA	BF 94/45 EIA	NA 94/7 III	BF 97/31 LIA	BF 96/22 MIA
GL	66.6	65.9	61.9	-	-
GB	21.1	21.0	20.1	18.5	18.4

os mt	BF 94/120 LIA
Bp	20.7

ph. 1	BF 94/120 LIA	BF 94/45 EIA	BF 94/45 EIA	BF 96/22 MIA	NA 92/2C subrec	BF 96/22 MIA	BF 94/120 LIA	BF 94/120 LIA	NA 92/2C subrec	BF 94/45 EIA
GLpe	51.5	44.7*	41.1	(40)	40.3	38.8	38.6	38.3	38.3	-
Bp	10.2	10.1	11.8	-	11.6	11.3	10.2	11.6	11.3	11.2
SD	8.3	8.4	9.0	9.3	9.6	9.4	8.8	9.0	9.0	9.9
Bd	9.9	10.0	9.2	10.5	10.5	9.6	-	10.1	10.2	10.1

ph. 1	BF 94/120 LIA	BF 94/45 EIA	BF 94/45 EIA	BF 94/45 EIA
Bp	9.7	8.9	-	-
SD	-	-	8.9	-
Bd	-	-	9.4	9.6

ph. 2	BF 94/45 EIA	BF 94/120 LIA	BF 96/22 MIA	BF 94/45 EIA	BF 97/31 LIA	BF 94/45 EIA	BF 97/31 LIA	BF 94/120 LIA	BF 94/120 LIA
GL	25.2	23.6	23.5	(23)	22.9	22.9	22.6	22.3	22.0
BP	10.0	9.5	9.4	(9)	9.8	9.6	10.1	9.5	8.9
SD	7.0	7.2	7.3	7.5	8.0	6.2	7.3	6.5	-
Bd	8.0	7.5	7.9	7.9	8.4	7.2	8.0	6.8	-

ph. 3	BF 94/120 LIA	BF 94/45 EIA	NA 92/2C subrec	BF 95/7 LIA	NA 99/75 IA	NA 97/26 IA	BF 94/45 EIA	BF 95/7 LIA	NA 93/45 II	BF 96/22 MIA
DLS	31.0	30.8	30.0	28.9	28.2	27.8	27.4	(26.5)	-	-
Ld	25.8	26.3	26.2	23.7	22.6	22.5	-	(21.5)	-	-
H	21.2	20.3	20.6	21.2	20.7	17.5	-	(18.5)	20.5	19.2
Bp	9.0	8.7	9.0	9.0	8.9	7.9	7.0	(7.5)	8.3	7.4

*unfused; ()=estimate

Table C.74: Measurements (mm) on redfronted gazelle bones

	BF 97/5 LSA	BF 94/45 EIA	BF 97/13 LIA	BF 97/30 LIA	BF 96/22 MIA	BF 97/31 LIA	BF 94/120 LIA	BF 95/7 LIA	NA 90/5BII Gaj IIc	NA 93/36 Gaj IIa	NA 99/65 Gaj IIc	NA 97/26 IA	NA 93/46 EIA	NA 94/7 I	NA 94/7 II	NA 94/7 IV	NA 93/45 I	NA 93/45 II	NA 93/45 IIIa	NA 93/45 IIIb	Blé A	Blé B	Blé C	Blé E	**All sites**
cranium	1	-	-	-	-	-	-	-	-	-	-	-	-	2	-	-	-	-	-	-	-	-	-	-	**3**
mandibulare	-	-	-	-	-	-	-	-	-	-	-	-	1	7	-	-	-	-	-	-	1	-	-	-	**9**
dentes mandibulare	-	-	-	-	-	-	-	-	-	-	-	-	2	1	-	-	-	1	-	-	-	-	-	-	**4**
dentes	1	-	-	-	-	-	-	-	-	-	-	-	-	-	-	-	-	-	-	-	-	-	-	-	**1**
atlas	-	-	-	-	-	-	-	-	-	-	-	-	-	5	-	-	-	-	-	-	-	-	-	-	**5**
axis	-	-	-	-	-	-	-	-	-	-	-	-	-	1	-	-	-	-	-	-	-	-	-	-	**1**
vertebrae thoracales	-	-	-	-	-	-	-	-	-	-	-	-	-	7	1	-	-	-	-	-	-	2	-	1	**11**
vertebrae lumbales	-	-	-	-	-	-	-	-	-	-	-	-	-	3	-	-	-	-	-	-	-	-	1	-	**4**
os sacrum	-	-	-	-	-	-	-	-	-	-	-	-	-	1	-	-	-	-	-	-	-	-	-	1	**2**
vertebrae caudales	-	-	-	-	-	-	-	-	-	-	-	-	-	3	-	-	-	-	-	-	-	-	-	-	**3**
vertebrae	-	-	-	-	-	-	-	-	-	-	-	-	-	-	-	-	-	-	-	-	-	-	-	1	**1**
costae	-	-	-	-	-	-	-	-	-	-	-	-	-	-	-	-	-	1	-	-	-	2	1	4	**8**
scapula	-	-	-	-	-	-	-	-	-	-	-	-	-	3	-	-	-	-	-	-	1	-	-	-	**4**
humerus	-	-	-	-	-	-	-	-	-	-	-	-	-	12	-	-	-	-	-	-	-	-	-	-	**12**
radius	-	-	-	-	-	-	-	-	-	-	-	-	2	4	-	-	-	-	-	-	-	-	-	-	**6**
ulna	-	1	-	-	-	-	1	-	1	-	-	-	-	3	-	-	-	-	-	-	-	-	-	-	**6**
ossa carpi	1	-	-	-	-	1	1	-	-	-	-	-	4	5	1	-	-	2	-	-	-	-	-	-	**15**
os centroquartale	1	-	-	-	-	-	-	-	-	-	-	-	-	-	-	-	-	-	-	-	-	-	-	-	**1**
os metacarpale III et IV	1	-	-	-	-	-	-	-	-	-	-	-	2	-	-	-	-	1	-	-	-	1	-	-	**5**
os coxae	-	-	-	-	-	1	-	-	-	-	-	-	2	7	-	-	-	-	-	-	-	-	2	-	**12**
os femoris	-	-	-	-	-	-	-	-	-	-	-	-	2	2	1	-	-	1	-	-	-	1	1	-	**8**
tibia	-	1	-	-	-	-	-	-	-	-	-	-	1	7	-	-	-	-	-	-	-	1	-	-	**10**
patella	-	-	-	-	-	-	-	-	-	-	-	-	-	1	-	-	-	-	-	-	-	-	-	-	**1**
talus	-	-	-	-	-	-	-	-	-	-	-	-	-	1	-	-	-	-	-	-	-	-	-	-	**1**
calcaneus	-	-	-	-	-	-	-	-	-	1	-	-	-	3	1	-	-	-	-	-	-	-	-	-	**5**
ossa tarsi	-	-	-	-	-	-	-	-	-	-	-	-	1	2	-	-	-	-	-	-	-	-	-	-	**3**
os metatarsale III et IV	-	1	-	-	-	1	-	-	-	-	-	-	2	3	-	-	1	1	-	-	-	-	-	-	**9**
ossa metapodalia	-	3	-	-	-	1	-	1	-	-	-	3	2	4	-	-	-	-	-	1	-	1	2	-	**18**
ossa sesamoidea	3	-	-	-	-	1	-	-	-	-	-	-	-	-	-	-	-	-	1	1	-	-	-	-	**6**
phalanx proximalis	-	1	1	1	-	-	-	-	-	-	1	1	2	3	-	-	1	1	-	-	-	-	-	-	**12**
phalanx media	1	1	-	-	1	1	-	-	-	-	-	-	-	3	-	1	-	1	-	-	-	1	-	-	**10**
phalanx distalis	3	2	-	-	-	-	-	-	-	-	-	-	-	2	-	-	-	-	-	-	-	-	-	-	**7**
Total	**12**	**10**	**1**	**1**	**1**	**6**	**2**	**1**	**1**	**1**	**1**	**4**	**23**	**96**	**4**	**1**	**2**	**9**	**1**	**2**	**2**	**9**	**7**	**7**	**204**
Tragelaphus scriptus	-	-	-	-	-	-	-	-	1	-	-	-	-	-	-	-	-	-	-	-	-	-	-	-	**1**
Redunca redunca	1	5	1	-	-	1	-	-	1	-	3	-	-	-	-	-	1	-	-	-	1	6	4	2	**26**
Kobus kob	3	1	-	-	-	-	1?	-	-	-	-	-	3	55	3	1	7	6	4	5	4	29	7	11	**139**
Gazella rufifrons	-	29	2	-	6	6	25	4	-	-	-	1	-	-	-	8	-	1	-	-	-	-	-	-	**82**

Table C.75: Skeletal element distribution of medium-sized antelopes by site and phase

site		unfused element	fusion age in *Ovis ammon* f. aries
BF 94/45	EIA	tibia prox.	36-42 months
BF 94/45	EIA	os metapodale III et IV dist.	18-28 months
NA 93/46	EIA	os femoris prox.	30-36 months
NA 93/46	EIA	phalanx proximalis	13-16 months

Table C.76: Unfused medium-sized antelope bones

		processus cornualis	cranium	mandibulare	scapula	humerus	radius	ulna	radio-ulna	ossa carpi	os metacarpale III et IV	os coxae	os femoris	talus	calcaneus	os metatarsale III et IV	ossa metapodalia	phalanx proximalis	phalanx media	phalanx distalis	**Total**
BF 94/45	EIA	-	-	-	-	-	-	-	-	-	1	-	-	-	-	4	-	2	-	-	**7**
BF 97/13	MIA	-	-	-	-	-	-	-	-	-	-	-	-	-	-	-	-	-	1	1	**2**
BF 97/13	LIA	1	-	-	-	-	1	-	-	1	-	-	-	-	-	-	1	-	4	1	**9**
BF 97/30	LIA	-	3	-	-	-	3	-	-	-	5	-	2	5	3	8	5	12	14	7	**67**
BF 96/22	MIA	-	-	-	-	-	-	-	-	-	-	-	-	-	-	-	-	-	2	1	**3**
BF 97/31	LIA	-	-	-	-	-	1	-	-	-	-	-	-	1	-	-	-	-	-	-	**2**
BF 94/120	LIA	-	-	-	-	2	-	-	-	2	3	-	-	4	2	3	1	5	4	3	**29**
BF 95/7	LIA	-	-	-	-	-	-	-	-	-	-	-	-	4	-	-	-	-	1	2	**7**
NA 90/5A	Gaj IIb	-	-	-	-	-	-	-	-	-	1	-	-	-	-	-	-	-	-	-	**1**
NA 90/5BI	Gaj IIa/b	-	-	-	-	-	-	1	-	-	-	-	-	-	-	-	-	-	-	-	**1**
NA 99/65	Gaj IIc	-	-	-	-	-	-	-	-	-	1	1	-	-	-	-	-	-	-	-	**2**
NA 97/37	Gaj III	-	-	-	-	-	-	-	-	-	-	-	-	-	-	-	-	1	-	-	**1**
NA 97/33	Gaj I	-	-	-	-	-	-	-	-	-	1	-	-	-	-	1	-	-	-	-	**2**
NA 97/33	Gaj IIa/b	-	-	-	-	-	-	-	-	-	-	-	-	-	-	-	-	1	-	-	**1**
NA 96/45	Gaj IIa/b	-	-	-	-	-	-	-	-	-	1	-	-	-	-	-	-	-	-	-	**1**
NA 99/75	IA	-	-	-	-	-	-	-	1	-	-	-	-	1	1	-	-	-	-	-	**3**
NA 92/2C	subrec	1	1	-	1	1	3	-	-	-	12	-	1	3	-	4	7	12	10	9	**65**
NA 93/46	EIA	-	1	-	-	-	-	-	-	-	1	-	-	-	-	-	1	-	3	-	**6**
NA 94/7	I	-	-	-	-	-	-	-	-	-	-	-	-	1	-	-	-	-	-	-	**1**
NA 93/45	I	-	-	-	-	-	-	-	-	-	-	-	-	-	-	-	-	-	2	1	**3**
NA 93/45	II	-	-	-	-	-	-	-	-	-	-	-	-	-	-	-	1	1	1	-	**3**
NA 93/45	IIIa	-	-	-	-	-	-	-	-	-	-	-	-	1	-	-	-	1	-	-	**2**
NA 93/45	IIIb	-	1	-	1	-	-	-	-	-	1	-	-	-	-	-	-	-	-	-	**3**
NA 93/45	IV	-	-	-	-	-	-	-	-	-	-	-	-	-	-	-	-	1	-	-	**1**
Blé B		-	-	1	-	-	-	-	-	-	1	-	-	-	-	-	-	-	-	-	**2**
Blé C		-	-	-	-	-	-	-	1	-	-	-	-	-	-	-	-	-	-	1	**2**
Blé E		-	-	-	-	-	-	-	-	-	1	-	-	-	-	-	-	-	-	-	**1**
All sites		**2**	**6**	**1**	**2**	**3**	**8**	**1**	**2**	**3**	**29**	**1**	**3**	**20**	**6**	**20**	**16**	**36**	**42**	**26**	**227**

Table C.77: Skeletal element distribution of sheep by site and phase

		processus cornualis	cranium	mandibulare	scapula	humerus	radius	ulna	radio-ulna	ossa carpi	os metacarpale III et IV	os coxae	os femoris	tibia	talus	calcaneus	os metatarsale III et IV	os centroquartale	ossa metapodalia	phalanx proximalis	phalanx media	phalanx distalis	**Total**
BF 94/45	EIA	-	2	-	1	-	-	-	-	1	1	-	-	-	5	1	2	-	1	5	4	-	**23**
BF 97/13	EIA	-	1	-	-	-	-	-	-	-	-	-	-	-	-	-	-	-	-	-	-	-	**1**
BF 97/13	MIA	-	1	-	-	-	1	-	-	3	-	-	-	-	-	2	-	-	1	2	1	-	**11**
BF 97/13	LIA	-	1	-	-	-	1	-	-	1	-	-	-	-	-	-	-	-	1	-	1	4	**9**
BF 97/30	LIA	-	1	-	-	1	3	-	3	-	3	-	-	1	5	4	2	-	7	12	9	1	**52**
BF 96/22	MIA	2	-	-	-	1	1	-	-	1	2	-	-	-	-	-	2	-	-	2	2	2	**15**
BF 97/31	LIA	-	-	-	-	-	-	-	-	5	1	-	-	-	1	1	-	-	-	1	2	-	**11**
BF 94/120	LIA	-	4	-	-	1	3	-	-	7	2	1	-	-	4	2	5	-	14	10	11	10	**74**
BF 95/7	LIA	-	-	-	-	-	-	-	-	-	-	-	-	-	1	-	5	-	-	2	1	2	**11**
NA 90/5C	Gaj IIa	-	-	-	-	-	1	-	-	-	-	-	-	-	-	-	-	-	-	-	-	-	**1**
NA 90/5A	Gaj IIb	2	-	-	1	4	7	3	-	1	5	-	-	2	8	6	4	1	2	13	8	6	**73**
NA 90/5BI	Gaj IIa/b	1	-	-	-	2	-	1	1	-	1	-	1	1	-	-	-	-	1	3	1	-	**13**
NA 90/5BII	Gaj IIc	-	-	-	2	1	1	-	-	-	1	1	-	-	1	-	-	-	-	-	-	1	**8**
NA 93/42	Gaj I	-	-	-	-	-	-	-	-	-	-	-	-	-	1	-	-	-	-	-	-	-	**1**
NA 99/65	Gaj IIb	-	1	-	-	-	-	-	-	-	-	-	-	-	1	1	-	-	-	1	-	-	**4**
NA 99/65	Gaj IIc	-	1	-	-	2	-	-	-	-	2	-	-	-	1	-	-	-	-	-	-	-	**6**
NA 97/37	Gaj III	-	-	-	-	-	-	-	-	-	-	-	-	-	-	-	1	-	-	1	-	-	**2**
NA 97/33	Gaj I	-	-	-	-	-	-	-	-	-	-	-	-	-	1	-	-	-	-	-	-	-	**1**
NA 97/33	Gaj IIa/b	1	-	-	-	-	-	-	-	-	-	-	-	-	-	-	-	-	-	-	-	-	**1**
NA 96/45	Gaj IIa/b	-	-	-	-	-	-	-	-	-	-	-	-	-	-	-	1	-	1	1	-	-	**3**
NA 95/1	Gaj IIa/b	-	1	-	-	-	2	-	-	-	-	-	-	-	2	-	-	-	1	5	-	1	**12**
NA 99/75	IA	-	1	-	-	-	-	-	-	-	-	-	-	-	-	-	-	-	-	-	-	-	**1**
NA 97/13	IA	-	-	-	-	-	-	-	-	-	-	-	-	-	1	-	-	-	-	-	-	-	**1**
NA 92/2C	subrec	-	-	1	1	3	2	-	-	-	3	-	-	-	2	-	5	-	9	8	5	1	**40**
NA 93/46	LSA	-	-	-	-	-	-	-	-	-	-	-	-	-	-	-	-	-	-	-	-	1	**1**
NA 93/46	EIA	-	2	-	-	1	-	-	-	-	-	-	-	-	1	-	-	-	-	4	2	-	**10**
NA 94/7	I	-	1	-	-	-	-	-	-	-	1	-	-	-	-	-	-	-	-	-	2	1	**5**
NA 94/7	II	-	-	-	-	-	-	-	-	-	-	-	-	-	-	-	-	-	-	-	1	-	**1**
NA 94/7	IV	-	-	-	-	-	-	-	-	-	-	-	-	-	-	1	-	-	-	-	-	-	**1**
NA 93/45	I	-	-	-	-	1	2	-	1	-	-	1	-	-	1	2	-	-	-	2	-	1	**11**
NA 93/45	II	-	1	1	-	-	-	-	-	-	3	-	-	-	-	-	-	-	1	1	-	1	**8**
NA 93/45	IIIa	-	-	-	-	-	-	-	-	-	-	-	-	-	1	-	-	-	-	1	-	-	**2**
NA 93/45	IIIb	-	-	-	-	-	-	-	-	-	-	-	-	-	-	-	-	-	-	1	-	-	**1**
NA 93/45	IV	1	-	-	1	-	-	-	-	-	1	-	-	-	-	-	-	-	-	-	-	2	**5**
Blé A		-	-	-	-	-	-	-	-	-	2	-	-	-	-	-	-	-	-	-	-	-	**2**
Blé B		-	1	-	-	-	-	-	-	-	-	-	-	-	-	-	-	-	-	-	-	-	**1**
Blé E		-	-	-	-	-	-	-	-	-	1	-	-	-	-	-	-	-	-	-	-	-	**1**
All sites		**7**	**19**	**2**	**6**	**17**	**24**	**4**	**5**	**19**	**29**	**3**	**1**	**4**	**37**	**20**	**27**	**1**	**39**	**75**	**50**	**34**	**423**

Table C.78: Skeletal element distribution of goat by site and phase

		processus cornualis	cranium	maxillare	dentes maxillare	mandibulare	dentes mandibulare	dentes	atlas	axis	vertebrae cervicales	vertebrae thoracales	os sacrum	costae	scapula	humerus	radius	ulna	radio-ulna	ossa carpi	os metacarpale III et IV	os coxae	os femoris	tibia	patella	malleolus lateralis	os centroquartale	os metatarsale III et IV	talus	calcaneus	os centroquartale	ossa tarsi	ossa metapodalia	ossa sesamoidea	phalanx proximalis	phalanx media	phalanx distalis	unidentified	**Total**
BF 94/45	EIA	-	-	-	9	4	18	3	-	-	-	-	-	-	1	2	5	-	-	4	3	3	-	5	-	2	-	-	-	-	-	-	5	1	12	6	1	-	**84**
BF 94/45	AB EIA	-	-	-	-	5	-	-	-	-	-	-	-	-	-	-	-	-	-	-	-	-	-	-	-	2	-	-	-	-	-	-	-	-	-	-	-	-	**7**
BF 97/13	EIA	1	-	-	-	-	-	-	1	-	-	-	-	-	-	-	1	-	-	1	2	-	-	-	-	-	-	-	-	-	-	-	-	1	-	-	-	-	**6**
BF 97/13	MIA	-	2	-	-	5	3	2	-	-	-	-	-	-	1	1	2	4	-	2	3	1	-	1	2	-	-	3	-	-	-	-	2	-	4	3	-	-	**41**
BF 97/13	LIA	-	-	-	3	2	6	2	-	-	-	-	-	-	1	-	-	-	-	11	-	-	-	-	-	1	1	1	1	-	-	1	4	-	14	5	1	-	**54**
BF 97/13	AB LIA	-	-	-	-	-	1	-	-	-	-	-	-	-	-	-	-	-	-	-	-	-	-	-	-	1	-	-	-	-	-	-	-	1	-	1	-	-	**4**
BF 97/30	LIA	-	9	3	9	18	19	12	1	1	-	-	-	-	3	4	13	9	1	19	12	6	5	11	5	1	-	15	3	4	6	1	7	5	28	9	8	17	**264**[1]
BF 96/22	MIA	1	8	-	16	3	22	2	-	-	-	-	-	-	2	2	3	1	-	4	1	-	-	1	1	-	-	4	1	1	4	-	1	-	21	8	2	-	**108**
BF 97/31	LIA	-	5	-	6	2	12	-	-	-	-	-	-	-	-	1	-	2	-	2	3	1	4	3	-	1	-	3	-	-	-	-	-	-	4	6	1	-	**56**
BF 94/120	LIA	-	11	-	26	20	59	31	-	1	-	-	1	-	2	4	5	5	1	12	15	6	-	7	3	2	-	16	5	4	9	3	6	1	20	23	6	-	**304**
BF 95/7	LIA	-	-	-	18	6	24	-	-	-	-	-	-	-	-	-	7	2	-	3	8	1	-	1	-	-	2	3	-	-	-	-	3	1	6	3	1	-	**89**
NA 90/5C	Gaj IIa	-	1	-	-	-	-	3	-	-	-	-	-	-	-	-	-	-	-	-	-	-	-	-	-	-	-	-	-	-	-	-	-	-	-	-	-	-	**4**
NA 90/5A	Gaj IIb	2	8	26	-	21	13	30	-	-	-	-	-	-	2	6	10	6	-	-	11	6	3	3	1	-	-	4	3	1	-	-	4	-	9	2	1	-	**170**
NA 90/5BI	Gaj IIa/b	-	-	-	1	3	3	-	-	1	1	-	-	-	1	-	-	-	-	-	1	-	-	-	-	-	-	-	-	-	-	-	-	-	-	2	-	-	**13**
NA 90/5BII	Gaj IIc	-	-	1	-	6	4	-	-	-	-	-	-	-	2	1	-	1	-	1	1	1	-	-	1	-	-	4	1	-	-	-	-	-	-	-	-	-	**24**[2]
NA 93/42	Gaj I	-	-	-	-	-	-	-	-	1	-	-	-	-	-	-	-	-	-	-	-	-	-	-	-	-	-	-	-	-	-	-	-	-	-	-	-	-	**1**
NA 99/65	Gaj IIb	-	-	-	3	-	3	-	-	-	-	-	-	-	1	-	-	-	-	1	-	1	1	1	-	-	-	2	-	-	-	-	-	-	1	-	-	-	**14**
NA 99/65	Gaj IIc	-	-	-	1	-	1	-	-	-	-	-	-	-	-	-	1	-	-	-	-	-	-	-	-	-	-	2	-	-	-	-	-	-	-	-	-	-	**5**
NA 93/10	Gaj IIc	-	-	-	-	-	-	-	-	-	-	-	-	-	-	1	-	-	-	-	1	-	-	1	-	-	-	-	-	1	-	-	-	-	-	-	-	-	**4**
NA 97/37	Gaj III	-	-	1	-	2	-	-	-	-	-	-	-	-	-	-	-	3	1	-	2	-	1	2	-	-	-	2	-	-	-	-	-	-	-	-	-	-	**14**
NA 97/33	Gaj I	-	2	1	2	3	4	-	-	-	-	-	-	-	2	5	2	1	-	-	3	3	1	1	-	-	-	3	-	-	-	-	2	-	1	-	-	1	**37**
NA 97/33	Gaj IIa/b	-	1	1	-	1	-	2	-	-	-	-	-	-	2	2	2	-	-	-	2	2	1	1	-	-	-	2	1	1	-	-	-	-	1	-	-	-	**22**
NA 96/45	Gaj IIa/b	-	-	2	-	6	-	3	-	-	-	-	-	-	2	-	-	-	-	-	1	5	-	2	-	-	-	-	1	-	-	-	-	-	-	-	-	-	**22**
NA 95/1	Gaj IIa/b	-	3	1	6	8	3	1	-	-	-	-	-	-	-	4	2	2	2	2	1	4	8	3	-	-	-	2	-	3	2	-	3	-	4	-	-	-	**64**
NA 97/26	IA	-	1	-	6	-	1	1	-	-	-	-	-	-	-	-	-	-	-	-	1	-	-	-	-	-	-	-	-	-	-	-	-	-	-	-	-	-	**10**
NA 99/75	IA	-	-	-	3	1	2	1	-	-	-	-	-	-	-	-	2	-	-	-	-	1	-	-	1	-	-	2	1	-	-	1	-	-	1	-	1	-	**16**
NA 97/13	IA	-	2	2	-	2	7	3	-	1	1	-	-	1	1	1	1	1	-	3	2	-	-	2	-	-	-	2	-	-	-	-	1	-	4	-	-	-	**37**[3]
NA 92/2C	subrec	-	11	1	6	11	1	4	-	2	-	-	-	-	14	3	4	2	-	5	6	16	18	13	-	-	-	5	3	3	9	2	4	-	5	1	3	-	**152**
NA 93/46	LSA	-	1	-	2	-	1	-	-	-	-	-	-	-	-	-	2	1	-	1	1	-	-	-	-	-	-	-	-	-	-	-	-	-	1	1	-	-	**11**
NA 93/46	EIA	-	4	-	17	7	24	-	-	-	-	-	-	-	1	2	7	4	-	5	5	3	5	6	1	-	-	7	-	3	1	2	5	-	9	2	-	-	**120**
NA 94/7	I	-	2	-	6	3	7	1	-	-	-	-	-	-	2	2	-	-	1	1	-	-	1	1	-	-	-	-	-	-	-	-	1	-	1	-	1	-	**30**
NA 94/7	II	-	-	-	1	-	1	-	-	-	-	-	-	-	-	-	1	-	-	-	-	-	-	-	-	-	-	-	-	-	-	-	-	-	1	-	-	-	**4**
NA 94/7	III	-	-	-	1	3	9	-	-	-	-	-	-	-	1	2	2	-	-	-	-	2	2	3	-	-	-	-	-	-	-	-	4	-	-	-	-	-	**29**[4]
NA 94/7	IV	-	-	-	3	1	-	-	-	-	-	-	-	-	-	-	1	-	-	-	-	1	-	-	-	-	-	-	-	-	-	1	-	-	-	-	-	-	**7**
NA 93/45	I	-	1	1	6	3	3	1	-	1	-	1	1	4	1	4	3	2	-	5	1	2	4	6	-	-	-	1	-	1	-	-	-	-	7	2	2	-	**63**[5]
NA 93/45	II	-	1	-	10	5	9	-	-	-	-	-	-	-	2	4	4	-	-	1	4	4	-	2	-	-	-	3	-	-	-	-	2	-	7	2	1	-	**61**
NA 93/45	IIIa	-	1	-	4	1	3	2	-	-	-	-	-	-	-	1	-	3	-	1	3	1	1	-	-	-	-	2	-	-	-	-	-	1	4	-	-	-	**28**
NA 93/45	IIIb	-	-	-	2	1	1	-	-	-	-	-	-	-	-	-	-	-	1	-	1	-	-	2	1	-	-	2	-	-	-	-	-	-	-	-	-	-	**11**
NA 93/45	IV	-	1	-	2	2	2	-	-	-	-	-	-	-	1	1	2	-	-	1	1	3	-	-	-	-	-	1	-	-	-	-	-	-	1	-	-	-	**18**
Blé A		-	-	-	1	1	-	1	-	-	-	-	-	-	-	-	1	-	2	-	1	2	1	2	-	-	-	1	-	1	-	-	1	-	1	1	-	-	**17**
Blé B		-	1	-	1	1	-	-	-	-	-	-	-	-	-	2	2	-	1	-	1	-	-	-	-	-	-	-	-	-	-	-	-	-	-	-	-	-	**9**
Blé C		-	-	-	-	-	-	-	-	-	-	-	-	-	2	-	-	-	-	-	-	2	1	2	-	-	-	1	-	-	-	-	-	-	-	-	-	-	**8**
Blé E		-	1	1	1	3	1	-	-	1	-	-	-	-	2	2	1	-	-	-	1	1	-	6	-	-	-	1	-	-	-	-	-	-	-	-	-	-	**22**
All sites		**4**	**77**	**41**	**172**	**160**	**267**	**105**	**2**	**9**	**2**	**1**	**2**	**5**	**49**	**57**	**86**	**49**	**10**	**85**	**98**	**78**	**57**	**88**	**15**	**10**	**3**	**94**	**20**	**23**	**31**	**11**	**55**	**11**	**167**	**77**	**29**	**18**	**2068**

[1]including 22 bones of min. 1 fœtus, [2]including 2 bones of min. 1 fœtus, [3]including 18 bones of min. 1 juvenile individual, [4]including 16 bones of min. 2 foetuses, [5]including 10 bones of min. 1 juvenile individual

Table C.79: Skeletal element distribution of sheep or goat by site and phase

fusion age (months)		6-8 scap.		10 hum. dist. rad. prox.		13-16 ph. 1 ph. 2			18-28 os mp dist.			18-24 os mc dist.		20-28 os mt dist.		30-36 fem. prox.		36 rad. dist.	
		NF	F	NF	F	NF	fus	F	NF	fus	F	NF	F	NF	F	NF	F	NF	F
BF 94/45	EIA	-	-	-	-	-	-	2	-	-	-	-	1	-	1	-	-	-	-
BF 97/13	MIA	-	-	-	-	-	-	1	-	-	-	-	-	-	-	-	-	-	-
BF 97/13	LIA	-	-	-	1	-	-	4	1	-	-	-	-	-	-	-	-	-	-
BF 97/30	LIA	-	-	-	-	-	3	20	2	1	2	1	1	1	3	-	2	2	1
BF 96/22	MIA	-	-	-	-	-	-	2	-	-	-	-	-	-	-	-	-	-	-
BF 97/31	LIA	-	-	-	1	-	-	-	-	-	-	-	-	-	-	-	-	-	-
BF 94/120	LIA	-	-	-	1	1	-	5	-	-	1	-	-	-	-	-	-	-	-
BF 95/7	LIA	-	-	-	-	-	-	1	-	-	-	-	-	-	-	-	-	-	-
NA 92/2C	subrec	-	1	-	2	4	3	9	4	-	3	-	2	-	1	1	-	1	1
Blé C		-	-	-	-	-	-	-	-	-	-	-	-	-	-	-	-	-	1

NF=not fused, fus=fusing, F=fused

Table C.80: Fusion data of sheep bones by site and phase

fusion age (months)		juv.	10 hum. dist. rad. prox.		13-16 ph. 1 ph. 2			18-24 os mc dist.		18-28 os mp dist.		20-28 os mt dist.		30-36 calc.		36 rad. dist. ulna prox.		36-42 hum. prox. tib. prox.	
			NF	F	NF	fus	F	NF	F	NF	F	NF	F	NF	F	NF	F	NF	F
BF 94/45	EIA	-	-	-	-	-	7	-	1	-	-	-	-	1	-	-	-	-	-
BF 97/13	MIA	-	-	2	1	-	2	-	-	-	1	-	-	2	-	-	-	-	-
BF 97/13	LIA	-	-	-	-	-	1	-	-	-	-	-	-	-	-	-	-	-	-
BF 97/30	LIA	-	-	7	-	-	19	2	-	3	-	-	1	1	2	-	1	-	1
BF 96/22	MIA	-	-	1	-	-	3	1	1	-	-	1	-	-	-	-	-	-	-
BF 97/31	LIA	-	-	-	-	-	3	-	1	-	-	-	-	-	1	-	-	-	-
BF 94/120	LIA	-	-	1	-	1	14	-	2	8	3	2	-	2	-	2	-	-	-
BF 95/7	LIA	-	-	-	-	-	3	-	-	-	-	1	-	-	-	-	-	-	-
NA 90/5A	Gaj IIb	-	-	2	7	-	10	3	-	-	-	-	-	-	-	5	1	2	1
NA 90/5B	Gaj IIa/b	-	-	-	-	-	-	-	-	-	-	-	-	1	3	-	-	-	-
NA 99/65	Gaj IIb	-	-	-	-	-	1	-	-	-	-	-	-	1	-	-	-	-	-
NA 99/65	Gaj IIc	-	-	-	-	-	-	-	1	-	-	-	-	-	-	-	-	-	-
NA 97/37	Gaj III	-	-	-	-	1	-	-	-	-	-	-	-	-	-	-	-	-	-
NA 96/45	Gaj II	-	-	-	-	-	-	-	-	-	-	-	-	-	-	-	-	-	-
NA 95/1	Gaj II	-	-	-	-	1	4	-	-	-	1	-	-	-	-	2	-	-	-
NA 92/2C	rec	2	-	3	5	-	6	-	-	4	-	1	1	-	-	1	-	-	-
NA 93/46	EIA	1	-	-	1	1	4	-	-	-	1	-	-	-	-	-	-	-	-
NA 93/45	I	-	-	1	-	-	2	-	-	-	-	-	-	1	1	2	-	-	-
NA 93/45	II	-	-	-	-	-	1	-	1	-	1	-	-	-	-	-	-	-	-
NA 93/45	IIIa	-	-	-	-	-	1	-	-	-	-	-	-	-	-	-	-	-	-
NA 93/45	IIIb	-	-	-	-	-	1	-	-	-	-	-	-	-	-	-	-	-	-

NF=not fused, fus=fusing, F=fused

Table C.81: Fusion data of goat bones by site and phase

fusion age (months)		foet.	juv.	6-10 os coxae		10 hum. dist. rad. prox.			13-16 ph. 1 ph. 2			18-24 os mc dist. tibia dist.			18-28 os mp dist.			20-28 os mt dist.			30-36 fem. prox. calc.		36 rad. dist. ulna prox.			36-42 hum. prox. fem. dist. tibia prox.		
				NF	F	NF	fus	F	NF	fus	F	NF	fus	F	NF	fus	F	NF	fus	F	NF	F	NF	fus	F	NF	fus	F
BF 94/45	EIA	-	-	-	-	-	-	5	3	-	12	5	-	1	1	-	-	-	-	-	-	-	-	-	1	-	-	1
BF 94/45	AB EIA	1	-	-	-	-	-	-	-	-	-	-	-	-	-	-	-	-	-	-	-	-	-	-	-	-	-	-
BF 97/13	EIA	1	-	-	-	-	-	-	-	-	-	-	-	-	-	-	-	-	-	-	-	-	-	-	-	-	-	-
BF 97/13	MIA	-	-	-	-	-	-	-	2	-	-	1	-	-	2	1	1	-	-	-	-	-	2	-	-	-	-	-
BF 97/13	LIA	-	-	-	-	-	-	-	3	-	8	-	-	-	-	-	-	-	-	-	-	-	-	-	-	-	-	-
BF 97/13	AB LIA	-	-	-	-	-	-	-	1	-	-	-	-	-	-	-	-	-	-	-	-	-	-	-	-	-	-	-
BF 97/30	LIA	22*	2	-	-	-	-	1	3	-	13	3	-	6	4	-	-	-	-	-	2	2	4	-	1	2	-	3
BF 96/22	MIA	-	-	-	-	-	-	-	1	-	19	-	-	-	1	-	-	1	-	-	1	-	4	-	-	-	-	-
BF 97/31	LIA	-	-	-	-	-	-	-	1	-	4	2	-	-	-	-	-	-	1	-	-	-	-	-	-	-	-	-
BF 94/120	LIA	1	-	-	-	-	-	2	7	1	20	3	-	3	4	-	-	-	-	-	1	2	2	-	1	-	-	-
BF 95/7	LIA	-	1	-	-	-	-	-	4	-	2	-	-	-	-	-	-	1	-	-	-	-	2	1	1	-	-	-
NA 90/5A	Gaj IIb	1	-	1	-	-	-	1	-	-	2	2	-	-	-	-	-	1	-	-	-	2	1	-	1	-	-	-
NA 90/5BI	Gaj IIa/b	-	-	-	-	-	-	3	-	4	-	1	-	-	1	-	1	-	-	-	-	1	-	-	-	-	-	-
NA 90/5BII	Gaj IIc	2*	-	-	-	-	-	-	-	-	-	-	-	-	-	-	-	1	-	-	-	-	-	-	1	-	-	-
NA 93/42	Gaj I	-	-	-	-	-	-	-	-	-	-	-	-	-	-	-	-	-	-	-	-	-	-	-	-	-	-	-
NA 99/65	Gaj IIb	1	1	1	-	-	-	-	-	-	-	-	-	-	-	-	-	-	-	-	-	-	-	-	-	1	-	-
NA 99/65	Gaj IIc	2	-	-	-	-	-	-	-	-	-	-	-	-	-	-	-	-	-	-	-	-	-	-	-	-	-	-
NA 97/37	Gaj III	-	-	-	-	-	-	-	-	-	-	2	-	-	-	-	-	-	-	-	-	-	2	-	-	-	-	-
NA 97/33	Gaj I	1	3	-	-	-	-	-	-	-	-	1	-	-	-	-	1	1	-	-	-	-	1	-	-	-	-	-
NA 97/33	Gaj IIa/b	-	-	-	-	-	-	1	1	-	-	1	-	1	-	-	-	-	-	-	1	-	-	-	-	-	-	2
NA 96/45	Gaj IIa/b	-	-	-	-	-	-	-	-	-	-	-	-	1	-	-	-	-	-	-	-	-	-	-	-	-	-	-
NA 95/1	Gaj IIa/b	1	-	-	-	-	-	1	1	-	1	-	-	1	-	-	-	-	-	-	3	2	-	-	1	-	-	-
NA 97/26	IA	1	-	-	-	-	-	-	-	-	-	-	-	-	-	-	-	-	-	-	-	-	-	-	-	-	-	-
NA 97/13	IA	-	18*	-	-	1	-	-	2	-	-	-	-	-	-	-	-	-	-	-	-	-	-	-	-	-	-	-
NA 92/2C	subrec	-	8	-	-	-	-	-	4	-	-	2	1	5	-	-	1	-	-	-	7	2	-	-	-	-	1	-
NA 93/46	LSA	-	-	-	-	-	-	-	1	-	-	-	-	-	-	-	-	-	-	-	-	-	1	-	-	-	-	-
NA 93/46	EIA	1	4	-	-	-	-	1	4	-	1	-	-	1	2	-	-	-	-	-	5	-	5	-	-	1	-	-
NA 94/7	I	-	-	-	-	-	-	-	-	-	-	-	-	1	-	-	-	-	-	-	-	-	-	-	-	-	-	-
NA 94/7	II	-	-	-	-	-	-	-	1	-	-	-	-	-	-	-	-	-	-	-	-	-	-	-	-	-	-	-
NA 94/7	III	16**	-	-	-	-	-	-	-	-	-	-	-	-	-	-	-	-	-	-	-	-	-	-	-	-	-	-
NA 93/45	I	-	10*	-	-	-	-	-	3	-	4	2	-	1	-	-	-	-	-	-	-	-	1	-	-	1	-	-
NA 93/45	II	-	3	-	-	1	1	-	-	-	4	-	-	-	-	-	-	-	-	-	-	-	-	-	-	-	-	-
NA 93/45	IIIa	1	2	-	-	-	-	-	1	-	-	-	-	-	-	-	-	-	-	-	1	-	1	-	-	-	-	-
NA 93/45	IIIb	-	1	-	-	-	-	-	-	-	-	-	-	-	-	-	-	-	-	-	-	-	-	-	-	-	-	-
NA 93/45	IV	2	1	-	-	-	-	-	-	-	-	-	-	-	-	-	-	-	-	-	-	-	-	-	-	-	-	-
Blé A		-	1	-	-	-	-	-	-	1	1	1	-	-	1	-	-	1	-	-	-	-	-	-	-	-	-	-
Blé B		-	-	-	-	-	-	1	-	-	-	-	-	-	-	-	-	-	-	-	-	-	-	-	-	-	-	-
Blé C		-	1	-	-	-	-	-	-	-	-	-	-	-	-	-	-	-	-	-	-	-	-	-	-	1	-	-
Blé E		-	5	-	-	-	-	-	-	-	-	1	-	-	-	-	-	-	-	-	-	-	-	-	-	-	-	-

NF=not fused, fus=fusing, F=fused; *MNI=1, **MNI=2

Table C.82: Fusion data of sheep or goat bones by site and phase

	Blé B		NA 92/2C	NA 93/45	BF 97/13		BF 94/120	NA 92/2C
mand.		**scapula**	subrec	IIIb	LIA	**hum.**	LIA	subrec
7	67.2	SLC	19.5	17.4	-	Bd	30.0	29.8
8	47.4	GLP	31.6	-	24.5	BT	28.7	-
9	19.9	LG	24.8	-	27.1			
L M3/B M3	23.2/9.2	BG	20.2	-	15.2			
12	66							
15b	26.7							
15c	14.9							

	BF 97/31	BF 97/13	NA 99/75	BF 97/30	Blé C	BF 97/30	BF 97/30		NA 90/5BI
rad.-uln.	LIA	LIA	IA	LIA		LIA	LIA	**ulna**	Gaj IIa/b
Bp	31.3	30.3	27.7	-	-	-	-	DPA	21.7
BFp	28.7	28.5	24.5	-	-	-	-		
SD	-	15	-	-	-	-	-		
Bd	-	-	-	26.4	25.2	23.2	22.5*		
BPC	-	-	19.3	-	-	-	-		

	NA 92/2C	Blé B	NA 92/2C	BF 94/120	NA 92/2C	BF 94/120	NA 92/2C	NA 92/2C	NA 92/2C	NA 92/2C	NA 92/2C
os mc	subrec		subrec	LIA	subrec	LIA	subrec	subrec	subrec	subrec	subrec
GL	150	135	-	-	-	-	-	-	-	-	-
Bp	22.6	22.2	26.0	24.4	24.4	23.7	23.4	23.4	23.4	22.6	21.4
SD	-	13.5	-	-	13.2	-	-	-	-	13.0	-
Bd	-	23.7	-	-	-	-	-	-	-	-	-

	Blé E	NA 92/2C	NA 90/5A	BF 94/120	BF 97/30	NA 99/65	NA 92/2C	NA 92/2C	BF 97/30	BF 94/45
os mc		subrec	Gaj IIb	LIA	LIA	Gaj IIc	subrec	subrec	LIA	EIA
Bp	21.0	20.4	20.0	19.6	18.4	17.2	-	-	-	-
SD	-	11.2	-	-	-	-	11.6	-	-	-
Bd	-	-	-	-	-	-	30.9	29.0	24.5	21.6

	NA 92/2C	BF 94/45	NA 92/2C	BF 94/45	BF 97/30	BF 97/30	NA 92/2C	BF 94/45	BF 94/120	BF 94/120	NA 92/2C
os mt	subrec	EIA	subrec	EIA	LIA	LIA	subrec	EIA	LIA	LIA	subrec
GL	156	-	-	-	-	-	-	-	-	-	-
Ll	149	-	-	-	-	-	-	-	-	-	-
Bp	20.0	23.1	21.4	21.3	18.6	18.6	18.6	18.0	17.7	17.4	-
SD	10.2	-	11.0	-	8.8	8.8	-	12.2	-	-	8.9
Bd	29.8	-	-	-	-	-	-	-	-	-	-

	BF 94/45	BF 97/30	BF 97/30		NA 99/65		BF 97/30
os mt	EIA	LIA	LIA	**os coxae**	Gaj IIc	**os fem.**	LIA
Bd	25.0	23.8	20.9	LA	26.3	Bp	31.5
						DC	13.1
						SD	13.6

	BF 94/120	BF 95/7	BF 95/7	BF 94/120	NA 92/2C	BF 95/7	BF 94/120	NA 92/2C	NA 93/45	NA 92/2C	BF 94/120
talus	LIA	LIA	LIA	LIA	subrec	LIA	LIA	subrec	IIIa	subrec	LIA
GLl	29.7	29.6	29.4	29.3	28.1	28.1	27.6	27.4	26.6	26.5	26.1
GLm	27.4	27.6	26.9	27.3	26.1	26.0	26.0	25.4	-	24.4	24.7
Dl	19.0	19.6	19.4	18.4	16.4	17.1	18.9	15.0	14.7	15.4	16.3
Bd	16.8	16.3	-	-	-	14.7	16.1	17.4	-	11.6	13.7

	BF 97/30	BF 97/30	NA 99/75	BF 97/30	BF 97/31	BF 97/30
talus	LIA	LIA	IA	LIA	LIA	LIA
GLl	25.9	24.5	24.0	23.7	22.9	-
GLm	24.5	23.9	23.6	25.5	21.5	24.8
Dl	14.8	19.0	13.9	14.0	14.9	14.8
Bd	16.1	16.8	16.0	15.7	12.1	18.9

*unfused

Table C.83: Measurements (mm) on sheep bones

	NA 92/2C	BF 94/120	BF 97/30	NA 92/2C	NA 92/2C	BF 97/30	BF 97/30	NA 92/2C	NA 92/2C	BF 94/45	NA 92/2C
ph. 1	subrec	LIA	LIA	subrec	subrec	LIA	LIA	subrec	subrec	EIA	subrec
GLpe	47.5	41.3	40.4	39.8	39.6	39.5	38.9	38.6	37.6	36.0	35.6
Bp	13.3	14.9	13.5	11.7	11.4	11.8	11.6	11.6	-	10.0	11.9
SD	10.4	13.7	11.0	8.8	7.9	9.0	6.5	9.6	10.0	7.0	8.6
Bd	12.2	14.0	-	11.1	10.4	11.5	9.8	9.7	10.4	10.1	10.9
	BF 94/45	NA 93/45	NA 97/37	BF 97/30	BF 97/30	BF 97/30	BF 97/30	NA 97/33	BF 94/120	BF 97/30	NA 93/45
ph. 1	EIA	II	Gaj III	LIA	LIA	LIA	LIA	Gaj IIa/b	LIA	LIA	IV
GLpe	35.1	34	33.8	33.8	33.3	32.6	32.5	32.2	30.6	-	-
Bp	9.9	10.6	11.8	10.5	10.1	10.6	10.2	12.0	9.6	9.9	-
SD	7.4	8.7	9.6	7.6	7.5	8.0	7.8	9.9	7.3	-	8.7
Bd	9.5	11.0	10.4	9.5	9.3	10.5	9.9	11.3	8.8	-	10.0
	BF 97/30	BF 97/30	NA 92/2C	BF 94/120	BF 97/30	BF 97/30	BF 94/120	BF 94/120			
ph. 1	LIA	LIA	subrec	LIA	LIA	LIA	LIA	LIA			
SD	7.6	-	-	-	-	-	-	-			
Bd	9.2	12.1	11.8	10.9	10.6	9.3	9.2	9.2			
	BF 97/30	NA 92/2C	BF 97/13	NA 92/2C	BF 97/30	BF 97/30	BF 97/30	NA 92/2C	NA 92/2C	BF 97/30	NA 92/2C
ph. 2	LIA	subrec	LIA	subrec	LIA	LIA	LIA	subrec	subrec	LIA	subrec
GL	29.4	25.1	24.3	24.2	24.0	23.9	23.8	23.4	23.4	23.2	23.0
Bp	12.4	12.1	-	8.2	9.0	13.3	8.9	11.2	10.8	9.5	12.0
SD	8.7	10.0	-	7.2	10.8	9.7	6.9	8.6	8.4	6.8	7.8
Bd	10.2	8.8	-	8.2	10.9	9.8	8.1	9.4	9.2	7.2	8.2
	NA 92/2C	BF 94/120	BF 97/30	BF 97/13	BF 97/13	NA 92/2C	BF 97/13	BF 94/120	NA 93/45	BF 97/30	NA 92/2C
ph. 2	subrec	LIA	LIA	LIA	LIA	subrec	LIA	LIA	I	LIA	subrec
GL	22.9	22.8	22.7	22.5	22.5	22.4	22.4	22.1	22.1	21.9	21.9
Bp	9.9	10.8	11.1	10.9	10.7	13.6	10	12.4	11.3	9.7	10.2
SD	7.7	7.6	7.3	7.4	-	7.7	7.8	9.0	8.7	6.9	7.0
Bd	-	7.6	9.2	7.8	8.4	-	8.8	9.1	9.0	9.7	7.4
	BF 97/30	NA 93/45	BF 94/120	BF 96/22	BF 97/30	BF 97/30	BF 96/22	BF 97/30	BF 97/13	BF 97/30	NA 93/45
ph. 2	LIA	II	LIA	MIA	LIA	LIA	MIA	LIA	MIA	LIA	I
GL	21.7	21.3	20.9	20.9	20.8	20.7	20.6	20.6	20.3	19.5	19.3
Bp	10.2	10.7	11.0	-	9.0	11.0	11.2	10.0	10.3	11.1	-
SD	6.8	8.0	7.8	-	6.6	8.1	8.1	7.3	7.3	8.0	9
Bd	8.7	8.5	8.6	10.0	-	9.5	8.0	7.8	7.4	7.8	8.4
	BF 97/30	BF 95/7	NA 93/46	NA 93/46	BF 97/30	NA 92/2C	NA 92/2C	BF 94/120	NA 92/2C	NA 93/46	
ph. 2	LIA	LIA	EIA	EIA	LIA	subrec	subrec	LIA	subrec	EIA	
GL	19.1	18.8	18.0	17.6	17.6	-	-	-	-	-	
Bp	9.1	9.7	9.6	9.4	10.3	10.6	10.1	-	-	-	
SD	6.3	6.7	6.4	7.2	7.2	-	7.2	12.4	6.4	5.9	
Bd	7.1	7.5	7.7	7.3	7.5	-	-	-	8.2	7.0	
	BF 97/30	Blé C	NA 93/45	NA 92/2C	NA 92/2C	NA 92/2C	NA 92/2C	NA 92/2C	BF 95/7	BF 97/30	BF 97/13
ph. 3	LIA		I	subrec	subrec	subrec	subrec	subrec	LIA	LIA	MIA
DLS	35.2	32.9	32.1	31.8	30.9	30.6	30.4	28.8	28.5	28.4	28.3
Ld	28.8	26.2	26.1	23.4	24.8	25.4	24.6	24.8	23.6	23.5	23.3
H	-	-	17.5	20.8	19.4	-	19.8	18.2	19.2	-	19.8
Bp	-	-	9.5	9.8	9.3	9.2	9.8	9.4	9.2	-	9.6
	NA 92/2C	BF 97/30	BF 97/30	BF 97/30	BF 94/120	BF 94/120	NA 92/2C	BF 96/22	NA 92/2C	BF 94/120	BF 95/7
ph. 3	subrec	LIA	LIA	LIA	LIA	LIA	subrec	MIA	subrec	LIA	LIA
DLS	28.2	25.1	24.7	24.5	23.4	20.6	-	-	-	-	-
Ld	21.5	20.3	20.8	18.9	-	15.6	19.6	-	-	-	-
H	20.4	-	-	-	-	11.6	20.2	20.8	18.4	12.8	-
Bp	9.0	-	-	-	7.1	6.4	8.6	9.2	9.4	-	9.2
	NA 92/2C	BF 97/13									
ph. 3	subrec	LIA									
Bp	9.2	8.5									

Table C.83 cont.

	BF 94/45		NA 92/2C	BF 97/31		BF 94/45	NA 93/45	NA 90/5BII	NA 92/2C
cranium	EIA	**mand.**	subrec	LIA	**scapula**	EIA	IV	Gaj IIc	subrec
32	74.0	7	74.4	-	GLP	30.5	28.1	23.3	-
40	9.4	8	51.4	-	LG	25.0	21.9	17.2	-
41	32.3	9	22.1	-	BG	18.8	16.0	15.8	-
42	23.1	L M3/B M3	22.4/7.4	-	SLC	-	17.9	-	17.0
43	8.5	12	61.0	-					
		14	-	20.0					
		15a	39.6	-					
		15b	22.1	-					
		15c	15.4	-					

	NA 92/2C	NA 90/5BI	NA 90/5BI	NA 92/2C	NA 93/46	NA 99/65	NA 93/45	BF 97/30	BF 96/22	NA 90/5A	NA 90/5A
hum.	subrec	Gaj IIa/b	Gaj IIa/b	subrec	EIA	Gaj IIc	I	LIA	MIA	Gaj IIb	Gaj IIb
SD	9.6	-	-	-	-	-	-	-	-	-	-
Bd	-	33.5	30.5	27.3	26.7	26.3	24.9	23.9	23.3	-	-
BT	25.8	31.9	30.3	-	25.8	27.6	24.2	23.4	23.9	30.0	29.5

	NA 90/5C	BF 97/30	NA 90/5A	NA 92/2C	BF 97/13	NA 93/45	BF 97/30	BF 97/13	BF 96/22	BF 97/30
radius	Gaj IIa	LIA	Gaj IIb	subrec	MIA	I	LIA	LIA	MIA	LIA
Bp	30.1	29.6	28.6	28.2	27.3	26.8	26.7	26.1	24.1	(23)
BFp	28.6	28.7	-	22.2	27.1	26.0	26.4	-	23	-

	BF 94/120	NA 90/5BI	BF 97/30	BF 94/120	NA 95/1	BF 94/120	NA 95/1		NA 93/45	BF 97/13	NA 90/5A
rad.-uln.	LIA	Gaj IIa/b	LIA	LIA	Gaj IIa/b	LIA	Gaj IIa/b	**ulna**	I	MIA	Gaj IIb
Bp	29.4	28.4	-	-	-	-	-	SDO	15.9	-	-
BFp	28.5	26.8	24.0	-	-	-	-	BPC	18.2	20.6	18.6
Bd	-	-	-	29.6*	27.8*	26.5*	25.0*				
DPA	-	-	19.9	-	-	-	-				

	NA 93/45	BF 97/30	NA 90/5A	NA 90/5BII	NA 99/65	NA 93/45	NA 92/2C	NA 97/37	NA 94/7	NA 99/65	BF 94/45
os mc	II	LIA	Gaj IIb	Gaj IIc	Gaj IIc	IIa/b	subrec	Gaj III	I	Gaj IIc	EIA
GL	92	-	-	-	-	-	-	-	-	-	-
Bp	20.0	23.2	20.8	20.3	20.3	19.1	19.0	16.6	16.0	-	-
SD	14.1	-	-	-	-	11.8	10.8	-	-	-	-
Bd	-	-	-	-	-	-	-	-	-	29.7	26.0

	BF 97/31	BF 97/30	BF 96/22	NA 90/5BI	BF 96/22	BF 94/120	BF 94/120	NA 90/5A	BF 97/30	NA 93/45
os mc	LIA	LIA	MIA	Gaj IIa/b	MIA	LIA	LIA	Gaj IIb	LIA	II
Bd	25.1	25.1*	24.1	23.9	23.3*	22.9	22.9	21.7	21.6*	20.9

	NA 90/5BI		BF 97/30	NA 90/5BI		NA 90/5A	BF 97/30	BF 97/30	BF 94/45	BF 94/45	BF 94/120
os fem.	Gaj IIa/b	**tibia**	LIA	Gaj IIa/b	**talus**	Gaj IIb	LIA	LIA	EIA	EIA	LIA
Bp	35.5	Bp	33.1	-	GLl	30.0	29.4	28.3	27.2	26.9	26.8
DC	17.5	Bd	-	32.2	GLm	26.7	27.3	26.4	26.5	26.4	25.6
					Bd	14.8	15.3	17.3	18.1	16.1	17.7
					Dl	18.9	-	15.3	14.1	13.8	15.5

	NA 90/5A	NA 90/5A	BF 97/30	BF 97/30	NA 95/1	NA 92/2C	BF 94/120	NA 99/65	BF 94/120	NA 93/45	NA 90/5A
talus	Gaj IIb	Gaj IIb	LIA	LIA	Gaj IIa/b	subrec	LIA	Gaj IIb	LIA	IIIa	Gaj IIb
GLl	26.8	26.7	26.6	26.5	26.5	26.4	26.0	26.0	25.9	25.8	25.8
GLm	24.8	24.5	26.1	25.6	24.2	24.9	24.5	24.2	22.2	24.6	24.0
Bd	13.7	14.1	16.9	16.7	13.4	15.1	16.2	-	16.0	13.3	13.2
Dl	18.4	16.5	15.3	15.0	16.6	16.4	13.2	18.3	12.0	17	16.7

*unfused

Table C.84: Measurements (mm) on goat bones

	NA 97/13	NA 95/1	BF 94/45	NA 90/5A	NA 97/33	BF 94/45	BF 97/30	NA 99/65	NA 93/46	NA 93/45	BF 94/120
talus	IA	Gaj IIa/b	EIA	Gaj IIb	Gaj I	EIA	LIA	Gaj IIc	EIA	I	LIA
GLl	25.5	25.5	25.2	25.0	25.0	25.0	24.4	24.1	23.6	22	-
GLm	24.8	23.9	23.9	23.5	23.4	22.9	25.6	22.5	22.6	20.9	24.0
Bd	16.6	12.9	16.1	12.9	15.8	15.8	17.9	-	12.6	11.3	17.2
Dl	14.1	14.9	13.4	15.0	12.8	10.5	13.9	12.2	15.0	14.1	13.0
	NA 93/42		BF 97/30	BF 97/31	NA 93/45	BF 97/30	BF 94/120	NA 90/5A	NA 90/5A	NA 90/5A	BF 94/45
talus	Gaj I	**calcaneus**	LIA	LIA	I	LIA	LIA	Gaj IIb	Gaj IIb	Gaj IIb	EIA
GLm	22.8	GL	63.8	59.9	52.7	51.2	50.7	49.5	46.9	46.0	-
		GB	18.5	17.7	19.2	14.2	16.5*	-	17.3	16.1	17.5*
	BF 97/13	BF 94/120	BF 97/30	NA 93/45		NA 90/5A	BF 95/7	NA 92/2C	BF 94/120	NA 92/2C	BF 95/7
calcaneus	MIA	LIA	LIA	I	**os mt**	Gaj IIb	LIA	subrec	LIA	subrec	LIA
GB	17.5	17.0*	(17)	13.2*	Bp	21.5	20.4	20.2*	20.1	19.4	18 5
					SD	-	-	9.6	-	-	-
	BF 95/7	BF 95/7	NA 90/5A	NA 92/2C	BF 94/120	BF 94/120	BF 94/120	NA 90/5A	BF 94/45	BF 96/22	BF 97/30
os mt	LIA	LIA	Gaj IIb	subrec	LIA	LIA	LIA	Gaj IIb	EIA	EIA	LIA
Bp	18.5	18.5	18.5	18.4	18.1	18.0	18.0	17.9	17.0	16.6	16.5
SD	-	-	-	9.0	-	-	-	-	12.2	-	-
	BF 96/22	NA 92/2C	BF 95/7		BF 94/45	BF 94/45	BF 97/30	NA 92/2C	NA 92/2C	NA 92/2C	NA 92/2C
os mt	MIA	subrec	LIA	**ph. 1**	EIA	EIA	LIA	subrec	subrec	subrec	subrec
Bp	18.7*	-	-	GLpe	42.8	41.9	41.6	39.8	38.7*	38.2	38.2
Bd	-	25.8	23.5*	Bp	12.4	14.7	13.3	14.8	12.8	12.8	12.7
				SD	9.9	12.0	10.8	10.1	10.0	9.6	10.2
				Bd	11.1	13.7	12.7	11.1	12.2	12.2	11.8
	BF 95/7	NA 90/5BI	NA 90/5BI	BF 97/30	BF 94/45	BF 97/31	NA 95/1	BF 97/13	BF 97/30	BF 94/120	BF 95/7
ph. 1	LIA	Gaj IIa/b	Gaj IIa/b	LIA	EIA	LIA	Gaj IIa/b	MIA	LIA	LIA	LIA
GLpe	38.2	38.1	37.4	37.3	37.0	37.0	36.8	35.9	35.4	35.1	35.0
Bp	12.5	-	-	11.4	11.5	11.5	13	11.5	12.1	11.3	11.4
SD	9.6	-	-	9.7	9.8	9.0	12.3	9.4	10.0	8.9	9.2
Bd	11.1	-	-	11.7	11.4	-	13.0	11.2	11.0	11.5	12.7
	BF 97/30	NA 93/45	NA 93/45	NA 97/37	NA 95/1	BF 94/120	BF 96/22	NA 99/65	NA 95/1	BF 94/120	NA 95/1
ph. 1	LIA	IIIb	IIIa	Gaj III	Gaj IIa/b	LIA	MIA	Gaj IIb	Gaj IIa/b	LIA	Gaj IIa/b
GLpe	34.3	34.2	34.2	34.2	33.5	33.1	32.8	32.7	32.6	32.6	32.5
Bp	10.9	12.2	10.6	10.4	13.6	9.9	11.2	10.4	12.4	-	11**
SD	9.6	10.2	10.2	9	10.4	8.4	8.6	9.2	10.5	8.1	8.9
Bd	10.3	11.3	11.8	10.4	12.7	9.3	10.3	-	12.0	10.2	10.4
	NA 93/45	NA 93/46	NA 93/45	BF 97/30	NA 96/45	BF 97/30	BF 94/120	NA 93/46	BF 97/30	NA 95/1	NA 93/46
ph. 1	I	EIA	II	LIA	Gaj IIa/b	LIA	LIA	EIA	LIA	Gaj IIa/b	EIA
GLpe	32.2	32.0	31.9	31.9	31.8	31.6	31.5	30.7	30.3	30.0	28.7
Bp	11.2	11.7	10.2	10.0	12.3	9.1	11.0	10.4	10.2	11.3	-
SD	10.6	9.3	9.0	7.9	10.2	8.7	9.3	7.9	7.3	9.8	8.7
Bd	10.8	11.5	10.7	9.6	11.3	9.3	10.4	9.6	8.8	10.5	9.4

*unfused, ** fusing; ()=estimate

Table C.84 cont. 1

	NA 92/2C	BF 97/30	BF 97/30	NA 93/45	BF 97/30	BF 94/45	NA 92/2C	NA 92/2C	BF 97/13	NA 92/2C	BF 94/45
ph. 1	subrec	LIA	LIA	I	LIA	EIA	subrec	subrec	MIA	subrec	EIA
Bp	15.0	12.3	11.3	10.9	10.7	-	-	-	-	-	-
SD	10.2	-	9.3	8.3	-	10.4	9.8	9.6	9.6	8.8	8.2
Bd	12.8	-	-	10.0	-	12.2	10.7	10.8	10.7	10.6	9.0
	BF 94/120	BF 94/120	BF 94/120	BF 97/30	BF 94/120	BF 96/22	BF 97/30				
ph. 1	LIA	LIA	LIA	LIA	LIA	MIA	LIA				
Bd	10.7	10.7	10.1	9.8	9.7	9.0	8.1				
	BF 97/30	BF 94/45	BF 94/45	BF 94/120	NA 92/2C	BF 94/120	BF 94/120	BF 94/45	BF 94/120	BF 95/7	BF 97/30
ph. 2	LIA	EIA	EIA	LIA	subrec	LIA	LIA	EIA	LIA	LIA	LIA
GL	26.0	25.2	23.9	23.5	23.4	23.4	23.4	22.6	22.6	22.4	22.4
Bp	12.5	13.0	12.7	10.7	12.1	11.4	11.1	9.4	11.1	11.9	11.5
SD	8.9	10.0	9.0	7.2	9.8	8.5	12.8	7.5	8.0	8.5	7.9
Bd	9.5	10.2	9.1	7.9	9.8	9.2	8.3	9.1	8.8	9.1	8.6
	NA 92/2C	BF 96/22	BF 97/31	BF 97/31	BF 97/13	BF 94/120	NA 94/7	BF 97/30	BF 94/120	BF 94/120	NA 94/7
ph. 2	subrec	MIA	LIA	LIA	MIA	LIA	I	LIA	LIA	LIA	II
GL	22.3	22.1	22.1	21.9	21.8	21.8	21.7	21.5	21.5	(21)	20.9
Bp	11.5	9.1	11.0	11.6	12.5	10.3	12.0	9.8	10.0	(11.5)	11.8
SD	8.8	6.1	8.4	8.0	8.7	7.2	8.5	6.5	7.1	8.7	7.3
Bd	9.3	7.4	8.9	9.3	8.9	7.5	8.9	8.8	7.6	-	9.0
	NA 94/7	BF 97/13	BF 94/120	BF 97/30	BF 96/22	BF 97/30	NA 93/46	NA 93/46	BF 97/30	BF 97/30	BF 97/30
ph. 2	I	EIA	LIA	LIA	MIA	LIA	EIA	EIA	LIA	LIA	LIA
GL	20.5	20.3	20.2	20.1	20.0	19.6	19.3	18.4	18.3	17.6	-
Bp	14.2	11.3	9.6	10.5	10.5	11.7	10.2	10.0	9.3	9.1	11.8
SD	8.8	7.4	7.1	7.4	7.5	8.0	7.3	7.1	6.2	6.3	8.0
Bd	10.3	8.6	8.2	8.2	8.6	8.8	8.0	6.5	6.5	6.6	-
	NA 92/2C	BF 94/45	BF 94/120	BF 97/30	NA 92/2C	BF 94/120		NA 90/5BII	NA 96/45	BF 94/120	BF 97/13
ph. 2	subrec	EIA	LIA	LIA	subrec	LIA	**ph. 3**	Gaj IIc	Gaj IIa/b	LIA	LIA
Bp	11.3	10.6	10.3	9.3	-	-	DLS	33.7	33.7	31.8	30.0
SD	7.4	-	7.5	-	8.2	-	Ld	28.5	28.5	26.9	-
Bd	7.7	-	-	-	9.8	9.7	H	19.7	19.7	12.5	-
							Bp	8.9	8.9	9.6	-
	NA 93/45	NA 93/45	NA 93/45	NA 95/1	BF 97/13	BF 97/13	NA 92/2C	NA 93/45	BF 96/22	NA 93/46	BF 94/120
ph. 3	IV	IV	I	Gaj IIa/b	LIA	LIA	subrec	II	MIA	LSA	LIA
DLS	28.6	28.2	27.8	27.3	26.6	25.7	24.4	23.7	-	-	-
Ld	8.1	22.4	-	-	-	-	19.4	-	13.8	-	-
H	-	18.4	-	-	-	-	18.8	-	-	18.5	17.9
Bp	-	8.6	-	-	7.6	-	8.2	-	7.2	8.0	7.6
	BF 94/120	BF 94/120	BF 94/120	BF 94/120	BF 94/120	BF 94/120	BF 95/7	BF 94/120	BF 94/120	BF 96/22	BF 95/7
ph. 3	LIA	LIA	LIA	LIA	LIA	LIA	LIA	LIA	LIA	MIA	LIA
H	17.4	17.4	-	-	-	-	-	-	-	-	-
Bp	8.9	8.4	10.4	10.1	10.1	8.5	8.4	7.9	7.8	7.7	6.0

()=estimate

Table C.84 cont. 2

	BF 97/31	NA 92/2C	BF 97/30	BF 94/120	BF 97/30	BF 94/120		BF 97/13			
mand.	LIA	subrec	LIA	LIA	LIA	LIA	**atlas**	EIA			
3	42.5	-	-	-	-	-	BFcr	18.6			
7	-	70.6	-	-	-	-					
8	-	50.8	-	-	-	-					
9	-	22.2	20.8	-	-	-					
13	56.0	-	-	-	-	-					
15a	36.5	34.4	-	-	-	-					
15b	-	18.4	16.9	-	-	-					
15c	-	12.2	12.1	15.2	14.9	11.9					
	NA 92/2C	NA 92/2C	NA 92/2C	BF 97/30	NA 92/2C	BF 97/30	BF 94/120	BF 96/22	NA 93/45		
scapula	subrec	subrec	subrec	LIA	subrec	LIA	LIA	MIA	II		
GLP	36.8	30.8	30.2	29.3	26.8	25.8	-	-	-		
LG	23.2	25.8	23.2	24.9	23.6	21.3	24.2	-	-		
BG	24.8	19.0	21	-	19.6	19.7	18.3	22.7	19.1		
	BF 94/45	BF 94/120	BF 94/45	BF 97/13	BF 97/30	NA 94/7	NA 93/45				
humerus	EIA	LIA	EIA	MIA	LIA	I	I				
Bd	27.0	26.3	24.8	-	-	-	-				
BT	25.5	-	-	25.8	25.3	-	-				
SD	-	-	-	-	-	13.5	10.9				
	BF 95/7	BF 94/45	BF 94/120	BF 94/45	BF 94/45	NA 94/7	BF 94/120	BF 96/22	BF 94/45	BF 96/22	BF 95/7
radius	LIA	EIA	LIA	EIA	EIA	I	LIA	MIA	EIA	MIA	LIA
GL	147	-	-	-	-	-	-	-	-	-	-
Bp	-	33.0	28.1	27.6	27.4	26.5	-	-	-	-	-
BFp	-	30.8	26.4	25.2	25.0	-	-	-	-	-	-
SD	14.3	-		-	-	-		-	-	-	-
Bd	25.0	-	-	-	-	-	28.9*	25.9*	24.0	22.7*	22.6
	NA 92/2C	NA 93/46	BF 94/120	NA 93/45	BF 97/31	BF 97/30	BF 97/30	BF 96/22	NA 93/45	NA 93/45	BF 97/30
ulna	subrec	LSA	LIA	I	LIA	LIA	LIA	MIA	IIIa	IIIa	LIA
LO	31.0	-	-	-	-	-	-	-	-	-	-
SDO	21.0	18.9	17.5	16.8	-	-	-	-	-	-	-
DPA	21.6	20.3	20.0	21.0	26.5	24.3	20.2	18.8	-	-	-
BPC	-	-	16.5	17.2	-	17.2	18.2	14.1	19.0	19.0	17.6
	BF 97/31	NA 97/13	BF 97/30		NA 93/45	BF 94/45	BF 97/31	NA 96/45	BF 95/7	Blé A	BF 95/7
ulna	LIA	IA	LIA	**os mc**	IV	EIA	LIA	Gaj IIa/b	LIA		LIA
BPC	17.4	15.0	14.5	Bp	27.1	25.2	23.6	22.6	21.3	20.0	19.7
	BF 97/31	BF 96/22	Blé A	BF 94/120	NA 97/13		BF 97/30	BF 95/7	NA 92/2C	BF 97/30	BF 97/30
os mc	LIA	MIA		LIA	IA	**os coxae**	LIA	LIA	subrec	LIA	LIA
Bp	19.6	18.7	17.0	-	-	SB	6.2	-	-	-	-
Bd	-	-	-	22.4	15.3	LA	28.5	31.0	28.4	(28)	26.8
	NA 92/2C	NA 93/45	NA 93/45	BF 94/120	Blé A		NA 99/65				
os coxae	subrec	II	II	LIA		**os fem.**	Gaj IIb				
LA	26.2	24	23.5	(22)	20	Bd	32.5				
	BF 96/22	BF 97/30	BF 97/30	BF 97/30	BF 94/120	BF 94/120	BF 97/30	NA 93/45	BF 97/13	BF 97/30	
patella	MIA	LIA	LIA	LIA	LIA	LIA	LIA	IIIb	MIA	LIA	
GL	(29)	(29)	28.3	26.1	25.5	(23.5)	22.7	(24)	(22)	-	
GB	(20)	21.1	19.4	-	21.4	(17.5)	-	(16)	(16)	15.9	

*unfused; ()=estimate

Table C.85: Measurements on sheep or goat bones

	NA 93/45	BF 97/30	BF 97/31	BF 97/30	NA 92/2C	NA 92/2C	BF 94/120	NA 92/2C	NA 96/45	BF 97/30	BF 97/30
tibia	I	LIA	LIA	LIA	subrec	subrec	LIA	subrec	Gaj IIa/b	LIA	LIA
Bp	35.2*	33.9	-	-	-	-	-	-	-	-	-
Bd	-	-	28.8	28.5	26.8	26.2	25.5	25.2*	24.8	24.6	24.2
	BF 97/30	BF 94/45	BF 94/120	NA 92/2C	NA 92/2C	BF 97/30	BF 97/30	BF 94/45	BF 94/45		
tibia	LIA	EIA	LIA	subrec	subrec	LIA	LIA	EIA	EIA		
Bd	24.1	23.6	23.2	23.1	23.0	22.8	21.7	21.4*	21.1*		
	BF 94/120	BF 94/45	BF 97/13	BF 94/120	BF 97/31	BF 94/45					
malleol.	LIA	EIA	EIA	LIA	LIA	AB EIA					
GD	13.6	12.3	12.3	11.8	11.6	11.5					
	NA 92/2C	BF 94/120	BF 94/120	NA 99/75	NA 92/2C	BF 94/120	BF 94/120	BF 94/120	NA 92/2C		
talus	subrec	LIA	LIA	IA	subrec	LIA	LIA	LIA	subrec		
GLl	30.6	26.9	26.7	26.5	26.1	-	-	-	-		
GLm	29.4	-	25.3	24.7	24.8	24.7	22.4	-	-		
BD	16.8	-	18.0	13.7	14.4	-	15.6	17.4	16.8		
Dl	19.8	13.8	14.0	17.0	16.9	11.7	12.8	-	19.4		
	BF 94/120	BF 96/22									
calc.	LIA	MIA									
GL	47.4	-									
GB	14.1	13.0									
	NA 92/2C	NA 92/2C	NA 92/2C	BF 94/120	NA 95/1	BF 97/30	BF 94/120	BF 94/120	NA 92/2C	NA 92/2C	BF 95/7
centroq.	subrec	subrec	subrec	LIA	Gaj IIa/b	LIA	LIA	LIA	subrec	subrec	LIA
GB	24.3	22.8	22.6	22.4	22.1	21.0	20.7	20.6	20.4	20.3	20.1
	BF 94/120	NA 90/5A	BF 97/30	NA 92/2C	BF 96/22	BF 94/120	BF 97/30	BF 97/30	BF 94/120	BF 96/22	BF 97/30
centroq.	LIA	Gaj IIb	LIA	subrec	MIA	LIA	LIA	LIA	LIA	MIA	LIA
GB	20.0	20.0	19.8	19.8	19.7	19.4	19.1	19.1	19.1	18.9	18.8
	NA 92/2C	BF 94/120	BF 95/7	BF 97/13	BF 97/30		NA 99/75	BF 97/30	BF 94/120		
centroq.	subrec	LIA	LIA	EIA	LIA	**os mt**	IA	LIA	LIA		
GB	18.8	18.3	18.3	17.9	17.9	Bp	18.1	16.9	-		
						SD	-	-	8.2		
	BF 94/120	BF 94/120	NA 90/5A	BF 97/30	NA 93/45	BF 96/22	NA 90/5A	BF 95/7	Blé A	BF 96/22	BF 96/22
ph. 1	LIA	LIA	Gaj IIb	LIA	I	MIA	Gaj IIb	LIA		MIA	MIA
GLpe	41.6	38.1	38.0	37.9	36.0	35.2	34.6	34.4	34.3	34.3	33.7
Bp	-	13.4	15.8	12.2	11.6	10.9	12.5	10.5	12.5	-	11.0
SD	-	10.4	12.1	10.5	9.0	9.1	9.9	8.6	10.7	7.2	8.1
Bd	-	12.6	15.0	12.4	10.0	9.9	11.4	9.7	11.8	8.5	9.2
	NA 93/45	BF 94/45	BF 96/22	NA 90/5A	BF 97/30	BF 96/22	NA 90/5A	BF 97/30	BF 94/45	BF 94/120	BF 97/30
ph. 1	IV	EIA	MIA	Gaj IIb	LIA	MIA	Gaj IIb	LIA	EIA	LIA	LIA
GLpe	(33)	32.2	32.1	32.0	31.5	31.1	-	-	-	-	-
Bp	(13)	10.0	10.2	11.3	10.6	9.4	12.8	12.5	12.5	12.4*	12.0
SD	(10)	7.0	7.5	9.8	8.0	7.3	-	-	-	-	-
Bd	(12)	9.4	8.8	11.0	9.8	-	-	-	-	-	-
	BF 97/13	BF 96/22	BF 94/45	BF 94/120	BF 96/22	BF 94/120	NA 93/45	BF 97/30	NA 95/1	BF 96/22	NA 93/45
ph. 1	LIA	MIA	EIA	LIA	MIA	LIA	II	LIA	Gaj IIa/b	MIA	I
Bp	12.0	11.8	11.6*	11.3*	11.2	11.2	11.2	11.1	11.1	10.9*	10.9*

*unfused; ()=estimate

Table C.85 cont. 1

	BF 94/45	BF 94/45	BF 96/22	BF 94/45	BF 96/22	BF 94/120	BF 94/120	BF 96/22	BF 94/45	BF 96/22	BF 94/45
ph. 1	EIA	EIA	MIA	EIA	MIA	LIA	LIA	MIA	EIA	MIA	EIA
Bp	10.8	10.8	10.7	10.6	10.4	10.4	10.3	10.1	10.0	(10)	9.7*
	BF 97/31	BF 95/7	BF 94/45	BF 97/30	NA 92/2C	BF 94/120	BF 97/30	NA 92/2C	BF 97/30	NA 92/2C	BF 97/13
ph. 1	LIA	LIA	EIA	LIA	subrec	LIA	LIA	subrec	LIA	subrec	LIA
Bp	9.7*	9.7*	9.6*	9.1	-	-	-	-	-	-	-
SD	-	-	-	-	11.2	10.9	9.6	9.1	8.0	6.3	-
Bd	-	-	-	-	12.8	12.2	-	11.0	-	8.4	12.7
	NA 90/5A	BF 94/120	BF 97/13	BF 97/30	NA 90/5A	BF 97/30	BF 97/30	NA 93/45	NA 93/45	BF 97/30	BF 94/45
ph. 1	Gaj IIb	LIA	MIA	LIA	Gaj IIb	LIA	LIA	IIIa	I	LIA	EIA
Bd	12.6	12.3	11.2	10.8	10.8	10.8	10.7	10.7	10.5	10.5	10.5
	BF 94/45	BF 97/13	BF 97/13	BF 97/13	BF 97/30	BF 94/120	BF 95/7	BF 97/13	BF 96/22	BF 97/31	BF 96/22
ph. 1	EIA	LIA	LIA	MIA	LIA	LIA	LIA	LIA	MIA	LIA	MIA
Bd	10.5	10.3	10.2	10.2	10.0	9.8	9.8	9.7	9.7	9.6	9.5
	NA 99/75	NA 99/65	BF 94/45	BF 96/22	BF 96/22	NA 97/13	BF 96/22	BF 97/13	BF 96/22	BF 95/7	
ph. 1	IA	Gaj IIb	EIA	MIA	MIA	IA	MIA	LIA	MIA	LIA	
Bd	9.3	9.2	9.2	9.2	9.0	8.8	8.8	8.4	8.3	8.0	
	BF 94/120	BF 94/120	BF 94/120	BF 97/13	BF 94/120	BF 97/30	NA 90/5A	BF 96/22	BF 94/120	BF 96/22	BF 94/120
ph. 2	LIA	LIA	LIA	LIA	LIA	LIA	Gaj IIb	MIA	LIA	MIA	LIA
GL	28.3	24.2	23.4	22.8	22.8	22.3	22.1	22.0	(22)	21.9	21.6
Bp	-	10.0	11.1	12.5	10.7	10.0	11.3	9.3	-	11.1	9.1
SD	9.4	7.6	7.9	9.2	7.9	6.9	7.9	6.6	8.4	8.2	6.7
Bd	10.3	-	-	-	8.4	7.1	8.5	7.4	9.0	8.3	7.2
	BF 96/22	BF 96/22	BF 94/120	NA 90/5A	NA 90/5A	BF 96/22	BF 94/45	BF 94/120	BF 97/30	BF 96/22	BF 94/45
ph. 2	MIA	MIA	LIA	Gaj IIb	Gaj IIb	MIA	EIA	LIA	LIA	MIA	EIA
GL	21.3	21.2	21.2	21.0	21.0	20.9	20.7	20.4	20.4	20.4	20.3
Bp	12.1	8.9	10.8	11.8	10.8	8.9	-	9.9	8.8	10.3	11.2
SD	9.2	6.6	7.7	8.7	8.1	6.6	9.0	6.7	6.1	7.0	6.9
Bd	9.7	7.5	8.5	8.8	7.5	7.5	-	-	6.7	7.0	7.1
	BF 97/31	BF 97/31	NA 93/45	Ble A	BF 97/13	BF 94/120	BF 95/7	BF 94/120	BF 94/120	BF 97/31	NA 93/45
ph. 2	LIA	LIA	I		LIA	LIA	LIA	LIA	LIA	LIA	I
GL	19.0	18.9	18.5	16.2	-	-	-	-	-	-	-
Bp	11.7	9.6	10.8	9.4	13.8	12.7	12.5*	12.1	11.3*	11.3	11.3
SD	9.2	6.0	8.5	6.9	9.2	-	-	-	-	-	-
Bd	9.5	-	8.8	7.6	-	-	-	-	-	-	-
	BF 94/120	BF 94/120	BF 94/45	BF 96/22	BF 96/22	BF 94/120	BF 94/120	BF 94/120	BF 97/31	BF 97/30	BF 94/120
ph. 2	LIA	LIA	EIA	MIA	MIA	LIA	LIA	LIA	LIA	LIA	LIA
Bp	10.9*	10.9	10.6	10.5	10.1	9.7	9.2*	9.1	8.5	7.9	-
SD	-	-	8.1	-	-	-	-	-	-	-	-
Bd	-	-	8.8	-	-	-	-	-	-	-	10.0
	BF 94/120	BF 94/45	BF 97/31	BF 94/45		NA 90/5A	NA 90/5A	NA 92/2C	BF 94/120	BF 94/120	BF 94/45
ph. 2	LIA	EIA	LIA	EIA	**ph. 3**	Gaj IIb	Gaj IIb	subrec	LIA	LIA	EIA
Bd	9.1	8.8	7.9	7.6	DLS	32.5	26.4	-	-	-	-
					Ld	26.8	22.4	18.4	-	-	-
					Bp	-	-	9.8	9.7	8.0	7.9

*unfused; ()=estimate

Table C.85 cont. 2

site			element	cut	chop
BF 94/45	EIA	O/C	tibia prox.		x
BF 94/45	EIA	O/C	phalanx proximalis	x	
BF 97/13	MIA	C	phalanx proximalis	x	
BF 97/13	MIA	O/C	phalanx media	x	
BF 97/13	LIA	C	phalanx media	x	
BF 97/13	LIA	O/C	phalanx media	x	
BF 97/30	LIA	O/C	os femoris diaph.		x
BF 97/30	LIA	O/C	os metatarsale III et IV diaph.	x	
BF 97/30	LIA	O	phalanx proximalis	x	
BF 97/30	LIA	C	phalanx proximalis	x	
BF 97/30	LIA	O	2 phalanges media	x	
BF 97/30	LIA	C	2 phalanges media	x	
BF 97/30	LIA	O/C	phalanx media	x	
BF 97/31	LIA	O/C	phalanx media	x	
BF 94/120	LIA	O	humerus dist.	x	
BF 94/120	LIA	C	os carpi	x	
BF 94/120	LIA	O/C	os metatarsale III et IV diaph.	x	
BF 94/120	LIA	C	talus	x	
BF 94/120	LIA	C	phalanx proximalis	x	
BF 94/120	LIA	O/C	phalanx proximalis	x	
BF 94/120	LIA	O/C	3 phalanx media	x	
BF 95/7	LIA	O/C	radius diaph.	x	
NA 97/37	Gaj III	O/C	ulna prox.	x	
NA 95/1	Gaj IIa/b	O/C	radio-ulna		?
NA 99/75	IA	O	talus	x	x
NA 92/2C	subrec	O/C	maxillare	x	
NA 92/2C	subrec	C	humerus dist.		x
NA 92/2C	subrec	O/C	axis	x	
NA 92/2C	subrec	O	os metacarpale III et IV	x	
NA 92/2C	subrec	C	os metacarpale III et IV		x
NA 92/2C	subrec	O/C	os femoris dist		x
NA 92/2C	subrec	O	tibia diaph.		x
NA 92/2C	subrec	O	phalanx proximalis	x	
NA 93/46	EIA	O/C	phalanx proximalis	x	
NA 93/45	I	O/C	phalanx media		x
NA 93/45	I	O/C	os carpi	x	
NA 93/45	I	C	phalanx proximalis	x	
NA 93/45	I	O	phalanx media	x	
NA 93/45	II	C	os metacarpale III et IV	x	
NA 93/45	II	O/C	radius diaph.		x
NA 93/45	II	O	phalanx media	x	
NA 93/45	IIIa	O/C	phalanx proximalis	x	
NA 93/45	IIIb	C	phalanx proximalis		
NA 93/45	IV	O/C	phalanx proximalis		
Blé A		C	os metacarpale III et IV prox.		x
Blé A		O/C	os coxae		x
Blé B		O	os metacarpale III et IV	x	
Blé B		O/C	radio-ulna dist.		x
Blé C		O/C	scapula		x
Blé C		O/C	scapula	x	
Blé C		O	radio-ulna		?
Blé E		O/C	scapula		x
Blé E		C	os metacarpale III et IV prox.	x	x
Blé E		O	os metacarpale III et IV prox.	x	x

O=*Ovis ammon* f. aries, C=*Capra aegagrus* f. hircus

Table C.86: Traces of butchery on ovicaprine bones

Site	Phase	processus cornualis	cranium	maxillare	dentes maxillare	mandibulare	dentes mandibulare	maxillare/mandibulare	dentes	hyoideus	atlas	axis	vertebrae cervicales	vertebrae thoracales	vertebrae lumbales	os sacrum	vertebrae caudales	vertebrae	costae	sternum	scapula	humerus	radius	ulna	radio-ulna	ossa carpi	os metacarpale III et IV	os coxae	os femoris	patella	tibia	malleolus lateralis	calcaneus	talus	ossa tarsi	os centroquartale	os metatarsale III et IV	ossa metapodalia	ossa sesamoidea	phalanx proximalis	phalanx media	phalanx distalis	phalanx	**Total**	Wild	Domestic
BF 94/133	LSA	-	-	-	-	-	-	-	-	-	-	-	-	-	-	-	-	-	-	-	1	-	-	-	-	-	-	-	-	-	-	-	-	-	-	-	-	-	-	-	-	-	-	**1**	0	0
BF 97/5	LSA	-	-	-	-	2	1	-	8	-	-	1	2	-	-	-	-	-	5	-	-	-	-	-	-	-	-	-	-	-	-	-	-	-	1	-	1	1	-	-	1	-	-	**23**	17	0
BF 97/5	EIA	-	1	-	-	-	1	-	-	-	-	-	1	-	-	-	-	-	-	-	-	1	-	-	-	-	1	-	-	-	-	-	-	-	-	-	-	-	-	-	-	-	-	**5**	0	0
BF 94/45	LSA	-	-	-	1	-	-	-	-	-	-	-	-	-	-	-	-	-	-	-	-	-	-	-	-	-	-	-	-	-	1	-	-	-	-	-	-	-	-	-	-	-	-	**2**	0	0
BF 94/45	EIA	-	2	-	-	7	1	-	8	1	-	1	2	2	1	-	-	-	8	-	8	8	9	1	2	9	5	11	1	-	5	-	-	3	1	1	8	16	2	7	5	-	-	**135**	63	114
BF 94/45	AB EIA	-	-	-	-	-	-	1	35	-	-	-	-	-	-	-	-	-	-	-	-	1	-	-	-	-	-	-	-	-	1	-	-	-	-	-	-	-	-	-	-	-	-	**38**	0	7
BF 97/13	EIA	-	1	-	-	-	-	-	10	-	-	-	-	-	-	-	-	-	1	-	-	-	2	1	-	-	-	-	1	-	-	-	2	-	-	-	-	2	-	-	-	-	-	**20**	0	8
BF 97/13	AB EIA	-	-	-	-	-	-	-	-	-	-	-	-	-	-	-	-	-	-	-	-	-	-	-	-	-	-	-	-	-	-	-	-	-	-	-	-	-	1	-	-	-	-	**1**	0	0
BF 97/13	MIA	1	4	-	-	2	1	-	33	2	-	-	-	1	-	-	-	-	8	2	6	3	3	-	-	3	3	4	1	-	7	-	2	4	6	1	7	3	1	2	1	-	-	**111**	3	54
BF 97/13	LIA	-	3	-	-	5	2	-	51	1	-	-	-	1	1	-	-	-	5	-	3	-	11	1	-	7	3	1	4	1	3	-	7	1	1	2	4	12	4	14	2	-	-	**150**	5	72
BF 97/13	AB LIA	-	-	-	-	-	-	-	16	-	-	-	-	-	-	-	-	-	1	-	-	-	-	-	-	2	-	-	-	-	-	-	-	-	-	2	-	-	1	-	1	-	-	**23**	1	4
BF 97/30	LIA	-	1	-	1	10	-	-	49	1	-	1	5	14	9	-	2	3	52	1	12	9	17	2	2	-	3	17	34	-	22	-	4	6	-	2	8	2	1	6	1	-	-	**297**	2	383
BF 96/22	MIA	1	1	-	2	7	3	-	63	1	-	-	2	3	-	-	-	2	11	-	6	3	10	3	2	2	1	4	2	1	4	1	4	2	1	-	6	4	-	4	1	1	-	**158**	9	127
BF 97/31	LIA	-	2	-	1	3	1	-	77	1	-	-	-	1	1	-	-	-	15	-	7	4	6	1	1	5	1	8	11	2	13	-	4	-	1	-	5	-	-	6	2	1	-	**180**	15	69
BF 94/120	LIA	-	3	-	1	14	3	-	227	-	-	1	2	3	8	-	1	1	45	-	11	10	21	8	2	10	17	8	7	1	12	-	11	10	4	1	15	15	9	19	10	9	-	**519**	40	407
BF 95/7	LIA	-	3	1	2	13	3	-	83	1	-	1	-	5	-	-	1	1	46	-	5	6	14	3	1	6	1	4	13	2	24	-	5	2	-	2	16	12	4	4	-	-	-	**284**	15	107
NA 90/5C	Gaj IIa	-	-	-	-	-	-	-	-	-	-	-	-	-	-	1	-	-	-	-	-	-	-	-	-	1	-	-	-	-	-	-	-	-	-	-	-	-	-	-	-	-	-	**2**	3	5
NA 90/5A	Gaj IIb	-	24	-	-	16	1	-	11	-	-	-	-	-	-	-	-	37	18	1	8	25	12	2	-	19	11	18	16	1	21	1	6	4	6	-	2	17	-	9	2	1	-	**289**	37	320
NA 90/5BI	Gaj IIa/b	-	4	-	1	4	1	-	-	-	-	-	-	-	-	-	-	3	-	-	3	9	5	1	-	5	1	4	1	1	-	-	-	1	1	-	-	3	-	1	2	-	-	**51**	13	27
NA 90/5BII	Gaj IIc	-	1	1	-	-	-	-	2	-	1	1	5	4	2	-	-	2	14	-	4	9	7	1	-	-	1	5	3	-	2	-	-	-	-	-	-	5	-	1	-	-	-	**71**	14	34
NA 93/42	Gaj I	-	-	-	-	-	-	-	1	-	-	-	-	-	-	-	-	2	-	-	-	-	-	-	-	-	-	1	2	-	-	-	-	-	-	-	-	-	-	-	-	-	-	**6**	0	2
NA 93/36	Gaj IIa	-	-	-	-	-	-	-	2	-	-	-	-	-	-	-	-	-	-	-	-	-	-	-	-	-	-	-	-	-	1	-	-	-	-	-	-	-	-	-	-	-	-	**3**	1	0
NA 99/65	Gaj IIb	-	-	-	-	1	-	-	15	-	-	-	-	2	-	-	-	-	2	-	3	2	1	-	-	-	-	1	-	-	-	-	-	-	-	-	1	-	-	-	-	-	-	**28**	2	18
NA 99/65	Gaj IIc	-	1	-	1	1	1	-	9	-	-	-	-	3	1	-	-	-	1	-	1	-	-	-	1	-	-	-	-	-	1	-	-	-	-	-	-	1	-	-	-	-	-	**22**	5	13
NA 97/37	Gaj III	-	1	-	-	-	-	-	1	-	-	1	-	1	-	-	-	-	3	-	1	-	-	-	-	-	-	2	2	-	1	-	1	-	-	-	-	-	-	-	-	-	-	**14**	0	17
NA 97/33	Gaj I	-	-	-	-	-	-	-	2	-	-	-	2	2	1	-	-	-	27	-	3	2	1	-	-	-	-	-	-	-	-	-	-	-	-	-	-	-	-	-	-	-	-	**40**	1	40
NA 97/33	Gaj IIa/b	-	-	-	-	-	-	-	2	-	-	1	3	1	1	-	-	1	14	-	-	-	-	1	-	-	2	-	-	-	1	-	-	-	-	-	-	-	-	-	-	-	-	**27**	11	24
NA 96/45	Gaj IIa/b	-	2	-	-	2	-	-	1	-	-	-	4	6	3	-	-	10	21	-	4	12	11	2	-	-	3	6	8	-	6	-	4	-	-	-	-	4	-	-	-	-	-	**109**	2	26
NA 95/1	Gaj IIa/b	-	3	-	-	1	-	-	13	-	-	1	3	-	3	-	-	2	17	-	-	1	1	-	1	-	-	1	3	-	3	-	-	-	-	-	-	-	-	-	-	-	-	**53**	30	76
NA 97/26	IA	-	-	-	-	-	-	-	21	-	-	-	-	-	-	-	-	-	1	-	-	1	4	-	1	-	2	2	3	-	2	-	4	-	-	-	3	-	-	1	-	-	-	**45**	5	10
NA 99/75	IA	-	-	-	-	1	-	-	9	-	-	-	-	-	-	-	-	-	7	-	2	1	-	-	-	-	-	-	-	-	1	-	-	-	-	-	-	1	-	-	-	-	-	**22**	5	20
NA 97/13	IA	-	1	-	-	-	-	-	9	-	-	-	1	1	5	-	-	-	1	-	-	2	-	-	-	-	-	-	-	1	1	-	1	-	-	-	-	-	-	1	-	-	-	**24**	1	38
NA 92/2C	subrec	-	11	-	2	6	-	-	16	3	2	3	11	26	20	7	2	19	126	1	8	16	14	4	-	-	4	10	19	1	25	1	2	1	2	3	2	4	3	1	2	-	2	**387***	10	257
NA 93/46	LSA	-	-	-	-	-	-	-	-	1	-	-	-	-	-	-	-	-	3	-	-	-	-	-	-	-	-	-	-	-	-	-	-	-	-	-	-	-	-	-	-	-	-	**4**	0	12
NA 93/46	EIA	-	3	-	-	4	1	-	43	1	2	-	1	-	-	-	-	4	23	-	5	5	-	1	-	5	-	1	8	-	5	-	5	-	1	-	-	1	4	2	1	3	-	**129**	31	136
NA 94/7	I	-	-	-	-	7	-	-	-	-	1	4	3	2	2	-	3	4	2	-	1	4	4	-	-	-	-	13	-	-	-	-	1	-	-	-	-	3	-	1	-	-	-	**55**	151	36
NA 94/7	II	-	-	-	-	-	-	-	-	-	1	-	-	1	-	-	2	-	2	-	-	-	-	-	-	-	-	-	-	-	-	-	-	-	-	-	1	1	-	-	-	-	-	**8**	11	5
NA 94/7	III	-	-	1	-	4	-	-	-	-	-	-	-	5	-	-	-	-	-	-	-	-	1	-	-	-	-	1	1	-	-	-	-	-	-	-	-	-	-	-	-	-	-	**13**	13	29
NA 94/7	IV	-	-	-	-	-	-	-	-	-	-	-	-	-	-	-	-	-	1	-	2	1	-	-	-	-	-	-	1	-	-	-	-	-	-	-	1	-	-	-	-	-	-	**6**	2	8
NA 93/45	I	-	1	-	-	-	-	-	1	1	-	-	2	3	2	-	1	-	16	-	1	-	-	-	-	2	1	2	2	-	4	-	1	1	-	-	1	2	-	1	-	-	-	**45**	10	77
NA 93/45	II	-	5	-	-	2	-	-	2	1	-	-	1	2	1	-	-	-	14	-	-	1	4	1	-	1	-	-	2	-	5	-	-	1	1	-	3	1	-	1	2	-	-	**51**	16	72
NA 93/45	IIIa	-	1	-	-	6	-	-	5	-	-	-	-	1	-	-	-	-	15	-	-	-	5	1	-	-	-	1	-	-	1	-	1	1	-	-	-	-	-	-	-	-	-	**38**	5	32
NA 93/45	IIIb	-	-	-	-	5	1	-	3	-	-	-	-	1	1	-	-	-	5	-	1	-	-	-	-	-	-	-	2	-	2	-	-	-	-	-	1	-	-	-	-	1	-	**23**	7	15
NA 93/45	IV	-	-	-	1	1	-	-	8	1	-	-	-	3	-	-	-	-	14	-	-	-	1	1	-	-	-	-	2	-	3	-	-	-	-	-	-	1	-	3	-	-	-	**39**	1	24
Blé A		-	-	-	-	-	-	-	-	1	-	-	-	-	-	-	-	-	-	-	2	1	-	-	-	-	-	-	-	-	1	-	-	-	-	-	1	-	-	-	-	-	-	**6**	7	19
Blé B		-	-	-	-	-	-	-	-	-	-	-	-	2	-	-	-	-	5	-	1	1	-	-	-	-	-	-	-	-	-	-	-	-	-	-	-	-	-	-	-	-	-	**9**	45	12
Blé C		-	-	-	-	-	-	-	2	-	-	-	-	-	-	-	-	3	8	-	-	-	-	-	-	-	-	1	2	-	1	-	-	-	-	-	-	-	-	-	-	-	-	**17**	18	10
Blé E		-	-	-	-	1	-	-	1	-	-	-	1	1	-	-	-	-	6	-	3	-	-	-	-	-	-	-	-	-	1	-	-	-	-	-	-	-	-	-	-	-	-	**14**	20	24
All sites		**2**	**79**	**3**	**13**	**125**	**21**	**1**	**839**	**17**	**7**	**16**	**51**	**97**	**62**	**8**	**12**	**94**	**563**	**5**	**112**	**138**	**164**	**35**	**13**	**77**	**60**	**126**	**151**	**11**	**180**	**3**	**65**	**37**	**26**	**14**	**86**	**111**	**30**	**84**	**33**	**16**	**2**	**3589**	647	2790

*including 9 bones of min. 1 juvenile individual

Table C.87: Skeletal element distribution of small bovids by site and phase

fusion age (months)		juv.	6-10 os coxae		10 hum. dist. rad. prox.		13-16 ph. 1 ph. 2		18-24 os mc dist.		18-28 os mp dist.		18-24 tibia dist.		30-36 fem. prox. calc.		36 rad. dist. ulna prox.		36-42 hum. prox. fem. dist. tibia prox.	
			NF	F	NF	F	NF	F	NF	F	NF	F	NF	F	NF	F	NF	F	NF	F
BF 97/5	EIA	1	-	-	-	-	-	-	-	-	-	-	-	-	-	-	-	-	-	-
BF 94/45	EIA	2	-	-	-	4	2	1	-	1	8	2	1	1	-	-	-	-	-	2
BF 94/45	AB EIA	1	-	-	-	-	-	-	-	-	-	-	-	-	-	-	-	-	-	-
BF 97/13	EIA	-	-	-	-	-	-	-	-	-	-	-	-	-	1	-	-	-	-	-
BF 97/13	MIA	-	-	-	-	-	1	-	-	-	1	-	-	-	-	-	-	-	1	-
BF 97/13	LIA	-	-	-	-	-	1	4	-	-	2	-	-	-	-	2	1	-	-	1
BF 97/13	AB LIA	-	-	-	-	-	1	-	-	-	-	-	-	-	-	-	-	-	-	-
BF 97/30	LIA	3	-	-	-	3	2	-	-	-	-	-	1	1	9	2	2	-	6	1
BF 96/22	MIA	-	-	-	-	3	-	1	1	-	2	-	-	-	2	-	-	-	1	-
BF 97/31	LIA	1	-	-	-	4	2	1	-	-	-	-	-	-	4	2	-	-	1	-
BF 94/120	LIA	1	-	-	-	1	9	8	-	-	3	2	2	3	8	-	2	-	1	-
BF 95/7	LIA	-	-	-	-	-	-	1	-	-	-	-	-	3	5	-	2	-	-	-
NA 90/5BII	Gaj IIc	-	-	-	1	1	4	1	-	-	-	1	1	-	-	-	-	2	-	1
NA 97/37	Gaj III	-	-	-	-	-	-	-	-	-	-	-	-	-	2	-	-	-	-	-
NA 97/33	Gaj I	2	-	-	-	-	-	-	-	-	-	-	-	-	-	-	-	-	1	-
NA 97/33	Gaj IIa/b	1	-	-	-	-	-	-	-	-	-	-	-	-	-	-	-	1	-	1
NA 96/45	Gaj IIa/b	-	2	-	-	-	-	-	-	-	-	-	1	-	1	-	-	-	2	-
NA 95/1	Gaj IIa/b	2	-	-	-	-	-	-	-	-	-	-	-	-	-	-	-	-	-	-
NA 97/26	IA	-	-	-	-	1	-	-	-	-	-	-	-	-	1	1	-	-	-	-
NA 97/13	IA	1	-	-	-	-	-	-	-	-	-	-	-	-	-	-	-	-	-	-
NA 92/2C	subrec	9*	-	-	-	-	-	-	-	-	-	-	-	-	1	-	-	-	-	-
NA 93/46	EIA	1	-	-	-	1	-	-	-	-	-	-	-	-	1	1	-	-	1	-
NA 94/7	I	-	-	-	-	-	-	-	-	-	-	-	-	-	1	-	-	-	-	-
NA 94/7	IV	1	-	-	-	-	-	-	-	-	-	-	-	-	-	-	-	-	-	-
NA 93/45	I	-	-	-	-	-	-	-	-	-	-	1	1	-	-	-	1	-	-	-
NA 93/45	II	1	-	-	-	1	-	1	-	-	-	-	-	-	-	-	1	-	1	-
NA 93/45	IIIa	-	-	-	-	1	-	-	-	-	-	-	-	-	-	-	-	-	-	-
NA 93/45	IIIb	-	-	-	-	-	-	-	-	-	-	-	-	-	-	-	-	-	1	-
NA 93/45	IV	-	-	-	1	-	-	-	-	-	-	-	-	1	1	-	-	-	-	-
Blé C		1	-	-	-	-	-	-	-	-	-	-	-	-	-	1	-	-	-	-

NF= not fused, fus=fusing, F=fused; *MNI=1

Table C.88: Fusion data of small bovid bones by site and phase

site		element	cut	chop
BF 94/45	EIA	vertebra thoracalis	x	
BF 94/45	EIA	radius diaph.		x
BF 97/13	LIA	tibia diaph.	x	
BF 97/13	LIA	os carpi	?	
BF 97/30	LIA	2 vertebrae lumbales	x	
BF 94/120	LIA	humerus diaph.		x
BF 94/120	LIA	costa	x	
BF 94/120	LIA	os femoris	x	
BF 94/120	LIA	phalanx media	x	
NA 95/1	Gaj IIa/b	costa	x	
NA 92/2C	subrec	axis		x
NA 92/2C	subrec	vertebra thoracalis	?	
NA 92/2C	subrec	20 costae	x	
NA 92/2C	subrec	7 costae		x
NA 92/2C	subrec	scapula	?	
NA 92/2C	subrec	2 humerus diaph.		x
NA 92/2C	subrec	radius diaph.		?
NA 92/2C	subrec	radius dist.	x	
NA 92/2C	subrec	2 ossa coxae	x	
NA 92/2C	subrec	tibia diaph.	x	
NA 93/45	II	2 costae		x
NA 93/45	IIIb	costa		x
NA 93/45	IV	3 costae		x
Blé A		scapula dist.		x
Blé C		costa		x
Blé E		mandibulare	x	
Blé E		tibia diaph.		x

Table C.89: Traces of butchery on small bovid bones

	NA 90/5BII		BF 94/120	BF 94/120
ph. 1	Gaj IIc	**ph. 3**	LIA	LIA
GLpe	79.4	DLS	60.6	57.2
Bd	26.5	Ld	52.4	50.1
		H	43.3	-
		Bp	21.4	21.4

Table C.90: Measurements (mm) on roan antelope bones

		radius	radio-ulna	os femoris	tibia	phalanx proximalis	phalanx media	**Total**
BF 94/45	LSA	-	-	-	-	-	1	**1**
BF 97/31	LIA	-	-	-	-	1	1	**2**
BF 94/120	LIA	-	-	1	-	-	-	**1**
NA 90/5BII	Gaj IIc	1	1	-	-	-	-	**2**
NA 99/65	Gaj IIc	-	-	-	-	1	-	**1**
NA 97/33	Gaj I	-	-	-	-	1	2	**3**
NA 96/45	Gaj IIa/b	-	-	-	1	-	-	**1**
All sites		**1**	**1**	**1**	**1**	**2**	**4**	**11**

Table C.91: Skeletal element distribution of hartebeest or topi by site and phase

	NA 99/65		NA 90/5BII	NA 90/5BII		NA 96/45
mand.	Gaj IIc	**radius**	Gaj IIc	Gaj IIc	**tibia**	Gaj IIa/b
9	18.0	Bp	60.3	-	Bd	48.4
15b	22.2	BFp	55.2	-		
L M3/B M3	16.6/7.3	Bd	-	45.7		

	NA 97/33	NA 99/65	BF 97/31		NA 97/33	BF 94/45	BF 97/31
ph. 1	Gaj I	Gaj IIc	LIA	**ph. 2**	Gaj I	EIA	LIA
GLpe	63.1	58.2	-	GL	37.1	34.9	34.4
Bp	20.9	16.9	19.5	Bp	18.6	17.1	16.7
SD	15.2	13.4	-	SD	14.9	14.2	13.9
Bd	18.9	15.2	-	Bd	17.1	15.6	14.5

Table C.92: Measurements (mm) on hartebeest or topi bones

		mandibulare	dentes mandibulare	ossa carpi	os coxae	os metatarsale III et IV	ossa sesamoidea	phalanx proximalis	**Total**	*Hippotragus equinus*	*Damaliscus lunatus/Alcelaphus buselaphus*
BF 97/13	LIA	-	-	1	-	-	-	-	**1**	-	-
BF 97/30	LIA	-	-	-	-	-	-	1	**1**	-	-
BF 97/31	LIA	1	-	-	-	-	-	-	**1**	-	1
BF 94/120	LIA	-	-	-	-	-	1	-	**1**	1	1
NA 90/5BII	Gaj IIc	-	1	-	1	2	3	-	**7**	1	2
NA 96/45	Gaj IIa/b	-	-	-	1	-	-	2	**3**	-	1
All sites		**1**	**1**	**1**	**2**	**2**	**4**	**3**	**14**	**2**	**5**

Table C.93: Skeletal element distribution of large antelopes by site and phase

element	n	element	n	element	n	element	n
cranium	13	vertebrae lumbales	4	os metacarpale III et IV	2	calcaneus	1
maxillare	4	vertebrae	4	phalanx proximalis manus	2	os centroquartale	1
mandibulare	2	costae	1	phalanx media manus	1	os metatarsale III et IV	2
dentes	6	hyoideus	1	os coxae	1	phalanx proximalis pedis	2
atlas	1	scapula	1	os femoris	3	phalanx media pedis	1
axis	1	humerus	2	tibia	3	metapodalia	1
vertebrae cervicales	1	radio-ulna	2	malleolus lateralis	1	phalanx proximalis	1
vertebrae thoracales	1	ossa carpi	2	talus	1	phalanx distalis	2

Table C.94: Buffalo bones from Gajiganna BII (NA 90/5BII)

	NA 90/5BII		NA 90/5BII		NA 90/5BII	NA 90/5BII		Blé E	NA 90/5BII	NA 90/5BII	
humerus	Gaj IIc	**radius**	Gaj IIc	**os mc**	Gaj IIc	Gaj IIc	**ph. 1 m.**		Gaj IIc	Gaj IIc	
Bd	90.5	Bd	88.8	Bp	66.4	-	GLpe	70	64	-	
BT	82.1			Bd	-	71	Bp	33.6	36.2	35.6	
							SD	30.3	34.5	-	
							Bd	31.8	33.5	-	
	NA 90/5BII	NA 90/5BII		NA 90/5BII		NA 90/5BII		NA 90/5BII	Blé E		NA 90/5BII
ph. 2 m.	Gaj IIc	Gaj IIc	**os fem.**	Gaj IIc	**malleolus**	Gaj IIc	**talus**	Gaj IIc		**calcaneus**	Gaj IIc
GL	47.4	46.4	Bd	111.5	GD	41.3	GLl	77.1	-	GL	145
Bp	35.2	34		NA 90/5BII			GLm	69.4	-	GB	51.5
SD	27.8	26.8	**tibia**	Gaj IIc			Dl	52.6	-		
Bd	30.2	28.4	Bd	76.6			Bd	41.9	47		
	NA 90/5BII		NA 90/5BII		NA 90/5BII	NA 90/5BII		NA 90/5BII		NA 90/5BII	NA 90/5BII
os mt	Gaj IIc	**ph. 1 p.**	Gaj IIc	**ph. 2 p.**	Gaj IIc	Gaj IIc	**ph. 1**	Gaj IIc	**ph. 3**	Gaj IIc	Gaj IIc
Bd	65.1	GLpe	(68)	GL	49.1	45.8	Bd	32.6	DLS	65.5	-
		Bp	35.4	Bp	33.4	33.1			Ld	50.6	-
		SD	33	SD	27.1	25.7			H	41.2	46.5
		Bd	32.5	Bd	27.4	26.9			Bp	20.9	28.3

()=estimate

Table C.95: Measurements (mm) on buffalo bones

		processus cornualis	cranium	maxillare	dentes maxillare	mandibulare	dentes mandibulare	dentes	hyoideus	atlas	axis	vertebrae cervicales	vertebrae thoracales	vertebrae lumbales	os sacrum	vertebrae caudales	vertebrae	sternum	costae	scapula	humerus	radius	ulna	radio- ulna	os metacarpale III et IV	ossa carpi	phalanx proximalis manus	phalanx media manus	phalanx distalis manus	os coxae	os femoris	tibia	malleolus lateralis	patella	os metatarsale I	os metatarsale III et IV	ossa metapodalia	os centroquartale	calcaneus	talus	os centroquartale	ossa tarsi	phalanx proximalis pedis	phalanx media pedis	phalanx distalis pedis	ossa metapodalia	ossa sesamoidea	phalanx proximalis	phalanx media	phalanx distalis	phalanx	unidentified	Total
BF 97/5	EIA	-	1	-	-	-	-	1	-	-	-	-	-	-	-	-	-	-	-	-	-	-	-	-	-	-	-	-	-	-	-	1	-	-	-	-	-	-	-	-	-	-	-	-	-	-	-	-	-	-	-	-	3
BF 94/45	EIA	1	1	-	8	2	20	2	-	-	-	5	2	1	-	-	-	-	18	3	5	1	1	1	3	4	1	-	1	1	-	1	-	1	1	7	2	1	-	1	-	1	1	-	-	-	3	5	2	1	-	-	108
BF 94/45	AB EIA	-	-	-	-	-	-	10	-	-	-	-	-	-	-	-	-	-	1	-	-	-	1	-	-	-	-	-	-	1	-	-	-	-	-	-	-	-	-	-	-	-	-	-	-	-	-	-	-	-	-	-	13
BF 97/13	EIA	-	-	-	-	-	1	6	-	-	-	-	-	-	-	-	-	-	-	-	-	-	-	-	-	-	-	-	-	-	-	-	-	-	-	-	-	-	-	-	-	1	-	-	-	-	-	-	-	-	-	-	8
BF 97/13	MIA	-	2	-	-	-	2	25	-	-	-	1	1	1	-	-	1	-	4	-	-	1	-	-	-	3	-	-	-	1	-	2	-	-	1	1	4	1	4	3	-	-	-	-	-	4	3	7	2	2	-	-	76
BF 97/13	LIA	2	1	-	5	3	19	49	-	-	-	1	-	1	-	-	-	-	-	-	-	-	-	-	-	2	-	-	-	-	-	-	-	-	-	-	-	2	1	1	-	2	-	-	-	5	3	4	1	-	1	-	103
BF 97/13	AB LIA	-	-	-	-	-	2	3	-	-	-	-	-	-	-	-	-	-	-	-	-	-	-	-	-	-	-	-	-	-	-	-	-	-	-	-	-	-	-	-	-	-	-	-	-	-	-	-	-	-	-	-	5
BF 97/30	LIA	1	4	2	3	6	9	8	-	1	1	1	1	1	-	1	2	-	5	3	6	8	2	-	3	4	2	4	-	-	2	3	-	1	-	5	-	-	1	1	-	1	-	-	-	4	5	6	5	1	-	-	113
BF 96/22	MIA	-	1	-	1	1	1	7	-	-	-	-	-	-	-	-	-	-	1	-	-	-	1	-	-	-	-	-	-	-	1	-	-	-	-	1	-	-	1	-	-	-	-	1	-	-	1	1	-	-	-	-	19
BF 97/31	LIA	-	1	-	5	-	-	9	-	-	-	-	-	-	-	-	-	-	3	-	2	1	-	-	1	1	-	-	-	-	-	1	-	-	-	2	-	-	-	-	-	-	-	1	-	-	1	2	1	1	-	-	32
BF 94/120	LIA	-	3	-	11	4	30	53	-	-	-	1	-	-	-	-	-	-	4	4	1	-	-	-	1	5	1	-	-	2	4	1	-	-	-	6	1	-	-	4	1	-	-	-	-	-	5	4	1	2	-	-	149
BF 95/7	LIA	-	-	-	2	1	-	13	-	-	-	-	1	-	-	-	-	-	5	2	2	3	-	-	-	1	-	-	-	1	1	6	1	-	-	3	-	-	-	1	-	-	-	-	-	-	1	1	1	-	-	-	46
NA 90/5C	Gaj IIa	-	1	-	-	-	-	2	-	-	-	-	2	-	-	-	-	-	2	-	-	-	1	-	-	1	-	-	-	-	-	-	-	-	-	-	-	-	-	-	-	-	-	-	-	-	1	-	-	-	-	-	10
NA 90/5A	Gaj IIb	1	46	-	36	44	66	90	1	-	-	-	-	-	-	-	94	-	47	1	24	19	14	-	14	37	-	-	-	14	22	22	7	3	-	45	-	-	9	14	-	17	-	-	-	46	35	41	24	22	-	-	855[1]
NA 90/5BI	Gaj IIa/b	4	29	1	31	35	29	35	3	-	-	-	-	-	-	-	60	-	17	5	11	24	7	-	6	5	-	-	-	14	8	12	1	2	-	6	-	-	10	6	-	6	-	-	-	12	5	24	12	6	-	-	426
NA 90/5BII	Gaj IIc	-	8	2	-	5	-	18	1	1	2	9	6	3	3	-	6	-	13	2	6	7	6	-	3	4	4	-	-	9	-	6	-	1	-	-	-	-	2	4	2	2	-	-	-	8	2	4	4	2	-	-	155
NA 91/1A	Gaj I	-	-	-	1	1	-	2	-	-	-	-	-	-	-	-	-	-	-	-	-	-	-	-	-	-	-	-	-	-	-	1	-	-	-	-	-	-	-	-	-	-	-	-	-	1	-	-	-	-	-	-	6
NA 93/42	Gaj I	-	1	-	-	-	1	7	-	-	-	-	1	-	-	-	6	-	1	-	-	1	-	-	-	-	-	-	-	-	1	-	-	-	-	1	1	-	-	-	-	-	-	1	-	-	-	3	2	4	-	-	31
NA 97/18	Gaj I	-	-	-	-	-	-	-	-	-	-	-	-	-	-	-	-	-	-	-	-	-	-	-	-	-	-	-	-	-	-	-	-	-	-	-	-	-	-	-	-	-	-	-	-	-	-	1	-	-	-	-	1
NA 97/24	Gaj I	-	1	1	1	1	-	1	-	-	-	-	-	-	-	2	-	-	-	-	-	-	-	-	-	-	-	-	-	-	3	-	-	-	-	-	-	-	-	-	-	-	-	-	-	-	-	-	1	3	-	-	14
NA 93/36	Gaj I	-	1	-	-	1	1	-	-	-	-	-	-	1	-	-	-	-	2	-	-	-	-	-	-	-	-	-	-	-	-	-	-	-	-	-	1	-	-	-	-	-	-	-	-	-	-	-	-	1	-	-	8
NA 99/65	Gaj IIa	-	-	-	-	-	-	-	-	-	-	-	1	-	-	-	-	-	-	-	-	-	-	-	-	-	-	-	-	-	-	-	-	-	-	-	-	-	-	-	-	-	-	-	-	-	-	-	-	-	-	-	1
NA 99/65	Gaj IIb	-	-	-	1	-	3	2	-	-	-	-	1	-	-	-	-	1	-	-	-	-	-	-	1	-	-	-	-	-	-	-	-	-	-	1	-	-	1	-	-	-	-	-	-	-	-	-	-	-	-	-	11
NA 99/65	Gaj IIc	-	-	-	3	1	5	4	-	-	-	-	1	-	-	-	-	-	1	-	-	-	-	-	-	2	-	-	-	-	-	-	-	-	-	1	1	-	-	-	-	-	-	-	-	-	1	1	2	-	-	-	23
NA 93/10	Gaj IIc	-	-	-	1	4	2	-	-	-	-	1	-	-	-	-	-	-	-	-	7	-	-	-	-	5	-	-	-	-	-	1	-	-	-	1	-	-	-	1	-	-	-	-	-	1	1	-	-	-	-	-	25
NA 97/37	Gaj III	1	5	-	6	10	6	8	-	2	-	3	1	1	-	-	1	-	6	2	3	3	1	-	1	4	1	-	1	7	9	1	-	-	-	7	4	-	1	1	-	-	3	2	4	-	1	5	5	3	-	-	119
NA 97/33	Gaj I	2	26	1	6	20	6	8	1	5	2	11	8	5	1	8	31	-	101	7	14	13	8	2	15	15	4	11	-	16	26	23	3	5	-	10	-	5	13	22	1	2	6	8	-	16	4	23	7	16	-	-	[2]
NA 97/33	Gaj II	1	8	-	6	13	8	9	-	1	1	10	5	-	1	1	7	-	20	-	11	6	1	-	6	10	3	1	-	2	6	11	1	2	-	12	-	-	4	4	3	2	2	3	-	5	3	10	1	2	-	-	202
NA 96/45	Gaj II	1	11	2	-	9	-	21	-	-	1	5	6	4	2	1	22	-	22	6	5	5	4	-	1	12	1	1	-	5	17	6	2	2	-	2	-	-	2	2	-	-	-	2	-	2	-	1	-	8	-	-	193
NA 95/1	Gaj II	-	20	-	10	10	41	32	-	1	1	13	12	8	-	2	9	-	44	3	8	7	12	-	14	20	3	4	-	9	8	17	2	2	-	11	-	-	14	9	5	8	2	11	-	22	11	17	7	14	1	1	445
NA 97/26	IA	3	-	-	1	-	4	3	-	-	-	-	-	-	-	-	-	-	-	-	-	-	-	-	-	1	-	-	-	-	1	2	-	-	-	-	-	-	-	-	-	-	-	-	-	-	-	-	-	-	-	-	15
NA 99/75	IA	-	1	-	1	-	3	6	-	-	-	-	-	-	-	-	-	-	3	-	-	-	-	-	-	-	-	-	-	-	-	-	-	-	-	1	-	-	-	-	-	-	-	-	-	-	1	1	-	-	-	-	17
NA 97/13	IA	1	-	-	-	-	1	-	-	-	-	-	-	-	1	-	-	-	1	-	-	-	-	-	-	-	-	-	-	-	-	-	-	-	-	-	1	-	-	-	-	-	-	-	-	-	-	1	-	-	-	-	6
NA 92/2C	subrec	-	15	-	6	10	5	17	-	1	-	3	4	6	-	5	18	1	65	-	9	6	3	-	2	8	1	1	1	1	8	4	1	-	-	12	-	-	2	3	1	2	3	3	-	14	18	13	4	1	-	3	280
NA 93/46	LSA	-	-	-	-	1	3	4	-	-	-	-	-	-	-	-	2	-	3	1	-	-	-	1	-	-	-	-	-	-	1	-	-	-	-	2	-	-	-	-	-	-	-	-	-	-	1	-	-	1	-	-	20
NA 93/46[3]	EIA	-	9	1	14	2	7	20	-	-	-	-	-	-	-	-	3	-	7	1	4	-	1	-	1	4	-	-	-	3	4	2	1	-	-	-	2	-	5	1	-	-	-	-	-	-	2	4	2	2	-	-	102
NA 94/7	I	5	40	-	53	81	55	29	-	8	2	15	15	20	2	6	3	-	23	-	25	-	5	-	2	17	-	-	-	18	20	25	-	5	-	8	-	-	15	10	-	3	-	-	-	4	-	18	10	5	-	-	547
NA 94/7	II	-	6	-	5	-	1	3	-	2	-	4	1	3	-	-	-	-	4	-	3	-	1	-	-	2	-	-	-	2	1	1	-	-	-	1	-	-	-	-	-	1	-	-	-	1	1	1	1	2	-	-	47
NA 94/7	III	-	-	-	1	6	2	-	-	-	-	-	-	-	-	-	-	-	1	-	1	-	-	-	-	-	-	-	-	-	2	1	-	-	-	-	-	-	-	-	-	-	-	-	-	-	-	-	1	1	-	-	16
NA 94/7	IV	1	1	-	-	-	1	2	-	-	-	2	1	1	-	-	-	-	-	-	-	-	1	-	-	1	-	-	-	-	-	1	-	-	-	-	-	-	2	-	-	-	-	-	-	-	-	-	-	-	-	-	14
NA 93/45	I	-	1	-	-	1	3	2	-	-	-	1	5	1	-	1	1	-	11	-	-	-	-	-	-	2	-	-	-	2	1	1	-	-	-	2	-	-	-	1	-	1	-	1	-	1	3	3	-	-	-	-	45
NA 93/45	II	1	1	-	1	3	3	4	-	-	-	2	2	-	-	2	1	-	8	-	1	1	1	-	1	3	-	1	-	1	2	1	-	1	-	2	-	-	-	1	1	1	-	1	1	1	3	-	-	1	-	-	53
NA 93/45	IIIa	-	3	-	1	1	2	5	2	-	-	1	-	1	-	-	1	-	9	-	-	2	1	-	-	2	-	-	-	1	1	2	-	-	-	-	-	-	-	5	1	-	-	-	-	1	2	1	1	1	-	-	47
NA 93/45	IIIb	-	1	-	3	8	6	2	-	-	-	2	2	2	1	1	1	-	15	2	2	3	1	-	2	3	2	-	-	2	4	6	2	-	-	1	-	-	-	1	1	-	1	1	-	2	3	1	1	-	-	-	85
NA 93/45	IV	-	4	1	4	5	2	4	1	-	-	2	4	4	1	-	1	-	14	1	1	2	-	-	2	1	-	-	1	-	-	2	-	-	-	3	2	-	-	-	-	-	-	-	-	-	-	2	-	-	-	-	64
Blé A		-	-	-	-	-	1	-	-	-	-	-	-	1	-	-	-	-	3	-	1	-	-	-	2	-	-	-	-	-	-	4	-	-	-	1	-	1	-	-	-	-	-	-	-	-	-	-	-	-	-	-	14
Blé B		-	-	-	-	-	2	-	-	-	-	-	1	-	-	-	1	-	-	-	-	1	-	-	-	-	-	-	-	-	1	-	-	-	-	-	1	-	-	-	-	-	-	-	-	-	-	-	-	-	-	-	7
Blé C		-	1	-	-	-	-	-	-	-	-	-	-	-	-	-	-	-	1	-	-	-	-	-	-	-	-	-	-	-	-	-	-	-	-	-	2	-	-	-	-	-	-	-	-	-	-	-	-	-	-	-	4
Blé E		-	-	-	1	2	1	-	-	-	-	1	-	-	-	-	2	-	5	2	-	-	-	-	-	-	-	-	-	-	1	3	-	-	-	-	-	-	1	1	-	-	-	1	-	-	-	-	-	1	-	-	22
All sites		**25**	**254**	**11**	**228**	**291**	**354**	**526**	**9**	**22**	**10**	**94**	**84**	**65**	**12**	**30**	**273**	**2**	**490**	**45**	**152**	**114**	**73**	**4**	**81**	**179**	**23**	**23**	**4**	**112**	**155**	**170**	**21**	**25**	**2**	**155**	**22**	**10**	**88**	**97**	**16**	**50**	**18**	**36**	**5**	**150**	**120**	**205**	**98**	**103**	**2**	**4**	**5142**

[1] [2]including 8 bones of min. 1 juvenile individual; [3]including 2 bones of min. 1 fœtus

Table C.96: Skeletal element distribution of cattle by site and phase

fusion ages (months)		foet.	juv.	7 scap. os coxae		12-18 hum. dist. radius prox.			18 ph. 1 ph. 2			24-30 os mc dist.		24-30 tibia dist.		24-36 os mp dist.		27-36 os mt dist.		36-42 calc.		42 fem. prox.			42-48 fem. dist. tibia prox. hum. prox. rad. dist. ulna prox.		
				NF	F	NF	fus	F	NF	fus	F	NF	F	NF	F	NF	F	NF	F	NF	F	NF	fus	F	NF	fus	F
BF 94/45	EIA	-	-	-	-	-	-	1	-	-	7	1	1	-	-	2	-	1	1	-	-	-	-	-	-	-	1
BF 94/45	AB EIA	-	-	-	-	1	-	-	-	-	-	-	-	-	-	-	-	-	-	-	-	-	-	-	-	-	-
BF 97/13	MIA	-	-	-	-	-	-	1	1	-	3	-	-	-	-	4	-	-	-	-	-	-	-	-	1	-	-
BF 97/13	LIA	-	-	-	-	-	-	-	-	-	2	-	-	-	-	-	-	-	-	-	-	-	-	-	-	-	-
BF 97/30	LIA	-	1	-	-	-	-	2	2	-	5	-	-	1	1	2	-	-	2	-	1	-	-	-	-	-	-
BF 96/22	MIA	-	-	-	-	-	-	-	-	-	1	-	-	-	-	-	-	-	-	-	-	-	-	-	-	-	-
BF 97/31	LIA	-	-	-	-	-	-	-	-	-	1	-	1	-	-	-	-	-	-	-	-	-	-	-	-	-	-
BF 94/120	LIA	-	1	-	-	-	-	-	-	-	2	-	-	-	-	-	-	-	-	-	-	-	-	-	-	-	-
BF 95/7	LIA	-	1	-	-	-	-	1	-	-	1	-	-	-	-	-	-	-	-	-	-	-	-	-	1	-	-
NA 90/5C	Gaj IIa	-	-	-	-	-	-	2	-	-	6	-	-	-	-	-	1	-	-	1	-	-	-	-	-	-	1
NA 90/5A	Gaj IIb	6*	-	-	1	3	-	9	4	-	27	-	-	2	2	12	7	-	-	3	-	1	-	3	11	-	5
NA 90/5BI	Gaj IIa/b	-	-	1	-	2	1	12	3	-	11	-	-	-	1	5	1	-	-	3	1	1	-	-	6	-	6
NA 90/5BII	Gaj IIc																										
NA 99/65	Gaj IIb	1	-	-	-	-	-	-	-	-	-	-	-	-	-	-	-	-	-	-	-	-	-	-	-	-	-
NA 97/37	Gaj III	-	2	-	-	-	-	1	-	1	8	-	-	-	-	2	1	-	1	1	-	-	-	-	-	1	1
NA 97/33	Gaj I	4	23**	-	-	-	-	3	8	2	35	2	2	3	8	3	2	-	-	1	2	2	3	3	2	-	8
NA 97/33	Gaj IIa/b	-	5	-	-	-	-	3	4	1	10	1	1	3	3	1	1	-	2	2	-	3	-	-	1	1	1
NA 96/45	Gaj IIa/b	-	-	-	-	-	-	1	-	-	4	-	-	-	-	-	-	-	-	-	-	-	-	-	-	-	2
NA 95/1	Gaj IIa/b	1	5	-	-	-	-	2	1	1	25	-	1	-	3	11	-	-	1	2	2	1	-	-	-	-	2
NA 99/75	IA	1	-	-	-	-	-	-	-	-	-	-	-	-	-	-	-	-	-	-	-	-	-	-	-	-	-
NA 97/13	IA	-	-	-	-	-	-	-	-	-	1	-	-	-	-	-	-	-	-	-	-	-	-	-	-	-	-
NA 92/2C	subrec	-	4	-	-	-	-	3	4	-	12	-	-	-	-	2	1	-	-	-	1	-	-	-	3	-	1
NA 93/46	EIA	-	2*	-	-	-	-	-	1	-	3	-	-	-	1	-	-	-	-	2	1	1	1	1	1	-	-
NA 94/7	I	1	3	-	-	-	-	-	-	-	11	-	-	-	1	-	-	-	-	-	1	-	-	-	-	-	-
NA 94/7	II	-	1	-	-	-	-	-	-	-	1	-	-	-	-	-	-	-	-	-	-	-	-	-	-	-	-
NA 94/7	III	-	-	-	-	-	-	-	-	-	1	-	-	-	-	-	-	-	-	-	-	-	-	-	-	-	-
NA 93/45	I	-	-	-	-	-	-	-	1	-	3	-	-	-	-	-	-	-	-	-	-	-	-	-	-	-	-
NA 93/45	II	-	-	-	-	-	-	-	-	-	1	-	-	-	-	-	-	-	-	-	-	-	-	-	1	-	-
NA 93/45	IIIa	-	1	-	-	-	-	-	-	-	1	-	-	-	-	-	1	-	-	-	-	-	-	-	-	-	-
NA 93/45	IIIb	-	1	-	-	-	-	1	1	-	4	-	-	1	-	1	-	-	1	-	-	-	-	-	2	-	-
NA 93/45	IV	-	1	-	-	1	-	-	-	-	1	-	1	2	-	-	-	-	1	-	-	-	-	-	-	-	1
Blé A		-	-	-	-	-	-	1	-	-	-	-	-	-	1	-	-	-	-	-	-	-	-	-	-	-	-
Blé B		-	-	-	-	-	-	-	-	-	-	-	-	-	-	-	1	-	-	-	-	-	-	1	1	-	-
Blé C		-	-	-	-	-	-	-	-	-	-	-	-	-	-	-	1	-	-	-	-	-	-	-	-	-	-
Blé E		-	-	-	-	-	-	-	-	-	-	-	-	-	1	-	-	-	-	1	-	-	-	-	-	-	-

NF=not fused, fus=fusing, F=fused; *MNI=1, **including 8 bones with MNI=1

Table C.97: Fusion data of cattle bones by site and phase

site		element	cut	chop	shave
BF 94/45	EIA	phalanx media	x		
BF 94/45	EIA	vertebra thoracalis	x		
BF 94/45	EIA	2 costae	x		
BF 97/13	MIA	phalanx proximalis	x		
BF 97/30	LIA	costa	x		
BF 97/30	LIA	humerus diaph.		x	
BF 97/30	LIA	os femoris diaph.		x	
BF 97/30	LIA	os metatarsale III et IV diaph.	x		
BF 97/31	LIA	radius diaph.		x	
NA 90/5A	Gaj IIb	costa	x		
NA 90/5BI	Gaj IIa/b	radio-ulna prox.		x	
NA 90/5BI	Gaj IIa/b	calcaneus	x		
NA 90/5BI	Gaj IIa/b	phalanx proximalis		x	
NA 90/5BII	Gaj IIc	2 ossa metapodalia		x	
NA 97/37	Gaj III	mandibulare		x	
NA 97/37	Gaj III	humerus dist.		x	
NA 97/37	Gaj III	ulna diaph.		x	
NA 97/37	Gaj III	2 ossa femoris diaph.		x	
NA 97/37	Gaj III	talus		x	
NA 97/37	Gaj III	costa	x	x	

Table C.98: Traces of butchery on cattle bones

site		element	cut	chop	shave
NA 97/33	Gaj I	os coxae		x	
NA 97/33	Gaj I	os femoris diaph.		x	
NA 97/33	Gaj I	os metapodale III et IV	x		
NA 95/1	Gaj IIa/b	mandibulare proc. art.	x		
NA 95/1	Gaj IIa/b	os metapodale III et IV dist.		x	
NA 92/2C	subrec	6 costae	x		
NA 92/2C	subrec	costa		x	
NA 92/2C	subrec	humerus dist.	x		
NA 92/2C	subrec	radius prox.		x	
NA 92/2C	subrec	os carpi		x	
NA 92/2C	subrec	os metatarsale III et IV prox.		x	
NA 92/2C	subrec	talus	x		
NA 92/2C	subrec	phalanx proximalis		x	
NA 93/46	EIA	costa		x	
NA 94/7	I	vertebra thoracalis	x		
NA 94/7	I	humerus		x	
NA 94/7	I	humerus			x
NA 94/7	III	costa		x	
NA 94/7	IV	ulna	x	x	
NA 93/45	I	costa		x	
NA 93/45	I	os coxae	x		
NA 93/45	I	os coxae	x		
NA 93/45	I	os sesamoideum	x		
NA 93/45	II	vertebra caudalis		x	
NA 93/45	II	vertebra thoracalis		x	
NA 93/45	II	os metatarsale III et IV diaph.		x	
NA 93/45	II	os metatarsale III et IV prox.		x	
NA 93/45	IIIa	vertebra		x	
NA 93/45	IIIa	talus		x	
NA 93/45	IIIb	costa		x	
NA 93/45	IIIb	vertebra lumbalis		x	
NA 93/45	IIIb	2 humeri dist.		x	
NA 93/45	IIIb	humerus		x	
NA 93/45	IIIb	os metacarpale III et IV prox.	x		
NA 93/45	IIIb	os metatarsale III et IV dist.	x		
NA 93/45	IIIb	radius prox.		x	
NA 93/45	IIIb	radius diaph.		x	
NA 93/45	IIIb	talus		x	
NA 93/45	IIIb	tibia diaph.		x	
NA 93/45	IV	vertebra		x	
NA 93/45	IV	costa		x	
NA 93/45	IV	os metatarsale III et IV dist.		x	
Blé A		costa		x	
Blé A		os metacarpale III et IV prox.			?
Blé B		os femoris prox.		x	
Blé B		vertebra thoracalis		x	
Blé E		mandibulare		x	
Blé E		costa	x		
Blé E		costa		x	
Blé E		2 scapulae		x	
Blé E		talus		x	
Blé E		phalanx distalis		x	

Table C.98 cont.

mand.	NA 90/5A Gaj IIb	NA 90/5BI Gaj IIa/b	NA 90/5BI Gaj IIa/b	NA 90/5BI Gaj IIa/b	NA 90/5BII Gaj IIc
7	131	123	-	-	-
L M3	-	35.0	36.0	33.0	-
15b	-	-	-	-	45.8
15c	-	-	-	-	30.7

axis	BF 97/30 LIA
BFcr	82.0

scapula	NA 97/33 Gaj I	BF 97/30 LIA	BF 94/45 EIA	BF 94/45 EIA	NA 96/45 Gaj IIa/b
GLP	73.1	59.1	55.4	55.2	49.5
LG	61.4	(53)	47.8	48.3	41.5
BG	45.4	(35)	41.8	42.2	32.7

hum.	NA 90/5BII Gaj IIc	NA 90/5A Gaj IIb	BF 94/45 EIA	NA 90/5BI Gaj IIa/b	NA 90/5A Gaj IIb	NA 97/33 Gaj I
Bd	64.8	-	-	-	-	-
BT	60.1	66	64.6	64	63	59.8

radius	NA 90/5BI Gaj IIa/b	NA 90/5BI Gaj IIa/b	NA 92/2C subrec	NA 90/5BI Gaj IIa/b	NA 90/5A Gaj IIb	NA 90/5A Gaj IIb	NA 96/45 Gaj IIa/b	NA 90/5A Gaj IIb	NA 90/5BI Gaj IIa/b	NA 90/5A Gaj IIb	NA 90/5A Gaj IIb
Bp	74.0	72.5	72	71.0	70.0	62.7	63.4	62.4	50.6*	-	-
BFp	71.0	66.0	68	67.5	63.1	53.6	57.1	59.5	47.3*	-	-
Bd	-	-	-	-	-	-	-	-	-	62	60.7

radius	NA 97/33 Gaj IIa/b	NA 93/45 I	NA 97/37 Gaj III
Bd	60.1*	60	45.8

ulna	NA 95/1 Gaj IIa/b	NA 90/5BII Gaj IIc	NA 92/2C subrec
BPC	35.0	-	-
DPA	-	60.2	-
SDO	-	-	41.5

os mc	NA 90/5A Gaj IIb	NA 93/45 I	NA 95/1 Gaj IIa/b	NA 97/33 Gaj I	Blé A	NA 90/5A Gaj IIb	BF 97/30 LIA	NA 97/33 Gaj I	NA 90/5BI Gaj IIa/b	NA 90/5BI Gaj IIa/b	NA 93/45 IIIb
GL	179	-	-	-	-	-	-	-	-	-	-
Bp	51.2	58.8	55.8	50.4	50.4	48.9	41.8	(37)*	(37)*	(35)*	-
SD	27.1	-	-	-	-	-	-	-	-	-	-
Bd	50.4	-	-	-	-	-	-	-	-	-	59.6

os mc	NA 97/33 Gaj IIa/b	NA 90/5A Gaj IIb	NA 97/33 Gaj I	NA 90/5A Gaj IIb	NA 97/33 Gaj I
Bd	57.2	56.6	49.0	46.5	46.2

ph. 1 m.	NA 92/2C subrec	NA 90/5A Gaj IIb	NA 97/33 Gaj IIa/b	NA 90/5A Gaj IIb	NA 93/45 IIIb	NA 93/45 IIIb	NA 90/5BII Gaj IIc	NA 97/33 Gaj IIa/b	NA 97/33 Gaj IIa/b	BF 97/30 LIA	BF 94/45 EIA
GLpe	61.6	56.3	55.5	55.5	55.4	53.7	53.2	(53)	(53)	(53)	52.0
Bp	30.0	27.1	27.1	25.8	-	24.4	26.5	(24)	(27)	(27)	26.0
SD	26.1	22.5	21.9	22.1	21.4	20.3	20.8	22.1	23.0	(23)	23.2
Bd	27.8	25.3	25.0	25.9	25	22.8	22.6	(24)	(24)	(24)	24.0

ph. 1 m.	NA 97/33 Gaj I	NA 90/5BII Gaj IIc	NA 95/1 Gaj IIa/b	NA 90/5BII Gaj IIc	NA 97/33 Gaj I	NA 96/45 Gaj IIa/b	NA 95/1 Gaj IIa/b	BF 97/30 LIA	NA 97/37 Gaj III	NA 97/37 Gaj III	NA 90/5BII Gaj IIc
GLpe	50.8	50.1	48.9	48	45.4	44.3	41.0	-	-	-	-
Bp	26.1	22.6	(23)	25.7	24.5	24.4	23.9	27.0	(27)	26	25.4
SD	21.2	21.2	20.0	18.9	20.1	19.7	-	24.8	-	22.3	21.5
Bd	23.5	24	-	22	22.1	23.6	-	-	-	24.5	-

*unfused; ()=estimate

Table C.99: Measurements (mm) on cattle bones

	NA 97/33	NA 95/1	BF 94/120		BF 97/30	NA 93/45	NA 95/1	NA 97/33	NA 90/5BI	NA 97/33	NA 97/33
ph. 1 m.	Gaj I	Gaj IIa/b	LIA	**ph. 2 m.**	LIA	II	Gaj IIa/b	Gaj IIa/b	Gaj IIa/b	Gaj I	Gaj I
Bp	25.4	-	-	GL	37.0	36.9	36.2	35.7	35.0	34.2	34.1
SD	-	20.6	-	Bp	27.0	29.8	25.9	26.9	26.7	25.6	25.9
Bd	-	-	27.8	SD	22.4	24.6	20.4	22.2	22.8	20.8	22.0
				Bd	22.8	24.3	22.0	23.0	23.1	22.7	22.2

	NA 95/1	NA 97/33	NA 95/1	NA 97/33	NA 97/33	NA 97/33	NA 97/33	NA 93/42	NA 97/33	NA 96/45	NA 97/33
ph. 2 m.	Gaj IIa/b	Gaj I	Gaj IIa/b	Gaj I	Gaj I	Gaj I	Gaj I	Gaj I	Gaj I	Gaj IIa/b	Gaj I
GL	33.0	33.0	32.7	32.1	31.7	31.7	31.7	31.3	(31)	30.9	-
Bp	26.5	24.0	26.6	24.8	24.2	23.9	23.8	23.9	22.7	24.5	24.0
SD	21.8	19.3	20.2	19.2	19.1	18.5	19.5	20.4	19.7	20.2	18.8
Bd	22.5	21.0	21.1	20.9	-	(20)	21.5	20.4	22.7	20.6	(20)

	NA 97/33	NA 95/1		NA 93/45	NA 92/2C	NA 97/37	BF 94/45	BF 94/45		NA 90/5BII	
ph. 2 m.	Gaj I	Gaj IIa/b	**ph. 3 m.**	IV	subrec	Gaj III	EIA	EIA	**os coxae**	Gaj IIc	
Bd	21.2	19.3	DLS	72.2	66.6	62.2	-	-	LA	62.1	
			Ld	51.7	47.4	-	-	-			
			H	47.7	42.2	-	43.9	36.4			
			Bp	24.9	22.4	-	20.5	22.7			

	NA 97/33	NA 90/5A	NA 90/5A	NA 90/5A	NA 97/33	NA 97/33	NA 90/5BI	NA 96/45
os fem.	Gaj I	Gaj IIb	Gaj IIb	Gaj IIb	Gaj I	Gaj I	Gaj IIa/b	Gaj IIa/b
DC	40.2	39.5	38.2	33.5	33.4	30.4	-	-
Bd	-	-	-	-	-	-	74.0	72.8

	NA 97/33	NA 90/5A	NA 95/1	NA 90/5BI	NA 90/5A	NA 97/33	NA 96/45
patella	Gaj IIa/b	Gaj IIb	Gaj IIa/b	Gaj IIa/b	Gaj IIb	Gaj I	Gaj IIa/b
GL	56.1	56	(55)	54	54	(50)	43.4
GB	47.1	-	48.3	-	44	45.0	-

	NA 90/5BI	BF 97/30	NA 90/5A	NA 93/45	NA 91/1A	NA 90/5A	NA 90/5BI	NA 94/7
tibia	Gaj IIa/b	LIA	Gaj IIb	IV	Gaj I	Gaj IIb	Gaj IIa/b	I
Bp	88.0	-	-	-	-	-	-	-
Bd	-	70.3	56.2	55.8*	53.5	51.8	51.5	41.7

	NA 93/46	NA 93/45	NA 95/1	NA 90/5A	NA 90/5A	NA 93/45	NA 90/5A	NA 97/33	NA 92/2C	NA 97/33
malleol.	EIA	IIIb	Gaj IIa/b	Gaj IIb	Gaj IIb	IIIb	Gaj IIb	Gaj IIa/b	subrec	Gaj I
GD	32.6	31.8	(31)	29.5	29.5	29.0	28.9	28.7	28.4	27.2

	BF 94/120	NA 90/5A	NA 90/5A	NA 93/46	NA 90/5BI	NA 93/45	NA 94/7	NA 90/5A	NA 97/33	NA 97/33	NA 96/45
talus	LIA	Gaj IIb	Gaj IIb	EIA	Gaj IIa/b	IIIa	I	Gaj IIb	Gaj I	Gaj I	Gaj IIa/b
GLl	65.5	65	64	(60)	62.0	61.8	61.4	59	58.8	58.6	57.6
GLm	-	57	58	-	57.0	58.0	54.9	53.5	51.9	-	-
Bd	40.7	41	42.5	(38)	42.2	39.7	39.4	-	32.9	-	32.6
Dl	36.0	34	35	(36)	34.5	34.5	33.1	-	35.1	32.2	-

	NA 90/5A	NA 93/45	NA 93/45	NA 97/33	NA 97/33	NA 97/33	NA 90/5BI	NA 97/33	NA 97/33	NA 97/33	NA 97/33
talus	Gaj IIb	IIIa	II	Gaj IIa/b	Gaj I	Gaj I	Gaj IIa/b	Gaj I	Gaj I	Gaj I	Gaj I
GLl	57	56.9	56.4	56.1	55.2	55.1	55.0	54.3	54.1	(54)	50.7
GLm	51	-	51.7	49.8	-	49.4	-	(50)	-	-	-
Bd	36	-	36.7	35.0	-	38.5	-	34.3	37.0	-	-
Dl	31	-	32.7	31.9	(30)	29.9	-	29.7	-	-	-

*unfused; ()=estimate

Table C.99 cont. 1

	NA 96/45	NA 92/2C	NA 94/7	BF 94/120	BF 94/120	NA 97/33	BF 94/120
talus	Gaj IIa/b	subrec	I	LIA	LIA	Gaj I	LIA
GLl	50.6	-	-	-	-	-	-
GLm	45.1	60.6	58.2	56.3	56.3	-	-
Bd	30.3	-	41.0	-	-	37.9	-
Dl	27.2	-	-	-	-	-	36.5

	NA 97/33	NA 90/5BI	NA 94/7	NA 92/2C	BF 97/30	NA 96/45	Blé E
calc.	Gaj I	Gaj IIa/b	I	subrec	LIA	Gaj IIa/b	
GL	131.5	128	128	127	-	-	-
GB	43.5	42	-	40	38.0	35.1	32

	NA 93/45	BF 94/45	NA 97/33	NA 97/33	NA 93/45	NA 95/1	NA 97/33	NA 90/5A	NA 90/5A	NA 90/5BII	NA 97/33
centroq.	II	EIA	Gaj I	Gaj IIa/b	IIIb	Gaj IIa/b	Gaj I	Gaj IIb	Gaj IIb	Gaj IIc	Gaj IIa/b
GB	58.5	57.9	55.9	51.9	51.7	51	47.4	46	46	44.7	(44)

	NA 90/5BII		NA 90/5A	NA 93/45	NA 97/33	NA 90/5A	NA 90/5BI	BF 97/30	NA 90/5A	BF 97/30	NA 90/5A
centroq.	Gaj IIc	**os mt**	Gaj IIb	IV	Gaj IIa/b	Gaj IIb	Gaj IIa/b	LIA	Gaj IIb	LIA	Gaj IIb
GB	41.7	Bp	49.5	43.9	43.1	42.2	-	-	-	-	-
		Bd	-	-	-	-	61.3	54.4	(53)	(52)	(50)

	BF 94/45	NA 95/1	NA 97/33	NA 97/33		NA 92/2C	NA 93/45	NA 90/5A	NA 92/2C	BF 94/45	NA 90/5BI
os mt	EIA	Gaj IIa/b	Gaj IIa/b	Gaj IIa/b	**ph. 1 p.**	subrec	IIIb	Gaj IIb	subrec	EIA	Gaj IIa/b
Bd	47.8	47.3	46.9	45.0	GLpe	62.8	60.4	60.0	60.0	58.0	57.1
					Bp	27.4	26.1	26.6	24.2	25.9	25.8
					SD	-	-	21.6	21.2	23.1	-
					Bd	-	-	25.5	23.8	23.8	24.1

	NA 90/5A	NA 90/5A	NA 90/5A	NA 90/5A	NA 90/5A	NA 97/33	BF 96/22	NA 90/5A	NA 95/1	NA 97/33	NA 97/33
ph. 1 p.	Gaj IIb	Gaj IIb	Gaj IIb	Gaj IIb	Gaj IIb	Gaj I	MIA	Gaj IIb	Gaj IIa/b	Gaj I	Gaj I
GLpe	57.0	56.0	56.0	55.4	55.0	55.0	54.4	54.0	53.3	53.1	52.2
Bp	26.0	26.6	-	26.0	23.5	-	-	23.8	23.1	25.0	23.1
SD	21.2	19.8	20.8	21.1	19.4	21.0	23.5	21.0	18.9	20.0	20.4
Bd	26.2	-	25.5	23.9	21.4	23.8	26.5	22.5	21.6	24.8	23.8

	NA 92/2C	NA 97/33	NA 97/33	NA 97/33	NA 95/1	NA 97/33		BF 97/31	BF 97/30	BF 97/30	BF 97/30
ph. 1 p.	subrec	Gaj I	Gaj I	Gaj I	Gaj IIa/b	Gaj IIa/b	**ph. 2 p.**	LIA	LIA	LIA	LIA
GLpe	50.0	50.0	-	-	-	-	GL	40.5	40.4	40.1	39.6
Bp	30.1	23.9	26.3	25.4	-	-	Bp	30.3	30.4	27.6	25.0
SD	25.0	19.0	-	-	20.0	19.1	SD	26.3	25.3	23.6	20.4
Bd	27.8	23.0	-	-	-	24.0	Bd	26.2	25.0	24.2	20.8

	BF 97/30	BF 97/30	Blé E	NA 97/33	NA 97/33	NA 97/33	NA 97/33	NA 97/33	NA 97/37	NA 97/37	NA 97/33
ph. 2 p.	LIA	LIA		Gaj IIa/b	Gaj IIa/b	Gaj I	Gaj I	Gaj I	Gaj III	Gaj III	Gaj I
GL	39.4	38.5	38.1	37.9	37.5	37.4	37.2	37.0	(37)	(37)	36.6
Bp	26.4	28.7	24.4	26.5	24.2	26.9	24.7	25.3	(24)	(23)	24.6
SD	21.5	23.6	19.2	22.3	19.4	21.7	19.9	20.3	20.1	(21)	19.6
Bd	22.3	26.0	21.0	23.1	21.4	22.6	20.8	21.4	(20)	(19)	20.8

	NA 97/33	NA 97/33	NA 97/33	NA 97/33	NA 97/33	NA 90/5A	NA 90/5A	NA 90/5A	NA 90/5BI	NA 90/5BI	NA 90/5BI
ph. 2 p.	Gaj I	Gaj I	Gaj I	Gaj IIa/b	Gaj I	Gaj IIb	Gaj IIb	Gaj IIb	Gaj IIa/b	Gaj IIa/b	Gaj IIa/b
GL	36.4	36.1	36.0	35.4	35.0	35.0	35.0	34.9	34.4	34.4	34.2
Bp	-	24.3	25.4	27.2	25.1	24.0	22.9	23.2	25.1	(23)	23.7
SD	-	19.8	20.0	21.2	20.1	20.1	19.0	18.0	19.3	19.7	20.1
Bd	22.5	19.9	22.5	22.2	21.9	20.2	19.5	19.9	20.3	20.5	(20)

()=estimate

Table C.99 cont. 2

	NA 90/5BI	NA 93/42	NA 96/45	NA 96/45	NA 92/2C	NA 92/2C	NA 97/37	NA 93/45	NA 97/37	NA 92/2C	NA 93/45
ph. 2 p.	Gaj IIa/b	Gaj I	Gaj IIa/b	Gaj IIa/b	subrec	subrec	Gaj III	II	Gaj III	subrec	I
GL	34.1	34	33.8	33.8	33.6	33.2	(33)	33	(33)	32.1	32
Bp	23.9	22	24.2	21.3	23.6	-	24.5	22.2	21.7	21.6	-
SD	18.9	19	19.3	21.2	19.0	18.2	20.3	17.1	18.0	17.7	18.6
Bd	20.5	18	19.8	23.2	19.6	19.1	(20)	18.8	-	-	-

	NA 92/2C	NA 93/42	NA 93/45	NA 95/1	NA 95/1	NA 95/1	NA 95/1	NA 95/1	NA 95/1	NA 95/1
ph. 2 p.	subrec	Gaj I	IIIb	Gaj IIa/b	Gaj IIa/b	Gaj IIa/b	Gaj IIa/b	Gaj IIa/b	Gaj IIa/b	Gaj IIa/b
GL	31.5	30.8	(30)	-	-	-	-	-	-	-
Bp	23.0	-	20.4	24.8	24.5	22.5	-	-	-	-
SD	17.7	-	16.5	19.8	18.8	-	23.9	18.7	-	-
Bd	18.4	-	19.0	21.1	-	-	-	18.4	(22)	18.7

	NA 93/45	NA 97/37	NA 97/37	NA 97/37
ph. 3 p.	II	Gaj III	Gaj III	Gaj III
DLS	70	(55)	-	-
Ld	52	(44)	-	-
MBS	-	-	-	-
H	40.2	38.7	30.7	-
Bp	22.7	17.8	19.4	22.2

	NA 92/2C	NA 94/7	NA 94/7	NA 94/7	BF 97/30	NA 94/7	NA 97/37	NA 96/45	NA 93/46	NA 94/7	BF 97/30
ph. 1	subrec	I	I	I	LIA	I	Gaj III	Gaj IIa/b	EIA	I	LIA
GLpe	67	64.5	(60)	-	-	-	-	-	-	-	-
Bp	28.4	-	(25)	34.0	(28)	27.9	27.6	27.1	27.0	(27)	26.2*
SD	26.1	-	22.4	30.7	-	24.2	-	21.6	-	-	-
Bd	30	24.7	25.6	33.2	-	27.1	-	24.2	-	-	-

	NA 95/1	BF 94/45	BF 94/45	BF 94/45	NA 97/33	BF 94/120	NA 97/13	NA 99/65	NA 94/7	NA 97/18	NA 94/7
ph. 1	Gaj IIa/b	EIA	EIA	EIA	Gaj I	LIA	IA	Gaj II	I	Gaj I	I
Bp	(24)	23.8	23.8	23.2	23.2	22.5	21.5	18.5	(26)	-	-
SD	19.5	-	-	-	-	-	-	-	-	24.2	22.8
Bd	-	-	-	-	-	-	-	-	-	20.7	25.1

	NA 94/7	NA 95/1	NA 99/75	NA 97/33	NA 97/37	NA 97/33	NA 97/37	NA 97/33
ph. 1	I	Gaj IIa/b	IA	Gaj I	Gaj III	Gaj I	Gaj III	Gaj I
SD	22.6	20.2	-	-	-	-	-	-
Bd	(27)	-	(28)	25.6	24.6	23.5	21.6	(19)

	NA 94/7	BF 94/120	NA 94/7	NA 94/7	NA 92/2C	NA 94/7	NA 94/7	BF 94/45	BF 97/30	BF 94/120	NA 94/7
ph. 2	I	LIA	I	III	subrec	II	I	EIA	LIA	LIA	I
GL	41.8	41.6	41.4	39.8	39.6	37.8	36.1	34.2	-	-	-
Bp	28.2	28.0	27.1	28.0	28.6	25.6	28.0	27.0	30.0	27.6	27.2
SD	23.1	22.4	22.6	23.1	23.2	20.1	22.7	22.3	-	-	23.6
Bd	24.0	23.1	23.2	25.1	-	19.9	23.7	-	-	-	(24)

	NA 97/37	BF 95/7	NA 97/37	BF 94/45	NA 90/5BII
ph. 2	Gaj III	LIA	Gaj III	EIA	Gaj IIc
Bp	26.1	25.7	(22)	-	-
SD	-	-	-	-	-
Bd	24.3	-	-	21.1	19.4

*unfused; ()=estimate

Table C.99 cont. 3

ph. 3	NA 93/45 II	NA 97/33 Gaj IIa/b	NA 90/5A Gaj IIb	NA 90/5A Gaj IIb	NA 97/33 Gaj I	NA 90/5BI Gaj IIa/b	NA 97/33 Gaj I	NA 94/7 II	NA 93/46 LSA	NA 90/5A Gaj IIb	BF 97/13 MIA
DLS	71.5	69.7	68	67	66	65	(65)	63.4	60.9	60	59.5
Ld	56.4	48.7	52	49	49.3	49	49.2	47.5	50.1	45	49
MBS	-	20.8	-	-	-	-	24.2	22.0	-	-	-
H	(47)	-	-	-	-	-	-	-	43.5	-	42.7
Bp	25.0	-	-	-	-	-	-	-	21.9	-	23.1

ph. 3	NA 92/2C subrec	NA 93/36 Gaj IIa	NA 90/5BI Gaj IIa/b	NA 90/5BII Gaj IIc	NA 90/5A Gaj IIb	NA 94/7 I	NA 97/33 Gaj I	NA 96/45 Gaj IIa/b	NA 96/45 Gaj IIa/b	NA 97/33 Gaj I	NA 97/37 Gaj III
DLS	58.4	57.1	57	56.1	55	(50)	-	-	-	-	-
Ld	44.8	-	44	41.1	46	-	(50)	45.7	45.6	-	-
MBS	-	-	-	-	-	-	-	-	-	20.2	-
H	42.2	-	-	37.2	-	-	-	-	-	-	46.4
Bp	22.4	-	-	-	-	-	-	-	24.2	-	23.7

ph. 3	NA 90/5BII Gaj IIc	BF 94/120 LIA	BF 97/31 LIA	NA 96/45 Gaj IIa/b
H	38.4	-	-	-
Bp	-	27.6	26.7	20.2

()=estimate

Table C.99 cont. 4

		cranium	dentes maxillare	mandibulare	dentes mandibulare	dentes	humerus	radius	radio-ulna	os metacarpale III et IV	ossa carpi	os coxae	tibia	os metatarsale III et IV	talus	os centroquartale	ossa tarsi	ossa metapodalia	ossa sesamoidei	phalanx proximalis	**Total**
BF 94/96	LSA	-	-	-	-	5	-	-	-	-	-	-	-	-	-	-	-	-	-	-	**5**
BF 97/5	LSA	-	-	1	-	2	-	-	-	-	-	-	1	-	-	-	-	-	-	-	**4**
BF 94/45	LSA	1	6	2	11	2	1	1	1	3	4	1	1	3	1	1	1	1	3	2	**46**
All sites		**1**	**6**	**3**	**11**	**9**	**1**	**1**	**1**	**3**	**4**	**1**	**2**	**3**	**1**	**1**	**1**	**1**	**3**	**2**	**55**

Table C.100: Skeletal element distribution of cattle or buffalo by site and phase

		cranium	dentes	vertebrae thoracales	**Total**
BF 94/133	LSA	-	7	1	**8**
BF 97/5	LSA	1	-	-	**1**
BF 94/45	LSA	-	1	-	**1**
All sites		**1**	**8**	**1**	**10**

Table C.101: Skeletal element distribution of large bovids by site and phase

Appendix D: Faunal lists by studied locality

Softshell turtle (Trionychidae)	1
Identified reptiles	**1**
Buffalo (*Syncerus caffer*)	1
Identified wild mammals	**1**
Small bovid	1
Large bovid	8
Identified wild or domestic mammals	**9**
Unidentified mammals	**792**
TOTAL	**803**

Table D.1: Faunal list of Tin Akof (BF 94/133)
Revised and adapted from Van Neer (Breunig *et al.*, 1996; Van Neer, 2002a)

Identified wild or domestic mammals: medium-sized carnivore	**1**

Table D.2: Faunal list of Dori 94/40 (BF 94/40)
Revised and adapted from Van Neer (unpublished data)

Identified wild or domestic mammals: cattle (*Bos primigenius* f. taurus) or buffalo (*Syncerus caffer*)	5
Unidentified mammals	**3**
TOTAL	**8**

Table D.3: Faunal list of Dori 94/96 (BF 94/96)
Revised and adapted from Van Neer (Breunig *et al.*, 1996; Van Neer, 2002a)

	LSA (trench II-VI)	LSA/EIA (trench VII)	EIA (trench VIII)
Identified molluscs: bivalve	**0**	**0**	**2**
Lungfish (*Protopterus annectens*)	-	-	9
Polypterus sp.	6	-	2
Heterotis niloticus	3	-	-
Mormyrid (Mormyridae)	1	-	-
Gymnarchus niloticus	24	-	-
Auchenoglanis sp.	34	-	-
Clarias sp.	-	-	1
Clariidae	11	-	346
Synodontis sp.	20	-	-
Unidentified catfish (Siluriformes)	383	-	1
Parachanna obscura	27	-	30
Nile perch (*Lates niloticus*)	25	-	-
Tilapia	364	-	459
Identified fish	**898**	**0**	**848**
Unidentified fish	**485**	**0**	**44**
Senegal flapshell turtle (*Cyclanorbis senegalensis*)	4	-	-
Adanson's mud turtle (*Pelusios adansonii*)	23	-	2
Crocodile (*Crocodylus* sp.)	3	-	-
Snake	1	-	-
Identified reptiles	**31**	**0**	**2**
Small rodent	3	-	1
Savannah cane rat (*Thryonomys swinderianus*)	7	-	-
Mongoose (Herpestidae), genet or civet (Viverridae)	1	-	-
Bohor reedbuck (*Redunca redunca*)	2	-	-
Kob (*Kobus kob*)	3	-	-
Medium-sized antelope	11	-	-
Identified wild mammals	**27**	**0**	**1**
Small bovid	23	-	5
Cattle (*Bos primigenius* f. taurus) or buffalo (*Syncerus caffer*)	4	-	3
Large bovid	1	-	-
Identified wild or domestic mammals	**28**	**0**	**8**
Unidentified mammals	**1574**	**6**	**68**
TOTAL	**3043**	**6**	**973**

Table D.4: Faunal list of Corcoba (BF 97/5)
Revised and adapted from Van Neer and Lambrecht (unpublished data)

				AB		
	LSA	EIA	**Total**	LSA	EIA	**Total**
Gastropod	1	47	**48**	-	9	**9**
Chambardia sp./*Spathopsis* sp.	-	6	**6**	-	1	**1**
Bivalve	9	162	**171**	-	-	**0**
Identified molluscs	**10**	**215**	**225**	**0**	**10**	**10**
Unidentified molluscs	**1**	**11**	**12**	**0**	**0**	**0**
Lungfish (*Protopterus annectens*)	-	281	**281**	-	47	**47**
Polypterus sp.	-	-	**0**	-	2	**2**
Tigerfish (*Hydrocynus* sp.)	-	-	**0**	-	1	**1**
Heterobranchus sp.	-	-	**0**	-	3	**3**
Clarias gariepinus	-	2	**2**	-	-	**0**
Clarias sp.	-	38	**38**	-	13	**13**
Clariidae	21	871	**892**	11	181	**192**
Synodontis sp.	-	7	**7**	1	3	**4**
Nile perch (*Lates niloticus*)	-	1	**1**	-	-	**0**
Tilapia	4	93	**97**	1	242	**243**
Identified fish	**25**	**1293**	**1318**	**13**	**492**	**505**
Unidentified fish	**5**	**92**	**97**	**2**	**129**	**131**
Identified amphibians: frog or toad (Anura)	**2**	**60**	**61**	**3**	**145**	**148**
Softshell turtle (Trionychidae)	-	2	**2**	-	-	**0**
Adanson's mud turtle (*Pelusios adansonii*)	-	87	**87**	-	3	**3**
Agama (*Agama* sp.)	-	2	**2**	-	14	**14**
Monitor lizard (*Varanus* sp.)	1	51	**52**	1	10	**11**
Snake	3	10	**13**	-	11	**11**
Identified reptiles	**4**	**150**	**154**	**1**	**38**	**39**
Heron (*Ardea* sp.)	-	1	**1**	-	-	**0**
Medium-sized duck or goose (Anatidae)	-	3	**3**	-	1	**1**
Large duck or goose (Anatidae)	8	6	**14**	-	2	**2**
Accipitridae	-	1	**1**	-	-	**0**
Quail (*Coturnix coturnix*)	-	1	**1**	-	-	**0**
Pigeon or dove (Columbidae)	1	-	**1**	-	1	**1**
Identified wild birds	**9**	**12**	**21**	**0**	**4**	**4**
Identified wild or domestic birds: large galliform	**0**	**12**	**12**	**0**	**2**	**2**
Unidentified birds	**4**	**22**	**26**	**0**	**8**	**8**
Eggshell ostrich (*Struthio camelus*)	-	12	12	-	1	1
Eggshell unidentified bird	1	36	37	2	32	34

Table D.5: Faunal list of Oursi (BF 94/45)
Revised and adapted from Lambrecht and Van Neer (unpublished data)

				AB		
	LSA	EIA	**Total**	LSA	EIA	**Total**
African hedgehog (*Atelerix albiventris*)	-	1	**1**	1	3	**4**
White-toothed shrew (*Crocidura* sp.)	1	-	**1**	-	-	**0**
Hare (*Lepus capensis/saxatilis*)	-	15	**15**	-	2	**2**
Striped ground squirrel (*Euxerus erythropus*)	-	-	**0**	-	1	**1**
Gerbil (*Gerbillus* sp.)	-	1	**1**	-	-	**0**
Multimammate rat (*Mastomys natalensis*)	-	2	**2**	-	-	**0**
Unstriped grass rat (*Arvicanthis niloticus*)	1	-	**1**	1	1	**2**
Small rodent	8	49	**57**	3	87	**90**
Giant pouched rat (*Cricetomys gambianus*)	-	1	**1**	-	-	**0**
Mongoose (Herpestidae), genet or civet (Viverridae)	-	4	**4**	-	-	**0**
Common warthog (*Phacochoerus africanus*)	-	2	**2**	-	-	**0**
Bush duiker (*Sylvicapra grimmia*)	-	2	**2**	-	-	**0**
Oribi (*Ourebia ourebi*)	-	1	**1**	-	-	**0**
Bush duiker (*Sylvicapra grimmia*) or oribi (*Ourebia ourebi*)	-	13	**13**	-	-	**0**
Bohor reedbuck (*Redunca redunca*)	-	5	**5**	-	-	**0**
Kob (*Kobus kob*)	-	1	**1**	-	-	**0**
Red-fronted gazelle (*Gazella rufifrons*)	-	29	**29**	-	-	**0**
Medium-sized antelope	-	10	**10**	-	-	**0**
Hartebeest (*Alcelaphus buselaphus*) or topi (*Damaliscus korrigum*)	-	1	**1**	-	-	**0**
Identified wild mammals	**10**	**137**	**147**	**5**	**94**	**99**
Domestic dog (*Canis lupus* f. familiaris)	-	1	**1**	-	-	**0**
Sheep (*Ovis ammon* f. aries)	-	7	**7**	-	-	**0**
Goat (*Capra aegagrus* f. hircus)	-	23	**23**	-	-	**0**
Sheep (*Ovis ammon* f. aries) or goat (*Capra aegagrus* f. hircus)	-	84	**84**	-	7	**7**
Cattle (*Bos primigenius* f. taurus)	-	108	**108**	-	13	13
Identified domestic mammals	**0**	**223**	**223**	**0**	**20**	**20**
African wild cat or domestic cat (*Felis silvestris* or *F.s.* f. catus)	-	10	**10**	-	1	**1**
Medium-sized carnivore	-	3	**3**	-	-	**0**
Small bovid	2	135	**137**	-	38	**38**
Cattle (*Bos primigenius* f. taurus) or buffalo (*Syncerus caffer*)	3	-	**3**	-	-	**0**
Identified wild or domestic mammals	**5**	**148**	**153**	**0**	**39**	**39**
Unidentified mammals	**290**	**8010**	**8300**	**43**	**862**	**905**
Human (*Homo sapiens sapiens*)	**0**	**2**	**2**	**0**	**0**	**0**
TOTAL	**366**	**10435**	**10800**	**69**	**1876**	**1945**

Table D.5 cont.

					AB			
	EIA	MIA	LIA	**Total**	EIA	MIA	LIA	**Total**
Identified molluscs: bivalve	**0**	**0**	**1**	**1**	**0**	**0**	**0**	**0**
Lungfish (*Protopterus annectens*)	10	2	3	**15**	-	-	-	**0**
Clarias sp.	2	4	7	**13**	-	-	1	**1**
Clariidae	40	35	41	**116**	6	-	10	**16**
Identified fish	**52**	**41**	**51**	**144**	**6**	**0**	**11**	**17**
Unidentified fish	**2**	**4**	**2**	**8**	**2**	**1**	**0**	**3**
Identified amphibians: frog or toad (Anura)	**5**	**2**	**1**	**8**	**14**	**9**	**13**	**36**
Adanson's mud turtle (*Pelusios adansonii*)	1	-	2	**3**	-	-	-	**0**
Agama (*Agama* sp.)	-	-	-	**0**	1	-	-	**1**
Monitor lizard (*Varanus* sp.)	1	3	8	**12**	-	-	-	**0**
Snake	1	-	-	**1**	-	-	-	**0**
Identified reptiles	**3**	**3**	**10**	**16**	**1**	**0**	**0**	**1**
Large duck or goose (Anatidae)	1	-	-	**1**	-	-	-	**0**
Helmeted guineafowl (*Numida meleagris*)	-	1	-	**1**	-	-	-	**0**
Identified wild birds	**1**	**1**	**0**	**2**	**0**	**0**	**0**	**0**
Identified domestic birds: domestic fowl (*Gallus gallus* f. domestica)	**0**	**0**	**0**	**0**	**0**	**0**	**1**	**1**
Identified wild or domestic birds: large galliform	**0**	**1**	**2**	**3**	**1**	**2**	**8**	**11**
Unidentified bird	**2**	**2**	**1**	**5**	**0**	**0**	**0**	**0**
Hare (*Lepus capensis/saxatilis*)	-	4	7	**11**	-	-	2	**2**
Multimammate rat (*Mastomys natalensis*)	3	1	-	**4**	1	-	-	**1**
Small rodent	38	5	5	**48**	18	5	5	**28**
Sandfox (*Vulpes pallida*)	-	1	-	**1**	-	-	-	**0**
Mongoose (Herpestidae), genet or civet (Viverridae)	2	-	2	**4**	-	1	-	**1**
Serval cat or caracal (*Felis serval/caracal*)	-	-	1	**1**	-	-	-	**0**
Common warthog (*Phacochoerus africanus*)	-	1	1	**2**	-	-	-	**0**
Bush duiker (*Sylvicapra grimmia*) or oribi (*Ourebia ourebi*)	-	1	1	**2**	-	-	-	**0**
Bohor reedbuck (*Redunca redunca*)	-	1	1	**2**	-	-	-	**0**
Red-fronted gazelle (*Gazella rufifrons*)	-	1	2	**3**	-	-	-	**0**
Medium-sized antelope	-	-	1	**1**	-	-	-	**0**
Large antelope	-	-	1	**1**	-	-	-	**0**
Identified wild mammals	**43**	**15**	**22**	**80**	**19**	**6**	**7**	**32**
Domestic dog (*Canis lupus* f. familiaris)	1	28	25	**54**	-	-	3	**3**
Horse (*Equus ferus* f. caballus) or donkey (*Equus africanus* f. asinus)	-	1?	2	**3**	-	-	-	**0**
Sheep (*Ovis aries* f. ammon)	-	2	9	**11**	-	-	-	**0**
Goat (*Capra aegagrus* f. hircus)	1	11	9	**21**	-	-	-	**0**
Sheep (*Ovis aries* f. ammon) or goat (*Capra aegagrus* f. hircus)	7	41	54	**102**	-	-	4	**4**
Cattle (*Bos primigenius* f. taurus)	8	76	103	**187**	-	-	5	**5**
Identified domestic mammals	**17**	**159**	**202**	**378**	**0**	**0**	**12**	**12**
Medium-sized carnivore	-	11	9	**20**	-	-	-	**0**
Small bovid	20	111	150	**281**	1	-	23	**24**
Identified wild or domestic mammals	**20**	**122**	**159**	**301**	**1**	**0**	**23**	**24**
Unidentified mammals	**464**	**3910**	**7938**	**12312**	**50**	**71**	**672**	**793**
TOTAL	**609**	**4260**	**8389**	**13258**	**94**	**89**	**747**	**930**

Table D.6: Faunal list of Oursi village (BF 97/13)

	hand	Sieve test (10 mm)	AB
Gastropod	1	-	-
Chambardia sp./*Spathopsis* sp.	11	-	-
Identified molluscs	**11**	**0**	**0**
Lungfish (*Protopterus annectens*)	3	-	-
Gymnarchus niloticus	1	-	-
Clarias sp.	-	1	-
Clariidae	72	4	1
Nile perch (*Lates niloticus*)	1	-	-
Identified fish	**77**	**5**	**1**
Unidentified fish	**1**	**0**	**0**
Identified amphibians: frog or toad (Anura)	-	1	-
Sahelian giant tortoise (*Geochelone sulcata*)	13[1]	-	-
Adanson's mud turtle (*Pelusios adansonii*)	1	-	-
Agama (*Agama* sp.)	1	-	-
Monitor lizard (*Varanus* sp.)	2	-	-
Snake	1	-	-
Crocodile (*Crocodylus* sp.)	6	-	-
Identified reptiles	**24**	**0**	**0**
Identified wild birds: pigeon or dove (Columbidae)	**1**	**0**	**0**
Identified domestic birds: chicken (*Gallus gallus* f. domestica)	**4**	**0**	**0**
Identified wild or domestic birds: large galliform	**24**	**0**	**0**
Unidentified birds	**20**	**0**	**0**
Eggshell unidentified bird	16	-	-
White-tooted shrew (*Crocidura* sp.)	1	-	-
Hare (*Lepus capensis/saxatilis*)	6	-	-
Striped ground squirrel (*Euxerus erythropus*)	1	-	-
Multimammate rat (*Mastomys natalensis*)	14	-	-
Small rodent	239	-	-
Giant pouched rat (*Cricetomys gambianus*)	1	-	-
Mongoose (Herpestidae), genet or civet (Viverridae)	11[2]	-	-
Serval cat or caracal (*Felis serval/caracal*)	5[1]	-	-
Aardvark (*Orycteropus afer*)	6[1]	-	-
African savannah elephant (*Loxodonta africana*)	1	-	-
Common warthog (*Phacohoerus africanus*)	1	-	-
Bush duiker (*Sylvicapra grimmia*) or oribi (*Ourebia ourebi*)	1	-	-
Medium antelope	1	-	-
Large antelope	1	-	-
Identified wild mammals	**289**	**0**	**0**
Dog (*Canis lupus* f. familiaris)	54	-	-
Horse (*Equus ferus* f. caballus) or donkey (*Equus africanus* f. asinus)	2	-	-
Sheep (*Ovis ammon* f. aries)	67	-	-
Goat (*Capra aegagrus* f. hircus)	52	-	1
Sheep (*Ovis ammon* f. aries) or goat (*Capra aegagrus* f. hircus)	262[3]	2	-
Cattle (*Bos primigenius* f. taurus)	114	1	-
Identified domestic mammals	**551**	**3**	**1**
Medium-sized carnivore	11	-	-
Small bovid	297	-	-
Identified wild or domestic mammals	**308**	**0**	**0**
Unidentified mammals	**6244**	**0**	**0**
Animal droppings: Donkey?	1	-	-
Ovicaprine	frequent	-	-
unidentified	1	-	-
Human (*Homo sapiens sapiens*)	**1**	**0**	**0**
TOTAL	**7573**	**8**	**2**

[1]MNI=1, [2]including 7 bones with MNI=1, [3]including 22 bones of min. 1 foetus

Table D.7: Faunal list of Oursi hu-beero (BF 97/30)
Revised and adapted from Linseele (in press a)

Gastropod	10
Bivalve	23
Identified molluscs	**33**
Lungfish (*Protopterus annectens*)	27
Clarias sp.	2
Clariidae	55
Synodontis sp.	1
Tilapia	4
Identified fish	**89**
Unidentified fish	**0**
Identified amphibians: frog or toad (Anura)	260
Adanson's mud turtle (*Pelusios adansonii*)	1
Agama (*Agama* sp.)	2
Monitor lizard (*Varanus* sp.)	57
Snake	4
Identified reptiles	**64**
Large duck or goose (Anatidae)	1
Coraciiformes	1
Identified wild birds	**2**
Identified wild or domestic birds: large galliform	**12**
Unidentified birds	**9**
Eggshell unidentified bird	84
African hedgehog (*Atelerix albiventris*)	3
Hare (*Lepus capensis/saxatilis*)	13
Tatera gerbil (*Tatera* sp.)	2
Multimammate rat (*Mastomys natalensis*)	3
Small rodent	26
Mongoose (Herpestidae), genet or civet (Viverridae)	1
Common warthog (*Phacochoerus africanus*)	1
Red-fronted gazelle (*Gazella rufifrons*)	6
Bush duiker (*Sylvicapra grimmia*) or oribi (*Ourebia ourebi*)	2
Medium-sized antelope	1
Identified wild mammals	**58**
Domestic dog (*Canis lupus* f. familiaris)	13
Sheep (*Ovis ammon* f. aries)	3
Goat (*Capra aegagrus* f. hircus)	10
Sheep (*Ovis ammon* f. aries) or goat (*Capra aegagrus* f. hircus)	109
Cattle (*Bos primigenius* f. taurus)	19
Identified domestic mammals	**141**
African wild cat or domestic cat (*Felis silvestris* or *F. s.* f. catus)	2
Medium-sized carnivore	6
Small bovid	158
Identified wild or domestic mammals	**461**
Unidentified mammals	**2627**
Human (*Homo sapiens sapiens*)	**8**
TOTAL	**3848**

Table D.8: Faunal list of Kissi 22 (BF 96/22)

Gastropod	3
Chambardia sp./*Spathopsis* sp.	3
Bivalve	20
Identified molluscs	**26**
Unidentified molluscs	**1**
Lungfish (*Protopterus annectens*)	20
Clarias sp.	4
Clariidae	93
Synodontis sp.	1
Tilapia	5
Identified fish	**123**
Unidentified fish	**2**
Identified amphibians: frog or toad (Anura)	47
Adanson's mud turtle (*Pelusios adansonii*)	4
Agama (*Agama* sp.)	11
Monitor lizard (*Varanus* sp.)	26
Snake	6
Identified reptiles	**47**
Unidentified reptiles	**3**
Identified wild birds: large duck or goose (Anatidae)	**1**
Identified wild or domestic birds: large galliform	**22**
Unidentified birds	**14**
Eggshell unidentified bird	27
Hare (*Lepus capensis/saxatilis*)	17
Multimammate rat (*Mastomys natalensis*)	4
Unstriped grass rat (*Arvicanthis niloticus*)	1
Small rodent	39
Giant pouched rat (*Cricetomys gambianus*)	3
Mongoose (Herpestidae), genet or civet (Viverridae)	1
Bush duiker (*Sylvicapra grimmia*) or oribi (*Ourebia ourebi*)	2
Bohor reedbuck (*Redunca redunca*)	1
Red-fronted gazelle (*Gazella rufifrons*)	6
Medium-sized antelope	6
Hartebeest (*Alcelaphus buselaphus*) or topi (*Damaliscus korrigum*)	2
Large antelope	1
Identified wild mammals	**83**
Dog (*Canis lupus* f. familiaris)	23
Sheep (*Ovis ammon* f. aries)	2
Goat (*Capra hircus* f. aegagrus)	11
Sheep (*Ovis ammon* f. aries) or goat (*Capra hircus* f. aegagrus)	56
Cattle (*Bos primigenius* f. taurus)	32
Identified domestic mammals	**124**
Medium-sized carnivore	8
Small bovid	180
Identified wild or domestic mammals	**188**
Unidentified mammals	**3815**
Human (*Homo sapiens sapiens*)	**1**
TOTAL	**4524**

Table D.9: Faunal list of Kissi 40 (BF 97/31)

Cowry (*Cypraea moneta/annulus*)	3
Lanistes varicus	5
Gastropod	31
Chambardia sp./*Spathopsis* sp.	55
Eupera ferruginea	1
Nile oyster (*Etheria elliptica*)	2
Bivalve	324
Identified molluscs	**421**
Lungfish (*Protopterus annectens*)	62
Clarias sp.	47
Clariidae	250
Synodontis sp.	9
Nile perch (*Lates niloticus*)	7
Tilapia	3
Identified fish	**378**
Unidentified fish	**14**
Identified amphibians: frog or toad (Anura)	44
Adanson's mud turtle (*Pelusios adansonii*)	11
Agama (*Agama* sp.)	1
Monitor lizard (*Varanus* sp.)	72
Snake	18
Crocodile (*Crocodylus* sp.)	50
Total reptiles	**152**
Cattle egret (*Bubulcus ibis*)	1
Medium-sized duck or goose (Anatidae)	2
Large duck or goose (Anatidae)	1
Francolin (*Francolinus* sp.)	1
Accipitridae	2
Helmeted guineafowl (*Numida meleagris*)	1
Pigeon or dove (Columbidae)	1
Identified wild birds	**9**
Identified domestic birds: domestic fowl (*Gallus gallus* f. domestica)	**3**
Identified wild or domestic birds: large galliform	**103**
Unidentified birds	**51**
Eggshell unidentified bird	85

Table D.10: Faunal list of Saouga 94/120 (BF 94/120)

African hedgehog (*Atelerix albiventris*)	1
Hare (*Lepus capensis/saxatilis*)	50
Striped ground squirrel (*Euxerus erythropus*)	1
Dwarf gerbil (*Desmodilliscus braueri*)	3
Multimammate rat (*Mastomys natalensis*)	4
Unstriped grass rat (*Arvicanthis niloticus*)	1
Small rodent	95
Giant pouched rat (*Cricetomys gambianus*)	3
Mongoose (Herpestidae), genet or civet (Viverridae)	7
Sandfox (*Vulpes pallida*)	1
Striped (*Hyaena hyaena*) or spotted hyaena (*Crocuta crocuta*)	1
Elephant (*Loxodonta africana*) or giraffe (*Giraffa camelopardalis*)	1
Common warthog (*Phacochoerus africanus*)	4
Bush duiker (*Sylvicapra grimmia*)	5
Oribi (*Ourebia ourebi*)	2
Bush duiker (*Sylvicapra grimmia*) or oribi (*Ourebia ourebi*)	5
Kob (*Kobus kob*)	1
Red-fronted gazelle (*Gazella rufifrons*)	25
Medium-sized antelope	2
Roan antelope (*Hippotragus equinus*)	3
Hartebeest (*Alcelaphus buselaphus*) or topi (*Damaliscus korrigum*)	1
Large antelope	1
Total wild mammals	**217**
Dog (*Canis lupus* f. familiaris)	128
Horse (*Equus ferus* f. caballus)	2
Sheep (*Ovis ammon* f. aries)	29
Goat (*Capra aegagrus* f. hircus)	74
Sheep (*Ovis ammon* f. aries) or goat (*Capra aegagrus* f. hircus)	304
Cattle (*Bos primigenius* f. taurus)	149
Total domestic mammals	**686**
African wild cat or domestic cat (*Felis silvestris* or *F.s.* f. catus)	4
Medium-sized carnivore	47
Small bovid	519
Total domestic or wild mammals	**570**
Unidentified mammals	**25000**
Coprolite domestic dog or jackal (*Canis* sp.)	**1**
Sheep (*Ovis ammon* f. aries) or goat (*Capra aegagrus* f. hircus) faeces	**5**
TOTAL	**27739**

Table D.10 cont.

Cowry (*Cypraea moneta/annulus*)	4
Gastropod	12
Chambardia sp./*Spathopsis* sp.	12
Nile oyster (*Etheria elliptica*)	1
Bivalve	89
Identified molluscs	**118**
Lungfish (*Protopterus annectens*)	33
Clarias sp.	10
Clariidae	88
Synodontis sp.	8
Tilapia	4
Identified fish	**143**
Unidentified fish	**5**
Identified fish: frog or toad (Anura)	**127**
Softshell turtle (Trionychidae)	1
Adanson's mud turtle (*Pelusios adansonii*)	29
Agama (*Agama* sp.)	1
Monitor lizard (*Varanus* sp.)	74
Snake	15
Crocodile (*Crocodylus* sp.)	1
Identified reptiles	**120**
Goliath heron (*Ardea goliath*)	1
Helmeted guineafowl (*Numida meleagris*)	1
Pigeon or dove (Columbidae)	2
Identified wild birds	**4**
Identified domestic birds: domestic fowl (*Gallus gallus* f. domestica)	**4**
Identified wild or domestic birds: large galliform	**56**
Unidentified birds	**17**
Eggshell unidentified bird	30
African hedgehog (*Atelerix albiventris*)	4
White-toothed shrew (*Crocidura* sp.)	1
Hare (*Lepus capensis/saxatilis*)	19
Unstriped grass rat (*Arvicanthis niloticus*)	2
Small rodent	76
Mongoose (Herpestidae), genet or civet (Viverridae)	2
Sandfox (*Vulpes pallida*)	2
Common warthog (*Phacochoerus africanus*)	2
Bush duiker (*Sylvicapra grimmia*)	1
Oribi (*Ourebia ourebi*)	1
Bush duiker (*Sylvicapra grimmia*) or oribi (*Ourebia ourebi*)	8
Red-fronted gazelle (*Gazella rufifrons*)	4
Medium-sized antelope	1
Identified wild mammals	**123**
Dog (*Canis lupus* f. familiaris)	176
Horse (*Equus ferus* f. caballus)	1
Horse (*Equus ferus* f. caballus) or donkey (*Equus africanus* f. asinus)	2
Sheep (*Ovis ammon* f. aries)	7
Goat (*Capra aegagrus* f. hircus)	11
Sheep (*Ovis ammon* f. aries) or goat (*Capra aegagrus* f. hircus)	89
Cattle (*Bos primigenius* f. taurus)	46
Identified domestic mammals	**332**
Medium-sized carnivore	77
Small bovid	284
Identified wild or domestic mammals	**361**
Unidentified mammals	**6400**
Human (*Homo sapiens sapiens*)	**8**
TOTAL	**7848**

Table D.11: Faunal list of Saouga 95/7 (BF 95/7)

Identified molluscs: *Pila wernei*	1
Stork (Ciconiidae)	1
Helmeted guineafowl (*Numida meleagris*)	1
Identified wild birds	**2**
Serval cat or caracal (*Felis serval/caracal*)	2
Bush duiker (*Sylvicapra grimmia*) or oribi (*Ourebia ourebi*)	2
Bohor reedbuck (*Redunca redunca*)	1
Identified wild mammals	**5**
Goat (*Capra aegagrus* f. hircus)	1
Sheep (*Ovis ammon* f. aries) or goat (*Capra aegagrus* f. hircus)	4
Cattle (*Bos primigenius* f. taurus)	10
Identified domestic mammals	**15**
Identified wild or domestic mammals: small bovid	**2**
Unidentified mammals	**165**
TOTAL	**190**

Table D.12: Faunal list of Gajiganna C (NA 90/5C)
Revised and adapted from Loonbeek (2001)

	Lower cultural layer (-1.6 to –1.8 m)	Upper cultural layer (-0.3 to –1.3 m)	All
Pila wernei	26	4	50
Chambardia sp./*Spathopsis* sp.	1	-	4
Limicolaria sp.	3	-	3
Identified molluscs	**4**	**30**	**57**
Polypterus sp.	1	-	1
Heterotis niloticus	1	-	1
Gymnarchus niloticus	2	-	2
Clarias gariepinus	-	-	1
Clarias anguillaris	-	-	9
Clarias sp.	-	-	14
Clariidae	224	135	361
Identified fish	**135**	**228**	**389**
Identified reptiles: Adanson's mud turtle (*Pelusios adansonii*)	**4**	**21**	**26**
Heron (*Ardea* sp.)	2	2	2
Stork (Ciconiidae)	-	1	2
Large duck or goose (Anatidae)	-	6	6
Accipitridae	-	1	1
Francolin (*Francolinus* sp.)	-	1	1
Helmeted guineafowl (*Numida meleagris*)	-	3	3
Identified wild birds	**14**	**1**	**15**
Unidentified birds	**11**	**2**	**8**
Savannah or patas monkey (*Cercopithecus aethiops/patas*)	-	4	4
Unstriped grass rat (*Arvicanthis niloticus*)	-	2	2
Small rodent	-	2	4
Giant pouched rat (*Cricetomys gambianus*)	-	1	1
Mongoose (Herpestidae), genet or civet (Viverridae)	1	9	10
African wild cat (*Felis silvestris*)	-	1	1
Serval cat or caracal (*Felis serval/caracal*)	-	4	4
Common warthog (*Phacochoerus africanus*)	2	17	19
Bush duiker (*Sylvicapra grimmia*)	1	1	2
Bohor reedbuck (*Redunca redunca*)	3	29	32
Identified wild mammals	**7**	**66**	**75**
Domestic dog (*Canis lupus* f. familiaris)	1	5	6
Sheep (*Ovis ammon* f. aries)	-	1	1
Goat (*Capra aegagrus* f. hircus)	10	53	73
Sheep (*Ovis ammon* f. aries) or goat (*Capra aegagrus* f. hircus)	17	132	172
Cattle (*Bos primigenius* f. taurus)	154	662	855*
Identified domestic mammals	**182**	**853**	**1107**
Identified wild or domestic mammals: small bovid	**26**	**243**	**289**
Unidentified mammals	**1110**	**8694**	**10338**
TOTAL	**1489**	**10117**	**12278**

*Including 6 bones of min. 1 foetus

Table D.13: Faunal list of Gajiganna A (NA 90/5A)
Revised and adapted from Van Neer (Breunig *et al.*, 1996)

	Lower cultural layer (-1.3 to –3.8 m)	Upper cultural layer (-0.8 to –1.3 m)	All
Identified molluscs: *Pila wernei*	**36**	**0**	**36**
Polypterus sp.	26	-	28
Heterotis niloticus	21	-	36
Clarias gariepinus	-	-	1
Clarias sp.	-	-	12
Clariidae	671	5	733
Identified fish	**718**	**5**	**810**
Unidentified fish	**2**	**0**	**2**
Adanson's mud turtle (*Pelusios adansonii*)	5	1	9
Monitor lizard (*Varanus* sp.)	-	-	1
Snake	2	7	22
Identified reptiles	**7**	**8**	**32**
Medium-sized duck or goose (Anatidae)	-	-	1
Bustard (Otididae)	1	-	1
Helmeted guineafowl (*Numida meleagris*)	-	-	1
Identified wild birds	**1**	**0**	**3**
Unidentified birds	**3**	**1**	**6**
Mongoose (Herpestidae), genet or civet (Viverridae)	1	-	1
Common warthog (*Phacochoerus africanus*)	2	-	3
Forest duiker (*Cephalophus* cf. *rufilatus*)	3	-	3
Oribi (*Ourebia ourebi*)	3	3	7
Bohor reedbuck (*Redunca redunca*)	2	-	6
Identified wild mammals	**9**	**3**	**14**
Domestic dog (*Canis lupus* f. familiaris)	1	-	3
Sheep (*Ovis ammon* f. aries)	-	-	1
Goat (*Capra aegagrus* f. hircus)	7	-	13
Sheep (*Ovis ammon* f. aries) or goat (*Capra aegagrus* f. hircus)	8	3	13
Cattle (*Bos primigenius* f. taurus)	245	29	426
Identified domestic mammals	**261**	**32**	**456**
Identified wild or domestic mammals: small bovid	**17**	**10**	**51**
Unidentified mammals	**1698**	**495**	**4096**
Coprolite domestic dog or jackal (*Canis* sp.)	**3**	**0**	**3**
TOTAL	**2757**	**555**	**5512**

Table D.14: Faunal list of Gajiganna BI (NA 90/5BI)
Revised and adapted from Van Neer (Breunig *et al.*, 1996)

Pila wernei	20
Gastropod	27
Large bivalve	3
Identified molluscs	**50**
Unidentified mollusc	**3**
Clarias sp.	2
Clariidae	4
Identified fish	**6**
Adanson's mud turtle (*Pelusios adansonii*)	2
Monitor lizard (*Varanus* sp.)	21
All reptiles	**23**
Large duck or goose (Anatidae)	5
Helmeted guineafowl (*Numida meleagris*)	34
Identified wild birds	**39**
Striped ground squirrel (*Euxerus erythropus*)	1
Small rodent	1
Mongoose (Herpestidae), genet or civet (Viverridae)	5
African wild cat (*Felis silvestris*)	4
Common warthog (*Phacochoerus africanus*)	8
Bush duiker (*Sylvicapra grimmia*) or oribi (*Ourebia ourebi*)	10
Bushbuck (*Tragelaphus scriptus*)	1
Bohor reedbuck (*Redunca redunca*)	1
Medium-sized antelope	2
Roan antelope (*Hippotragus equinus*)	1
Hartebeest (*Damaliscus korrigum*) or topi (*Alcelaphus buselaphus*)	2
Large antelope	7
Buffalo (*Syncerus caffer*)	71[1]
Identified wild mammals	**111**
Goat (*Capra aegagrus* f. hircus)	8
Sheep (*Ovis ammon* f. aries) or goat (*Capra aegagrus* f. hircus)	26[2]
Cattle (*Bos primigenius* f. taurus)	155
Identified domestic mammals	**189**
Identified wild or domestic mammals: small bovid	**71**
Unidentified mammals	**1512**
Coprolite domestic dog or jackal (*Canis* sp.)	**3**
TOTAL	**2007**

[1]MNI=1, [2]including 2 bones of min. 1 foetus

Table D.15: Faunal list of Gajiganna BII (NA 90/5BII)
Revised and adapted from Loonbeek (2001)

Identified mammals: cattle (*Bos primigenius* f. taurus)	**6**
Unidentified mammals	**1**
TOTAL	7

Table D.16: Faunal list of NA 91/1A

	Cultural layer 2	Pit	**Total**
Pila wernei	-	1	**1**
Gastropod	-	3	**3**
Identified molluscs	**0**	**4**	**4**
Polypterus sp.	10	3	**13**
Heterotis niloticus	-	11	**11**
Gymnarchus niloticus	1	1	**2**
Clariidae	22	18	**40**
Identified fish	**33**	**33**	**66**
Adanson's mud turtle (*Pelusios adansonii*)	3	2	**5**
Monitor lizard (*Varanus* sp.)	4	-	**4**
Identified reptiles	**7**	**2**	**9**
Unidentified birds	**1**	**0**	**1**
Mongoose (Herpestidae), genet or civet (Viverridae)	1	-	**1**
Common warthog (*Phacochoerus africanus*)	1	-	**1**
Identified wild mammals	**2**	**0**	**2**
Goat (*Capra aegagrus* f. hircus)	1	-	**1**
Sheep (*Ovis ammon* f. aries) or goat (*Capra aegagrus* f. hircus)	-	1	**1**
Cattle (*Bos primigenius* f. taurus)	25	6	**31**
Identified domestic mammals	**26**	7	**33**
Identified wild or domestic mammals: small bovid	**4**	**2**	**6**
Unidentified mammals	**645**	**23**	**668**
TOTAL	**718**	**71**	**789**

Table D.17: Faunal list of NA 93/42.
Revised and adapted from Van Neer (unpublished data)

Polypterus sp.	15
Clarias sp.	1
Clariidae	9
Unidentified catfish (Siluriformes)	1
Identified fish	**26**
Identified reptiles: Adanson's mud turtle (*Pelusios adansonii*)	**6**
Identified mammals: cattle (*Bos primigenius* f. taurus)	**1**
Unidentified mammals	**20**
TOTAL	**53**

Table D.18: Faunal list of Dumge A (NA 97/18)
Revised and adapted from Lambrecht (unpublished data)

Identified molluscs: *Pila wernei*	**1**
Identified mammals: cattle (*Bos primigenius* f. taurus)	**14**
Unidentified mammals	**1**
TOTAL	**16**

Table D.19: Faunal list of Labe (NA 97/24)
Revised and adapted from Lambrecht (unpublished data)

Pila wernei	2
Gastropod	4
Identified molluscs	**6**
Identified fish: Clariidae	**6**
Unidentified fish	**1**
Identified reptiles: Adanson's mud turtle (*Pelusios adansonii*)	**2**
Identified wild mammals: medium-sized antelope	**1**
Identified domestic mammals: cattle (*Bos primigenius* f. taurus)	**8**
Identified wild or domestic mammals: small bovid	**3**
Unidentified mammals	**284**
Human ***(Homo sapiens sapiens)***	**4**
TOTAL	**315**

Table D.20: Faunal list of NA 93/36

	Gaj IIa	Gaj IIb	Gaj IIc	**Total**
Identified molluscs: bivalve	**0**	**0**	**1**	**1**
Clarias sp.	-	3	-	**3**
Clariidae	-	55	5	**60**
Identified fish	**0**	**58**	**5**	**63**
Unidentified fish	**0**	**2**	**0**	**2**
Adanson's mud turtle (*Pelusios adansonii*)	-	1	-	**1**
Monitor lizard (*Varanus* sp.)	-	1	-	**1**
Identified reptiles	**0**	**2**	**0**	**2**
Egret (*Egretta* sp.)	-	1	-	**1**
Accipitridae	-	-	1	**1**
Large duck or goose (Anatidae)	-	1	-	**1**
Helmeted guineafowl (*Numida meleagris*)	-	1	-	**1**
Identified wild birds	**0**	**3**	**1**	**4**
Identified wild or domestic birds: large galliform	**0**	**2**	**0**	**2**
Unidentified birds	**0**	**4**	**2**	**6**
Eggshell unidentified bird	-	1	-	1
Olive baboon (*Papio anubis*)	-	-	2	**2**
Savannah or patas monkey (*Cercopithecus aethiops/patas*)	-	2	3	**5***
Unstriped grass rat (*Arvicanthis niloticus*)	-	2	-	**2**
Small rodent	-	4	1	**5**
Striped (*Hyaena hyaena*) or spotted hyaena (*Crocuta crocuta*)	-	-	1	**1**
Forest duiker (*Cephalophus* cf. *rufilatus*)	-	1	-	**1**
Bush duiker (*Sylvicapra grimmia*)	-	1	1	**2**
Bohor reedbuck (*Redunca redunca*)	-	-	3	**3**
Medium-sized antelope	-	-	1	**1**
Hartebeest (*Alcelaphus buselaphus*) or topi (*Damaliscus korrigum*)	-	-	1	**1**
Identified wild mammals	**0**	**10**	**11**	**21**
Domestic dog (*Canis lupus* f. familiaris)	-	-	1	**1**
Sheep (*Ovis ammon* f. aries)	-	-	2	**2**
Goat (*Capra aegagrus* f. hircus)	-	4	6	**10**
Sheep (*Ovis ammon* f. aries) or goat (*Capra aegagrus* f. hircus)	-	14	5	**19**
Cattle (*Bos primigenius* f. taurus)	1	11	23	**35**
Identified domestic mammals	**1**	**29**	**37**	**67**
Medium-sized carnivore	-	1	4	**5**
Small bovid	-	28	22	**50**
Identified wild or domestic mammals	**0**	**29**	**26**	**55**
Unidentified mammals	**24**	**1283**	**1326**	**2633**
Human (*Homo sapiens sapiens*)	**1**	**2**	**8**	**9**
TOTAL	**26**	**1425**	**1417**	**2866**

*MNI=1

Table D.21: Faunal list of Gilgila (NA 99/65)

Sheep (*Ovis ammon* f. aries) or goat (*Capra aegagrus* f. hircus)	4
Cattle (*Bos primigenius* f. taurus)	25
Identified domestic mammals	**29**
Unidentified mammals	**223**
TOTAL	**252**

Table D.22: Faunal list of NA 93/10.
Revised and adapted from Lambrecht (unpublished data)

	Z1	Z3	Z4	Z5	Z5 (pit 1)	Z5 (pit 2)	Z5 - all	Z6	Z6 (pit 1)	Z6 (pit 2)	Z6 (pit 3)	Z6-total	**Total**
Pila wernei	-	-	1	-	5	-	5	4	-	1	-	5	**11**
Gastropod	-	-	-	1	1	-	2	5	1	-	-	6	**8**
Identified molluscs	**0**	**0**	**1**	**1**	**6**	**0**	**7**	**9**	**1**	**1**	**0**	**11**	**19**
Lungfish (*Proptopterus annectens*)	-	-	-	1	-	-	1	39	-	-	-	39	**40**
Heterotis niloticus	-	1	-	-	-	-	0	7	7	-	-	14	**8**
Gymnarchus niloticus	-	-	-	-	-	-	0	1	-	-	-	1	**1**
Clarias gariepinus	-	-	-	-	-	-	0	1	-	-	-	1	**1**
Clarias anguillaris	-	-	-	-	-	-	0	1	-	-	-	1	**1**
Clarias sp.	-	-	1	1	-	1	2	2	1	-	-	3	**6**
Clariidae	-	-	4	7	2	84	93	42	7	-	-	49	**146**
Identified fish	**0**	**1**	**5**	**9**	**2**	**85**	**96**	**93**	**15**	**0**	**0**	**108**	**203**
Unidentified fish	**0**	**0**	**0**	**1**	**0**	**30**	**31**	**10**	**1**	**0**	**0**	**11**	**42**
Identified amphibians: frog or toad (Anura)	0	0	0	9	0	0	9	10	0	0	0	10	**19**
Senegal or Aubrey's flapshell turtle (*Cyclanorbis senegalensis/Cycloderma aubryi*)	-	-	-	-	-	-	0	1	-	-	-	1	**1**
Adanson's mud turtle (*Pelusios adansonii*)	-	-	-	1	-	-	1	15	212*	-	-	227	**228**
Monitor lizard (*Varanus* sp.)	-	-	-	2	-	-	2	9	1	-	-	10	**12**
Crocodile (*Crocodylus* sp.)	-	-	-	-	-	-	0	4	-	-	-	4	**4**
Identified reptiles	**0**	**0**	**0**	**3**	**0**	**0**	**3**	**29**	**213**	**0**	**0**	**242**	**245**
Egret (*Egretta* sp.)	-	-	-	1	-	-	1	-	-	-	-	0	**1**
Medium-sized duck or goose (Anatidae)	-	-	-	1	-	-	1	7	-	-	-	7	**8**
Large duck or goose (Anatidae)	-	-	1	5	3	-	8	17	-	2	-	19	**28**
Identified wild birds	**0**	**0**	**1**	**7**	**3**	**0**	**10**	**24**	**0**	**2**	**0**	**26**	**37**
Unidentified birds	**0**	**0**	**2**	**4**	**2**	**0**	**6**	**7**	**0**	**4**	**0**	**11**	**17**
Eggshell unidentified bird	-	-	-	1	-	-	1	-	-	-	-	-	1
Small rodent	-	-	-	-	-	-	0	7	-	-	-	7	**7**
African wild cat (*Felis silvestris*)	-	-	-	-	-	-	0	1	-	-	-	1	**1**
Mongoose (Herpestidae), genet or civet (Viverridae)	-	-	-	1	-	-	1	1	-	-	-	1	**2**
Identified wild mammals	**0**	**0**	**0**	**1**	**0**	**0**	**1**	**9**	**0**	**0**	**0**	**9**	**10**
Domestic dog (*Canis lupus* f. familiaris)	-	-	-	1	-	-	1	2	-	-	-	2	**3**
Sheep (*Ovis ammon* f. aries)	-	-	-	-	-	-	0	1	-	-	-	1	**1**
Goat (*Capra aegagrus* f. hircus)	-	1	-	-	-	-	0	1	-	-	-	1	**2**
Sheep (*Ovis ammon* f. aries) or goat (*Capra aegagrus* f. hircus)	-	1	1	2	-	-	2	9	1	-	-	10	**14**
Cattle (*Bos primigenius* f. taurus)	-	1	3	28	10	2	40	66	8	-	1	75	**119**
Identified domestic mammals	**0**	**3**	**4**	**31**	**10**	**2**	**43**	**79**	**9**	**0**	**1**	**89**	**139**
Medium-sized carnivore	-	-	-	-	-	-	0	1	-	-	-	1	**1**
Small bovid	-	-	-	-	3	-	3	11	-	-	-	11	**14**
Identified wild or domestic mammals	**0**	**0**	**0**	**0**	**3**	**0**	**3**	**12**	**0**	**0**	**0**	**12**	**15**
Unidentified mammals	**1**	**3**	**60**	**307**	**47**	**3**	**357**	**535**	**20**	**0**	**0**	**555**	**976**
TOTAL	**1**	**7**	**73**	**374**	**73**	**120**	**567**	**817**	**259**	**7**	**1**	**1084**	**1723**

*MNI=5

Table D.23: Faunal list of Zilum (NA 97/37)

	Gaj I	Gaj II	**Total**
Pila wernei	29	40	**69**
Limicolaria sp.	-	1	**1**
Chambarida sp./*Spathopsis* sp.	-	2	**2**
Bivalve	9	10	**19**
Identified molluscs	**38**	**53**	**91**
Unidentified molluscs	**1**	**1**	**2**
Lungfish (*Protopterus annectens*)	1	-	**1**
Clarias sp.	-	2	**2**
Clariidae	3	3	**6**
Identified fish	**4**	**5**	**9**
Senegal flapshell turtle (*Cyclanorbis senegalensis*)	-	1	**1**
Adanson's mud turtle (*Pelusios adansonii*)	-	1	**3**
Monitor lizard (*Varanus* sp.)	2	-	**2**
Identified reptiles	**2**	**2**	**6**
Giant pouched rat (*Cricetomys gambianus*)	2	-	**2**
Common warthog (*Phacochoerus africanus*)	2	-	**2**
Bush duiker (*Sylvicapra grimmia*)	-	2	**2**
Bush duiker (*Sylvicapra grimmia*) or oribi (*Ourebia ourebi*)	1	10	**11**
Hartebeest (*Alcelaphus buselaphus*) or topi (*Damaliscus korrigum*)	3	-	**3**
Identified wild mammals	**8**	**12**	**20**
Sheep (*Ovis ammon* f. aries)	2	1	**3**
Goat (*Capra aegagrus* f. hircus)	1	1	**2**
Sheep (*Ovis ammon* f. aries) or goat (*Capra aegagrus* f. hircus)	37	22	**59**
Cattle (*Bos primigenius* f. taurus)	537*	202	**739**
Identified domestic animals	**577**	**226**	**803**
Identified wild or domestic mammals: small bovid	**40**	**27**	**67**
Unidentified mammals	**1875**	**939**	**2814**
TOTAL	**2545**	**1265**	**3812**

*including 8 bones of min. 1 juvenile individual

Table D.24: Faunal list of Bukarkurari (NA 97/33)
Revised and adapted from Lambrecht (unpublished data)

Pila wernei	1
Gastropod	1
Chambarida sp./*Spathopsis* sp.	5
Bivalve	9
Identified molluscs	**16**
Unidentified molluscs	**5**
Clarias sp.	3
Clariidae	33
Synodontis sp.	1
Tilapia	3
Identified fish	**40**
Identified wild birds: helmeted guineafowl (*Numida meleagris*)	1
Common warthog (*Phacochoerus africanus*)	1
Bush duiker (*Sylvicapra grimmia*) or oribi (*Ourebia ourebi*)	2
Hartebeest (*Alcelaphus buselaphus*) or topi (*Damaliscus korrigum*)	1
Large antelope	3
Identified wild mammals	**7**
Sheep (*Ovis ammon* f. aries)	1
Goat (*Capra aegagrus* f. hircus)	3
Sheep (*Ovis ammon* f. aries) or goat (*Capra aegagrus* f. hircus)	22
Cattle (*Bos primigenius* f. taurus)	193
Identified domestic mammals	**219**
Identified wild or domestic mammals: small bovid	**109**
Unidentified mammals	**637**
TOTAL	**1034**

Table D.25: Faunal list of Kelumeri (NA 96/45)
Revised and adapted from Loonbeek (2001)

Polypterus sp.	6
Heterotis niloticus	1
Gymnarchus niloticus	1
Clarias sp.	3
Clariidae	22
Nile perch (*Lates niloticus*)	2
Identified fish	**35**
Unidentified fish	**1**
Identified reptiles: Adanson's mud turtle (*Pelusios adansonii*)	**19**
Identified wild birds: large cuckoo (*Cuculus canorus*)	**1**
Identified wild or domestic birds: large galliform	**2**
Small rodent	2
Serval cat or caracal (*Felis serval/caracal*)	1
Aardvark (*Orycteropus afer*)	1
Common warthog (*Phacochoerus africanus*)	3
Bush duiker (*Sylvicapra grimmia*)	12
Bush duiker (*Sylvicapra grimmia*) or oribi (*Ourebia ourebi*)	17
Bohor reedbuck (*Redunca redunca*)	1
Identified wild mammals	**37**
Goat (*Capra aegagrus* f. hircus)	12
Sheep (*Ovis ammon* f. aries) or goat (*Capra aegagrus* f. hircus)	64
Cattle (*Bos primigenius* f. taurus)	447
Identified domestic mammals	**523**
Medium-sized carnivore	4
Small bovid	53
Identified wild or domestic mammals	**57**
Unidentified mammals	**4444**
Human (*Homo sapiens sapiens*)	**15**
TOTAL	**5134**

Table D.26: Faunal list of Kariari C (NA 95/1)
Revised and adapted from Lambrecht (unpublished data)

Pila wernei	4
Lymnaea natalensis	1
Limicolaria sp.	34
Gastropod	33
Identified molluscs	**72**
Unidentified molluscs	**8**
Lungfish (*Protopterus annectens*)	43
Clarias sp.	1
Clariidae	10
Identified fish	**54**
Identified amphibians: frog or toad (Anura)	**54**
Adanson's mud turtle (*Pelusios adansonii*)	4
Monitor lizard (*Varanus* sp.)	9
Snake	14
Identified reptiles	**27**
Identified wild birds: small Passeriformes	**2**
Unidentified birds	**4**
Hare (*Lepus capensis/saxatilis*)	1
Multimammate rat (*Mastomys natalensis*)	1
Unstriped grass rat (*Arvicanthis niloticus*)	4
Small rodent	15
Mongoose (Herpestidae), genet or civet (Viverridae)	2
Red-fronted gazelle (*Gazella rufifrons*)	1
Medium-sized antelope	4
Identified wild mammals	**28**
Domestic dog (*Canis lupus* f. familiaris)	1
Sheep (*Ovis ammon* f. aries) or goat (*Capra aegagrus* f. hircus)	10
Cattle (*Bos primigenius* f. taurus)	15
Identified domestic mammals	**26**
Identified wild or domestic mammals: small bovid	**45**
Unidentified mammals	**648**
Human (*Homo sapiens sapiens*)	**3**
TOTAL	**971**

Table D.27: Faunal list of Labe Kanuri (NA 97/26)

	Lower cultural layer (-1.7/1.8 to –2.7 m)	Middle cultural layer (-1.2 to –1.6 m)	Upper cultural layer (-1.1 to –1.2 m)	**All**
Pila wernei	4	1	-	7
Cleopatra bulimoides	1	-	-	1
Limicolaria sp.	-	-	1	1
Gastropod	8	-	-	9
Bivalve	1	1	-	2
Identified molluscs	**13**	**2**	**1**	**20**
Lungfish (*Protopterus annectens*)	7	1	2	13
Polypterus sp.	1	-	-	1
Heterotis niloticus	1	1	-	2
Clarias sp.	3	1	-	4
Clariidae	26	3	-	43
Identified fish	**38**	**6**	**2**	**63**
Unidentified fish	**8**	**0**	**0**	**13**
Adanson's mud turtle (*Pelusios adansonii*)	13	-	-	14
Agama (*Agama* sp.)	-	1	-	1
Monitor lizard (*Varanus* sp.)	2	2	1	5
Snake	1	-	-	1
Identified reptiles	**16**	**3**	**1**	**21**
Identified amphibians: frog or toad (Anura)	**7**	**0**	**1**	**12**
Identified wild birds: large duck or goose (Anatidae)	**4**	**0**	**1**	**6**
Identified wild or domestic birds: large galliform	**2**	**0**	**0**	**3**
Unidentified birds	**3**	**0**	**1**	**6**
Eggshell unidentified bird	4	-	-	4
Unstriped grass rat (*Arvicanthis niloticus*)	-	-	-	1
Small rodent	10	-	4	16
Giant pouched rat (*Cricetomys gambianus*)	7	-	-	7
Mongoose (Herpestidae), genet or civet (Viverridae)	1	-	-	1
Forest duiker (*Cephalophus* cf. *rufilatus*)	-	-	3*	3
Bohor reedbuck (*Redunca redunca*)	1	-	-	1
Red-fronted gazelle (*Gazella rufifrons*)	-	-	1	1
Identified wild mammals	**9**	**0**	**8**	**30**
Domestic dog (*Canis lupus* f. familiaris)	-	1	-	1
Sheep (*Ovis ammon* f. aries)	-	3	-	3
Goat (*Capra aegagrus* f. hircus)	-	-	1	1
Sheep (*Ovis ammon* f. aries) or goat (*Capra aegagrus* f. hircus)	4	11	7	22
Cattle (*Bos primigenius* f. taurus)	7	11	9	17
Identified domestic mammals	**11**	**26**	**17**	**44**
Identified wild or domestic mammals: small bovid	**23**	**29**	**0**	**52**
Unidentified mammals	**120**	**106**	**52**	**324**
Human (*Homo sapiens sapiens*)	**19**	**31**	**0**	**52**
TOTAL	**277**	**203**	**84**	**650**

*MNI=1

Table D.28: Faunal list of Elkido North (NA 99/75)

Gastropod	4
Chambardia sp./*Spathopsis* sp.	2
Bivalve	3
Identified molluscs	**9**
Identified amphibians: frog or toad (Anura)	**13**
Monitor lizard (*Varanus* sp.)	3
Snake	1
Identified reptiles	**4**
Identified wild birds: helmeted guineafowl (*Numida meleagris*)	**1**
Identified wild or domestic birds: large galliform	**5**
Hare (*Lepus capensis/saxatilis*)	1
Multimammate rat (*Mastomys natalensis*)	7
Small rodent	32
Bohor reedbuck (*Redunca redunca*)	1
Identified wild mammals	**41**
Domestic dog (*Canis lupus* f. familiaris)	2
Goat (*Capra aegagrus* f. hircus)	1
Sheep (*Ovis ammon* f. aries) or goat (*Capra aegagrus* f. hircus)	37*
Cattle (*Bos primigenius* f. taurus)	6
Identified domestic animals	**44**
Medium-sized carnivore	2
Small bovids	24
Identified wild or domestic mammals	**116**
Unidentified mammals	**794**
Human (*Homo sapiens sapiens*)	**8**
TOTAL	**1035**

*including 18 bones of min. 1 juvenile individual

Table D.29: Faunal list of Dorota (NA 97/13)

Pila wernei	5
Gastropod	9
Bivalve	5
Identified molluscs	**19**
Unidentified molluscs	**2**
Lungfish (*Protopterus annectens*)	4
Heterotis niloticus	4
Gymnarchus niloticus	1
Distichodontidae or Citharinidae	6
Clariidae	37
Synodontis sp.	27
Nile perch (*Lates niloticus*)	33
Tilapia	1
Identified fish	**113**
Unidentified	**11**
Identified amphibians: frog or toad (Anura)	39
Identified reptiles: monitor lizard (*Varanus* sp.)	39
Stork (Ciconiidae)	11
Sacred ibis (*Threskiornis aethiopica*)	1
Medium-sized duck or goose (Anatidae)	20
Francolin (*Francolinus* sp.)	2
Bustard (Otididae)	1
Helmeted guineafowl (*Numida meleagris*)	2
Identified wild birds	**37**
Identified domestic birds: domestic fowl (*Gallus gallus* f. domestica)	**3**
Identified wild or domestic birds: large galliform	**30**
Unidentified birds	**63**
Eggshell ostrich (*Struthio camelus*)	2
Eggshell unidentified bird	4
Hare (*Lepus capensis/saxatilis*)	7
Multimammate rat (*Mastomys natalensis*)	1
Small rodent	2
Giant pouched rat (*Cricetomys gambianus*)	1
Mongoose (Herpestidae), genet or civet (Viverridae)	1
Common warthog (*Phacochoerus africanus*)	1
Bush duiker (*Sylvicapra grimmia*) or oribi (*Ourebia ourebi*)	1
Red-fronted gazelle (*Gazella rufifrons*)	9
Identified wild mammals	**23**
Domestic dog (*Canis lupus* f. familiaris)	2
Sheep (*Ovis ammon* f. aries)	65
Goat (*Capra aegagrus* f. hircus)	40
Sheep (*Ovis ammon* f. aries) or goat (*Capra aegagrus* f. hircus)	152
Cattle (*Bos primigenius* f. taurus)	282
Identified domestic mammals	**539**
Wild cat or domestic cat (*Felis silvestris* or *F. s.* f. catus)	3
Small bovid	387*
Identified wild or domestic mammals	**1470**
Unidentified mammal	**2622**
TOTAL	**5016**

*including 9 bones of min. 1 juvenile individual

Table D.30: Faunal list of Galaga (NA 92/2C) (squares A1 and A4)
Revised and adapted from De Cock (2002)

				AB		
	LSA	EIA	**Total**	LSA	EIA	**Total**
Pila wernei	1	-	**1**	-	-	**0**
Gastropod	-	6	**6**	-	-	**0**
Bivalve	-	2	**2**	-	-	**0**
Identified molluscs	**1**	**8**	**9**	**0**	**0**	**0**
Unidentified molluscs	**0**	**24**	**24**	**0**	**0**	**0**
Lungfish (*Protopterus annectens*)	-	7	7	-	-	**0**
Polypterus sp.	122	3500	**3622**	53	457	**510**
Heterotis niloticus	16	557	**573**	-	3	**3**
Mormyrid (Mormyridae)	1	11	**12**	10	2	**12**
Gymnarchus niloticus	32	208	**240**	-	1	**1**
Tigerfish (*Hydrocynus* sp.)	-	4	**4**	-	-	**0**
Distichodontidae or Citharinidae	1	15	**16**	-	2	**2**
Cyprinid (Cyprinidae)	-	3	**3**	-	-	**0**
Bagrus bajad	-	5	**5**	-	-	**0**
Auchenoglanis sp.	-	13	**13**	-	-	**0**
Heterobranchus sp.	-	1	**1**	-	-	**0**
Clarias gariepinus	2	26	**28**	-	1	**1**
Clarias anguillaris	-	4	**4**	-	1	**1**
Clarias sp.	22	424	**446**	-	1	**1**
Clariidae	391	9924	**10315**	56	295	**351**
Synodontis sp.	-	36	**36**	-	-	**0**
Unidentified catfish (Siluriformes)	-	53	**53**	-	-	**0**
Parachanna obscura	8	600	**608**	-	4	**4**
Nile perch (*Lates niloticus*)	18	451	**469**	-	-	**0**
Tilapia	11	1028	**1039**	2	53	**55**
Identified fish	**624**	**16870**	**17494**	**121**	**820**	**941**
Unidentified fish	**281**	**7292**	**7573**	**365**	**1489**	**1854**
Identified amphibians: frog or toad (Anura)	**0**	**19**	**19**	**0**	**0**	**0**
African softshell turtle (*Trionyx triunguis*)	-	3	**3**	-	-	**0**
Senegal or Aubrey's flapshell turtle (*Cyclanorbis senegalensis/Cycloderma aubryi*)	-	8	**8**	-	-	**0**
Adanson's mud turtle (*Pelusios adansonii*)	23	769	**792**	-	3	**3**
Agama (*Agama* sp.)	-	1	**1**	-	1	**1**
Monitor lizard (*Varanus* sp.)	-	4	**4**	-	1	**1**
Snake	4	4	**8**	1	1	**2**
Crocodile (*Crocodylus* sp.)	-	6	**6**	-	-	**0**
Identified reptiles	**27**	**795**	**822**	**1**	**6**	**7**

Table D.31: Faunal list of Kursakata (NA 93/46)

				AB		
	LSA	EIA	**Total**	LSA	EIA	**Total**
Birds						
Reed cormorant (*Phalacrocorax africanus*)	-	1	**1**	-	-	**0**
Small duck or goose (Anatidae)	-	19	**19**	-	-	**0**
Large duck or goose (Anatidae)	-	8	**8**	-	-	**0**
Charadriiformes	-	2	**2**	-	-	**0**
Identified wild birds	**0**	**30**	**30**	**0**	**0**	**0**
Unidentified birds	**2**	**28**	**30**	**0**	**0**	**0**
White-toothed shrew (*Crocidura* sp.)	-	1	**1**	-	-	**0**
Multimammate rat (*Mastomys natalensis*)	-	4	**4**	-	-	**0**
Unstriped grass rat (*Arvicanthis niloticus*)	-	13	**13**	-	-	**0**
Small rodent	-	53	**53**	-	5	**5**
Hippopotamus (*Hippopotamus amphibius*)	-	1	**1**	-	-	**0**
Common warthog (*Phacochoerus africanus*)	-	1	**1**	-	-	**0**
Oribi (*Ourebia ourebi*)	-	1	**1**	-	-	**0**
Bush duiker (*Sylvicapra grimmia*) or oribi (*Ourebia ourebi*)	-	4	**4**	-	-	**0**
Kob (*Kobus kob*)	-	2	**2**	-	-	**0**
Medium-sized antelope	-	23	**23**	-	-	**0**
Identified wild mammals	**0**	**103**	**103**	**0**	**5**	**5**
Sheep (*Ovis ammon* f. aries)	-	6	**6**	-	-	**0**
Goat (*Capra aegagrus* f. hircus)	1	10	**11**	-	-	**0**
Sheep (*Ovis ammon* f. aries) or goat (*Capra aegagrus* f. hircus)	11	120	**131**	-	-	**0**
Cattle (*Bos primigenius* f. taurus)	20	102*	**122**	-	1	**1**
Identified domestic mammals	**32**	**238**	**270**	**0**	**1**	**1**
Identified wild or domestic mammals: small bovid	**4**	**129**	**133**	**0**	**0**	**0**
Unidentified mammals	**712**	**12322**	**13034**	**4**	**21**	**25**
Human (*Homo sapiens sapiens*)	**0**	**1**	**1**	**0**	**0**	**0**
TOTAL	**1683**	**37859**	**39542**	**491**	**2342**	**2833**

*including 2 bones of min. 1 foetus

Table D.31 cont.

	I	II	III	IV	Total
Molluscs					
Limicolaria sp.	-	-	-	1	**1**
Pila sp.	-	-	1	-	**1**
Gastropod	-	3	5	3	**11**
Bivalve	-	1	2	7	**10**
Identfied molluscs	**0**	**4**	**8**	**11**	**23**
Unidentified molluscs	**0**	**1**	**5**	**1**	**7**
Polypterus sp.	57	6	1	-	**64**
Heterotis niloticus	25	32	5	8	**70**
Mormyrid (Mormyridae)	-	4	-	-	**4**
Gymnarchus niloticus	10	19	1	-	**30**
Auchenoglanis sp.	7	3	1	1	**12**
Heterobranchus sp.	-	-	1	-	**1**
Clarias gariepinus	-	-	2	-	**2**
Clariidae	1534	375	383	309	**2601**
Synodontis sp.	6	2	1	1	**10**
Parachanna obscura	-	2	2	-	**4**
Nile perch (*Lates niloticus*)	8	8	3	4	**23**
Tilapia	4	83	15	5	**107**
Identified fish	**1651**	**534**	**415**	**328**	**2928**
Unidentified fish	**437**	**251**	**289**	**140**	**1117**
Identified amphibians: frog or toad (Anura)	0	0	0	4	**4**
Adanson's mud turtle (*Pelusios adansonii*)	173	69	2	-	**244**
Monitor lizard (*Varanus* sp.)	-	-	4	2	**6**
Identified reptiles	**173**	**69**	**6**	**2**	**250**

Table D.32: Faunal list of Mege (NA 94/7) (square B2)
Revised and adapted from Lambrecht (1997)

	I	II	III	IV	Total
Birds					
Small duck or goose (Anatidae)	1	-	-	-	**1**
Medium-sized duck or goose (Anatidae)	-	-	-	1	**1**
Large duck or goose (Anatidae)	1	-	2	-	**3**
Helmeted guineafowl (*Numida meleagris*)	-	-	3	-	**3**
Identified wild birds	**2**	**0**	**5**	**1**	**8**
Identified domestic birds: domestic fowl (*Gallus gallus* f. domestica)	**0**	**0**	**1**	**0**	**1**
Identifed wild or domestic birds: large galliform	**0**	**0**	**3**	**2**	**5**
Unidentified bird	**7**	**1**	**2**	**16**	**26**
Eggshell unidentified bird	-	-	5	2	7
Multimammate rat (*Mastomys natalensis*)	-	1	-	4	-
Small rodent	-	2	5	9	**16**
Mongoose (Herpestidae), genet or civet (Viverridae)	-	-	-	1	**1**
Common warthog (*Phacochoerus africanus*)	1	-	-	-	**1**
Bush duiker (*Sylvicapra grimmia*) or oribi (*Ourebia ourebi*)	1	4	4	1	**10**
Kob (*Kobus kob*)	55	3	1	-	**59**
Red-fronted gazelle (*Gazella rufifrons*)	-	-	8	-	**8**
Medium-sized antelope	95	4	-	1	**100**
Identified wild mammals	**152**	**13**	**18**	**12**	**195**
Domestic dog (*Canis lupus* f. familiaris)	1	-	-	-	**1**
Sheep (*Ovis aries* f. ammon)	1	-	-	-	**1**
Goat (*Capra aegagrus* f. hircus)	5	1	-	1	**7**
Sheep (*Ovis ammon* f. aries) or goat (*Capra aegagrus* f. hircus)	30	4	29*	7	**70**
Cattle (*Bos primigenius* f. taurus)	547	47	16	14	**624**
Identified domestic mammals	**584**	**52**	**45**	**22**	**703**
African wild or domestic cat (*Felis silvestris* or *F. s.* f. catus)	-	-	2	-	**2**
Small bovid	55	8	13	6	**82**
Identified wild or domestic mammals	**1223**	**112**	**105**	**50**	**1490**
Unidentified mammals	**7348**	**891**	**679**	**584**	**9502**
TOTAL	**11577**	**1928**	**1586**	**1175**	**16266**

*Including 16 bones of min. 2 foetuses

Table D.32 cont.

	I	II	IIIa	IIIb	IV	**Total**
Identified molluscs: *Pila wernei*	**8**	**6**	**3**	**2**	**4**	**23**
Lungfish (*Protopterus annectens*)	7	17	6	6	5	41
Polypterus sp.	29	64	20	2	8	123
Heterotis niloticus	24	26	14	12	13	89
Hyperopisus bebe	1	2	-	-	-	3
Gymnarchus niloticus	18	23	9	5	4	59
Cyprinid (Cyprinidae)	2	-	-	-	-	2
Bagrus sp.	2	-	1	-	3	6
Auchenoglanis sp.	2	-	1	1	1	5
Heterobranchus sp.	1	-	-	-	-	1
Clarias gariepinus	-	1	-	-	-	1
Clarias anguillaris	-	-	-	-	1	1
Clarias sp.	10	18	11	8	9	56
Clariidae	208	617	251	221	266	1563
Synodontis sp.	6	6	5	5	4	26
Unidentified catfish (Siluriformes)	-	1	1	-	1	3
Parachanna obscura	15	19	1	2	3	40
Nile perch (*Lates niloticus*)	49	85	30	8	18	190
Tilapiini	23	14	11	6	10	64
Identified fish (quadrant A3 & C1)	**397**	**893**	**361**	**276**	**346**	**2273**
Unidentified fish (quadrant A3 & C1)	**137**	**235**	**100**	**34**	**51**	**557**
Estimate total fish all quadrants	**2045**	**4359**	**1808**	**1187**	**1528**	**10927**
Identified amphibians: frog or toad (Anura)	**3**	**18**	**0**	**3**	**2**	**26**
African softshell turtle (*Trionyx triunguis*)	-	1	-	-	-	1
Senegal flapshell turtle (*Cyclanorbis senegalensis*)	2	11	2	2	2	19
Softshell turtle (Trionychidae)	1	-	-	1	-	2
Adanson's mud turtle (*Pelusios adansonii*)	40	117	31	71	81	340
Monitor lizard (*Varanus* sp.)	11	17	5	2	4	39
Snake	1	1	-	-	-	2
Crocodile (*Crocodylus* sp.)	-	7	4	2	-	13
Identified reptiles	**55**	**147**	**38**	**76**	**87**	**403**

Table D.33: Faunal list of Ngala (NA 93/45)
Revised and adapted from Linseele (2005)

	I	II	IIIa	IIIb	IV	**Total**
Heron (*Ardea* sp.)	-	-	-	-	1	1
Sacred ibis? (*Threskiornis aethiopica*)	-	1	-	-	-	1
Small duck or goose (Anatidae)	-	2	-	-	1	3
Medium-sized duck or goose (Anatidae)	1	3	-	2	2	8
Large duck or goose (Anatidae)	9	8	2	3	3	25
Charadriiformes	-	1	-	-	-	1
Small passeriformes	1	-	-	-	-	1
Identified wild birds	**11**	**15**	**2**	**5**	**7**	**40**
Identified domestic birds: domestic fowl (*Gallus gallus* f. domestica)	**1**	**5**	**0**	**0**	**1**	**7**
Identified wild or domestic birds: large galliform	**19**	**35**	**5**	**3**	**8**	**70**
Unidentified birds	**23**	**18**	**5**	**2**	**10**	**58**
Hare (*Lepus capensis/saxatilis*)	-	1	-	-	-	1
White-toothed shrew (*Crocidura* sp.)	1	-	-	-	-	1
Multimammate rat (*Mastomys natalensis*)	9	1	1	-	-	11
Unstriped grass rat (*Arvicanthis niloticus*)	2	5	1	-	-	8
Small rodent	41	19	3	2	5	70
Serval cat or caracal (*Felis serval/caracal*)	-	2	-	-	-	2
Hippopotamus (*Hippopotamus amphibius*)	-	-	1	-	-	1
Elephant (*Loxodonta africana*) or hippo (*Hippopotamus amphibius*)	-	-	1	-	-	1
Warthog (*Phacochoerus africanus*)	-	-	2	-	-	2
Bohor reedbuck (*Redunca redunca*)	1	-	-	-	-	1
Kob (*Kobus kob*)	7	6	4	5	1	23
Red-fronted gazelle (*Gazella rufifrons*)	-	1	-	-	-	1
Medium-sized antelope	2	9	1	2	-	14
Identified wild mammals	**63**	**44**	**14**	**9**	**6**	**136**
Domestic dog (*Canis lupus* f. familiaris)	3	-	-	-	-	3
Sheep (*Ovis ammon* f. aries)	3	3	2	3	1	12
Goat (*Capra aegagrus* f. hircus)	11	8	2	1	5	27
Sheep (*Ovis ammon* f. aries) or goat (*Capra aegagrus* f. hircus)	63*	61	28	11	18	181
Cattle (*Bos primigenius* f. taurus)	45	53	47	85	64	295
Identified domestic mammals	**80**	**72**	**32**	**15**	**24**	**223**
Medium-sized carnivore	1	-	-	-	-	1
Small bovid	45	51	38	23	39	196
Identified domestic or wild mammals	**46**	**51**	**39**	**22**	**39**	**197**
Unidentified mammals	**143**	**431**	**776**	**679**	**565**	**2594**
Human (*Homo sapiens sapiens*)	**35**	**11**	**1**	**0**	**0**	**0**
TOTAL	**2532**	**5212**	**2721**	**2005**	**2281**	**14704**

*including 10 bones of min. one juvenile individual

Table D.33 cont.

	OH XV (0-20 cm)	OH XIV (20-40 cm)	OH XIII (60-80 cm)	OH XII (80-120 cm)	OH X (140-160 m)	OH VI (220-260 cm)	OH IV (300-340 cm)	OH III (340-380 cm)	OH II (380-420 cm)	**Total**
Pila wernei	3	1	-	4	-	-	-	-	-	**8**
Bivalve	2	-	-	-	-	-	-	-	-	**2**
Identified molluscs	**5**	**1**	**0**	**4**	**0**	**0**	**0**	**0**	**0**	**10**
Lungfish (*Protopterus annectens*)	-	-	-	2	13	-	-	-	1	**16**
Heterotis niloticus	-	1	-	1	-	-	-	-	-	**2**
Cyprinid (Cyprinidae)	-	1	-	1	-	-	-	-	-	**2**
Clarias sp.	2	-	-	-	-	-	-	-	-	**2**
Clariidae	11	2	1	3	29	-	-	1	-	**47**
Synodontis sp.	10	5	-	3	10	-	-	-	-	**28**
Nile perch (*Lates niloticus*)	1	2	-	2	2	-	-	1	-	**8**
Tilapia	-	-	-	-	1	-	-	-	-	**1**
Identified fish	**24**	**11**	**1**	**12**	**55**	**0**	**0**	**2**	**1**	**106**
Unidentified fish	**4**	**2**	**0**	**0**	**12**	**0**	**0**	**1**	**0**	**19**
Adanson's mud turtle (*Pelusios adansonii*)	1	1	1	-	15*	-	-	-	-	**18**
Monitor lizard (*Varanus* sp.)	-	-	-	-	3	-	-	-	-	**3**
Identified reptiles	**1**	**1**	**1**	**0**	**18**	**0**	**0**	**0**	**0**	**21**
Heron (*Ardea* sp.)	-	1	-	-	-	-	-	-	-	**1**
Large duck or goose (Anatidae)	3	-	-	2	-	1	-	-	-	**6**
Identified wild birds	**3**	**1**	**0**	**2**	**0**	**1**	**0**	**0**	**0**	**7**
Identified wild or domestic birds: large galliform	**2**	**0**	**0**	**0**	**0**	**1**	**0**	**0**	**0**	**3**
Unidentified birds	**1**	**0**	**0**	**0**	**0**	**0**	**0**	**0**	**0**	**1**
Serval cat or caracal (*Felis serval/caracal*)	1	-	-	-	-	-	-	-	-	**1**
Common warthog (*Phacochoerus africanus*)	-	1	-	-	-	-	-	-	-	**1**
Bohor reedbuck (*Redunca redunca*)	-	-	-	1	-	-	-	-	-	**1**
Kob (*Kobus kob*)	1	1	-	1	-	1	-	-	-	**4**
Medium-sized antelope	1	-	-	-	-	-	-	1	-	**2**
Identified wild mammals	**3**	**2**	**0**	**2**	**0**	**1**	**0**	**1**	**0**	**9**
Goat (*Capra aegagrus* f. hircus)	1	-	-	-	-	-	-	1	-	**2**
Sheep (*Ovis ammon* f. aries) or goat (*Capra aegagrus* f. hircus)	2	2	-	2	-	2	2	6	1	**17**
Cattle (*Bos primigenius* f. taurus)	2	3	-	1	-	-	-	5	1	**12**
Identified domestic mammals	**5**	**5**	**0**	**3**	**0**	**2**	**2**	**12**	**2**	**31**
Medium-sized carnivore	1	-	-	-	-	-	-	-	-	**1**
Small bovid	4	-	-	1	-	-	-	-	1	**6**
Identified wild or domestic mammals	**5**	**0**	**0**	**1**	**0**	**0**	**0**	**0**	**1**	**7**
Unidentified mammals	**51**	**26**	**14**	**28**	**9**	**4**	**9**	**48**	**17**	**206**
TOTAL	**104**	**49**	**16**	**52**	**94**	**9**	**11**	**64**	**21**	**420**

*MNI=1

Table D.34: Faunal list of Blé A
Revised and adapted from Lambrecht and Van Neer (unpublished data)

	OH XII (0-40 cm)	OH XI (40-80 cm)	OH IX (100-120 cm)	OH VIII (120-140 cm)	OH VII (160-180 cm)	OH V (220-240 cm)	OH III (260-280 cm)	**Total**
Pila wernei	2	-	-	-	-	-	-	**2**
Chambardia sp./*Spathopsis* sp.	1	-	-	-	-	-	-	**1**
Bivalve	3	-	1	-	-	-	-	**4**
Identified bivalve	**6**	**0**	**1**	**0**	**0**	**0**	**0**	**7**
Lungfish (*Protopterus annectens*)	-	5	1	-	-	-	-	**6**
Polypterus sp.	1	-	-	-	-	-	-	**1**
Heterotis niloticus	1	2	-	-	-	-	1	**4**
Gymnarchus niloticus	-	3	-	-	-	-	-	**3**
Distichodontidae or Citharinidae	-	6	-	-	1	-	-	**7**
Cyprinid (Cyprinidae)	7	14	-	-	-	-	-	**21**
Auchenoglanis sp.	2	-	-	-	-	-	-	**2**
Clarias sp.	-	1	-	-	-	-	-	**1**
Clariidae	9	58	5	-	3	4	33	**112**
Synodontis sp.	10	59	1	-	-	1	-	**71**
Parachanna obscura	1	4	-	-	-	-	-	**5**
Nile perch (*Lates niloticus*)	6	7	1	1	-	6	1	**22**
Tilapia	-	8	-	-	-	-	-	**8**
Identified fish	**37**	**167**	**8**	**1**	**4**	**11**	**35**	**263**
Unidentified fish	**22**	**47**	**2**	**0**	**1**	**6**	**10**	**88**
Senegal flapshell turtle (*Cyclanorbis senegalensis*)	-	1	-	-	-	-	-	**1**
Adanson's mud turtle (*Pelusios adansonii*)	8	2	5	-	-	-	1	**16**
Identified reptiles	**8**	**3**	**5**	**0**	**0**	**0**	**1**	**17**
Identified wild birds: large duck or goose (Anatidae)	**0**	**0**	**1**	**0**	**0**	**0**	**0**	**1**
Identified domestic birds: domestic fowl (*Gallus gallus* f. domestica)	**0**	**0**	**0**	**0**	**0**	**1**	**0**	**1**
Identified wild or domestic birds: large galliform	**0**	**0**	**0**	**0**	**0**	**1**	**0**	**1**
Unidentified birds	**0**	**0**	**0**	**0**	**0**	**1**	**0**	**1**
Bush duiker (*Sylvicapra grimmia*) or oribi (*Ourebia ourebi*)	-	-	-	1	-	-	-	**1**
Bohor reedbuck (*Redunca redunca*)	5	-	1	-	-	-	-	**6**
Kob (*Kobus kob*)	10	13	3	-	1	2	-	**29**
Medium-sized antelope	4	1	-	-	1	1	2	**9**
Identified wild mammals	**19**	**14**	**4**	**1**	**2**	**3**	**2**	**45**
Sheep (*Ovis ammon* f. aries)	-	-	1	1	-	-	-	**2**
Goat (*Capra aegagrus* f. hircus)	-	1	-	-	-	-	-	**1**
Sheep (*Ovis ammon* f. aries) or goat (*Capra aegagrus* f. hircus)	1	5	-	-	-	1	2	**9**
Cattle (*Bos primigenius* f. taurus)	-	1	4	-	-	1	1	**7**
Identified domestic mammals	**1**	**7**	**5**	**1**	**0**	**2**	**3**	**19**
Identified wild or domestic mammals: small bovid	**3**	**4**	**0**	**0**	**1**	**0**	**1**	**9**
Unidentified mammals	**176**	**221**	**57**	**15**	**36**	**99**	**19**	**623**
TOTAL	**272**	**463**	**83**	**18**	**44**	**124**	**71**	**1075**

Table D.35: Faunal list of Blé B
Revised and adapted from Lambrecht and Van Neer (unpublished data)

	OH VIII (0-20 cm)	OH VII (20-40 cm)	OH VI (40-80 cm)	OH V (80-100 cm)	OH IV (100-120 cm)	OH III (140-200 cm)	OH II (200-220 cm)	OH I (220-280 cm)	**Total**
Pila wernei	4	-	1	-	1	-	-	-	**6**
Chambardia sp./*Spathopsis* sp.	-	-	-	-	-	-	1	-	**1**
Nile oyster (*Etheria elliptica*)	-	-	1	-	-	-	-	-	**1**
Bivalve	-	-	2	-	-	-	-	-	**2**
Identified molluscs	**4**	**0**	**4**	**0**	**1**	**0**	**1**	**0**	**10**
Lungfish (*Protopterus annectens*)	-	2	2	-	-	-	-	-	**4**
Polypterus sp.	-	1	-	-	-	-	-	-	**1**
Heterotis niloticus	-	5	20	-	-	-	-	-	**25**
Mormyrids (Mormyridae)	-	1	1	-	-	-	-	-	**2**
Gymnarchus niloticus	-	2	3	-	-	-	-	-	**5**
Distichodontidae or Citharinidae	-	4	-	-	-	-	-	-	**4**
Cyprinid (Cyprinidae)	5	6	6	-	1	-	-	-	**18**
Bagrus sp.	-	-	1	-	-	-	-	-	**1**
Auchenoglanis sp.	-	2	1	-	-	-	-	-	**3**
Clariidae	14	19	52	-	1	2	2	-	**90**
Synodontis sp.	13	24	66	-	1	-	-	-	**104**
Nile perch (*Lates niloticus*)	-	6	8	3	1	1	-	-	**19**
Tilapia	-	2	22	-	-	1	-	-	**25**
Identified fish	**32**	**63**	**156**	**3**	**4**	**4**	**2**	**0**	**301**
Unidentified fish	**16**	**30**	**46**	**0**	**0**	**0**	**1**	**0**	**93**
Senegal flapshell turtle (*Cyclanorbis senegalensis*)	1	-	2	-	-	-	-	-	**3**
Softshell turtle (Trionychidae)	-	-	-	-	-	1	-	-	**1**
Adanson's mud turtle (*Pelusios adansonii*)	-	10	2	-	-	-	-	-	**12**
Crocodile (*Crocodylus* sp.)	-	1	-	-	-	-	-	-	**1**
Identified reptiles	**1**	**11**	**4**	**0**	**0**	**1**	**0**	**0**	**17**
Identified wild birds: large duck or goose (Anatidae)	**2**	**7**	**3**	**0**	**1**	**1**	**0**	**0**	**14**
Identified wild or domestic birds: large galliform	**0**	**1**	**0**	**0**	**0**	**0**	**0**	**0**	**1**
Unidentified birds	**0**	**1**	**1**	**0**	**0**	**0**	**0**	**0**	**2**
Serval cat or caracal (*Felis serval/caracal*)	-	-	1	-	-	-	-	-	**1**
Bohor reedbuck (*Redunca redunca*)	-	4	-	-	-	-	-	-	**4**
Kob (*Kobus kob*)	-	7	-	-	-	-	-	-	**7**
Medium-sized antelope	-	6	-	-	-	1	-	-	**7**
Identified wild mammals	**0**	**17**	**1**	**0**	**0**	**1**	**0**	**0**	**19**
Sheep (*Ovis ammon* f. aries)	-	-	-	-	-	1	1	-	**2**
Sheep (*Ovis ammon* f. aries) or goat (*Capra aegagrus* f. hircus)	-	6	-	1	-	1	-	-	**8**
Cattle (*Bos primigenius* f. taurus)	-	3	-	-	-	-	1	-	**4**
Identified domestic mammals	**0**	**9**	**0**	**1**	**0**	**2**	**2**	**0**	**14**
African wild cat or domestic dog (*Felis silvestris* or *F. s.* f. catus)	-	1	-	-	-	-	-	-	**1**
Small bovid	1	12	1	-	-	2	1	-	**17**
Identified wild or domestic mammals	**1**	**13**	**1**	**0**	**0**	**2**	**1**	**0**	**18**
Unidentified mammals	**29**	**202**	**138**	**4**	**12**	**4**	**5**	**0**	**394**
TOTAL	**85**	**354**	**354**	**8**	**18**	**15**	**12**	**0**	**883**

Table D.36: Faunal list of Blé C
Revised and adapted from Lambrecht and Van Neer (unpublished data)

	OH V (0-40 cm)	OH IV (40-120 cm)	OH II-III (120-180 cm)	OH I (180-280 cm)	**Total**
Pila wernei	3	8	1	1	**13**
Limicolaria sp.	-	-	1	-	**1**
Chambardia sp./*Spathopsis* sp.	2	2	2	1	**7**
Mutela nilotica	-	-	-	1	**1**
Nile oyster (*Etheria elliptica*)	-	-	1	-	**1**
Bivalve	2	3	1	7	**13**
Identified molluscs	**7**	**13**	**6**	**10**	**36**
Lungfish (*Protopterus annectens*)	4	6	-	1	**11**
Polypterus sp.	8	4	3	-	**15**
Heterotis niloticus	19	21	4	-	**44**
Gymnarchus niloticus	-	4	1	-	**5**
Distichodontidae or Citharinidae	9	5	1	-	**15**
Cyprinid (Cyprinidae)	13	4	-	-	**17**
Auchenoglanis sp.	2	3	1	-	**6**
Schilbe sp.	-	-	1	-	**1**
Clariidae	47	84	27	2	**160**
Synodontis sp.	94	80	1	-	**175**
Parachanna obscura	2	1	1	-	**4**
Nile perch (*Lates niloticus*)	50	33	12	7	**102**
Tilapia	18	46	2	2	**68**
Identified fish	**266**	**291**	**54**	**12**	**623**
Unidentified fish	**103**	**80**	**8**	**1**	**192**
Identified amphibians: frog or toad (Anura)	**1**	**0**	**0**	**0**	**1**
Senegal flapshell turtle (*Cyclanorbis senegalensis*)	-	2	-	-	**2**
Adanson's mud turtle (*Pelusios adansonii*)	8	20	3	1	**32**
Monitor lizard (*Varanus* sp.)	1	1	-	-	**2**
Identified reptiles	**9**	**23**	**3**	**1**	**36**
Large duck or goose (Anatidae)	-	4	1	-	**5**
Identified wild birds	**0**	**4**	**1**	**0**	**5**
Identified domestic birds: domestic fowl (*Gallus gallus* f. domestica)	**0**	**0**	**0**	**1**	**1**
Identified domestic or wild birds: large galliform	**2**	**0**	**0**	**1**	**3**
Common warthog (*Phacochoerus africanus*)	-	-	-	1	**1**
African savannah elephant (*Loxodonta africana*)	-	-	-	1	**1**
Bohor reedbuck (*Redunca redunca*)	-	-	-	2	**2**
Kob (*Kobus kob*)	1	4	1	5	**11**
Medium-sized antelope	-	1	2	4	**7**
Buffalo (*Syncerus caffer*)	-	2	1	-	**3**
Identified wild mammals	**1**	**7**	**4**	**13**	**25**
Sheep (*Ovis ammon* f. aries)	1	-	-	-	**1**
Goat (*Capra aegagrus* f. hircus)	-	-	1	-	**1**
Sheep (*Ovis ammon* f. aries) or goat (*Capra aegagrus* f. hircus)	2	10	4	6	**22**
Cattle (*Bos primigenius* f. taurus)	1	7	2	12	**22**
Identified domestic mammals	**4**	**17**	**7**	**18**	**46**
Identified wild or domestic mammals: small bovid	**4**	**8**	**0**	**2**	**14**
Unidentified mammals	**145**	**270**	**49**	**87**	**551**
Human (*Homo sapiens sapiens*)	**0**	**0**	**1**	**0**	**1**
TOTAL	**542**	**713**	**133**	**146**	**1534**

Table D.37: Faunal list of Blé E
Revised and adapted from Lambrecht and Van Neer (unpublished data)

CAMBRIDGE MONOGRAPHS IN AFRICAN ARCHAEOLOGY

No 1 BAR S75, 1980 **The Niger Delta** *Aspects of its Prehistoric Economy and Culture* by Nwanna Nzewunwa. ISBN 0 86054 083 9

No 2 BAR S89, 1980 **Prehistoric Investigations in the Region of Jenne, Mali** *A Study in the Development of Urbanism in the Sahel* by Susan Keech McIntosh and Roderick J. McIntosh ISBN 0 86054 103 7

No 3 BAR S97, 1981 **Off-Site Archaeology and Human Adaptation in Eastern Africa** *An Analysis of Regional Artefact Density in the Amboseli, Southern Kenya* by Robert Foley. ISBN 0 86054 114 2

No 4 BAR S114, 1981 **Later Pleistocene Cultural Adaptations in Sudanese Nubia** by Yousif Mukhtar el Amin. ISBN 0 86054 134 7

No 5 BAR S119, 1981 **Settlement Patterns in the Iron Age of Zululand** *An Ecological Interpretation* by Martin Hall. ISBN 0 86054 143 6

No 6 BAR S139, 1982 **The Neolithic Period in the Sudan, c. 6000-2500 B.C**. by Abbas S. Mohammed-Ali. ISBN 0 86054 170 3

No 7 BAR S195, 1984 **History and Ethnoarchaeology in Eastern Nigeria** *A Study of Igbo-Igala relations with special reference to the Anambra Valley* by Philip Adigwe Oguagha and Alex Ikechukwu Okpoko. ISBN 0 86054 249 1

No 8 BAR S197, 1984 **Meroitic Settlement in the Central Sudan** *An Analysis of Sites in the Nile Valley and the Western Butana* by Khidir Abdelkarim Ahmed. ISBN 0 86054 252 1

No 9 BAR S201, 1984 **Economy and Technology in the Late Stone Age of Southern Natal** by Charles Cable. ISBN 0 86054 258 0

No 10 BAR S207, 1984 **Frontiers** *Southern African Archaeology Today* edited by M. Hall, G. Avery, D.M. Avery, M.L. Wilson and A.J.B. Humphreys. ISBN 0 86054 268 8. £23.00.

No 11 BAR S215, 1984 **Archaeology and History in Southern Nigeria** *The ancient linear earthworks of Benin and Ishan* by P.J. Darling. ISBN 0 86054 275 0

No 12 BAR S213, 1984 **The Later Stone Age of Southernmost Africa** by Janette Deacon. ISBN 0 86054 276 9

No 13 BAR S254, 1985 **Fisher-Hunters and Neolithic Pastoralists in East Turkana, Kenya** by John Webster Barthelme. ISBN 0 86054 325 0

No 14 BAR S285, 1986 **The Archaeology of Central Darfur (Sudan) in the 1st Millennium A.D.** by Ibrahim Musa Mohammed. ISBN 0 86054 367 6.

No 15 BAR S293, 1986 **Stable Carbon Isotopes and Prehistoric Diets in the South-Western Cape Province, South Africa** by Judith Sealy. ISBN 0 86054 376 5.

No 16 BAR S318, 1986 **L'art rupestre préhistorique des massifs centraux sahariens** by Alfred Muzzolini.. ISBN 0 86054 406 0

No 17 BAR S321, 1987 **Spheriods and Battered Stones in the African Early and Middle Stone Age** by Pamela R. Willoughby. ISBN 0 86054 410 9

No 18 BAR S338, 1987 **The Royal Crowns of Kush** *A study in Middle Nile Valley regalia and iconography in the 1st millennia B.C. and A.D.* by Lázló Török.. ISBN 0 86054 432 X

No 19 BAR S339, 1987 **The Later Stone Age of the Drakensberg Range and its Foothills** by H. Opperman. ISBN 0 86054 437 0

No 20 BAR S350, 1987 **Socio-Economic Differentiation in the Neolithic Sudan** by Randi Haaland. ISBN 0 86054 453 2

No 21 BAR S351, 1987 **Later Stone Age Settlement Patterns in the Sandveld of the South-Western Cape Province, South Africa** by Anthony Manhire. ISBN 0 86054 454 0

No 22 BAR S365, 1987 **L'art rupestre du Fezzan septentrional (Libye) Widyan Zreda et Tarut (Wadi esh-Shati)** by Jean-Loïc Le Quellec. ISBN 0 86054 473 7

No 23 BAR S368, 1987 **Archaeology and Environment in the Libyan Sahara** *The excavations in the Tadrart Acacus*, 1978-1983 edited by Barbara E. Barich. ISBN 0 86054 474 5

No 24 BAR S378, 1987 **The Early Farmers of Transkei, Southern Africa Before A.D. 1870** by J.M. Feely. ISBN 0 86054 486 9

No 25 BAR S380, 1987 **Later Stone Age Hunters and Gatherers of the Southern Transvaal** *Social and ecological interpretation* by Lyn Wadley. ISBN 0 86054 492 3

No 26 BAR S405, 1988 **Prehistoric Cultures and Environments in the Late Quaternary of Africa** edited by John Bower and David Lubell. ISBN 0 86054 520 2

No 27 BAR S418, 1988 **Zooarchaeology in the Middle Nile Valley** *A Study of four Neolithic Sites near Khartoum* by Ali Tigani El Mahi. ISBN 0 86054 539 3

No 28 BAR S422, 1988 **L'Ancienne Métallurgie du Fer à Madagascar** by Chantal Radimilahy. ISBN 0 86054 544 X

No 29 BAR S424, 1988 **El Geili The History of a Middle Nile Environment, 7000 B.C.-A.D. 1500** edited by I. Caneva. ISBN 0 86054 548 2

No 30 BAR S445, 1988 **The Ethnoarchaeology of the Zaghawa of Darfur (Sudan) Settlement and Transcience** by Natalie Tobert. ISBN 0 86054 574 1

No 31 BAR S455, 1988 **Shellfish in Prehistoric Diet Elands Bay, S.W. Cape Coast, South Africa** by W.F. Buchanan. ISBN 0 86054 584 9

No 32 BAR S456, 1988 **Houlouf I** *Archéologie des sociétés protohistoriques du Nord-Cameroun* by Augustin Holl. ISBN 0 86054 586 5

No 33 BAR S469, 1989 **The Predynastic Lithic Industries of Upper Egypt** by Liane L. Holmes. ISBN 0 86054 601 2 (two volumes)

No 34 BAR S521, 1989 **Fishing Sites of North and East Africa in the Late Pleistocene and Holocene** *Environmental Change and Human Adaptation* by Kathlyn Moore Stewart. ISBN 0 86054 662 4

No 35 BAR S523, 1989 **Plant Domestication in the Middle Nile Basin** *An Archaeoethnobotanical Case Study* by Anwar Abdel-Magid. ISBN 0 86054 664 0

No 36 BAR S537, 1989 **Archaeology and Settlement in Upper Nubia in the 1st Millennium A.D.** by David N. Edwards. ISBN 0 86054 682 9

No 37 BAR S541, 1989 **Prehistoric Settlement and Subsistence in the Kaduna Valley, Nigeria** by Kolawole David Aiyedun and Thurstan Shaw. ISBN 0 86054 684 5

No 38 BAR S640, 1996 **The Archaeology of the Meroitic State** *New perspectives on its social and political organisation* by David N. Edwards. ISBN 0 86054 825 2

No 39 BAR S647, 1996 **Islam, Archaeology and History** *Gao Region (Mali) ca. AD 900 - 1250* by Timothy Insoll. ISBN 0 86054 832 5

No 40 BAR S651, 1996 **State Formation in Egypt**: *Chronology and society* by Toby A.H. Wilkinson. ISBN 0 86054 838 4

No 41 BAR S680, 1997 **Recherches archéologiques sur la capitale de l'empire de Ghana** *Etude d'un secteur d'habitat à Koumbi Saleh, Mauritanie. Campagnes II-III-IV-V (1975-1976)-(1980-1981)* by S. Berthier. ISBN 0 86054 868 6

No 42 BAR S689, 1998 **The Lower Palaeolithic of the Maghreb** *Excavations and analyses at Ain Hanech, Algeria* by Mohamed Sahnouni. ISBN0 86954 875 9

No 43 BAR S715, 1998 **The Waterberg Plateau in the Northern Province, Republic of South Africa, in the Later Stone Age** by Maria M. Van der Ryst. ISBN 0 86054 893 7

No 44 BAR S734, 1998 **Cultural Succession and Continuity in S.E. Nigeria** *Excavations in Afikpo* by V. Emenike Chikwendu. ISBN 0 86054 921 6

No 45 BAR S763, 1999 **The Emergence of Food Production in Ethiopia** by Tertia Barnett. ISBN 0 86054 971 2

No 46 BAR S768, 1999 **Sociétés préhistoriques et Mégalithes dans le Nord-Ouest de la République Centrafricaine** by Étienne Zangato. ISBN 0 86054 980 1

No 47 BAR S775, 1999 **Ethnohistoric Archaeology of the Mukogodo in North-Central Kenya** *Hunter-gatherer subsistence and the transition to pastoralism in secondary settings* by Kennedy K. Mutundu. ISBN 0 86054 990 9

No 48 BAR S782, 1999 **Échanges et contacts le long du Nil et de la Mer Rouge dans l'époque protohistorique (IIIe et IIe millénaires avant J.-C.)** *Une synthèse préliminaire* by Andrea Manzo. ISBN 1 84171 002 4

No 49 BAR S838, 2000 **Ethno-Archaeology in Jenné, Mali** *Craft and status among smiths, potters and masons* by Adria LaViolette. ISBN 1 84171 043 1

No 50 BAR S860, 2000 **Hunter-Gatherers and Farmers** *An enduring Frontier in the Caledon Valley, South Africa* by Carolyn R. Thorp. ISBN 1 84171 061 X

No 51 BAR S906, 2000 **The Kintampo Complex** *The Late Holocene on the Gambaga Escarpment, Northern Ghana* by Joanna Casey. ISBN 1 84171 202 7

No 52 BAR S964, 2000 **The Middle and Later Stone Ages in the Mukogodo Hills of Central Kenya** *A Comparative Analysis of Lithic Artefacts from Shurmai (GnJm1) and Kakwa Lelash (GnJm2) Rockshelters* by G-Young Gang. ISBN 1 84171 251 5

No 53 BAR S1006, 2001 **Darfur (Sudan) In the Age of Stone Architecture c. 1000 - 1750 AD** *Problems in historical reconstruction* by Andrew James McGregor. ISBN 1 84171 285 X

No 54 BAR S1037, 2002 **Holocene Foragers, Fishers and Herders of Western Kenya** by Karega-Mũnene. ISBN 1 84171 1037

No 55 BAR S1090, 2002 **Archaeology and History in Ìlàrè District (Central Yorubaland, Nigeria) 1200-1900 A.D.** by Akinwumi O. Ogundiran. ISBN 1 84171 468 2

No 56 BAR S1133, 2003 **Ethnoarchaeology in the Zinder Region, Republic of Niger: the site of Kufan Kanawa** by Anne Haour. ISBN 1 84171 506 9

No 57 BAR S1187, 2003 **Le Capsien typique et le Capsien supérieur** *Évolution ou contemporanéité. Les données technologiques* by Noura Rahmani. ISBN 1 84171 553 0

No 58 BAR S1216, 2004 **Fortifications et urbanisation en Afrique orientale** by Stéphane Pradines. ISBN 1 84171 576 X

No 59 BAR S1247, 2004 **Archaeology and Geoarchaeology of the Mukogodo Hills and Ewaso Ng'iro Plains, Central Kenya** by Frederic Pearl. ISBN 1 84171 607 3

No 60 BAR S1289, 2004 **Islamic Archaeology in the Sudan** by Intisar Soghayroun Elzein. ISBN 1 84171 639 1.

No 61 BAR S1308, 2004 **An Ethnoarchaeological Study of Iron-Smelting Practices among the Pangwa and Fipa in Tanzania** by Randi Barndon. ISBN 1 84171 657 X.

No 62 BAR S1398, 2005 **Archaeology and History in North-Western Benin** by Lucas Pieter Petit. ISBN 1 84171 837 8.

No 63 BAR S1407, 2005 **Traditions céramiques, Identités et Peuplement en Sénégambie** *Ethnographie comparée et essai de reconstitution historique* by Moustapha Sall. ISBN 1 84171 850 5

No 64 BAR S1446, 2005 **Changing Settlement Patterns in the Aksum-Yeha Region of Ethiopia: 700 BC – AD 850** by Joseph W. Michels. ISBN 1 84171 882 3.

No 65 BAR S1454, 2006 **Safeguarding Africa's Archaeological Past** *Selected papers from a workshop held at the School of Oriental and African Studies, University of London, 2001* edited by Niall Finneran. ISBN 1841718920

No 66 BAR -S1537, 2006 **Excavations at Kasteelberg, and the Origins of the Khoekhoen in the Western Cape, South Africa** by Andrew B. Smith. ISBN 1 84171 969 2.

No 67 BAR –S1549, 2006 **Archéologie du Diamaré au Cameroun Septentrional** *Milieux et peuplements entre Mandara, Logone, Bénoué et Tchad pendant les deux derniers millénaires* by Alain Marliac ISBN 1 84171 978 1.

No 68 BAR –S1602, 2007 **Chasse et élevage dans la Corne de l'Afrique entre le Néolithique et les temps historiques** by Joséphine Lesur. ISBN 978 1 4073 0019 1.

No 69 BAR –S1617, 2007 **The Emergence of Social and Political Complexity in the Shashi-Limpopo Valley of Southern Africa, AD 900 to 1300** *Ethnicity, class, and polity* by John Anthony Calabrese ISBN 978 1 4073 0029 0.